NEW DEVELOPMENTS IN FUNDAMENTAL INTERACTION THEORIES

Related Titles from AIP Conference Proceedings

564 Quantum Electrodynamics and Physics of the Vacuum: QED 2000, Second Wkshp.
Edited by Giovanni Cantatore, May 2001, 0-7354-0000-8

562 Particles and Fields: Ninth Mexican School
Edited by Lukas Nellen and Gerardo Herrera Corral, April 2001, 1-56396-998-X

555 Cosmology and Particle Physics: CAPP 2000
Edited by Ruth Durrer, Juan Garcia-Bellido, and Mikhail Shaposhnikov, March 2001, 1-56396-986-6

541 Theoretical High Energy Physics: MRST 2000
Edited by C. R. Hagen, November 2000, 1-56396-966-1

539 Symmetries in Subatomic Physics: 3rd International Symposium
Edited by X.-H. Guo, A. W. Thomas, and A. G. Williams, October 2000, 1-56396-964-5

531 Particles and Fields: Seventh Mexican Workshop
Edited by Alejandro Ayala, Guillermo Contreras, and Gerardo Herrera, July 2000, 1-56396-954-8

488 High Energy Physics at the Millennium: MRST '99
Edited by Pat Kalyniak, Stephen Godfrey, and B. Kamal, October 1999, 1-56396-902-5

453 Particles, Fields, and Gravitation
Edited by Jakub Rembieliński, December 1998, 1-56396-837-1

423 Fundamental Particles and Interactions: Frontiers in Contemporary Physics
Edited by Robert S. Panvini and Thomas J. Weiler, February 1998, 1-56396-725-1

To learn more about these titles, or the AIP Conference Proceedings Series, please visit the webpage **http://www.aip.org/catalog/aboutconf.html**

NEW DEVELOPMENTS IN FUNDAMENTAL INTERACTION THEORIES

37th Karpacz Winter School of Theoretical Physics

Karpacz, Poland 6–15 February 2001

EDITORS

Jerzy Lukierski

University of Wroclaw
Wroclaw, Poland

Jakub Rembieliński

University of Lodz
Lodz, Poland

AMERICAN INSTITUTE OF PHYSICS

Melville, New York, 2001

AIP CONFERENCE PROCEEDINGS ■ VOLUME 589

Editors:

Jerzy Lukierski
Director, Inst. of Theoretical Physics
University of Wrocław
pl. Maxa Borna 9
50-204 Wrocław
POLAND

E-mail: lukier@ift.uni.wroc.pl

Jakub Rembieliński
Director, Institute of Physics
University of Łódź
ul. Pomorska 149/153
90-236 Łódź
POLAND

E-mail: jaremb@krysia.uni.lodz.pl

L.C. Catalog Card No. 2001094641
ISBN 0-7354-0029-6
ISSN 0094-243X
Printed in the United States of America

CONTENTS

*Italicized name indicates the author who presented the paper.

FIELD-THEORETIC FRAMEWORK: GENERAL FORMALISM AND MODELS

*Italicized name indicates the author who presented the paper.

RELATED TOPICS

*Italicized name indicates the author who presented the paper.

Preface

The Karpacz Schools started in 1964 and since then have been held every year except 1982, the year of martial law in Poland. The organizers of the School have always followed two principles: first, to choose front-line research subjects for the School; and second, to provide the grounds for the meeting of physicists from the West and the East. This School, I believe, follows that pattern.

The subject of this School was recent developments in fundamental interaction theories. However, basic notions as superstrings, p-branes, M-branes, which are very far from experimental particle physics, are really the direction new developments in research are taking. We shall read here about all these ideas which determine the latest trend of fundamental interaction theories: new supersymmetric extensions, quantum deformations of symmetries and dynamical models, new approach to multidimensional Kaluza-Klein formalism, and many others. Quite a remarkable number of prominent theoreticians who lectured on these exciting subjects gathered in Karpacz.

This year, the School was organized jointly by the Institute for Theoretical Physics, the University of Wrocław, and the Institute for Physics at Łódź University.

The School has been financially supported by several institutions: as usual, by the University of Wrocław, the University of Łódź, the Ministry of Education, the Physics Committee of Polish Academy of Sciences, and the Foundation of Karpacz Schools. We would like to thank our sponsors.

The main organizers of the School were Jerzy Lukierski (University of Wrocław), Ziemowit Popowicz (University of Wrocław), and Jakub Rembieliński (University of Łódź and the College of Computer Science (Łódź)). The members of the Organizing Committee included Arkadiusz Błaut (University of Wrocław), Andrzej Frydryszak (University of Wrocław), and Cezary Gonera (University of (Łódź).

We hope that both lecturers and students enjoyed their stay in Karpacz, but this was also a period of hard work. This volume consists of the documentation of the important content of all the lectures. Hopefully, some of the results described here will take an important place in the future development of Theoretical Physics.

Jerzy Lukierski
Jakub Rembieliński

Editors

SUPERSYMMETRY AND FUNDAMENTAL INTERACTION THEORIES

Superalgebra Cohomology, the Geometry of Extended Superspaces and Superbranes

J.A. de Azcárraga* and J. M. Izquierdo†

*Departamento de Física Teórica, Valencia Univ. and IFIC (CSIC-UVEG), 46100-Burjassot, Spain
email:j.a.de.azcarraga@ific.uv.es
†Departamento de Física Teórica, Valladolid University, 47011-Valladolid, Spain
email:izquierd@fta.uva.es

Abstract. We present here a cohomological analysis of the new spacetime superalgebras that arise in the context of superbrane theory. They lead to enlarged superspaces that allow us to write D-brane actions in terms of fields associated with the additional superspace variables. This suggests that there is an extended superspace/worldvolume fields democracy for superbranes.

INTRODUCTION

Due to the development of the new string theory, it has become clear that the supersymmetry algebra contains new bosonic tensorial generators, which are central if one does not consider the Lorentz part of the complete algebra. The study of the structure of the supersymmetry algebra goes back to [1], and tensorial charges were already considered in [2] (see also [3, 4]). An enlarged superalgebra with an additional 'central' fermionic generator was introduced in [5]; other, more general algebras were considered in [6], where it was proved that to every super-p-brane of the branescan in [7], corresponds a new spacetime superalgebra, generalizing the results of [5, 8]. The point of view in [6, 9] constitutes the Lie superalgebra counterpart of the Chevalley-Eilenberg (CE) supersymmetry algebra cohomology (see [10, 11]) analysis of the scalar branes previously done in [12]. We report here on a recent work [13] which leads to a systematic cohomological construction of all these algebras, with both bosonic and fermionic generators, as well as their associated extended superspace $\tilde{\Sigma}$ *groups*. These contain additional coordinates, besides those (x^μ, θ^α) of the standard superspace Σ, which can be used to construct manifestly invariant super-p-brane Wess-Zumino (WZ) terms [8, 6, 9, 13]. Indeed, it is well known (see *e.g.* [10]) that the quasi-invariance of Lagrangians (invariance but for a total derivative) exhibits the non-trivial cohomology of the symmetry group, and that they can be rendered manifestly invariant by using the additional variables associated with the extended group. This is reflected in the realization of the symmetries in terms of Noether currents and charges, and we shall provide their general expression for superbranes. As one might expect, the new variables on $\tilde{\Sigma}$ appear trivially (in total derivatives) in the action of the scalar super-p-branes. In the case of D-branes, however, some new variables appear non-trivially since the worldvolume fields can be constructed as pull-backs of suitably enlarged superspaces that correspond to new superalgebras with additional fermionic generators.

CP589, *Fundamental Interaction Theories*, edited by J. Lukierski and J. Rembieliński

EXTENDED SUPERSPACES GIVEN BY CENTRAL EXTENSIONS OF THE SUPERTRANSLATION GROUP

Standard superspace itself provides the simplest example of our point of view. Consider the abelian, odd, *supertranslation group* sTr_D, of group law $\theta''^{\alpha} = \theta'^{\alpha} + \theta^{\alpha}$ (generically, we denote a group law as $g'' = g'g \equiv L_{g'}g = R_g g'$; L and R are the left and right actions of the group on itself). Associated with sTr_D is the trivial Maurer-Cartan (MC) equation $d\Pi^{\alpha} = 0$ ($\Pi^{\alpha} = d\theta^{\alpha}$), which is the dual version of the corresponding abelian Lie superalgebra $\{D_{\alpha}, D_{\beta}\}=0$ [1]. Now, let θ^{α} be Majorana. Then, $(C\Gamma^{\mu})_{\alpha\beta}\Pi^{\alpha}\wedge\Pi^{\beta}$ defines a *Tr_D-valued non-trivial CE two-cocycle* since a) it is closed and left-invariant (LI) *i.e.*, it is a CE cocycle and b) it is not d of a LI form (not a coboundary). Therefore, it is consistent to extend $d\Pi^{\alpha} = 0$ by a one-form $\tilde{\Pi}^{\mu}$ ($\tilde{\Pi}^{\mu} \equiv (1/2)(C\Gamma^{\mu})_{\alpha\beta}\theta^{\alpha}d\theta^{\beta}$, say) so that

$$d\tilde{\Pi}^{\mu} = (1/2)(C\Gamma^{\mu})_{\alpha\beta}\Pi^{\alpha}\Pi^{\beta} \quad . \tag{1}$$

Clearly, $\tilde{\Pi}^{\mu}$ and Π^{α} define a free differential algebra (FDA) but they are not the MC one-forms of a *Lie* algebra since *$\tilde{\Pi}^{\mu}$ is not LI*. To remedy this, we introduce a *new group coordinate x^{μ} – Minkowski space Tr_D–* and define instead

$$\Pi^{\mu} = dx^{\mu} + \tilde{\Pi}^{\mu} = dx^{\mu} + (1/2)(C\Gamma^{\mu})_{\alpha\beta}\theta^{\alpha}d\theta^{\beta} \quad , \quad \mu = 0,1,\ldots,D-1 \quad . \tag{2}$$

Obviously, $d\Pi^{\mu} = d\tilde{\Pi}^{\mu}$ and we may now choose the transformation law for x^{μ} so that Π^{μ} is LI,

$$x''^{\mu} = x'^{\mu} + x^{\mu} - (1/2)(C\Gamma^{\mu})_{\alpha\beta}\theta'^{\alpha}\theta^{\beta} \quad . \tag{3}$$

This gives [14, 10] *rigid superspace* Σ as a *group* parametrized by $(\theta^{\alpha}, x^{\mu})$, with group law now given by $\theta''^{\alpha} = \theta'^{\alpha} + \theta^{\alpha}$ and (3): *supersymmetry is the result of the non-trivial cohomology of the odd supertranslation group.*

The philosophy behind this simple superspace example, that 'fermions (θ's) are first' and that *rigid superspaces are group extensions*, may be extended by considering other types of two-cocycles (*i.e*, valued on more general spaces than Tr_D) on the sTr_D algebra. Explicitly, given a particular sTr_D algebra to be extended, one

a) looks for a non-trivial CE two-cocycle of the desired Lorentz-covariant nature. This means searching for Lorentz-tensor-valued LI closed two-forms that are not d(LI one-form).

b) Introduces a new LI one-form, the differential of which is the two-cocycle. Then,

c) the left invariance of the new one-form is achieved by fixing the transformation properties of the *new group coordinate*. This defines in general an *extended superspace (super)group* manifold $\tilde{\Sigma}$.

d) The new LI one-form together with the MC equations automatically define by (LI one-forms/LI vector fields) duality an extended Lie algebra.

---

[1] We use D_{α} (covariant derivatives) rather than Q_{α} (supersymmetry generators) because we deal with LI (hence, supersymmetry invariant) forms and vector fields, but this is unessential: the left and right algebras have the same structure constants but for an overall sign that may be conveniently ignored here.

e) Since the required Lorentz group symmetry is implicit in the process, the extension cocycles must be covariant under the action of $Spin(1,D-1)$.

The extension procedure described above can be applied more than once.

Consider the *general case of 'central' bosonic extensions* of sTr$_D$ (they are really tensorial, since the Lorentz generators do not commute with the 'central' ones). As in the above superspace example, we may look at the problem from the Lie algebra $\mathcal{G}$ or the group G point of view:

1) *Lie algebra extension point of view*
When described in terms of LI forms, the algebra extensions require the existence of higher order (α,β)-*symmetric* Lorentz tensors $(C\Gamma^{\mu_1\ldots\mu_p})_{\alpha\beta}$ of rank p

$$d\Pi^{\mu_1\ldots\mu_p} \equiv (1/2)(C\Gamma^{\mu_1\ldots\mu_p})_{\alpha\beta}\Pi^\alpha\Pi^\beta\,, \tag{4}$$

$(\Pi^\alpha\Pi^\beta \equiv \Pi^\alpha \wedge \Pi^\beta = \Pi^\beta \wedge \Pi^\alpha$; we omit wedge products). The corresponding generators $Z_{\mu_1\ldots\mu_p}$ are all $(D_\alpha-)$*central*, as the translation generator $X_\mu = P_\mu$ itself, and are associated with new central charges.

The LI of the new forms in (4) *requires* new group parameters $\varphi^{\mu_1\ldots\mu_p}$ so that

$$\Pi^{\mu_1\ldots\mu_p} = d\varphi^{\mu_1\ldots\mu_p} + (1/2)(C\Gamma^{\mu_1\ldots\mu_p})_{\alpha\beta}\theta^\alpha\Pi^\beta\,, \tag{5}$$

is LI. These new parameters $\varphi^{\mu_1\ldots\mu_p}$ generalize the spacetime parameters x^μ, and their associated generators $Z_{\mu_1\ldots\mu_p}$ may be considered as *generalised momenta*. There are no (two-cocycle) restrictions coming from the Jacobi identity since the r.h.s of (4) is trivially consistent with $d(d\Pi^{\mu_1\ldots\mu_p}) \equiv 0$.

2) *Group extension point of view*
The closedness of the r.h.s. of (4) means that the Lorentz tensor-valued two-cocycle on sTr$_D$, $\xi^{\mu_1\ldots\mu_p}(\theta',\theta) = \theta'^\alpha(C\Gamma^{\mu_1\ldots\mu_p})_{\alpha\beta}\theta^\beta$, satisfies also (trivially) the *two-cocycle condition*

$$\xi(\theta,\theta')^{\mu_1\ldots\mu_p} + \xi(\theta+\theta',\theta'')^{\mu_1\ldots\mu_p} = \zeta(\theta,0'+\theta'')^{\mu_1\ldots\mu_p} + \xi(\theta',\theta'')^{\mu_1\ldots\mu_p}\,. \tag{6}$$

The symmetry of $(C\Gamma^{\mu_1\ldots\mu_p})_{\alpha\beta}$ is needed to prevent the above two-cocycle from being trivial, since the possible function $\eta^{\mu_1\ldots\mu_p}(\theta)$ on sTr$_D$ that might generate the two-coboundary $(\xi_{cob}^{\mu_1\ldots\mu_p}(\theta',\theta) = \eta^{\mu_1\ldots\mu_p}(\theta+\theta') - \eta^{\mu_1\ldots\mu_p}(\theta') - \eta^{\mu_1\ldots\mu_p}(\theta))$ is zero: $\eta^{\mu_1\ldots\mu_p}(\theta) = \theta^\alpha(C\Gamma^{\mu_1\ldots\mu_p})_{\alpha\beta}\theta^\beta \equiv 0$. Hence, *The problem of finding all* central *extensions of the sTr$_D$ algebra $\{D_\alpha, D_\beta\} = 0$ is reduced to finding a basis of the symmetric space $\Pi^{(\alpha}\otimes\Pi^{\beta)}$ in terms of p-Lorentz tensors $(C\Gamma^{\mu_1\ldots\mu_p})_{\alpha\beta}$ symmetric in (α,β).*

The answer for different spacetime dimensions D depends on the properties of their respective Γ matrices, since they determine the existence of non-trivial cocycles (see [13] for a table). We shall only consider here the example of the

D=11, M-Theory Extended Superspace

The maximally centrally extended FDA is obtained by adding Π^{μ}, $\Pi^{\mu_1\mu_2}$, $\Pi^{\mu_1\ldots\mu_5}$ to $\Pi^{\alpha}=d\theta^{\alpha}$ (θ^{α} Majorana, $\alpha=1,\ldots,32$) satisfying

$$d\Pi^{\alpha}=0\,,\; d\Pi^{\mu}=\frac{1}{2}(C\Gamma^{\mu})_{\alpha\beta}\Pi^{\alpha}\Pi^{\beta}\quad, \\ d\Pi^{\mu_1\mu_2}=\frac{1}{2}(C\Gamma^{\mu_1\mu_2})_{\alpha\beta}\Pi^{\alpha}\Pi^{\beta}\,,\; d\Pi^{\mu_1\ldots\mu_5}=\frac{1}{2}(C\Gamma^{\mu_1\ldots\mu_5})_{\alpha\beta}\Pi^{\alpha}\Pi^{\beta}\quad. \tag{7}$$

There are no *one*-forms $\Pi^{\mu_1\mu_2}$, $\Pi^{\mu_1\ldots\mu_5}$ LI on Σ. They can be made LI by introducing *new* 'central' (tensorial) coordinates $\varphi^{\mu_1\mu_2}$, $\varphi^{\mu_1\ldots\mu_5}$. These *define* the $D=11, N=1$ extended superspace $\tilde{\Sigma}(\theta^{\alpha},x^{\mu},\varphi^{\mu_1\mu_2},\varphi^{\mu_1\ldots\mu_5})$. In terms of the central generators $X_{\mu}=\partial/\partial x^{\mu}$, $Z_{\mu_1\mu_2}=\partial/\partial\varphi^{\mu_1\mu_2}$, $Z_{\mu_1\ldots\mu_5}=\partial/\partial\varphi^{\mu_1\ldots\mu_5}$, the $D=11$ *supersymmetry M-algebra* dual to (7) is

$$\{D_{\alpha},D_{\beta}\}=(C\Gamma^{\mu})_{\alpha\beta}X_{\mu}+(C\Gamma^{\mu_1\mu_2})_{\alpha\beta}Z_{\mu_1\mu_2}+(C\Gamma^{\mu_1\ldots\mu_5})_{\alpha\beta}Z_{\mu_1\ldots\mu_5}\quad. \tag{8}$$

This is usually referred to [15] as the *M-theory superalgebra*. The group law of the extended superspace $\tilde{\Sigma}(\theta^{\alpha},x^{\mu},\varphi^{\mu_1\mu_2},\varphi^{\mu_1\ldots\mu_5})$ is obtained easily as the simplest Σ case (*cf.* (2)).

For a recent discussion of the M-algebra, in which the tensorial central charges are considered as bilinears of spinors, see [16].

NON-CENTRAL ADDITIONAL GENERATORS AND THEIR EXTENDED SUPERSPACES

The above are *central* extensions of the basic odd abelian algebra $\{D_{\alpha},D_{\beta}\}=0$ by *bosonic* tensorial generators. But there are also extensions by fermionic generators that make non-abelian *e.g.*, the $[X_{\mu},D_{\alpha}]=0$ commutator. The CE cohomology analysis is also useful here. Let us start from a centrally extended superspace $\tilde{\Sigma}(\theta^{\alpha},x^{\mu},\varphi_{\mu_1\ldots\mu_p})$, p fixed, and LI one-forms Π^{μ}, Π^{α}, $\Pi_{\mu_1\ldots\mu_p}$, satisfying the MC eqs.

$$d\Pi^{\mu}=a_s(C\Gamma^{\mu})_{\alpha\beta}\Pi^{\alpha}\Pi^{\beta}\quad,\quad d\Pi_{\mu_1\ldots\mu_p}\equiv a_0(C\Gamma_{\mu_1\ldots\mu_p})_{\alpha\beta}\Pi^{\alpha}\Pi^{\beta}\quad, \tag{9}$$

where a_s, a_0 are not fixed for convenience. A non-trivial CE two-cocycle with p indices has to be of the type $(\mu_1\ldots\mu_{p-1}\alpha_1)$ and, hence, the only available LI *two*-forms (in this case, fermionic) are

$$\rho^{(1)}_{\mu_1\ldots\mu_{p-1}\alpha_1}=(C\Gamma_{\nu\mu_1\ldots\mu_{p-1}})_{\beta\alpha_1}\Pi^{\nu}\Pi^{\beta}\,,\; \rho^{(2)}_{\mu_1\ldots\mu_{p-1}\alpha_1}=(C\Gamma^{\nu})_{\beta\alpha_1}\Pi_{\nu\mu_1\ldots\mu_{p-1}}\Pi^{\beta}\quad. \tag{10}$$

For $p=1$, both are closed. For $p\geq 2$, the condition $d(\rho^{(1)}+\lambda_2\rho^{(2)})=0$ fixes $\lambda_2=a_s/a_0$ provided

$$(C\Gamma^{\nu})_{(\alpha\beta}(C\Gamma_{\nu\mu_1\ldots\mu_{p-1}})_{\gamma\delta)}=0\quad, \tag{11}$$

which holds for the (D,p) of the scalar branescan [7]. Condition (11) is a *new* feature of the *'non-central' case*; in the central (bosonic two-cocycles) case, the closedness was *trivially* satisfied and hence there was *no condition on D; only p* (the rank p of the Lorentz tensor, see below (5)) was restricted by the $(\alpha\beta)$ symmetry of the tensor. We may introduce now a new one–form $\Pi_{\mu_1\ldots\mu_{p-1}\alpha_1}$ with

$$d\Pi_{\mu_1\ldots\mu_{p-1}\alpha_1} = a_1\left((C\Gamma_{\nu\mu_1\ldots\mu_{p-1}})_{\beta\alpha_1}\Pi^\nu\Pi^\beta + \frac{a_s}{a_0}(C\Gamma^\nu)_{\beta\alpha_1}\Pi_{\nu\mu_1\ldots\mu_{p-1}}\Pi^\beta)\right) \quad (12)$$

(for $p = 1$ the coefficient of the second term can be arbitrary). This MC equation implies that both $[D_\alpha, X_\mu]$ and $[D_\alpha, Z^{\mu_1\ldots\mu_p}]$ are modified by a term proportional to $Z^{\mu_1\ldots\mu_{p-1}\alpha_1}$, the latter being the only central generator at this stage ($Z^{\mu_1\ldots\mu_{p-1}\alpha_1}$ is central because, by construction, $\Pi_{\mu_1\ldots\mu_{p-1}\alpha_1}$ cannot appear at the r.h.s. of a MC equation expressing the differential of a LI form).

The general features of the extensions with non-central fermionic generators are:

a) The extension two-cocycles (two-forms) may be *fermionic* (eqs. (10)). This leads to non-zero [bosonic,fermionic] commutators.

b) At any stage in the chain of extensions, the only *central* generator present is the one introduced in the *last* extension.

c) Successive central extensions substitute one spinorial index for a vectorial one. This leads to one-forms of the type

$$\Pi_{\mu_1\ldots\mu_{p-k}\alpha_1\ldots\alpha_k} \equiv \Pi_{\rho_k}\,, \quad \rho_k \equiv (\mu_1\ldots\mu_{p-k}\alpha_1\ldots\alpha_k) \quad , \quad (13)$$

where ρ_k labels the additional coordinates of the extended superspace $\tilde{\Sigma}$.

d) The procedure ends when the p vector indices have become spinorial ones so that $\Pi_{\rho_k} \to \Pi_{\rho_p} \equiv \Pi_{\alpha_1\ldots\alpha_p}$.

e) For a given p, there are consistency conditions that restrict the spacetime dimension D; for instance, the Green algebra exists for D=3,4,6 and 10 only [5].

All the extended superspaces have a natural fibre bundle structure that is inherited from their group extension character; we refer to [13] (see also [14]) for details.

Two Applications: the GS Superstring and the Supermembrane

Consider the Green-Schwarz superstring case (p=1, D=10, N=1). We shall denote by φ_μ the additional vector parameter and by Z^μ, $\Pi^{(\varphi)}_\mu$ the associated generator and LI form [13]. The MC eqs. are

$$\begin{aligned} d\Pi^\alpha &= 0\,, \quad d\Pi^\mu = (1/2)(C\Gamma^\mu)_{\alpha\beta}\Pi^\alpha\Pi^\beta\,, \\ d\Pi^{(\varphi)}_\mu &= (1/2)(C\Gamma_\mu)_{\alpha\beta}\Pi^\alpha\Pi^\beta\,,\ d\Pi_\alpha = (C\Gamma_\mu)_{\alpha\beta}\Pi^\mu\Pi^\beta + (C\Gamma^\mu)_{\alpha\beta}\Pi^{(\varphi)}_\mu\Pi^\beta; \end{aligned} \quad (14)$$

$\mu = 0,\ldots,9$, and all spinors here are MW ($\theta^\alpha \equiv \mathcal{P}_+\theta^\alpha$, $\Pi^\alpha \equiv \mathcal{P}_+ d\theta^\alpha$; notice that Π^α and Π_α are unrelated). The two terms in the r.h.s. of the last of (14) are individually closed ($d(d\Pi_\alpha) = 0$ follows from (11) for $p = 1$, *i.e.* by $(C\Gamma^\mu)_{(\alpha\beta}(C\Gamma_\mu)_{\gamma\delta)} = 0$) and hence their relative normalization cannot be fixed by requiring $d(d\Pi_\alpha) = 0$.

The corresponding Lie superalgebra contains an additional *fermionic* central generator, Z^{β}, and is given by

$$\begin{aligned}\{D_{\alpha}, D_{\beta}\} &= (C\Gamma^{\mu})_{\alpha\beta}X_{\mu} + (C\Gamma^{\mu})_{\alpha\beta}Z^{\mu}, \\ [D_{\alpha}, X_{\mu}] &= (C\Gamma^{\mu})_{\alpha\beta}Z^{\beta}, \quad [D_{\alpha}, Z^{\mu}] = (C\Gamma^{\mu})_{\alpha\beta}Z^{\beta} \quad ;\end{aligned} \tag{15}$$

if one omits Z^{μ}, it reduces to the *Green algebra* [5][2]. Note that X_{μ} *is no longer central* due to the presence of Z^{β}. The associated group manifold is the *GS superstring extended superspace* $\tilde{\Sigma}(\theta^{\alpha}, x^{\mu}, \varphi_{\mu}, \varphi_{\alpha})$; its group law is given and discussed in [13].

As mentioned, the extended superspaces are suitable to define manifestly invariant WZ terms. For instance, using the LI forms on $\tilde{\Sigma}(\theta^{\alpha}, x^{\mu}, \varphi_{\mu}, \varphi_{\alpha})$, one obtains the *manifestly invariant WZ term for the GS superstring* (*cf.* [8])

$$S_{WZ} = \int_W \phi^*(\tilde{b}) = \int_W \phi^*(\Pi^{(\varphi)}_{\mu}\Pi^{\mu} + \frac{1}{2}\Pi_{\alpha}\Pi^{\alpha}) \quad , \tag{16}$$

$d\tilde{b} = db = h = (C\Gamma^{\mu})_{\alpha\beta}\Pi^{\mu}\Pi^{\alpha}\Pi^{\beta}$ and hence $\phi^*(\tilde{b})$ and the standard WZ term $\phi^*(b)$ are equivalent; b and $\tilde{b}$ *differ only by an exact form.*

Similarly, it is also possible to write a manifestly invariant D=11 membrane (p=2) WZ term. It exists on the $D = 11$ *supermembrane extended superspace group* $\tilde{\Sigma}(\theta^{\alpha}, x^{\mu}, \varphi_{\mu\nu}, \varphi_{\mu\alpha}, \varphi_{\alpha\beta})$, and is found to be

$$\tilde{b} = (2/3)\Pi_{\mu\nu}\Pi^{\mu}\Pi^{\nu} - (3/5)\Pi_{\mu\alpha}\Pi^{\mu}\Pi^{\alpha} - (2/15)\Pi_{\alpha\beta}\Pi^{\alpha}\Pi^{\beta}\,, \tag{17}$$

as given in [6]. Again, $\tilde{b}$ *depends on the additional variables* φ *through total differentials* since $d\tilde{b} = db = h = (C\Gamma_{\mu\nu})_{\alpha\beta}\Pi^{\mu}\Pi^{\nu}\Pi^{\alpha}\Pi^{\beta}$. The non-WZ part of the action does not depend on the additional variables of extended superspace, and remains the standard one. As we shall see, this situation will change for D-branes.

NEW NOETHER CURRENTS AND CHARGES

Let us now give the general expression for the Noether currents for the additional symmetries $j^i_{\sigma_l}$ (see (13) for the notation) using the manifestly invariant WZ forms $\tilde{L}_{WZ}$ defined on the various extended superspaces $\tilde{\Sigma}$. The Lagrangian $\tilde{L}_{WZ}(\xi)$ on the worldvolume W (of coordinates ξ^i, $i = 0, 1, \ldots, p$) is the pull-back $\phi^*(\tilde{L}_{WZ}) = \tilde{L}_{WZ}(\xi)d^{p+1}\xi$ of the (p+1)-form $\tilde{L}_{WZ}$ on $\tilde{\Sigma}$ by the map $\phi : W \longrightarrow \tilde{\Sigma}$. The charges that correspond to the current densities $j^i_{\sigma_l}(\xi)$ appear on the r.h.s. of the supersymmetry algebra.

The WZ part of the action is $\int \tilde{L}_{WZ}(\xi)d^{p+1}\xi$ and, *since only* $\tilde{L}_{WZ}$ *depends on the additional variables of the extended superspace* $\tilde{\Sigma}$ (different from (x^{μ}, θ^{α}) of Σ), we

---

[2] This algebra may be viewed as a 'stabilising deformation' of Σ [17]. In this context, stability is achieved by exhausting the second Lie algebra cohomology group (the non-trivial two-cocycle space) *i.e.*, by extending maximally under certain conditions, as done in [13]. This means, *e.g.*, including *both* generators $Z_{\mu\nu}$ and $Z_{\mu_1\ldots\mu_5}$ in (8) if only bosonic ones are considered (*cf.* [9]), and similarly for the other cases.

shall focus on $\tilde{\mathcal{L}}_{WZ}$. We find first the general expression for the Noether currents and then apply it to the simple cases of the GS superstring and the supermembrane.

We start by writing the manifestly invariant density $\tilde{\mathcal{L}}_{WZ}(\xi)$ as

$$\tilde{\mathcal{L}}_{WZ}(\xi) \equiv \Pi_{\rho_k i}(\xi)\,\Lambda^{\rho_k i}(\xi) \quad , \quad \xi^i = (\tau,\sigma) \quad , \quad i = 0,1,\ldots,p \quad , \tag{18}$$

(see (13)) where Λ^{ρ_k} is defined by (18) and denotes the LI $(p\text{-})$form

$$\Lambda^{\rho_k} \equiv \Lambda^{\mu_1\ldots\mu_{p-k}\alpha_1\ldots\alpha_k} = a_k \Pi^{\mu_1}\ldots\Pi^{\mu_{p-k}}\Pi^{\alpha_1}\ldots\Pi^{\alpha_k} \quad , \tag{19}$$

$\Pi_{\rho_k i}$=$(\phi^*(\Pi_{\rho_k})_i$ and $\Lambda^{\rho_k i}$ corresponds to $a_k\varepsilon^{i j_1\ldots j_p}\Pi^{\mu_1}_{j_1}\ldots\Pi^{\mu_{p-k}}_{j_{p-k}}\Pi^{\alpha_1}_{j_{p-k+1}}\ldots\Pi^{\alpha_k}_{j_p}$ (the constants a_k are fixed by $d\tilde{b} = h$).

Given the group law $g'' = g'g$ $(g''^A = g''^A(g',g), A = (\alpha,\mu;\rho_k))$ of $\tilde{\Sigma}$, the LI one-forms $\Pi^A(g)$ and the RI vector fields $Z_A(g)$ are given by

$$\Pi^A(g) = \Pi^A_B(g)\,dg^B = \left.\frac{\partial g''^A(g',g)}{\partial g^B}\right|_{g'=g^{-1}} dg^B\ , \quad Z_A(g) = \left.\frac{\partial g''^D(g',g)}{\partial g'^A}\right|_{g'=e}\frac{\partial}{\partial g^D} \tag{20}$$

(see, *e.g.*, [10]). The Z_A generate the left g^A-translations of $\tilde{\Sigma}$.

The $\tilde{\mathcal{L}}_{WZ}(\xi)$ *contribution to* $j^i_A(\xi) = j^i_{A(kin)}(\xi) + j^i_{A(WZ)}(\xi)$ is

$$j^i_{A(WZ)}(\xi) = (\delta_A g^B)\frac{\partial\tilde{\mathcal{L}}_{WZ}}{\partial g^B{}_{,i}} \equiv (Z_A.g^B)\frac{\partial\tilde{\mathcal{L}}_{WZ}}{\partial g^B{}_{,i}} \quad . \tag{21}$$

Let the extended superspace index refer to a new coordinate, $A = \sigma_l$, and let us compute $j^i_{\sigma_l}$. Since only $\tilde{\mathcal{L}}_{WZ}$ depends on the new coordinates, $j^i_{\sigma_l(kin)} = 0$. The B summation in (21) is reduced to a summation over the additional coordinates index η_k since the vector fields Z_{σ_l} do not have $\partial/\partial x^\mu$, $\partial/\partial\theta^\alpha$ components and thus $Z_{\sigma_l}.g^B = 0$ for $g^B = (\theta^\alpha, x^\mu)$. Moreover, since $\Lambda^{\rho_k} = \Lambda^{\rho_k}(\Pi^\mu,\Pi^\alpha)$ and Π^μ,Π^α are defined on the standard Σ, the $\Lambda^{\rho_k i}$ part does not depend on φ_{η_k} (g^{η_k} in (21)),

$$j^i_{\sigma_l} = (Z_{\sigma_l}.g^{\eta_k})\left(\frac{\partial}{\partial g^{\eta_k}{}_{,i}}\Pi_{\rho_k i}\right)\Lambda^{\rho_k i} \quad . \tag{22}$$

Using Eq. (20),

$$\begin{aligned} j^i_{\sigma_l} &= (Z_{\sigma_l}.g^{\eta_k})\left\{\frac{\partial}{\partial g^{\eta_k}{}_{,i}}\left(\left.\frac{\partial g''_{\rho_k}(g',g)}{\partial g^B}\right|_{g'=g^{-1}} g^B{}_{,i}\right)\right\}\Lambda^{\rho_k i} \\ &= (Z_{\sigma_l}.g^{\eta_k})\left.\frac{\partial g''_{\rho_k}(g',g)}{\partial g^{\eta_k}}\right|_{g'=g^{-1}}\Lambda^{\rho_k i}, \end{aligned} \tag{23}$$

since $g'' \neq g''(g_{,i})$. This gives the *general expression for the Noether currents associated with the additional generators*:

$$j^i_{\sigma_l} = (Z_{\sigma_l}.g''_{\rho_k}(g',g)|_{g'=g^{-1}})\Lambda^{\rho_k i} \equiv T_{\sigma_l\rho_k}\Lambda^{\rho_k i} \quad , \tag{24}$$

where T corresponds to the adjoint representation $Ad(g^{-1})$ and depends on ξ through $g(\xi)$ (notice that if X^R is RI and Π^L as a LI one-form, $i_{X^R}\Pi^L = Ad(g^{-1})X^R$). Since for $A=\sigma_l$ we may restrict D to η_k ($Z^{\mu,\alpha}_{\sigma_l}(g)=0$), eq. (24) may also be written as

$$j^i_{\sigma_l}(\xi) = \left(\frac{\partial g'^{\eta_k}(g',g)}{\partial g'^{\sigma_l}}\bigg|_{g'=e} \frac{\partial g''_{\rho_k}(g',g)}{\partial g^{\eta_k}}\bigg|_{g'=g^{-1}} \right) \Lambda^{\rho_k i} \tag{25}$$

using (20); the bracketed term is determined by the group $\tilde{\Sigma}$ only, and $\Lambda^{\rho_k i}$ by $\tilde{\mathcal{L}}_{WZ}(\xi)$.

a) *D=10, N=1 superstring*:

Using expression (16) for (18), we find that the conserved Noether currents are

$$j^{\mu i}_{(\varphi)} = \varepsilon^{ij}\partial_j x^\mu \quad , \quad j^{\alpha i} = (1/2)\varepsilon^{ij}\partial_j\theta^\alpha \,, \tag{26}$$

and the charges [18]

$$Z^\mu = \oint d\sigma j^{\mu 0}_{(\varphi)} = \oint d\sigma \frac{\partial x^\mu}{\partial \sigma} \,,\quad Z^\alpha = \oint d\sigma j^{\alpha 0} = \oint d\sigma \frac{1}{2}\frac{\partial\theta^\alpha}{\partial\sigma} = 0 \,, \tag{27}$$

assuming that θ is periodic in σ (*cf.* [19]). It is clear that, in general, the integral of j^0 (as, *e.g.*, for $j^{\mu 0}_{(\varphi)}$) leads to a non-zero result if the topology is nontrivial (the loop is not contractible).

b) *D=11 2-brane*:

It can be shown from eq. (17) that the currents can be written as the worldvolume duals of the *current two-forms*

$$\begin{aligned} J^{\mu\nu} &= d\left(\frac{2}{3}x^{[\mu}dx^{\nu]} + \frac{1}{15}\theta^\alpha x^{[\mu}(C\Gamma^{\nu]})_{\alpha\beta}d\theta^\beta\right) \quad , \\ J^{\kappa\alpha} &= d\left(\frac{3}{5}dx^\kappa\theta^\alpha - \frac{1}{30}(C\Gamma^\kappa)_{\beta\gamma}\theta^\beta\theta^\alpha d\theta^\gamma\right) , \quad J^{\beta\gamma} = d(-\frac{2}{15}\theta^\beta d\theta^\gamma) \quad ; \end{aligned} \tag{28}$$

current conservation follows from $dJ=0$. For periodic θ's the charges $Z^{\kappa_1\alpha_1}$, $Z^{\beta_1\gamma_1}$ turn out to be zero, but not $Z^{\mu_1\nu_1}$ for a non-trivial closed two-cycle [18] (in the general p-case, the integrals are over non-trivial de Rham p-cycles; we refer to [18] for details on topological charges). Thus, the above assumptions provide a realization of the extended algebra where only the bosonic $Z^{\mu\nu}$ generator is realized non-trivially.

THE CASE OF D-BRANES

Consider first a bosonic background such that the action of the Dp-brane [20, 21] reduces to

$$I = \int d^{p+1}\xi \sqrt{-\det(\partial_i x^\mu \partial_j x_\mu + F_{ij})} \quad , \tag{29}$$

where $F=dA$ and $A(\xi) = A_i(\xi)d\xi^i$ is the worldvolume Born-Infeld (BI) field.

Let us look for a *manifestly supersymmetric generalisation*. This means substituting first Π_i^μ for $\partial_i x^\mu$, $F_{ij} = \partial_{[i}A_{j]}$ by $\mathcal{F} = dA - B$, and then adding a WZ term b, $db = h$. A previous analysis [12] of the WZ terms of the scalar branescan [7] showed that *WZ terms may be characterized and classified by CE-(p+2)cocycles*. The same philosophy is successful for the Dp-branes. The result is that Dp-branes may also be be characterized (see below and [13] for details and further references) by means of non-trivial CE $(p+2)$-cocycles, recovering Polchinski's consistency conditions [20] (p even/odd for IIA/IIB). In the case of D-branes, however, and due to the presence of F_{ij} in the kinetic term (29) the situation turns out to be different from that of the previous p-branes: the new variables will appear in the action *non-trivially*, not as total derivatives.

Example: the D2-Brane Defined on its Extended Superspace

Consider the D2-brane. The starting point is now the IIA-type FDA plus the $d\mathcal{F}$ equation *i.e.*

$$
\begin{aligned}
d\Pi^\alpha &= 0, & d\Pi^\mu &= \tfrac{1}{2}(C\Gamma^\mu)_{\alpha\beta}\Pi^\alpha\Pi^\beta, \\
d\Pi &= \tfrac{1}{2}(C\Gamma_{11})_{\alpha\beta}\Pi^\alpha\Pi^\beta, & d\Pi_{\mu\nu} &= \tfrac{1}{2}(C\Gamma_{\mu\nu})_{\alpha\beta}\Pi^\alpha\Pi^\beta, \\
d\Pi_\mu^{(z)} &= \tfrac{1}{2}(C\Gamma_\mu\Gamma_{11})_{\alpha\beta}\Pi^\alpha\Pi^\beta, & d\mathcal{F} &= (C\Gamma_\mu\Gamma_{11})_{\alpha\beta}\Pi^\mu\Pi^\alpha\Pi^\beta,
\end{aligned}
\tag{30}
$$

($\mu=0,\ldots 9, \alpha=1,\ldots 32$). This is justified *e.g* by the fact that the dual of the first 5 eqs. is the algebra obtained when one computes the algebra of Noether charges for the type IIA D2-brane [22]. The next step is extending this algebra with the generators obtained by replacing vector indices by spinorial ones, as outlined after (12). In the case of the D2-brane this is not difficult to do because, apart from the equation for $d\mathcal{F}$, the FDA above is actually the dimensional reduction to D=10 of the D=11 one (eq. (7) with generators with one or two vector indices since p=2),

$$
d\Pi^{\tilde\mu} = (1/2)(C\Gamma^{\tilde\mu})_{\alpha\beta}\Pi^\alpha\Pi^\beta, \qquad d\Pi_{\tilde\mu\tilde\nu} = (1/2)(C\Gamma_{\tilde\mu\tilde\nu})_{\alpha\beta}\Pi^\alpha\Pi^\beta, \tag{31}
$$

where ($\tilde\mu = (\mu, 10) = 0, 1, \ldots 10$), and in which one sets $\Pi^{\tilde\mu} \equiv (\Pi^\mu, \Pi^{10} \equiv \Pi)$, $\Pi_{\tilde\mu\tilde\nu} \equiv (\Pi_{\mu\nu}, \Pi_{\mu 10} \equiv \Pi_\mu^{(z)})$. This $D = 11$ FDA may be extended. The $D = 10$ dimensional reduction of the extended algebra gives

$$
\begin{aligned}
d\Pi^\alpha &= 0, \quad d\Pi^\mu = \frac{1}{2}(C\Gamma^\mu)_{\alpha\beta}\Pi^\alpha\Pi^\beta, \quad d\Pi = \frac{1}{2}(C\Gamma_{11})_{\alpha\beta}\Pi^\alpha\Pi^\beta, \\
d\Pi_{\mu\nu} &= \frac{1}{2}(C\Gamma_{\mu\nu})_{\alpha\beta}\Pi^\alpha\Pi^\beta, \quad d\Pi_\mu^{(z)} = \frac{1}{2}(C\Gamma_\mu\Gamma_{11})_{\alpha\beta}\Pi^\alpha\Pi^\beta, \\
d\Pi_{\mu\alpha} &= (C\Gamma_{\nu\mu})_{\alpha\beta}\Pi^\nu\Pi^\beta + (C\Gamma_{11}\Gamma_\mu)_{\alpha\beta}\Pi\Pi^\beta + (C\Gamma^\nu)_{\alpha\beta}\Pi_{\nu\mu}\Pi^\beta - (C\Gamma_{11})_{\alpha\beta}\Pi_\mu^{(z)}\Pi^\beta \quad , \\
d\Pi_\alpha^{(z)} &= (C\Gamma_\nu\Gamma_{11})_{\alpha\beta}\Pi^\nu\Pi^\beta + (C\Gamma^\nu)_{\alpha\beta}\Pi_\nu^{(z)}\Pi^\beta \quad ,
\end{aligned}
$$

$$
\begin{aligned}
d\Pi_{\alpha\beta} \;=\;& -\frac{1}{2}(C\Gamma_{\mu\nu})_{\alpha\beta}\Pi^{\mu}\Pi^{\nu} - (C\Gamma_{\mu}\Gamma_{11})_{\alpha\beta}\Pi^{\mu}\Pi - \frac{1}{2}(C\Gamma^{\mu})_{\alpha\beta}\Pi_{\mu\nu}\Pi^{\nu} \\
& +\frac{1}{2}(C\Gamma_{11})_{\alpha\beta}\Pi^{(z)}_{\mu}\Pi^{\mu} - \frac{1}{2}(C\Gamma^{\mu})_{\alpha\beta}\Pi^{(z)}_{\mu}\Pi + \frac{1}{4}(C\Gamma^{\mu})_{\alpha\beta}\Pi_{\mu\delta}\Pi^{\delta} \\
& +\frac{1}{4}(C\Gamma_{11})_{\alpha\beta}\Pi^{(z)}_{\delta}\Pi^{\delta} + 2\Pi_{\mu(\beta}(C\Gamma^{\mu})_{\alpha)\delta}\Pi^{\delta} + 2(C\Gamma_{11})_{\delta(\alpha}\Pi^{(z)}_{\beta)}\Pi^{\delta} \quad . \qquad (32)
\end{aligned}
$$

Using the new forms it is possible to find a manifestly invariant WZ form $\tilde{b}$, $d\tilde{b} = h$; h is given by

$$
h = (C\Gamma_{\mu\nu})_{\alpha\beta}\Pi^{\mu}\Pi^{\nu}\Pi^{\alpha}\Pi^{\beta} - (C\Gamma_{11})_{\alpha\beta}\Pi^{\alpha}\Pi^{\beta}\mathcal{F} \quad , \qquad (33)
$$

and the *manifestly invariant WZ term for the type IIA D2-brane* by

$$
\tilde{b} = \frac{2}{3}\Pi_{\mu\nu}\Pi^{\mu}\Pi^{\nu} + \frac{4}{3}\Pi^{(z)}_{\mu}\Pi^{\mu}\Pi - \frac{2}{15}\Pi_{\alpha\beta}\Pi^{\alpha}\Pi^{\beta} - \frac{3}{5}\Pi_{\mu\alpha}\Pi^{\mu}\Pi^{\alpha} - \frac{3}{5}\Pi^{(z)}_{\alpha}\Pi\Pi^{\alpha} - 2\Pi\mathcal{F} \; . \qquad (34)
$$

We expect that this analysis also holds true for the other values of p.

The extended free differential algebra is not the dual of a Lie algebra because it includes the equation for the *three*-form $d\mathcal{F}$. However,

$$
d(\frac{1}{2}\Pi^{\alpha}\Pi^{(z)}_{\alpha} - \Pi^{\mu}\Pi^{(z)}_{\mu}) = (C\Gamma_{\mu}\Gamma_{11})_{\alpha\beta}\Pi^{\mu}\Pi^{\alpha}\Pi^{\beta} \qquad (35)
$$

so that, on the extended superspace $\tilde{\Sigma}(\theta^{\alpha}, x^{\mu}, \varphi_{\mu}, \varphi_{\alpha})$ we may set

$$
\mathcal{F} = (1/2)\Pi^{\alpha}\Pi^{(z)}_{\alpha} - \Pi^{\mu}\Pi^{(z)}_{\mu} \quad . \qquad (36)
$$

Since $\mathcal{F} = dA - B$ *and* B *is defined on* Σ, it follows that dA may be written on $\tilde{\Sigma}$. Making use of the explicit form of the LI one-forms in terms of the extended superspace variables, it is easy to identify A *as the one-form on* $\tilde{\Sigma}$

$$
A = \varphi_{\mu}dx^{\mu} + (1/2)\varphi_{\alpha}d\theta^{\alpha} \quad . \qquad (37)
$$

In the present approach, the customary BI worldvolume field $A_i(\xi)d\xi^i$ becomes $\phi^*(A)$; we might even say that the existence of the BI field is a consequence of supersymmetry. We now check the consistency of the replacement (37).

a) The Euler Lagrange equations are still the same. Let $I[x^{\mu}(\xi), \theta^{\alpha}(\xi), A_i(\xi)]$ be the action before making the substitution. The EL equations are

$$
\delta I/\delta x^{\mu} = 0, \qquad \delta I/\delta\theta^{\alpha} = 0, \qquad \delta I/\delta A_j = 0. \qquad (38)
$$

When the substitution is made,

$$
\begin{aligned}
\int d\xi'^{p+1}\frac{\delta I}{\delta A_j(\xi')}\frac{\delta A_j(\xi')}{\delta x^{\mu}(\xi)} + \frac{\delta I}{\delta x^{\mu}} &= 0, & \frac{\delta I}{\delta\varphi_{\mu}} = \frac{\delta I}{\delta A_j}\partial_j x^{\mu} &= 0, \\
\int d\xi'^{p+1}\frac{\delta I}{\delta A_j(\xi')}\frac{\delta A_j(\xi')}{\delta\theta^{\alpha}(\xi)} + \frac{\delta I}{\delta\theta^{\alpha}} &= 0, & \frac{\delta I}{\delta\varphi_{\alpha}} = \frac{1}{2}\frac{\delta I}{\delta A_j}\partial_j\theta^{\alpha} &= 0.
\end{aligned} \qquad (39)
$$

We see that to avoid the collapse of one or more worldvolume dimensions we must have $\delta I/\delta A_j = 0$ which implies eqs. (38). This also follows from the fact that $\delta I/\delta\varphi_\mu = 0$ implies $(\delta I/\delta A_j)g_{ij} = 0$, where $g_{ij} \equiv \Pi_i^\mu \Pi_{\mu j} = \partial_i x^\mu \partial_j x_\mu +$ (nilpotent terms) is the induced worldvolume metric. Thus, we must have $\delta I/\delta A_j = 0$ to prevent g_{ij} from being degenerate. As a result, $\delta I/\delta\varphi_\alpha = 0$ is satisfied identically and it is a Noether identity.

b) The gauge transformations of $A_i(\xi)$ can be reinterpreted in the new language. If one defines $\delta\varphi_\mu = \partial_\mu\lambda$ and $\delta\varphi_\alpha = 2\partial_\alpha\lambda$, by means of a superfield λ such that $\phi^*\lambda(x^\mu,\theta^\alpha) = \Lambda(\xi)$, then $\phi^*(A)$ behaves as expected: $\delta(\phi^*[\varphi_\mu dx^\mu + \frac{1}{2}\varphi_\alpha d\theta^\alpha]) = \partial_i\Lambda$.

c) The number of worldvolume degrees of freedom remains the same. Let us first note that, since $\delta I/\delta\varphi_\alpha = 0$ is a Noether identity, the second Noether theorem tells us that there exists a gauge symmetry that can be used to set $\varphi_\alpha = 0$. Thus, the 'physical' part of A is contained in $\varphi_\mu dx^\mu$. The identification (37) is therefore equivalent to replacing $A_i(\xi)$ by $\varphi_\mu(\xi)\partial_i x^\mu(\xi)$. We now notice that the D equations $\delta I/\delta\varphi_\mu = 0$ produce only $(p+1)$ independent ones, $\delta I/\delta A_j = 0$; the remaining $D-(p+1)$ equations are Noether identities that reflect the existence of further gauge symmetries. To check explicitly the degrees of freedom we first adopt the gauge $(x^0(\xi) = \tau, x^1(\xi) = \xi^1, \ldots, x^p(\xi) = \xi^p)$. Then,

$$(\phi^*A)_i = \varphi_\mu(\xi)\partial_i x^\mu(\xi) = \varphi_i(\xi) + \varphi_K(\xi)\partial_i x^K(\xi)\,, \quad K = p+1,\ldots,D-1\,. \tag{40}$$

We see that, apparently, we are describing the $(p+1)$ components A_i of the BI field using D functions (φ_i, φ_K). The mismatch in the number of degrees of freedom is sorted out by the existence of the bosonic gauge symmetries that allow us to remove the additional $(D-p-1)$ functions. Futhermore, since the components φ_μ enter the action non-trivially only through $(\phi^*A)_i$, any local transformation of $\varphi_\mu(\xi)$ that leaves $(\phi^*A)_i$ unchanged will be a gauge symmetry of the action. Consider then

$$\delta\varphi_i(\xi) = -\alpha_K(\xi)\partial_i x^K(\xi)\,, \quad \delta\varphi_K(\xi) = \alpha_k(\xi)\,. \tag{41}$$

This specific transformation has the property $\delta(\phi^*A)_i = 0$, so it is a gauge symmetry that can be used to set $\varphi_K = 0$ by taking $\alpha_k = \varphi_K$, so $\phi^*(A)_i = \varphi_i$. Hence, we may identify $A_i = \phi^*(A)_i$.

For the IIB Dp-brane, an analysis similar to that in this section (for odd p) can be made. In fact, the origin of $A(\xi)$ in the p=1 IIB D-string case was discussed in [23] (see also [24]) by introducing an appropriate extended group manifold. We may conclude, then, that *the different worldvolume fields are introduced naturally through the pull-back of coordinates (forms) of (defined on) suitably extended superspaces.*

NOETHER CHARGES AND D-BRANE ACTIONS

The worldvolume field $A(\xi)$ that appears in the D2-brane action may be written in terms of the variables of the superstring extended superspace $\tilde{\Sigma}(x^\mu, \theta^\alpha, \varphi_\mu, \varphi_\alpha)$. The D2-WZ term, which is quasi-invariant in these coordinates, can be made strictly invariant by further extending the previous superspace to $\tilde{\Sigma} = (x^\mu, \theta^\alpha, \varphi_\mu, \varphi_\alpha, \varphi_{\mu\nu}, \varphi_{\mu\alpha}, \varphi_{\alpha\beta}, \varphi)$. In this way, the whole action is invariant. The canonical commutators of the charges generating the symmetries of the action (denoted by a hat) give a realization of the 'right' version of the 'left' Lie algebra dual to (32).

Consider the $\{Q_\alpha, Q_\beta\}$ commutator, that we shall write as

$$\{Q_\alpha, Q_\beta\} = (C\Gamma^\mu)_{\alpha\beta}P_\mu + (C\Gamma_\mu\Gamma_{11})_{\alpha\beta}\hat{Z}^\mu + (C\Gamma_{\mu\nu})_{\alpha\beta}\hat{Z}^{\mu\nu} + (C\Gamma_{11})_{\alpha\beta}\hat{Z} \quad . \tag{42}$$

With $A = A(\xi)$, the $C\Gamma_{\mu\nu}$ and $C\Gamma_{11}$ contributions would come from the quasi-invariance of the WZ Lagrangian, while $C\Gamma_{11}\Gamma_\mu$ would be the result of the contribution of the $A(\xi)$ field to the Noether current [22] (see also [25]). This is because the supersymmetry transformations *do not close on* A, and this produces an additional term by a mechanism similar to the one in the standard quasi-invariance case.

These modifications become transparent by formulating the action on the extended superspace [13]. Consider the formulation of the D2-brane on the extended superspace with quasi-invariant WZ term $b = b(x^\mu, \theta^\alpha, \varphi_\mu, \varphi_\alpha)$. The conserved Noether currents then *have to* include a term coming from the quasi-invariance of the WZ piece: if we wrongly ignored this the algebra of the charges would be

$$\{Q_\alpha, Q_\beta\} = (C\Gamma^\mu)_{\alpha\beta}P_\mu + (C\Gamma_\mu\Gamma_{11})_{\alpha\beta}\hat{Z}^\mu \tag{43}$$

rather than (42). Alternatively, we may find the correct algebra by replacing the quasi-invariant WZ term b by $\tilde{b} = \tilde{b}(x^\mu, \theta^\alpha, \varphi_\mu, \varphi_\alpha, \varphi_{\mu\nu}, \varphi_{\mu\alpha}, \varphi_{\alpha\beta}, \varphi)$, which is *manifestly invariant* since the transformation properties of the additional variables $(\varphi_{\mu\nu}, \varphi_{\mu\alpha}, \varphi_{\alpha\beta}, \varphi)$ remove the quasi-invariance of the WZ term b. Hence, the algebra of charges reproduces (42), and the contributions to $\hat{Z}^{\mu\nu}$ and $\hat{Z}$ are entirely due to the contribution of the additional variables $\varphi_{\mu\nu}$, $\varphi_{\mu\alpha}$, $\varphi_{\alpha\beta}$, φ in the WZ term $\tilde{b}$ (or to the quasi-invariance of $b(x^\mu, \theta^\alpha, \varphi_\mu, \varphi_\alpha)$ if we used b instead).

HIGHER ORDER TENSORS: THE CASE OF THE M5-BRANE

Consider the $D = 11$ M5-brane, which contains a worldvolume two-form field $A(\xi)$. As before, the supersymmetric action is obtained in two steps:

a) First, $H = dA - C$ where C is such that $dC = -(C\Gamma_{\mu\nu})_{\alpha\beta}\Pi^\mu\Pi^\nu\Pi^\alpha\Pi^\beta$, and the transformation properties of A are fixed so that H is invariant;

b) Secondly, a WZ term is added to obtain κ-symmetry.

The FDA generated by the LI one-forms Π^α, Π^μ and the three-form H is

$$d\Pi^\alpha = 0 \quad , \quad d\Pi^\mu = (1/2)(C\Gamma^\mu)_{\alpha\beta}\Pi^\alpha\Pi^\beta \quad , \quad dH = (C\Gamma_{\mu\nu})_{\alpha\beta}\Pi^\mu\Pi^\nu\Pi^\alpha\Pi^\beta . \tag{44}$$

(Note that $ddH \equiv 0$ implies $(C\Gamma^{\mu\nu})_{(\alpha\beta}(C\Gamma_\nu)_{\gamma\delta)} \equiv 0$ which is satisfied for $D = 11$). To find the nontrivial CE $(p+2)$-cocycles for the FDA (44) one may impose the closure condition for h on a general $(p+2)$-form with the correct dimensions. This gives two possible expressions for h. One of them is proportional to $(C\Gamma_{\mu\nu})_{\alpha\beta}\Pi^\mu\Pi^\nu\Pi^\alpha\Pi^\beta = dH$, so it is exact. The other is found to be

$$h \propto (C\Gamma_{\mu_1\ldots\mu_5})_{\alpha\beta}\Pi^{\mu_1}\ldots\Pi^{\mu_5}\Pi^\alpha\Pi^\beta - (15/2)(C\Gamma_{\mu_1\mu_2})_{\alpha\beta}\Pi^{\mu_1}\Pi^{\mu_2}\Pi^\alpha\Pi^\beta H \ , \tag{45}$$

which turns out to be not CE-exact. Hence, there is no solution unless $p = 5$: *the M5-brane, $p = 5$, is characterized by the only non-trivial D=11 (5+2)-CE-cocycle.*

H may be defined as a LI three-form on the extended superspace group of coordinates $\tilde{\Sigma}(\theta^\alpha, x^\mu, \varphi_{\mu\nu}, \varphi_{\mu\alpha}, \varphi_{\alpha\beta})$, namely

$$H = \frac{2}{3}\Pi^\mu\Pi^\nu\Pi_{\mu\nu} + \frac{3}{5}\Pi^\mu\Pi^\alpha\Pi_{\mu\alpha} - \frac{2}{15}\Pi_{\alpha\beta}\Pi^\alpha\Pi^\beta \quad . \tag{46}$$

Moreover, it may be shown that there exists a LI $\tilde{b}$ such that $h = d\tilde{b}$ on a suitably extended superspace [9]. By using the explicit form of the LI one-forms appearing in (46), we may replace the worldvolume two-form $A(\xi)$ by a two-form A on the extended superspace. Also, the gauge transformation $\delta A(\xi) = d\Lambda(\xi)$ is achieved by the one-form $\lambda = \lambda_\mu dx^\mu + \lambda_\alpha d\theta^\alpha$, $\phi^*(\lambda) = \Lambda(\xi)$. Then, defining $\delta\varphi_{\mu\nu}$, $\delta\varphi_{\mu\alpha}$, $\delta\varphi_{\alpha\beta}$ conveniently one obtains $\delta\phi^*(A) = d\Lambda(\xi)$.

The EL equations derived from $I[x^\mu(\xi), \theta^\alpha(\xi), A_{ij}(\xi)]$ are equivalent to the ones corresponding to the new action in which $A(\xi)$ is the pull-back $\phi^*(A)$. In fact, in parallel with the D2 brane case, it is found that $\delta I/\delta\varphi_{\alpha\beta} = 0$ and $\delta I/\delta\varphi_{\mu\alpha} = 0$ are identically satisfied (they are Noether identities) and that only $\delta I/\delta\varphi_{\mu\nu} = 0$ contains a non-trivial part, $\delta I/\delta A_{ij} = 0$. Thus, there remain $\binom{D}{2} - \binom{p+1}{2}$ Noether identities. Consider then $(\phi^*(\varphi_{\mu\nu}dx^\mu dx^\nu))_{ij} = A_{ij}(\xi) = \varphi_{\mu\nu}\partial_i x^\mu \partial_j x^\nu$. Again, the election $x^0 = \tau$, $x^1(\xi) = \xi^1, \ldots, x^5(\xi) = \xi^5$, gives

$$(\phi^* A)_{ij} = \varphi_{\mu\nu}\partial_i x^\mu \partial_j x^\nu = \varphi_{ij}(\xi) + \varphi_{iK}(\xi)\partial_j x^K - \varphi_{jK}\partial_i x^K + \varphi_{KL}\partial_i x^K \partial_j x^L \quad , \tag{47}$$

where $K, L = ((p+1) = 6, \ldots, D-1 = 10)$. The additional degrees of freedom associated with φ_{iK} and φ_{KL} may be removed by suitable gauge transformations. Indeed,

$$\begin{aligned} \delta_\alpha\varphi_{ij} &= 0 \quad , \quad \delta_\alpha\varphi_{iK} = \frac{1}{2}\alpha_{KL}\partial_i x^L \quad , \quad \delta_\alpha\varphi_{KL} = \alpha_{KL} \quad , \\ \delta_\beta\varphi_{ij} &= -\beta_{iK}\partial_j x^K + \beta_{jK}\partial_i x^K \quad , \quad \delta_\beta\varphi_{iK} = \beta_{iK} \quad , \quad \delta_\beta\varphi_{KL} = 0 \quad , \end{aligned} \tag{48}$$

leave $(\phi^* A)_{ij}$ invariant, δ_α removes φ_{KL} (by choosing $\alpha_{KL} = -\varphi_{KL}$), and δ_β sets φ_{iK} equal to zero (for $\beta_{iK} = -\varphi_{iK}$).

The previous discussions of the degrees of freedom for the D2 and M5 worldvolume fields set the pattern for other possible cases.

CONCLUSIONS

In view of the results described here, it seems natural to conclude that there exists an extended superspace origin for all the worldvolume fields appearing in the various super-p-brane actions: all worldvolume fields may be considered as pull-backs to W for the map $\phi : W \longrightarrow \tilde{\Sigma}$. In other words, *there exists a field/extended superspaces democracy* by which *all superbrane worldvolume fields may be seen as the pullbacks ϕ^* to W of some target extended superspace $\tilde{\Sigma}$ coordinates.*

The appropriate extended superspace $\tilde{\Sigma}$ of the specific theory being considered is determined by an extension of its associated *basic* sTr_D *fermionic group* and, using $\tilde{\Sigma}$, the action of the super-p-brane can be constructed in a manifestly invariant form. In fact, *in this field/extended superspace democracy context, the invariance properties and the non-trivial cocycles of the CE cohomology appear to characterise essentially the different superbranes and their actions* [13] (we might also say that they are *perfect* in the sense of [26]).

Are these extra 'dimensions' *necessary* or just *convenient* for a more geometrical and unified description of superbranes? We already saw that spacetime itself (x^μ) *is* a *consequence* of the *non-triviality* of the D-Minkowski space-valued second cohomology group of the abelian odd translation group sTr_D. Thus, it is reasonable to conclude that supersymmetry algebras and superspace groups going beyond the standard ones (see *e.g.* [18, 5, 6, 23, 27, 28, 29, 9, 30, 31]) are *required* for a suitable description of the various superstring and superbrane theories and that, as in the superspace case, Nature makes use of the extension possibilities offered by the non-trivial cohomology groups of sTr_D.

ACKNOWLEDGMENTS

This work has been partially supported by the DGICYT research grant PB 96-0756 and the Junta de Castilla y León research grant C02/199. The authors wish to thank Igor Bandos and Paul Townsend for helpful discussions.

REFERENCES

1. Haag, R., J. T. Lopuszański, J. T., and Sohnius, M., *Nucl. Phys.*, **B88**, 257–274 (1975).
2. van Holten, J.W., and Van Proeyen, A., *J. Phys.*, **A15**, 3763–3783 (1982).
3. D'Auria, R., and Fré, P., *Nucl. Phys.*, **B201**, 101-140 (1982); (E.: *ibid.* **B206**, 496 (1982)).
4. Zizzi, P.A., *Phys. Lett.*, **137B**, 57–61 (1984); *ibid.* **149B**, 333–336 (1984).
5. Green, M.B., *Phys. Lett.*, **B223**, 157–164 (1989).
6. Bergshoeff, E., and Sezgin, E., *Phys. Lett.*, **B354**, 256–263 (1995); see also *Phys. Lett.*, **B232**, 96-103 (1989).
7. Achúcarro, A., Evans, J.M., Townsend, P.K., and Wiltshire, D.L., *Phys. Lett.*, **198B**, 441–446 (1987).
8. Siegel, W., *Phys. Rev.*, **D50**, 2799–2805 (1994).
9. Sezgin, E., *Phys. Lett.*, **B392**, 323–331 (1997).
10. de Azcárraga, J.A., and Izquierdo, J.M., *Lie groups, Lie algebras and some applications in physics*, Camb. Univ. Press (1995).
11. de Azcárraga, J.A., and Izquierdo, J.M., *Chevalley-Eilenberg complex and extended objects*, to appear in the *Concise encyclopedia of supersymmetry*, J. Bagger et al. Eds., Kluwer, Dordrecht.
12. de Azcárraga, J.A., and Townsend, P.K., *Phys. Rev. Lett.*, **62**, 2579–2512 (1989).
13. Chryssomalakos, C., de Azcárraga, J.A., Izquierdo, J.M., and Pérez Bueno, J.C., *Nucl. Phys.*, **B567**, 293-330 (2000), `hep-th/990413`.
14. Aldaya, V., and de Azcárraga, J.A., *J. Math. Phys.*, **26**, 1818-1821 (1985).
15. Townsend, P.K., *p–brane democracy*, in *Particles, strings and cosmology*, J. Bagger, G. Domokos, A. Falk and A. Kovesi-Domokos (eds.), pp. 271–285 (World Sci., 1996), `hep-th/9507048`.
16. Bandos, I.A., de Azcárraga, J.A., Izquierdo. J.M., and Lukierski, J., *BPS states in M-theory and twistorial constituents*, to appear in *Phys. Rev. Lett.*, `hep-th/0101113`.
17. Chryssomalakos, C., *Stability of Lie superalgebras and branes*, `hep-th/0102134`.
18. de Azcárraga, J.A., Gauntlett, J.P., Izquierdo, J.M., and Townsend, P.K., *Phys. Rev. Lett.*, **63**, 2443–2446 (1989).
19. Hatsuda, M., and Sakaguchi, M., *Nucl. Phys.*, **B577**, 183-193 (2000).

20. Polchinski, J., *Phys. Rev. Lett.*, **75**, 4724-4727 (1995).
21. Cederwall, M., von Gussich, A., Nilsson, B.E.W., Sundell, P., and Westerberg, A., *Nucl. Phys.*, **B490**, 179-201 (1997); Aganagic, M., Popescu, C., and Schwarz, J.H., *Nucl. Phys.*, **B495**, 99-126 (1997); Bergshoeff, E., and Townsend, P.K., *Nucl. Phys.*, **B490**, 145-162 (1997).
22. Hammer, H., *Nucl. Phys.*, **B521**, 503-546 (1998).
23. Sakaguchi, M., *Phys. Rev.*, **D59**, 046007 (1999).
24. Sakaguchi, M., *JHEP*, 0004, 019 (2000); see also Abe, M., Hatsuda, M., Kamimura, K., and Tokunaga, T., *Nucl. Phys.*, **B553**, 305-316 (1999).
25. Bergshoeff, E., and Townsend, P.K., *Nucl. Phys.*, **B531**, 226-238 (1998).
26. Gayduk, A.V., Romanov, V.N., and Schwarz, A.S., *Commun. Math. Phys.*, **79**, 507–528 (1981).
27. Curtright, T., *Phys. Rev. Lett.*, **60**, 393-396 (1988).
28. de Azcárraga, J.A., Izquierdo, J.M., and Townsend, P.K., *Phys. Lett.*, **B267**, 366–373 (1991).
29. Bars, I., *Phys. Rev.*, **D54**, 5202–5210 (1996); see also *Survey of Two-Time Physics*, hep-th/0008164.
30. Deriglazov, A., and Galajinsky, A., *Mod. Phys. Lett.*, **A12**, 1517–1529 (1997).
31. Bars, I., *S–theory*, *Phys. Rev.*, **D55**, 2373–2381 (1997).

2T-Physics 2001

I. Bars [1]

CIT-USC Center for Theoretical Physics & Department of Physics
University of Southern California, Los Angeles, CA 90089-2535, USA,
email: bars@physics1.usc.edu

Abstract. The physics that is traditionally formulated in one–time-physics (1T-physics) can also be formulated in two-time-physics (2T-physics). The physical phenomena in 1T or 2T physics are not different, but the spacetime formalism used to describe them is. The 2T description involves two extra dimensions (one time and one space), is more symmetric, and makes manifest many hidden features of 1T-physics. One such hidden feature is that families of apparently different 1T-dynamical systems in d dimensions holographically describe the same 2T system in $d+2$ dimensions. In 2T-physics there are two timelike dimensions, but there is also a crucial gauge symmetry that thins out spacetime, thus making 2T-physics effectively equivalent to 1T-physics. The gauge symmetry is also responsible for ensuring causality and unitarity in a spacetime with two timelike dimensions. What is gained through 2T-physics is a unification of diverse 1T dynamics by making manifest hidden symmetries and relationships among them. Such relationships is the evidence for the presence of the higher dimensional spacetime structure. 2T-physics could be viewed as a device for gaining a better understanding of 1T-physics, but beyond this, 2T-physics offers new vistas in the search of the unified theory while raising deep questions about the meaning of spacetime. In these lectures, the recent developments in the powerful gauge field theory formulation of 2T-physics will be described after a brief review of the results obtained so far in the more intuitive worldline approach.

WORLDLINE APPROACH

A crucial element in the formulation of 2T-physics [1]-[10] is an Sp$(2,R)$ gauge symmetry in phase space. All new phenomena in 2T-physics can be traced to the presence of this gauge symmetry and its generalizations.

Spinless Particle and Interactions with Backgrounds

An elementary approach for understanding 2T-physics is offered by the worldline description of a spinless particle and its interactions. The action[2] has the form [6] [8]

$$I_Q = \int d\tau \left[\dot{X}^M P_M - \frac{1}{2} A^{ij}(\tau)\, Q_{ij}(X,P)\right], \tag{1}$$

[1] This research was partially supported by the US Department of Energy under grant number DE-FG03-84ER40168.

[2] This action is a generalization of the familiar elementary worldline action $\int d\tau \left[\dot{X}^M P_M - \frac{1}{2} e(\tau)\, P^2\right]$ for a free massless particle, as seen by specializing to $A_{11} = A_{12} = 0$, $A_{22} = e$ and $Q_{22} = P^2$.

CP589, *Fundamental Interaction Theories*, edited by J. Lukierski and J. Rembieliński

where the symmetric $A_{ij}=A_{ji}$ for $i=1,2$, denotes three Sp$(2,R)$ gauge fields, and the symmetric $Q_{ij}=Q_{ji}$ are three sp$(2,R)$ generators constructed from the phase space of the particle on the worldline $\left(X^M(\tau),P^M(\tau)\right)$. An expansion of $Q_{ij}(X,P)$ in powers of P_M in some local domain, $Q_{ij}(X,P)=\sum_s\left(f_{ij}(X)\right)^{M_1\cdots M_s}P_{M_1}\cdots P_{M_s}$, defines all the possible background fields in configuration space $\left(f_{ij}(X)\right)^{M_1\cdots M_s}$ that the particle can interact with. The local sp$(2,R)$ gauge transformations are

$$\delta X^M=-\omega^{ij}(\tau)\frac{\partial Q_{ij}}{\partial P_M},\quad \delta P_M=\omega^{ij}(\tau)\frac{\partial Q_{ij}}{\partial X^M},\quad \delta A^{ij}=\partial_\tau\omega^{ij}(\tau)+[A,\omega(\tau)]^{ij}. \tag{2}$$

The action I_Q is gauge invariant, with local parameters $\omega^{ij}(\tau)$, provided the $Q_{ij}(X,P)$ satisfy the sp$(2,R)$ Lie algebra under Poisson brackets. This is equivalent to a set of differential equations that must be satisfied by the background fields $\left(f_{ij}(X)\right)^{M_1\cdots M_s}$ [6][8]. The simplest solution is the free case denoted by $Q_{ij}=q_{ij}$ (no background fields, only the flat metric η_{MN})

$$q_{ij}=X_i^M X_j^N\eta_{MN}:\quad q_{11}=X\cdot X,\quad q_{12}=X\cdot P,\quad q_{22}=P\cdot P. \tag{3}$$

The general solution with $d+2$ dimensional background fields (see Eqs.(9-12) below) describes all interactions of the spinless particle with arbitrary electromagnetic, gravitational and higher spin gauge fields in d dimensions [8].

Two timelike dimensions is not an input, it is a result of the gauge symmetry Sp$(2,R)$ on phase space $\left(X^M,P^M\right)$. This gauge symmetry imposes the constraints $Q_{ij}(X,P)=0$ on phase space as a result of the equations of motion of the gauge field $A_{ij}(\tau)$. The meaning of the constraints is that the physical subspace of phase space should be gauge invariant under Sp$(2,R)$. There is non-trivial content in such phase space provided spacetime has $d+2$ dimensions, including two timelike dimensions [2]. The solution of the constraints is a physical phase space in two less dimensions, that is $(d-1)$ spacelike and 1 timelike dimensions. The non-trivial aspect is that there are many ways of embedding d dimensional phase space in a given $d+2$ dimensional phase space while satisfying the gauge invariance constraints. Each d dimensional solution represents a different dynamical system in 1T-physics, but all d dimensional solutions holographically represent the same higher dimensional $d+2$ theory in 2T-physics [2].

The Sp$(2,R)$ gauge symmetry is responsible for the effective holographic reduction of the $d+2$ dimensional spacetime to (a collection of) d dimensional spacetimes with one-time. Evidently, in the worldline formalism there is a single proper time τ, but the particle position $X^M(\tau)$ has two timelike dimensions $X^0(\tau),X^{0'}(\tau)$ and the particle momentum $P_M(\tau)$ has the corresponding two timelike components, $P_0(\tau),P_{0'}(\tau)$. Sp$(2,R)$ has 3 gauge parameters and 3 constraints. Two gauge parameters and two constraints can be used to eliminate one timelike and one spacelike dimensions from the coordinates and momenta. The third gauge parameter and constraint are equivalent to those associated with τ reparametrization, which is familiar in the 1T-physics worldline formulation (as in the footnote). In making the three gauge choices one must ask which combination of the two timelike dimensions is identified with τ? Evidently there are many possibilities, and once this choice is made, the 1T-Hamiltonian of the system, which will emerge from the solution of the constraints, will be the canonical momentum that is conjugate to this gauge choice of time. The many gauge choices correspond to different looking Hamiltonians with different 1T dynamical content. In this way, by various gauge fixing, the same 2T system (with the same fixed set of background fields) can be made to look like diverse 1T systems. Each 1T system in d dimensions holographically captures all the information of the 2T system in $d+2$ dimensions. Therefore there is a family of 1T systems that are in some sense dual to each other. Explicit examples of this holography/duality have been produced in the simplest 2T model [2], namely the free 2T particle in $d+2$ dimensions. The free 2T particle of Eq.(3) produces the following 1T holographic pictures in d dimensions: massless relativistic particle, massive relativistic particle, massive non-relativistic particle, particle in anti-de-Sitter space AdS_d, particle in $\mathrm{AdS}_{d-k}\times\mathrm{S}^k$ space for all $k\leq d-2$, non-relativistic Hydrogen atom ($1/r$ potential), non-relativistic harmonic oscillator in one less dimension, some examples of black holes (for d=2,3), and more ...

The higher symmetry of the $d+2$ system is present in all of the d dimensional holographic pictures. This symmetry is a global symmetry that commutes with $\mathrm{Sp}(2,R)$, and therefore is gauge invariant. Therefore it is a symmetry of the action (not of the Hamiltonian) for any choice of gauge. Before making a gauge choice the symmetry is realized in $d+2$ dimensions. After making a gauge choice the same symmetry is non-linearly realized on the fewer d dimensions, and therefore it is harder to detect in many 1T-dynamical systems, although it is present. A special example is provided by the free 2T particle of Eq.(3) which evidently has a linearly realized $\mathrm{SO}(d,2)$ Lorentz symmetry. This symmetry is interpreted in various ways from the point of view of 1T dynamics in the d dimensional holographic pictures: conformal symmetry for the free massless relativistic particle, dynamical $\mathrm{SO}(d,2)$ symmetry for the H-atom, $\mathrm{SO}(d,2)$ symmetry for the particle in AdS_d (n.b. larger than $\mathrm{SO}(d-1,2)$), etc.. The first two cases were familiar, although they were not usually thought of as having a relation to higher dimensions. All the other holographic cases mentioned above, including the non-relativistic massive particle, harmonic oscillator, $\mathrm{AdS}_{d-k}\times\mathrm{S}^k$, etc. all have the same hidden $\mathrm{SO}(d,2)$ symmetry which was understood for the first time in the context of 2T-physics. The generators of the symmetry have been explicitly constructed for all the cases mentioned [2]. What is more, the symmetry is realized in the same unitary representation as characterized by the eigenvalues of the Casimir operators of $\mathrm{SO}(d,2)$. In the classical theory, in which orders of phase space quantum operators are neglected, all the Casimir operators for $\mathrm{SO}(d,2)$ seem to vanish (this is a non-trivial representation for the non-compact group). However, when quantum ordering is taken into account, the Casimir eigenvalues do not vanish, but take on some special values corresponding to a special representation. The ordering of phase space operators is in general difficult, but it can be implemented explicitly in a few cases (conformal, H-atom, $\mathrm{AdS}_{d-k}\times\mathrm{S}^k$, harmonic oscillator). For all these cases the quantum quadratic Casimir operator is the same $C_2(SO(d,2))=1-d^2/4$, and similarly one obtains some fixed number for all higher Casimir operators. This unitary representation is the singleton/doubleton representation (name depends on d). This very specific representation, which is common to all the 1T dynamical models mentioned, corresponds to the free 2T particle. This fact is already part of the evidence of the duality that points to the existence of the unifying $d+2$ or 2T structure underlying these 1T systems.

In [8] the general system with background fields was studied. It was shown that all possible interactions of a point particle with background electromagnetic, gravitational and higher-spin fields in d dimensions emerges from the 2T-physics worldline theory in Eq.(3). The general $\mathrm{Sp}(2,R)$ algebraic relations of the $Q_{ij}(X,P)$ govern the interactions, and determine equations that the background fields of any spin must obey. The constraints were solved for a certain 2T to 1T holographic image which describes a relativistic particle interacting with background fields of any spin in $(d-1)+1$ dimensions. Two disconnected branches of solutions exist, which seem to have a correspondence with massless states in string theory, one containing low spins in the zero Regge slope limit, and the other containing high spins in the infinite Regge slope limit.

The same kind of holography/duality phenomena that exist in the free case should, in principle, be expected in the presence of background fields. This includes holographic capture of the $d+2$ dimensional dynamics in various forms in d dimensions, duality relations (analogs of same Casimir, and other related (dual) quantities) among many 1T dynamical systems which have the same background fields in $d+2$ dimensions, and hidden symmetries in d dimensions which become manifest in $d+2$ dimensions. Such higher symmetries include global, local, and reparametrization symmetries inherited from $d+2$ dimensions. In principle it is possible to construct numerous duality relationships as "experimental" evidence of the underlying higher dimensional structure. There is much detail of this type yet to be explored in the worldline theory. This should be a fruitful area of investigation in 2T-physics.

Spin, Supersymmetry

The worldline theory for the spinless particle has been generalized in several directions. One generalization is to spinning particles through the use of worldline supersymmetry, in which case the gauge group is $\mathrm{OSp}(n|2)$ instead of $\mathrm{Sp}(2,R)$ [3]. This case has also been generalized by the inclusion of some background fields [7], but the most general case analogous to Eq.(3) (including all powers of the fermion field) although straightforward, has not been investigated yet.

Another generalization involves spacetime supersymmetry, which has been obtained for the free particle (i.e. for $Q_{ij} \to q_{ij}$ as in Eq.(3)) [4][1]. In this case, in addition to the local $\mathrm{Sp}(2,R)$, there is local kappa supersymmetry as part of a local supergroup symmetry. The action is

$$S = \int d\tau \left[\dot{X}^M P_M - \frac{1}{2} A^{ij} \left(X_i \cdot X_j \right) - Str \left(L \left(\partial_\tau g g^{-1} \right) \right) \right], \tag{4}$$

where $g \in G$ is a supergroup element, and $L = L^{MN} \Gamma_{MN}$ is a coupling of the Cartan form $\partial_\tau g g^{-1}$ to the orbital $\mathrm{SO}(d,2)$ Lorentz generators $L^{MN} = X^M P^N - X^N P^M$ via the spinor representation Γ_{MN} of $\mathrm{SO}(d,2)$. The supergroup element g contains fermions Θ that are now coupled to phase space X^M, P_M.

The supergroups $\mathrm{OSp}(N|4)$, $\mathrm{SU}(2,2|4)$, $\mathrm{F}(4)$, $\mathrm{OSp}(6,2|N)$, contain $\mathrm{SO}(d,2)$ in the spinor representation for $d = 3,4,5,6$ respectively. The 2T free superparticle of [4] based on these supergroups has a holographic reduction from $d+2$ to d dimensions which produces from Eq.(4) the superparticle in $d = 3,4,5,6$,

$$S = \int d\tau \left[\dot{x}^\mu p_\mu - \frac{1}{2} e(\tau) p^2 + \dot{\theta} \gamma^\mu \theta p_\mu \right]. \tag{5}$$

In these special dimensions the 2T approach of Eq.(4) makes manifest the hidden superconformal symmetry of the superparticle action which precisely given by the corresponding supergroup. For other supergroups that contain the bosonic subgroup $SO(d,2)$ in the *spinor* representation, the 2T supersymmetric model of Eq.(4) includes p-brane degrees of freedom. In general, the approach of [4] shows how to formulate any of these systems in terms of twistors and supertwistors (instead of particle phase space) by simply choosing gauges, and thus obtaining the spectrum of the system by using oscillator methods developed a long time ago [12] (see [11] for a related twistor approach). For example, $OSp(1|8)$, which contains $SO(4,2) = SU(2,2)$ (with $\mathbf{4}+\mathbf{4}^*$ spinors), is used to construct an action in $4+2$ dimensions as in Eq.(4); this produces a holographic picture in $3+1$ dimensions for a superparticle together with 2-brane degrees of freedom (the 4D superalgebra containing the maximal 2-brane extension). There is enough gauge symmetry in the system to remove ghosts associated with timelike dimensions of the 2-brane. The physical, and unitary, quantum states of this system correspond to a particle-brane BPS realization of the supersymmetry $OSp(1|8)$. A closely related case is the toy M-model in 11+2 dimensions based on $OSp(1|64)$ [4][1]. This produces a holographic picture that includes the 11-dimensional 2-brane and 5-brane degrees of freedom in addition to the 11-dimensional superparticle phase space (with the maximally extended superalgebra in 11D). The spectrum of the toy M-model consists of 2^8 bosons and 2^8 fermions with the quantum numbers of the 11D supergravity multiplet, but with a BPS relation among the brane charges and particle momentum. Another interesting variation of Eq.(4) is a superparticle model in 10+2 dimensions with $SU(2,2|4)$ supersymmetry, in which the coupling L lives both in $SO(4,2) = SU(2,2)$ and $SO(6) = SU(4)$ [4][1]. This produces an anti-de-Sitter holographic picture that describes the complete Kaluza-Klein towers of states that emerge in the $AdS_5 \times S^5$ compactification of IIB-supergravity. A generalization of the latter, including brane degrees of freedom, is achieved by using $OSp(8|8)$. The methods and partial details of these constructions are given in [4][1], and the full details will appear in the near future.

The generalization of the 2T superparticle with background fields is a challenging problem that remains to be investigated.

FIELD THEORETIC FORMULATION OF 2T-PHYSICS

The local symmetry $Sp(2,R)$ is at the heart of the worldline formalism and the physical results of 2T-physics. This is a phase space symmetry as seen from Eq.(2). How can one implement such a local symmetry in field theory? The answer is naturally found in noncommutative field theory [9]. In fact, a beautiful and essentially unique gauge theory formulation of 2T-physics based on noncommutative $u_\star(1,1)$ [10] emerged soon after these lectures were delivered, and therefore it will be included as part of this summary of 2T-physics.

Field theory emerges from the first quantization of the worldline theory. There is a phase space approach to first quantization developed in the old days by Weyl-Wigner-Moyal and others [13]-[15]. Instead of using wavefunctions in configuration space $\psi(X)$, this approach uses wavefunctions in phase space $\phi(X,P)$, which are equivalent to functions of operators X,P by the Weyl correspondence. The correct quantum results are produced provided the phase space wavefunctions are always multiplied with each other using the Moyal star product

$$(\phi_1 \star \phi_2)(X,P) = \exp\left(\frac{i\hbar}{2}\frac{\partial}{\partial X^M}\frac{\partial}{\partial \tilde{P}_M} - \frac{i\hbar}{2}\frac{\partial}{\partial P_M}\frac{\partial}{\partial \tilde{X}^M}\right)\phi_1(X,P)\phi_2(\tilde{X},\tilde{P})\Big|_{X=\tilde{X},P=\tilde{P}}. \tag{6}$$

Then non-commutative field theory becomes a natural setting for implementing the local symmetries of 2T-physics.

Noncommutative Fields from First Quantization

A noncommutative field theory in phase space introduced recently [9] confirmed the worldline as well as the configuration space field theory [7] results of 2T-physics, and suggested some far reaching insights. As shown in [9], first quantization of the worldline theory is described by the noncommutative field equations

$$\left[Q_{ij}, Q_{kl}\right]_\star = i\left(\varepsilon_{jk}Q_{il} + \varepsilon_{ik}Q_{jl} + \varepsilon_{jl}Q_{ik} + \varepsilon_{il}Q_{jk}\right), \tag{7}$$

$$Q_{ij} \star \phi = 0, \tag{8}$$

where the Moyal star product appears in all products. The first equation is the Sp$(2,R)$ commutation relations which promote the Poisson brackets relations of the worldline theory to commutators to all orders of $\hbar$. According to the Weyl correspondence, we may think of ϕ as a projection operator in Hilbert space $\phi \sim |\psi\rangle\langle\chi|$. The ϕ equations are equivalent to sp$(2,R)$ singlet conditions in Hilbert space, $Q_{ij}|\psi\rangle = 0$, whose solutions are physical states that are gauge invariant under sp$(2,R)$. These equations were explicitly solved in several stages in [7][8][9]. The solution space is non-empty and is unitary only when spacetime has precisely two timelike dimensions, no less and no more [9]. The solution space of these equations confirm the same physical picture conveyed by the worldline theory, including the holography/duality and hidden higher dimensional symmetries, but now in a field theoretical setting [7][9]. Up to canonical transformations of (X,P) the general solution of Eq.(7) is given by

$$Q_{11} = X^M X^N \eta_{MN}, \quad Q_{12} = X^M\left(P_M + A_M(X)\right), \tag{9}$$

$$Q_{22} = G_0(X) + G_2^{MN}(X)\left(P+A(X)\right)_M\left(P+A(X)\right)_N \tag{10}$$

$$+\sum_{s=3}^{\infty} G_s^{M_1\cdots M_s}(X)\ \left(P+A(X)\right)_{M_1}\cdots\left(P+A(X)\right)_{M_s}, \tag{11}$$

where η^{MN} is the flat metric in $d+2$ dimensions, $A_M(X)$ is the Maxwell gauge potential, $G_0(X)$ is a dilaton, $G_2^{MN}(X) = \eta^{MN} + h^{MN}(X)$ is the gravitational metric, and the symmetric tensors $(G_s(X))^{M_1\cdots M_s}$ for $s \geq 3$ are high spin gauge fields. The sp$(2,R)$ closure condition in Eq.(28) requires these fields to be orthogonal to X^M and to be homogeneous of degree $(s-2)$

$$X \cdot \partial G_s = (s-2)\, G_s, \qquad X_{M_1} h_2^{M_1 M_2} = X_{M_1} G_{s\geq 3}^{M_1 \cdots M_s} = 0, \qquad X^M F_{MN} = 0, \tag{12}$$

where $F_{MN} = (\partial_M A_N - \partial_N A_M)$ is the Maxwell field strength. There is remaining canonical symmetry which, when expanded in powers of P, contains the gauge transformation parameters for all of these background gauge fields [8]. The background fields $A, G_0, G_2, G_{s\geq 3}$ determine all other background fields $\left(f_{ij}(X)\right)^{M_1\cdots M_s}$ up to canonical transformations. The solution of the $d+2$ dimensional equations (12) is given in [8] in terms of d dimensional background fields for Maxwell $A_\mu(x)$, dilaton $g(x)$, metric $g_{\mu\nu}(x)$ and higher spin fields $g^{\mu_1\cdots\mu_s}(x)$.

The solution to the matter field equation $\phi \star \phi = \phi$ is given by a Wigner distribution function constructed by Fourier transform from a wavefunction $\psi(X_1) = < X_1|\psi >$ in configuration space

$$\begin{aligned} \phi(X,P) &= \int d^D Y\, \psi(X) \star e^{-iY^M P_M} \star \chi^*(X) && (13)\\ &= \int d^D Y\; \psi\left(X - \frac{Y}{2}\right) e^{-iY^M P_M}\, \chi^*\left(X + \frac{Y}{2}\right), && (14) \end{aligned}$$

where the wavefunction satisfies $Q_{ij}|\psi\rangle = 0$. This equation has a non-empty solution space of sp$(2,R)$ gauge invariant and positive norm wavefunctions only when there are two timelike dimensions. The complete set of solutions $\psi_n(X)$ in $d+2$ dimensional configuration space is holographically given explicitly in 1T spacetime in terms of a complete set of wavefunctions in d dimensional configuration space. Thus, the equations (7,8) correctly represent the 1T physics of a particle in d dimensions interacting with background gauge fields, including the Maxwell, Einstein, and high spin fields [9].

Using the complete set of physical states, $\psi_n(X)$, one may construct a complete set of physical fields $\phi_{mn}(X,P)$ in noncommutative 2T quantum phase space that correspond to $\phi_m^n \sim |\psi_m\rangle\langle\chi_n|$. It can be shown explicitly that in noncommutative space these complete set of physical fields satisfy a closed algebra [9]

$$\phi_{n_1}^{m_1} \star \left(\phi^\dagger\right)_{m_2}^{n_2} \star \phi_{n_3}^{m_3} = \delta_{m_2}^{m_1}\, \delta_{n_3}^{n_2}\, \phi_{n_1}^{m_3}. \tag{15}$$

The positive norm (unitarity) of the physical states is captured by the δ_{nk} on the right hand side.

$\mathbf{u}_\star(1,1)$ Gauge Principle and Interactions

The goal of the field theory approach is to find a field theory, and appropriate gauge principles, from which the free Eqs.(7,8) follow as classical field equations of motion, much in the same way that the Klein-Gordon field theory arises from satisfying τ-reparametrization constraints ($p^2 = 0$), or string field theory emerges from satisfying Virasoro constraints, etc. The field theory approach, combined with gauge principles is expected to provide non-linear field interactions in 2T-physics.

The desired fundamental gauge symmetry principle that fulfill these goals is based on noncommutative $\mathrm{u}_\star(1,1)$ in phase space [10]. There is no non-commutative $\mathrm{su}(1,1)$ without the extra $\mathrm{u}(1)$ in noncommutative space, and therefore to include $\mathrm{sp}(2,R)$=$\mathrm{su}(1,1)$ one must take $\mathrm{u}_\star(1,1)$ as the smallest candidate symmetry (a smaller candidate $\mathrm{sp}_\star(2,R)$ [20] which also has a $\mathrm{u}_\star(1)$, is eliminated on other grounds [10]). The apparently extra noncommutative $\mathrm{u}_\star(1)$ is related to canonical transformations and plays an important role in the overall scheme.

The $\mathrm{u}_\star(1,1)$ gauge principle completes the formalism of [9] into an elegant and concise theory which beautifully describes 2T-physics in field theory in $d+2$ dimensions. The resulting theory has deep connections to standard d dimensional gauge field theories, gravity and the theory of high spin fields. There is also a finite matrix formulation of the theory in terms of $\mathrm{u}(N,N)$ matrices, such that the $N \to \infty$ limit becomes the $\mathrm{u}_\star(1,1)$ gauge theory.

The 4 noncommutative parameters of $\mathrm{u}_\star(1,1)$ can be written in the form of a 2×2 matrix, $\Omega_{ij} = \omega_{ij} + i\omega_0\varepsilon_{ij}$, whose symmetric part $\omega_{ij}(X,P)$ becomes $\mathrm{sp}(2,R)$ when it is global, while its antisymmetric part generates the local subgroup $\mathrm{u}_\star(1)$ with local parameter $\omega_0(X_1,X_2)$. The indices are raised with the $\mathrm{sp}(2,R)$ metric ε^{ij}, therefore in matrix form we have

$$\Omega_k^{\ l} = \omega_k^{\ l} - i\omega_0\delta_k^{\ l} = \begin{pmatrix} \omega_{12} - i\omega_0 & \omega_{22} \\ -\omega_{11} & -\omega_{12} - i\omega_0 \end{pmatrix}. \tag{16}$$

This matrix satisfies the following hermiticity conditions, $\Omega^\dagger = \varepsilon\Omega\varepsilon$. Such matrices close under matrix-star commutators to form $\mathrm{u}_\star(1,1)$.

We introduce a 2×2 matrix $\mathcal{J}_{ij} = J_{ij} + iJ_0\varepsilon_{ij}$ that parallels the form of the parameters Ω_{ij}. There will be a close relation between the fields $J_{ij}(X,P)$ and Q_{ij} as we will see soon. When one of the indices is raised, the matrix $\mathcal{J}$ takes the form

$$\mathcal{J}_i^{\ j} = \begin{pmatrix} J_{12} - iJ_0 & J_{22} \\ -J_{11} & -J_{12} - iJ_0 \end{pmatrix}. \tag{17}$$

Next we consider matter fields that transform under the noncommutative group $\mathrm{U}_\star^L(1,1) \times \mathrm{U}_\star^R(1,1)$. In this notation $\mathcal{J}$ transforms as the adjoint under $\mathrm{U}_\star^L(1,1)$ and is a singlet under $\mathrm{U}_\star^R(1,1)$, thus it is in the $(1,0)$ representation, which means its gauge transformations are defined by the matrix-star products in the form $\delta\mathcal{J} = \mathcal{J}\star\Omega^L - \Omega^L\star\mathcal{J}$. For the matter field we take the $\left(\frac{1}{2},\frac{1}{2}\right)$ representation given by a 2×2 complex matrix $\Phi_i^{\ \alpha}(X_1,X_2)$. This field is equivalent to a complex symmetric tensor Z_{ij} and a complex scalar φ. We define $\bar{\Phi} = \varepsilon\Phi^\dagger\varepsilon$. The $\mathrm{U}_\star^L(1,1) \times \mathrm{U}_\star^R(1,1)$ transformation rules for this field are $\delta\Phi = -\Omega^L\star\Phi + \Phi\star\Omega^R$, where Ω^L, Ω^R are the infinitesimal parameters for $\mathrm{U}_\star^L(1,1) \times \mathrm{U}_\star^R(1,1)$.

We now construct an action that will give the noncommutative field theory equations (7,8) in a linearized approximation and provide unique interactions in its full form. The action has a resemblance to the Chern-Simons type action introduced in [9], but now there is one more field, J_0, and the couplings among the fields obey a higher gauge symmetry

$$S_{J,\Phi} = \int d^{2D}X\, Tr\left(-\frac{i}{3}\mathcal{J}\star\mathcal{J}\star\mathcal{J} - \mathcal{J}\star\mathcal{J} + i\mathcal{J}\star\Phi\star\bar{\Phi} - V_\star\left(\Phi\star\bar{\Phi}\right)\right). \tag{18}$$

The invariance under the local $U^L_\star(1,1)\times U^R_\star(1,1)$ transformations is evident. $V(u)$ is a potential function with argument $u{=}\Phi\star\bar{\Phi}$.

The form of this action is unique as long as the maximum power of $\mathcal{J}$ is 3. We have not imposed any conditions on powers of Φ or interactions between $\mathcal{J},\Phi$, other than obeying the symmetries. A possible linear term in $\mathcal{J}$ can be eliminated by shifting J by a constant, while the relative coefficients in the action are all absorbed into a renormalization of $\mathcal{J},\Phi$. A term of the form $Tr\left(\mathcal{J}\star\mathcal{J}\star f\left(\Phi\star\bar{\Phi}\right)\right)$ that is allowed by the gauge symmetries can be eliminated by shifting $\mathcal{J}\to\left(\mathcal{J}-\frac{1}{3}f\left(\Phi\star\bar{\Phi}\right)\right)$. This changes the term $i\bar{\Phi}\star\mathcal{J}\star\Phi$ by replacing it with interactions of $\mathcal{J}$ with any function of $\bar{\Phi},\Phi$ that preserves the gauge symmetries. However, one can do field redefinitions to define a new Φ so that the interaction with the linear $\mathcal{J}$ is rewritten as given, thus shifting all complications to the function $V_\star\left(\bar{\Phi}\star\Phi\right)$. When the maximum power of J is cubic we have the correct link to the first quantized worldline theory. Therefore, with the only assumption being the cubic restriction on $\mathcal{J}$, this action explains the first quantized worldline theory, and generalizes it to an interacting theory based purely on a gauge principle.

The equations of motion are

$$\mathcal{J}\star\mathcal{J} - 2i\mathcal{J} - \Phi\star\bar{\Phi} = 0, \quad \left(\mathcal{J}+iV'\right)\star\Phi = 0, \tag{19}$$

where $V'(u) = \partial V/\partial u$. It is shown in [10] that one can choose gauges for Φ to simplify these equations, such that J_0 is fully solved, while the remaining fields satisfy

$$J_{ij}\star\phi = 0 \tag{20}$$

and

$$[J_{11},J_{12}]_\star = i\left\{J_{11},\sqrt{1-C_2(J)}\right\}_\star, \tag{21}$$

$$[J_{11},J_{22}]_\star = 2i\left\{J_{12},\sqrt{1-C_2(J)}\right\}_\star, \tag{22}$$

$$[J_{11},J_{22}]_\star = i\left\{J_{22},\sqrt{1-C_2(J)}\right\}_\star, \tag{23}$$

where $\Phi\star\bar{\Phi} = -\lambda\phi\mathbf{1}$, and the expression

$$C_2(J) = \frac{1}{2}J_{kl}\star J^{kl} = \frac{1}{2}J_{11}\star J_{22} + \frac{1}{2}J_{22}\star J_{11} - J_{12}\star J_{12}, \tag{24}$$

looks like a Casimir operator. But this algebra is not a Lie algebra, and in general one cannot show that $C_2(J)$ commutes with J_{ij}. However, assuming no anomalies in the associativity of the star product, the Jacobi identities $[J_{11},[J_{12},J_{22}]_\star]_\star + cyclic = 0$ require

$$\left[J_{ij},\left[J^{ij},\sqrt{1-C_2(J)}\right]_\star\right]_\star = 0, \tag{25}$$

but generally this is a weaker condition than the vanishing of $[J_{ij},C_2(J)]_\star$.

To understand the content of the nonlinear gauge field equations (21-23) we setup a perturbative expansion around a background solution

$$J_{ij} = J_{ij}^{(0)} + gJ_{ij}^{(1)} + g^2 J_{ij}^{(2)} + \cdots, \tag{26}$$

such that $J_{ij}^{(0)}$ is an exact solution, and then analyze the full equation perturbatively in powers of g. For the exact background solution we assume that $\frac{1}{2}J_{kl}^{(0)} \star J^{(0)kl}$ commutes with $J_{ij}^{(0)}$, therefore the background solution satisfies a Lie algebra. Then we can write the exact background solution to Eqs.(21-23) in the form

$$J_{ij}^{(0)} = Q_{ij} \star \frac{1}{\sqrt{1+\frac{1}{2}Q_{kl}\star Q^{kl}}}, \tag{27}$$

where Q_{ij} satisfies the sp$(2,R)$ algebra with the normalization of Eq.(7)

$$[Q_{11},Q_{12}]_\star = 2iQ_{11}, \quad [Q_{11},Q_{22}]_\star = 4iQ_{12}, \quad [Q_{12},Q_{22}]_\star = 2iQ_{22}, \tag{28}$$

and $\frac{1}{2}Q_{kl}\star Q^{kl}$ is a Casimir operator that commutes with all Q_{ij} that satisfies the sp$(2,R)$ algebra. Then the background $Q_{ij}(X_1,X_2)$ has the form of Eqs(9-12) up to a u$_\star(1)$ subgroup gauge symmetry. The square root is understood as a power series involving the star products and can be multiplied on either side of Q_{ij} since it commutes with the Casimir operator. For such a background, the matter field equations reduce to

$$Q_{ij}\star\phi = 0 = \phi\star Q_{ij}, \quad \phi\star\phi = \phi. \tag{29}$$

Summarizing, we have shown that our action $S_{J,\Phi}$ has yielded precisely what we had hoped for. The linearized equations of motion (0th power in g) in Eqs.(28,29) are exactly those required by the first quantization of the worldline theory as given by Eqs.(7,8). There remains to understand the propagation and self interactions of the fluctuations of the gauge fields $gJ_{ij}^{(1)} + g^2 J_{ij}^{(2)} + \cdots$, which are not included in Eqs.(28,29). However, the full field theory includes all the information uniquely, in particular the expansion of Eqs.(21-23) around the background solution $J_{ij}^{(0)}$ of Eq.(27) should determine both the propagation and the interactions of the fluctuations involving photons, gravitons, and high spin fields. At the linearized level in lowest order, it is shown in [10] that these fields satisfy the Klein-Gordon type equation.

REMARKS AND PROJECTS

We have learned that we can consistently formulate a worldline theory as well as a field theory of 2T-physics in $d+2$ dimensions based on basic gauge principles. The equations, compactly written in $d+2$ dimensional phase space in the form of Eq.(19), yield a unified description of various gauge fields in configuration space, including Maxwell, Einstein, and high spin gauge fields interacting with matter and among themselves in d dimensions.

All results follow from the field theory action in Eq.(18), which is essentially unique save for the assumption of maximum cubic power of $\mathcal{J}$. At this time it is not known what would be the consequences of relaxing the maximum cubic power of $\mathcal{J}$.

It appears that our approach provides for the first time an action principle that should contribute to the resolution of the long studied but unfinished problem of high spin fields [17][18][8][19]. The nature and detail of the interactions can in principle be extracted from our $d+2$ dimensional theory. The details of the interactions remain to be worked out, and a comparison to the perturbative equations in [17] is desirable.

For spinning particles the worldline theory introduced $osp(n|2)$ in place of $sp(2,R)$ [3]. For the field theory counterpart, we may guess that the appropriate gauge group for the supersymmetric noncommutative field theory would be $u_\star(n|1,1)$.

In the case of spacetime supersymmetry, the worldline theory with background fields remains to be constructed. We expect this to be a rather interesting and rewarding exercise, because kappa supersymmetry is bound to require the background fields to satisfy dynamical equations of motion, as it does in 1T physics [22]. The supersymmetric field equations thus obtained in $d+2$ dimensions should be rather interesting as they would include some long sought field theories in $d+2$ dimensions, among them super Yang-Mills and supergravity theories. Perhaps one may also attempt directly the spacetime supersymmetrization of the field theory approach, bypassing the background field formulation of the worldline theory. Based on the arguments given in [4][1] and [23] we expect the supersymmetry $osp(1|64)$ to play a crucial role in the relation of this work to M-theory ($osp(1|64)$ has also appeared in [24][25]).

The noncommutative field theory can be reformulated as a matrix theory in a large N limit [10]. It is conceivable that these methods would lead to a formulation of a covariant version of M(atrix) theory [21]. Using matrix methods one could relate to the 2T-physics formulation of strings and branes.

A different formulation of 2T-physics for strings (or p-branes) on the worldsheet (or worldvolume) was initiated in [5]. Tensionless strings (or p-branes) were described in the 2T approach. However, tensionful strings did not emerge yet in the formulation. The systematics of the 2T formulation for strings or branes is not as well understood as the particle case. The suspicion is that either the action in [5] was incomplete or the correct gauge choice (analogous to the massive particle) remains to be found. This is still a challenge for 2T-physics.

So far the field theory has been analyzed at the classical level. The action can now be taken as the starting point for a second quantized approach to 2T-physics. The technical aspects of this are open.

It would be interesting to consider phenomenological applications of the noncommutative field theory approach of 2T-physics, including spinning particles, and non-Abelian gauge groups.

ACKNOWLEDGMENTS

I would like to thank the organizers of the Karpacz Winter School for their support and for providing a stimulating and enjoyable environment.

REFERENCES

1. Bars, I., "Survey of Two-Time Physics", hep-th/0008164.
2. Bars, I., Deliduman, C., and Andreev, O., *Phys. Rev.*, **D58**, 066004 (1998), hep-th/9803188;
Bars, I., *Phys. Rev.*, **D58**, 066006 (1998), hep-th/9804028;
Vongehr, S., hep-th/9907077;
Bars, I., hep-th/9809034;
Bars, I., *Phys. Rev.*, **D59**, 045019 (1999), hep-th/9810025.
3. Bars, I., and Deliduman, C., *Phys. Rev.*, **D58**, 106004 (1998), hep-th/9806085.
4. Bars, I., Deliduman, C., and Minic, D., *Phys. Rev.*, **D59**, 125004 (1999), hep-th/9812161; *Phys. Lett.*, **B457**, 275 (1999), hep-th/9904063.
Bars, I., *Phys. Lett.*, **B483**, 248 (2000), hep-th/0004090; "$AdS_5\times S^5$ Supersymmetric Kaluza-Klein Towers as a 12-dimensional Theory in 2T-Physics", in preparation (partial results in [1]); "A Toy M-model", in preparation (partial results in hep-th/9904063 and [1]).
5. Bars, I., Deliduman, C., and Minic, D., *Phys. Lett.*, **B466**, 135 (1999), hep-th/9906223.
6. Bars, I., *Phys. Rev.*, **D62**, 085015 (2000), hep-th/0002140.
7. Bars, I., *Phys. Rev.*, **D62**, 046007 (2000), hep-th/0003100.
8. Bars, I., and Deliduman, C., "High spin gauge fields and Two-Time Physics", hep-th/0103042.
9. Bars, I., and Rey, S.J., "Noncommutative Sp(2,R) Gauge Theories - A Field Theory Approach to Two-Time Physics", hep-th/0104135.
10. Bars, I., "$U_\star(1,1)$ Noncommutative Gauge Theory of 2T-Physics ", hep-th/0105013.
11. Bandos, I., de Azcarraga, J.A., Izquierdo, J.M., and Lukierski, J., "BPS states in M theory and twistorial constituents", hep-th/0101113.
12. Bars, I., and Günaydin, M., *Comm. Math. Phys.*, **91** , 31 (1983);
Günaydin, M., and Marcus, N., *Class. and Quant. Grav.*, **2**, L11 (1985);
Günaydin, M., Minic, D., and Zagermann, M., *Nucl. Phys.*, **B534** , 96 (1998); *Nucl. Phys.*, **B544**, 737 (1999);
Günaydin, M., and Minic, D., *Nucl. Phys.*, **B523**, 145 (1998).
13. Weyl, H., *Zeit. für Phys.*, **46**, 1 (1927).
14. Wigner, E., *Phys. Rev.*, **40**, 749 (1932); in *Perspectives in Quantum Theory*, Eds. W. Yourgrau and A. van de Merwe (MIT Press, Cambridge, 1971).
15. Moyal, J., *Proc. Camb. Phil. Soc.*, **45**, 99 (1949).
16. Dirac, P.A.M., *Ann. Math.*, **37**, 429 (1936);
Kastrup, H.A., *Phys. Rev.*, **150**, 1183 (1966);
Mack, G., and Salam, A., *Ann. Phys.*, **53**, 174 (1969);
Preitschopf, C.R., and Vasiliev, M.A., hep-th/9812113.
17. Vasiliev, M.A., *Int. J. Mod. Phys.*, **D5**, 763 (1996); "Higher spin gauge theories, star products and AdS space", hep-th/9910096.
18. Segal, A. Yu., "Point particle in general background fields and generalized equivalence principle", hep-th/0008105; "A Generating formulation for free higher spin massless fields", hep-th/0103028.
19. Sezgin, E., and Sundell, P., "Doubletons and higher spin gauge theories", hep-th/0105001.
20. Bars, I., Sheikh-Jabbari, M.M., Vasiliev, M.A., "Noncommutative o*(N) and usp*(2N) algebras and the corresponding gauge field theories", hep-th/0103209.
21. Banks, T., Fischler, W., Shenker, S.H., and Susskind, L., *Phys. Rev.*, **D55**, 5112 (1997), hep-th/9610043.
Ishibashi, N., Kawai, H., Kitazawa, Y., and Tsuchiya, A., *Nucl. Phys.*, **B498**, 467 (1997), hep-th/9612115.

22. Witten, E., "Twistor-like transform in ten-dimensions", *Nucl. Phys.*, **B266**, 245 (1986).
23. Bars, I., *Phys. Rev.*, **D55**, 2373 (1997), hep-th/9607112; "Algebraic structure of S theory", hep-th/9608061.
24. West, P., **JHEP**, 0008:007 (2000), hep-th/0005270.
25. D'Auria, R., Ferrara, S., and Lledo, M.A., "On the embedding of space-time symmetries into simple superalgebras", hep-th/0102060.

Brane Plus Bulk Supersymmetry in Ten Dimensions

E. Bergshoeff [1] *, R. Kallosh†, T. Ortín**, D. Roest* and A. Van Proeyen‡

*University of Groningen, Nijenborgh 4, 9747 AG Groningen, The Netherlands
†Stanford University, Stanford, California 94305, USA
**Universidad Autónoma de Madrid, E-28049-Madrid, Spain
‡Katholieke Universiteit Leuven, Celestijnenlaan 200D B-3001 Leuven, Belgium

Abstract. We discuss a generalized form of IIA/IIB supergravity depending on *all* R-R potentials $C^{(p)}$ $(p = 0, 1, \dots 9)$ as the effective field theory of Type IIA/IIB superstring theory. For the IIA case we explicitly break this R-R democracy to either $p \leq 3$ or $p \geq 5$ which allows us to write a new bulk action that can be coupled to $N = 1$ supersymmetric brane actions.

The case of 8-branes is studied in detail using the new bulk & brane action. The supersymmetric negative tension branes without matter excitations can be viewed as orientifolds in the effective action. These D8-branes and O8-planes are fundamental in Type I′ string theory. A BPS 8-brane solution is given which satisfies the jump conditions on the wall. As an application of our results we derive a quantization of the mass parameter and the cosmological constant in string units.

MOTIVATION

Our purpose is to construct supersymmetric domain walls of string theory in $D = 10$ which may shed some light on the stringy origin of the brane world scenarios. In the process of pursuing this goal we have realized that all descriptions of the effective field theory of Type IIA/B string theory available in the literature are inefficient for our purpose. This has led us to introduce new versions of the effective supergravities corresponding to Type IIA/B string theory.

The standard IIA massless supergravity includes the $C^{(1)}$ and $C^{(3)}$ R-R potentials and the corresponding $G^{(2)}$ and $G^{(4)}$ gauge-invariant R-R forms. Type IIB supergravity includes the $C^{(0)}$, $C^{(2)}$ and $C^{(4)}$ R-R potentials and the corresponding $G^{(1)}$, $G^{(3)}$ and (self-dual) $G^{(5)}$ gauge-invariant R-R forms. On the other hand, string theory has all Dp-branes, odd and even, including the exotic ones, like 8-branes in IIA and 7-branes in IIB theory. These branes, of co-dimension 1 and 2, are special objects which are different in many respects from the other BPS-extended objects like the p-branes with $0 \leq p \leq 6$, which have co-dimension greater than or equal to 3. The basic difference is in the behavior of the form fields at large distance, $G^{(p+2)} \sim r^{p-8}$. For example the $G^{(10)}$ R-R form of the 8-brane does not fall off at infinity but takes a constant value there. It is believed that such extended objects can not exist independently but only in connection

[1] email: E.A.Bergshoeff@phys.rug.nl

CP589, *Fundamental Interaction Theories*, edited by J. Lukierski and J. Rembieliński

with orientifold planes [1]. However, the realization of the total system in supergravity is rather obscure.

It has been realized a while ago [2] that massive IIA supergravity, discovered by Romans [3], was the key to understand the spacetime picture of the 8-branes, which are domain walls in $D = 10$. A significant progress towards the understanding of the 8-brane solutions was made in [4, 5], where the bulk supergravity solution was found. Also, in [5], the description of the cosmological constant via a 9-form potential, based upon the work of [6, 7], was discussed. In [8] a standard 8-brane action coupling to this 9-form potential has been shown to be the appropriate source for the second Randall–Sundrum scenario [9]. Solutions for the coupled bulk & brane action system automatically satisfy the jump conditions and so they are consistent, at least from this point of view. A major unsolved problem was to find an explicitly supersymmetric description of coupled bulk & brane systems like it was done in [10]. Such a description should allow us to find out some important properties of the domain walls like the distance between the planes, the status of unbroken supersymmetry in the bulk and on the brane etc. We expect that realizing such a bulk & brane construction will lead to a better insight into the fundamental nature of extended objects of string theory.

The string backgrounds that we want to describe using an explicitly supersymmetric bulk & brane action are one-dimensional orbifolds obtained by modding out the circle S^1 by a reflection $\mathbb{Z}_2$. The orbifold direction is the transverse direction of the branes that fill the rest of the spacetime. Now, the orbifold $S^1/\mathbb{Z}_2$ being a compact space, we cannot place a single charged object (a D8-brane, say) in it, but we have to have at least two oppositely charged objects. However, this kind of system cannot be in supersymmetric equilibrium unless their tensions also have opposite signs. We are going to identify these negative-tension objects with O8-planes and we will propose an O8-plane action to be coupled to the bulk supergravity action. O8-planes can only sit at orbifold points because they require the spacetime to be mirror symmetric in their transverse direction and, thus, they can sit in any of the two endpoints of the segment $S^1/\mathbb{Z}_2$. We are going to place the other (positive tension, opposite R-R charge) brane at the other endpoint. Clearly we can, from the effective action point of view, identify the positive tension brane as a combination of O8-planes and D8-branes with positive total tension and the negative tension brane as a combination of O8-planes and D8-branes with negative total tension.

Our strategy will be to generalize the 5-dimensional construction of the supersymmetric bulk & brane action, proposed in [10]. The construction of [10] allowed to find a supersymmetric realization of the brane-world scenario of Randall and Sundrum [9]. We will repeat the construction of [10] in $D = 10$ with the aim to get a better understanding of branes and planes in string theory.

To solve the discrepancy between the bulk actions with limited field content (lower-rank R-R forms) and the wide range of brane actions that involve all the possible R-R forms, we have constructed a new formulation of IIA/IIB supergravity up to quartic order in fermions. In particular, the new formulation gives an easy control over the exotic $G^{(0)}$ and $G^{(10)}$ R-R forms associated with the mass and cosmological constant of the $D = 10$ supergravity. This in turn allows a clear study of the D8–O8 system describing a pair of supersymmetric domain walls which are fundamental objects of the Type I′ string theory. The quantization of the mass parameter and cosmological constant in stringy units are simple consequences of the theory. Apart from being a tool to understand the supersymmetric domain walls we were interested in, it can be expected that the new effective theories of $D = 10$ supersymmetry will have more general applications in the future.

In this talk we will summarize the results of [11].

A NEW DUAL FORMULATION OF D=10 SUPERGRAVITY

The standard formulation of $D = 10$ IIA (massless [12, 13, 14] and massive [3]) and IIB [15, 16] supergravity has the following field content

$$\begin{aligned} \text{IIA} \ &: \quad \left\{g_{\mu\nu}, B_{\mu\nu}, \phi, C^{(1)}_{\mu}, C^{(3)}_{\mu\nu\rho}, \psi_{\mu}, \lambda\right\}, \\ \text{IIB} \ &: \quad \left\{g_{\mu\nu}, B_{\mu\nu}, \phi, C^{(0)}, C^{(2)}_{\mu\nu}, C^{(4)}_{\mu\nu\rho\sigma}, \psi_{\mu}, \lambda\right\}. \end{aligned} \tag{1}$$

In the IIA case, the massive theory contains an additional mass parameter $G^{(0)} = m$. In the IIB case, an extra self-duality condition is imposed on the field strength of the four-form. It turns out that one can realize the N=2 supersymmetry on the R-R gauge fields of higher rank as well. These are usually incorporated via duality relations. To treat the R-R potentials democratically we propose a new formulation based upon a pseudo-action. This democratic formulation describes the dynamics of the bulk supergravity in the most elegant way. However, it turns out that this formulation is not well suited for our purposes. For the IIA case, we therefore give a different formulation where the constant mass parameter has been replaced by a field.

The Democratic Formulation

To explicitly introduce the democracy among the R-R potentials we propose a pseudo-action whose equations of motion are supplemented by duality constraints (see below). Of course this enlarges the number of degrees of freedom. Since a p- and an $(8-p)$-form potential carry the same number of degrees of freedom, the introduction of the dual potentials doubles the R-R sector. Including the highest potential $C^{(9)}$ in IIA does not alter this, since it carries no degrees of freedom. This 9-form potential can be seen as the potential dual to the constant mass parameter $G^{(0)} = m$. The doubling of number of degrees of freedom will be taken care of by a constraint, relating the lower- and higher-rank potentials. This new formulation of supersymmetry is inspired by the bosonic construction of [17], and, in the case of IIB supergravity, is related to the pseudo-action construction of [18].

A pseudo-action [18] can be used as a mnemonic to derive the equations of motion. It differs from a usual action in the sense that not all equations of motion follow from varying the fields in the pseudo-action. To obtain the complete set of equations of motion, an additional constraint has to be substituted by hand into the set of equations of motion that follow from the pseudo-action. The constraint itself does not follow from the pseudo-action. The construction we present here generalizes the pseudo-action construction of [17, 18] in the sense that our construction (i) treats the IIA and IIB case in a unified way, introducing all R-R potentials in the pseudo-action, and (ii) describes also the massive IIA case via a 9-form potential $C^{(9)}$ and a constant mass parameter $G^{(0)} = m$.

Our pseudo-action has the extended field content

$$\begin{aligned}
\text{IIA} \quad &: \quad \left\{g_{\mu\nu}, B_{\mu\nu}, \phi, C^{(1)}_{\mu}, C^{(3)}_{\mu\nu\rho}, C^{(5)}_{\mu\cdots\rho}, C^{(7)}_{\mu\cdots\rho}, C^{(9)}_{\mu\cdots\rho}, \psi_\mu, \lambda\right\}, \\
\text{IIB} \quad &: \quad \left\{g_{\mu\nu}, B_{\mu\nu}, \phi, C^{(0)}, C^{(2)}_{\mu\nu}, C^{(4)}_{\mu\cdots\rho}, C^{(6)}_{\mu\cdots\rho}, C^{(8)}_{\mu\cdots\rho}, \psi_\mu, \lambda\right\}. \qquad (2)
\end{aligned}$$

It is understood that in the IIA case the fermions contain both chiralities, while in the IIB case they satisfy

$$\Gamma_{11}\psi_\mu = \psi_\mu\,, \qquad \Gamma_{11}\lambda = -\lambda\,, \qquad (\text{IIB}). \qquad (3)$$

In that case they are doublets, and we suppress the corresponding index. The explicit form of the pseudo-action is given by[2]

$$\begin{aligned}
S_{\text{Pseudo}} = -\frac{1}{2\kappa_{10}^2}\int d^{10}x\sqrt{-g}\Big\{&e^{-2\phi}\Big[R(\omega(e)) - 4(\partial\phi)^2 + \tfrac{1}{2}H\cdot H + \\
&- 2\partial^\mu\phi\chi^{(1)}_\mu + H\cdot\chi^{(3)} + 2\bar\psi_\mu\Gamma^{\mu\nu\rho}\nabla_\nu\psi_\rho - 2\bar\lambda\Gamma^\mu\nabla_\mu\lambda + 4\bar\lambda\Gamma^{\mu\nu}\nabla_\mu\psi_\nu\Big] + \\
&+ \sum_{n=0,1/2}^{5,9/2} \tfrac{1}{4}G^{(2n)}\cdot G^{(2n)} + \tfrac{1}{2}G^{(2n)}\cdot\Psi^{(2n)}\Big\} + \text{quartic fermionic terms}.
\end{aligned} \qquad (4)$$

It is understood that the summation in the above pseudo-action is over integers ($n = 0, 1, \ldots, 5$) in the IIA case and over half-integers ($n = 1/2, 3/2, \ldots, 9/2$) in the IIB case. In the summation range we will always first indicate the lowest value for the IIA case, before the one for the IIB case. Furthermore,

$$\frac{1}{2\kappa_{10}^2} = \frac{g^2}{2\kappa^2} = \frac{2\pi}{(2\pi\ell_s)^8}\,, \qquad (5)$$

where κ^2 is the physical gravitational coupling, g is the string coupling constant and $\ell_s = \sqrt{\alpha'}$ is the string length. For notational convenience we group all potentials and field strengths in the formal sums

$$\mathbf{G} = \sum_{n=0,1/2}^{5,9/2} G^{(2n)}\,, \qquad \mathbf{C} = \sum_{n=1,1/2}^{5,9/2} C^{(2n-1)}\,. \qquad (6)$$

---

[2] We use the notation and conventions of [11].

The bosonic field strengths are given by

$$H = dB\,, \qquad \mathbf{G} = d\mathbf{C} - dB{\wedge}\mathbf{C} + G^{(0)}\mathbf{e}^B\,, \tag{7}$$

where it is understood that each equation involves only one term from the formal sums (6) (only the relevant combinations are extracted). The corresponding Bianchi identities then read

$$dH = 0\,, \qquad d\mathbf{G} - H{\wedge}\mathbf{G} = 0\,. \tag{8}$$

In this subsection $G^{(0)} = m$ indicates the constant mass parameter of IIA supergravity. In the IIB theory all equations should be read with vanishing $G^{(0)}$. The spin connection in the covariant derivative ∇_μ is given by its zehnbein part: $\omega_\mu{}^{ab} = \omega_\mu{}^{ab}(e)$. The bosonic fields couple to the fermions via the bilinears $\chi^{(1,3)}$ and $\Psi^{(2n)}$, which read

$$\begin{aligned}
\chi^{(1)}_\mu &= -\,2\bar\psi_\nu\Gamma^\nu\psi_\mu - 2\bar\lambda\Gamma^\nu\Gamma_\mu\psi_\nu\,,\\
\chi^{(3)}_{\mu\nu\rho} &= \tfrac{1}{2}\bar\psi_\alpha\Gamma^{[\alpha}\Gamma_{\mu\nu\rho}\Gamma^{\beta]}\mathcal{P}\psi_\beta + \bar\lambda\Gamma_{\mu\nu\rho}{}^\beta\mathcal{P}\psi_\beta - \tfrac{1}{2}\bar\lambda\mathcal{P}\Gamma_{\mu\nu\rho}\lambda\,,\\
\Psi^{(2n)}_{\mu_1\cdots\mu_{2n}} &= \tfrac{1}{2}e^{-\phi}\bar\psi_\alpha\Gamma^{[\alpha}\Gamma_{\mu_1\cdots\mu_{2n}}\Gamma^{\beta]}\mathcal{P}_n\psi_\beta + \tfrac{1}{2}e^{-\phi}\bar\lambda\Gamma_{\mu_1\cdots\mu_{2n}}\Gamma^\beta\mathcal{P}_n\psi_\beta +\\
&\quad - \tfrac{1}{4}e^{-\phi}\bar\lambda\Gamma_{[\mu_1\cdots\mu_{2n-1}}\mathcal{P}_n\Gamma_{\mu_{2n}]}\lambda\,.
\end{aligned} \tag{9}$$

We have used the following definitions:

$$\begin{aligned}
\mathcal{P} &= \Gamma_{11} \quad (\text{IIA}) \quad \text{or} \quad -\sigma^3\ (\text{IIB})\,,\\
\mathcal{P}_n &= (\Gamma_{11})^n\ (\text{IIA}) \quad \text{or} \quad \sigma^1\ (\text{n}+1/2\ \text{even}),\ \text{i}\sigma^2\ (\text{n}+1/2\ \text{odd})\ (\text{IIB})\,.
\end{aligned} \tag{10}$$

Note that the fermions satisfy

$$\Psi^{(2n)} = (-)^{\text{Int}[n]+1}\star\Psi^{(10-2n)}\,. \tag{11}$$

Due to the appearance of all R-R potentials, the number of degrees of freedom in the R-R sector has been doubled. Each R-R potential leads to a corresponding equation of motion:

$$d\star(G^{(2n)} + \Psi^{(2n)}) + H{\wedge}\star(G^{(2n+2)} + \Psi^{(2n+2)}) = 0\,. \tag{12}$$

Now, one must relate the different potentials to get the correct number of degrees of freedom. We therefore by hand impose the following duality relations

$$G^{(2n)} + \Psi^{(2n)} = (-)^{\text{Int}[n]}\star G^{(10-2n)}\,, \tag{13}$$

in the equations of motion that follow from the pseudo-action (4). It is in this sense that the action (4) cannot be considered as a true action. Instead, it should be considered as a mnemonic to obtain the full equations of motion of the theory. As usual, the Bianchi identities and equations of motions of the dual potentials correspond to each other when employing the duality relation. For the above reason the democratic formulation can be viewed as self-dual, since (13) places constraints relating the field content (2).

The pseudo-action (4) is invariant under supersymmetry provided we impose the duality relations (13) after varying the action. The supersymmetry rules read (here given modulo cubic fermion terms):

$$
\begin{aligned}
\delta_\varepsilon e_\mu{}^a &= \bar{\varepsilon}\Gamma^a\psi_\mu\,,\\
\delta_\varepsilon\psi_\mu &= \Big(\partial_\mu + \tfrac{1}{4}\,\not{\omega}_\mu + \tfrac{1}{8}\mathcal{P}\,\not{H}_\mu\Big)\varepsilon + \tfrac{1}{16}e^{\phi}\sum_{n=0,1/2}^{5,9/2}\frac{1}{(2n)!}\,\not{G}^{(2n)}\Gamma_\mu\mathcal{P}_n\varepsilon\,,\\
\delta_\varepsilon B_{\mu\nu} &= -\,2\bar{\varepsilon}\Gamma_{[\mu}\mathcal{P}\psi_{\nu]}\,,\\
\delta_\varepsilon C^{(2n-1)}_{\mu_1\cdots\mu_{2n-1}} &= -\,e^{-\phi}\bar{\varepsilon}\Gamma_{[\mu_1\cdots\mu_{2n-2}}\mathcal{P}_n\Big((2n-1)\psi_{\mu_{2n-1}]} - \tfrac{1}{2}\Gamma_{\mu_{2n-1}]}\lambda\Big) +\\
&\quad + (n-1)(2n-1)C^{(2n-3)}_{[\mu_1\cdots\mu_{2n-3}}\delta_\varepsilon B_{\mu_{2n-2}\mu_{2n-1}]}\,,\\
\delta_\varepsilon\lambda &= \Big(\not{\partial}\phi + \tfrac{1}{12}\not{H}\mathcal{P}\Big)\varepsilon + \tfrac{1}{8}e^{\phi}\sum_{n=0,1/2}^{5,9/2}(-)^{2n}\frac{5-2n}{(2n)!}\,\not{G}^{(2n)}\mathcal{P}_n\varepsilon\,,\\
\delta_\varepsilon\phi &= \tfrac{1}{2}\bar{\varepsilon}\lambda\,,
\end{aligned}
\qquad (14)
$$

where ε is a spinor similar to ψ_μ, i.e. in IIB: $\Gamma_{11}\varepsilon = \varepsilon$. Note that for n half-integer (the IIB case) these supersymmetry rules exactly reproduce the rules given in eq. (1.1) of [19].

Secondly, the pseudo-action (4) is also invariant under the usual bosonic NS-NS and R-R gauge symmetries with parameters Λ and $\Lambda^{(2n)}$ respectively:

$$
\delta_\Lambda B = d\Lambda\,, \qquad \delta_\Lambda \mathbf{C} = (d\mathbf{L} - G^{(0)}\Lambda)_\wedge \mathbf{e}^B\,, \qquad \text{with } \mathbf{L} = \sum_{n=0,1/2}^{4,7/2}\Lambda^{(2n)}\,. \qquad (15)
$$

Finally, there is a number of $\mathbb{Z}_2$-symmetries. However, in the IIA case these $\mathbb{Z}_2$-symmetries are *only valid for* $G^{(0)} = m = 0$. Below we show how these symmetries of the action act on supergravity fields. For both massless IIA and IIB there is a fermion number symmetry $(-)^{F_L}$ given by

$$
\begin{aligned}
&\{\phi, g_{\mu\nu}, B_{\mu\nu}\} \to \{\phi, g_{\mu\nu}, B_{\mu\nu}\}\,,\\
&\{C^{(2n-1)}_{\mu_1\cdots\mu_{2n-1}}\} \to -\{C^{(2n-1)}_{\mu_1\cdots\mu_{2n-1}}\}\,,\\
&\{\psi_\mu, \lambda, \varepsilon\} \to +\mathcal{P}\{\psi_\mu, -\lambda, \varepsilon\}\,, \quad \text{(IIA)},\\
&\{\psi_\mu, \lambda, \varepsilon\} \to +\mathcal{P}\{\psi_\mu, \lambda, \varepsilon\}\,, \quad \text{(IIB)}.
\end{aligned}
\qquad (16)
$$

In the IIB case there is an additional worldsheet parity symmetry Ω given by

$$
\begin{aligned}
&\{\phi, g_{\mu\nu}, B_{\mu\nu}\} \to \{\phi, g_{\mu\nu}, -B_{\mu\nu}\}\,,\\
&\{C^{(2n-1)}_{\mu_1\cdots\mu_{2n-1}}\} \to (-)^{n+1/2}\{C^{(2n-1)}_{\mu_1\cdots\mu_{2n-1}}\}\,,\\
&\{\psi_\mu, \lambda, \varepsilon\} \to \sigma^1\{\psi_\mu, \lambda, \varepsilon\}\,,
\end{aligned}
\qquad (17)
$$

In the massless IIA case there is a similar $I_9\Omega$-symmetry involving an additional parity transformation in the 9-direction. Writing $\mu = (\underline{\mu}, \dot{9})$, the rules are given by

$$\begin{aligned} x^{\dot{9}} &\to -x^{\dot{9}}, \\ \{\phi, g_{\underline{\mu}\underline{\nu}}, B_{\underline{\mu}\underline{\nu}}\} &\to \{\phi, g_{\underline{\mu}\underline{\nu}}, -B_{\underline{\mu}\underline{\nu}}\}, \\ \{C^{(2n-1)}_{\underline{\mu}_1\cdots\underline{\mu}_{2n-1}}\} &\to (-)^{n+1}\{C^{(2n-1)}_{\underline{\mu}_1\cdots\underline{\mu}_{2n-1}}\}, \\ \{\psi_{\underline{\mu}}, \lambda, \varepsilon\} &\to +\Gamma^9\{\psi_{\underline{\mu}}, -\lambda, \varepsilon\}. \end{aligned} \tag{18}$$

The parity of the fields with one or more indices in the $\dot{9}$-direction is given by the rule that every index in the $\dot{9}$-direction gives an extra minus sign compared to the above rules.

In both IIA and IIB there is also the obvious symmetry of interchanging all fermions by minus the fermions, leaving the bosons invariant.

The $\mathbb{Z}_2$-symmetries are used for the construction of superstring theories with sixteen supercharges, see [20]. $(-)^{F_L}$ gives a projection to the $E_8 \times E_8$ heterotic superstring (IIA) or the SO(32) heterotic superstring theory (IIB). Ω is used to reduce the IIB theory to the SO(32) Type I superstring, while the $I_9\Omega$-symmetry reduces the IIA theory to the Type I′ $SO(16) \times SO(16)$ superstring theory.

One might wonder at the advantages of the generalized pseudo-action (4) above the standard supergravity formulation. At the cost of an extra duality relation we were able to realize the R-R democracy in the action. Note that only kinetic terms are present; by allowing for a larger field content the Chern–Simons term is eliminated. Under T-duality all kinetic terms are easily seen to transform into each other [21]. The same goes for the duality constraints. This formulation is elegant and comprises all potentials. However, it is impossible to construct a proper action in this formulation due to the doubling of the degrees of freedom. Therefore, to add brane actions to the bulk system, the democratic formulation is not suitable. This is due to two reasons. First, the $I_9\Omega$ symmetry is only valid for $G^{(0)} = 0$, but we will need this symmetry in our construction of the bulk & 8-brane system. Secondly, to describe a charged domain wall, we would like to have opposite values for $G^{(0)}$ at the two sides of the domain wall, i.e. we want to allow for a mass parameter that is only piecewise constant. The R-R democracy has to be broken to accommodate for an action and this will be discussed in the next subsection.

The Dual Formulation of IIA

We will present here the new dual formulation with action, available for the IIA case only. A proper action will be constructed in this formulation. It is this formulation that we will apply in our construction of the bulk & brane system. We will call this the dual formulation.

The independent fields in this formulation are

$$\left\{e^a_\mu, B_{\mu\nu}, \phi, G^{(0)}, G^{(2)}_{\mu\nu}, G^{(4)}_{\mu_1\cdots\mu_4}, A^{(5)}_{\mu_1\cdots\mu_5}, A^{(7)}_{\mu_1\cdots\mu_7}, A^{(9)}_{\mu_1\cdots\mu_9}, \psi_\mu, \lambda\right\}. \tag{19}$$

The bulk action reads

$$
\begin{aligned}
S_{\rm bulk} = & -\frac{1}{2\kappa_{10}^2}\int d^{10}x\sqrt{-g}\Big\{e^{-2\phi}\big[R(\omega(e)) - 4(\partial\phi)^2 + \tfrac{1}{2}H\cdot H - 2\partial^\mu\phi\chi_\mu^{(1)} + H\cdot\chi^{(3)} \\
& + 2\bar\psi_\mu\Gamma^{\mu\nu\rho}\nabla_\nu\psi_\rho - 2\bar\lambda\Gamma^\mu\nabla_\mu\lambda + 4\bar\lambda\Gamma^{\mu\nu}\nabla_\mu\psi_\nu\big] \\
& + \sum_{n=0,1,2} \tfrac{1}{2}G^{(2n)}\cdot G^{(2n)} + G^{(2n)}\cdot\Psi^{(2n)} \\
& - \star\big[\tfrac{1}{2}G^{(4)}G^{(4)}B - \tfrac{1}{2}G^{(2)}G^{(4)}B^2 + \tfrac{1}{6}G^{(2)2}B^3 + \tfrac{1}{6}G^{(0)}G^{(4)}B^3 - \tfrac{1}{8}G^{(0)}G^{(2)}B^4 \\
& + \tfrac{1}{40}G^{(0)2}B^5 + \mathbf{e}^{-B}\mathbf{G}d(A^{(5)} - A^{(7)} + A^{(9)})\big]\Big\} + \text{quartic fermionic terms}, \qquad (20)
\end{aligned}
$$

where all $\wedge$'s have been omitted in the last two lines. In the last term a projection on the 10-form is understood. Here $\mathbf{G}$ is defined as in (6) but where $G^{(0)}$, $G^{(2)}$ and $G^{(4)}$ are now independent fields (which we will call black boxes) and are no longer given by (7). Note that their Bianchi identities are imposed by the Lagrange multipliers $A^{(9)}$, $A^{(7)}$ and $A^{(5)}$. The NS-NS three-form field strength is given by (7). Note that the standard action for IIA supergravity can be obtained by integrating out the dual potentials in (20).

The symmetries of the action are similar to those of the democratic formulation with some small changes. In the supersymmetry transformations of gravitino and gaugino, the sums now extend only over $n = 0,1,2$:

$$
\begin{aligned}
\delta_\varepsilon e_\mu{}^a &= \bar\varepsilon\Gamma^a\psi_\mu\,, \\
\delta_\varepsilon\psi_\mu &= \Big(\partial_\mu + \tfrac{1}{4}\not{\omega}_\mu + \tfrac{1}{8}\Gamma_{11}\not{H}_\mu\Big)\varepsilon + \tfrac{1}{8}e^\phi\sum_{n=0,1,2}\frac{1}{(2n)!}\not{G}^{(2n)}\Gamma_\mu(\Gamma_{11})^n\varepsilon\,, \\
\delta_\varepsilon B_{\mu\nu} &= -2\bar\varepsilon\Gamma_{[\mu}\Gamma_{11}\psi_{\nu]}\,, \\
\delta_\varepsilon\lambda &= \Big(\not\partial\phi - \tfrac{1}{12}\Gamma_{11}\not{H}\Big)\varepsilon + \tfrac{1}{4}e^\phi\sum_{n=0,1,2}\frac{5-2n}{(2n)!}\not{G}^{(2n)}(\Gamma_{11})^n\varepsilon\,, \\
\delta_\varepsilon\phi &= \tfrac{1}{2}\bar\varepsilon\lambda\,, \\
\delta_\varepsilon\mathbf{A} &= \mathbf{e}^{-B}\wedge\mathbf{E}\,, \\
\delta_\varepsilon\mathbf{G} &= d\mathbf{E} + \mathbf{G}\wedge\delta_\varepsilon B - H\wedge\mathbf{E}\,, \\
\text{with } E^{(2n-1)}_{\mu_1\cdots\mu_{2n-1}} &\equiv -e^{-\phi}\bar\varepsilon\Gamma_{[\mu_1\cdots\mu_{2n-2}}(\Gamma_{11})^n\Big((2n-1)\psi_{\mu_{2n-1}]} - \tfrac{1}{2}\Gamma_{\mu_{2n-1}]}\lambda\Big)\,. \qquad (21)
\end{aligned}
$$

The transformation of the black boxes $\mathbf{G}$ follow from the requirement that $\mathbf{e}^{-B}\mathbf{G}$ transforms in a total derivative. Here the formal sums

$$
\mathbf{A} = \sum_{n=1}^{5} A^{(2n-1)}\,, \qquad \mathbf{E} = \sum_{n=1}^{5} E^{(2n-1)}\,, \qquad \mathbf{G} = \sum_{n=0}^{5} G^{(2n)}\,, \qquad (22)
$$

have been used. Note that the first formal sum in (22) contains fields, $A^{(1)}$ and $A^{(3)}$, that do not occur in the action. The same applies to $\mathbf{G}$, which contains the extra fields $G^{(6)}, G^{(8)}$ and $G^{(10)}$. Although these fields do not occur in the action, one can nevertheless show that the supersymmetry algebra is realized on them. To do so one must use the supersymmetry rules of (21) and the equations of motion that follow from the action (20).

The gauge symmetries with parameters Λ and $\Lambda^{(2n)}$ are

$$\begin{aligned} \delta_\Lambda B &= d\Lambda\,, \qquad \delta_\Lambda \mathbf{A} = d\mathbf{L} - G^{(0)}\Lambda - d\Lambda_\wedge \mathbf{A}\,, \\ \delta_\Lambda \mathbf{G} &= d\Lambda_\wedge\big(\mathbf{G} - \mathbf{e}^B{}_\wedge(d\mathbf{A} + G^{(0)})\big) + \mathbf{e}^B{}_\wedge \Lambda_\wedge dG^{(0)}\,. \end{aligned} \tag{23}$$

Note that, with respect to the R-R gauge symmetry, the $\mathbf{A}$ potentials transform as a total derivative while the black boxes are invariant.

Finally, there are $\mathbb{Z}_2$-symmetries, $(-)^{F_L}$ and $I_9\Omega$, which leave the action invariant. In contrast to the democratic formulation these two $\mathbb{Z}_2$-symmetries are valid symmetries even for $G^{(0)} \neq 0$. The $(-)^{F_L}$-symmetry is given by

$$\begin{aligned} \{\phi, g_{\mu\nu}, B_{\mu\nu}\} &\to \{\phi, g_{\mu\nu}, B_{\mu\nu}\}\,, \\ \{G^{(2n)}_{\mu_1\cdots\mu_{2n}}, A^{(2n-1)}_{\mu_1\cdots\mu_{2n-1}}\} &\to -\{G^{(2n)}_{\mu_1\cdots\mu_{2n}}, A^{(2n-1)}_{\mu_1\cdots\mu_{2n-1}}\}\,, \\ \{\psi_\mu, \lambda, \varepsilon\} &\to +\Gamma_{11}\{\psi_\mu, -\lambda, \varepsilon\}\,, \end{aligned} \tag{24}$$

while the second $I_9\Omega$-symmetry reads

$$\begin{aligned} x^9 &\to -x^9\,, \\ \{\phi, g_{\underline{\mu\nu}}, B_{\underline{\mu\nu}}\} &\to \{\phi, g_{\underline{\mu\nu}}, -B_{\underline{\mu\nu}}\}\,, \\ \{G^{(2n)}_{\underline{\mu_1\cdots\mu_{2n}}}, A^{(2n-1)}_{\underline{\mu_1\cdots\mu_{2n-1}}}\} &\to (-)^{n+1}\{G^{(2n)}_{\underline{\mu_1\cdots\mu_{2n}}}, A^{(2n-1)}_{\underline{\mu_1\cdots\mu_{2n-1}}}\}\,, \\ \{\psi_{\underline{\mu}}, \lambda, \varepsilon\} &\to +\Gamma^9\{\psi_{\underline{\mu}}, -\lambda, \varepsilon\}\,. \end{aligned} \tag{25}$$

ADDING THE BRANE ACTIONS

Having established supersymmetry in the bulk, we now turn to supersymmetry on the brane. As mentioned in the introduction, our main interest is in one-dimensional orbifold constructions with 8-branes at the orbifold points. Using the techniques of the three-brane on the orbifold in five dimensions [10], we want to construct an orientifold using a $\mathbb{Z}_2$-symmetry of the bulk action. On the fixed points we insert brane actions, which will turn out to be invariant under the reduced ($N = 1$) supersymmetry. For the moment we will not restrict to domain walls (in this case eight-branes) since our brane analysis is similar for orientifolds of lower dimension. In the previous section we have seen that our bulk action possesses a number of symmetries, among which a parity operation. To construct an orientifold, the relevant $\mathbb{Z}_2$-symmetry must contain parity operations in the transverse directions. Furthermore, in order to construct a charged domain wall, we want for a p-brane the $(p+1)$-form R-R potential to be even. For the 8-brane the $I_9\Omega$ symmetry satisfies the desired properties. For the other p-branes, it would seem natural to use the $\mathbb{Z}_2$-symmetry

$$I_{9,8,\ldots,p+1}\Omega \equiv (I_9\Omega)(I_8\Omega)\cdots(I_{p+1}\Omega)\,, \tag{26}$$

where $I_q\Omega$ is the transformation (18) with 9 replaced by q, and I_q and Ω commute. However, for some p-branes ($p = 2,3,6,7$) the corresponding $C^{(p+1)}$ R-R-potential is odd under this $\mathbb{Z}_2$-symmetry. To obtain the correct parity one must include an extra $(-)^{F_L}$ transformation in these cases, which also follows from T-duality [24]. This leads for each p-brane to the $\mathbb{Z}_2$-symmetry indicated in Table 1.

TABLE 1. The $\mathbb{Z}_2$-symmetries used in the orientifold construction of an Op-plane. The T-duality transformation from IIA to IIB in the lower dimension induces each time a $(-)^{F_L}$.

p	IIB	IIA
9	Ω	-
8	-	$I_9\Omega$
7	$(-)^{F_L}I_{9,8}\Omega$	-
6	-	$(-)^{F_L}I_{9,8,7}\Omega$
5	$I_{9,8,\ldots,6}\Omega$	-
4	-	$I_{9,8,\ldots,5}\Omega$
3	$(-)^{F_L}I_{9,8,\ldots,4}\Omega$	-
2	-	$(-)^{F_L}I_{9,8,\ldots,3}\Omega$
1	$I_{9,8,\ldots,2}\Omega$	-
0	-	$I_{9,8,\ldots,1}\Omega$

Thus the correct $\mathbb{Z}_2$-symmetry for a general IIA Op-plane is given by

$$((-)^{F_L})^{p/2} I_{9,8,\ldots,p+1}\Omega\,. \tag{27}$$

The effect of this $\mathbb{Z}_2$-symmetry on the bulk fields reads (the underlined indices refer to the worldvolume directions, i.e. $\mu = (\underline{\mu}, p+1,\ldots,9)$

$$\begin{aligned}
\{x^{p+1},\ldots,x^9\} &\to -\{x^{p+1},\ldots,x^9\}\,,\\
\{\phi, g_{\underline{\mu\nu}}, B_{\underline{\mu\nu}}\} &\to \{\phi, g_{\underline{\mu\nu}}, -B_{\underline{\mu\nu}}\}\,,\\
\{A^{(5)}_{\underline{\mu_1\cdots\mu_5}}, A^{(9)}_{\underline{\mu_1\cdots\mu_9}}, G^{(2)}_{\underline{\mu\nu}}\} &\to (-)^{\frac{p}{2}}\{A^{(5)}_{\underline{\mu_1\cdots\mu_5}}, A^{(9)}_{\underline{\mu_1\cdots\mu_9}}, G^{(2)}_{\underline{\mu\nu}}\}\,,\\
\{A^{(7)}_{\underline{\mu_1\cdots\mu_7}}, G^{(0)}, G^{(4)}_{\underline{\mu_1\cdots\mu_4}}\} &\to (-)^{\frac{p}{2}+1}\{A^{(7)}_{\underline{\mu_1\cdots\mu_7}}, G^{(0)}, G^{(4)}_{\underline{\mu_1\cdots\mu_4}}\}\,,\\
\{\psi_{\underline{\mu}}, \varepsilon\} &\to -\alpha\Gamma^{p+1\cdots 9}(-\Gamma_{11})^{\frac{p}{2}}\{\psi_{\underline{\mu}}, \varepsilon\}\,,\\
\{\lambda\} &\to +\alpha\Gamma^{p+1\cdots 9}(+\Gamma_{11})^{\frac{p}{2}}\{\lambda\}\,,
\end{aligned} \tag{28}$$

and for fields with other indices there is an extra minus sign for each replacement of a worldvolume index $\underline{\mu}$ by an index in a transverse direction. We have left open the possibility of combining the symmetry with the sign change of all fermions. This possibility introduces a number $\alpha = \pm 1$ in the above rules. This symmetry will be used for the orientifold construction.

For this purpose we choose spacetime to be $\mathcal{M}^{p+1} \times T^{9-p}$ with radii $R^{\bar{\mu}}$ of the torus that may depend on the world-volume coordinates. All fields satisfy

$$\Phi(x^{\bar{\mu}}) = \Phi(x^{\bar{\mu}} + 2\pi R^{\bar{\mu}}), \tag{29}$$

with $\bar{\mu} = (p+1, \ldots, 9)$. The parity symmetry (27) relates the fields in the bulk at $x^{\bar{\mu}}$ and $-x^{\bar{\mu}}$. At the fixed point of the orientifolds, however, this relation is local and projects out half the fields. This means that we are left with only $N = 1$ supersymmetry on the fixed points, where the branes will be inserted. Consider for example a nine-dimensional orientifold. The projection truncates our bulk $N = 2$ supersymmetry to $N = 1$ on the brane; only half of the 32 components of ε are even under (28). The original field content, a $D = 10$, $(128 + 128)$, $N = 2$ supergravity multiplet, gets truncated on the brane to a reducible $D = 9$, $(64 + 64)$, $N = 1$ theory consisting of a supergravity plus a vector multiplet. One may further restrict to a constant torus. This particular choice of spacetime then projects out a $N = 1$ $(8 + 8)$ vector multiplet (containing $e_{\dot{9}}{}^{9}$), leaving us with the irreducible $D = 9$, $(56 + 56)$, $N = 1$ supergravity multiplet. Similar truncations are possible in lower dimensional orientifolds, on which the $(64 + 64)$ $N = 1$ theory also consists of a number of multiplets.

We propose the p-brane action ($p = 0, 2, 4, 6, 8$) to be proportional to

$$\begin{aligned} \mathcal{L}_p &= -e^{-\phi}\sqrt{-g_{(p+1)}} - \alpha \frac{1}{(p+1)!}\varepsilon^{(p+1)}C^{(p+1)}, \\ \text{with } \varepsilon^{(p+1)}C^{(p+1)} &\equiv \varepsilon^{(p+1)}_{\underline{\mu_0 \cdots \mu_p}} C^{(p+1)\underline{\mu_0 \cdots \mu_p}}, \end{aligned} \tag{30}$$

with $\varepsilon^{(p+1)\,\underline{\mu_0 \cdots \mu_p}} = \varepsilon^{(10)\,\underline{\mu_0 \cdots \mu_p}\,\dot{p}\dot{+}\dot{1}\ldots\dot{9}}$, which follows from $e_{\underline{\mu}}{}^{\bar{a}} = 0$ (being odd). Here the underlined indices are $(p+1)$-dimensional and refer to the world-volume. The parameter α is the same that appears in (28) and takes the values $\alpha = +1$ for *branes*, which are defined to have tension and charge with the same sign in our conventions, and $\alpha = -1$ for *anti-branes*, which are defined to have tension and charge of opposite signs. Note that due to the vanishing of B on the brane the potentials $C^{(p+1)}$ and $A^{(p+1)}$ are equal. The p-brane action can easily be shown to be invariant under the appropriate $N = 1$ supersymmetry:

$$\delta_\varepsilon \mathcal{L}_p = -e^{-\phi}\sqrt{-g_{(p+1)}}\,\bar{\varepsilon}\big(1 - \alpha\Gamma^{p+1\cdots 9}(\Gamma_{11})^{\frac{p}{2}}\big)\,\Gamma^{\underline{\mu}}\big(\psi_{\underline{\mu}} - \tfrac{1}{18}\Gamma_{\underline{\mu}}\lambda\big). \tag{31}$$

The above variation vanishes due to the projection under (28) that selects branes or anti-branes depending on the sign of α ($+1$ or -1 respectively). In the following discussions we will assume $\alpha = 1$ but the other case just amounts to replacing branes by anti-branes.

By truncating our theory we are able to construct a brane action that only consists of bosons and yet is separately supersymmetric. Having these at our disposal, we can introduce source terms for the various potentials. In general there are 2^{9-p} fixed points. The compactness of the transverse space implies that the total charge must vanish. Thus the total action will read (we take the special case that all branes are equally distributed over all 2^{9-p} fixed points)

$$\begin{aligned} \mathcal{L} &= \mathcal{L}_{\text{bulk}} + k_p \mathcal{L}_p \Delta_p\,, \\ \text{with } \Delta_p &\equiv \big(\delta(x^{p+1}) - \delta(x^{p+1} - \pi R^{p+1})\big) \cdots \big(\delta(x^9) - \delta(x^9 - \pi R^9)\big)\,, \end{aligned} \qquad (32)$$

where the branes at all fixed points have a tension and a charge proportional to $\pm k_p$, a parameter of dimension $1/[\text{length}]^{p+1}$. Since anti-branes do not satisfy the supersymmetry condition (31), we need both positive and negative tension branes to accomplish vanishing total charge. As explained in the introduction we are going to interpret the negative tension branes as O-planes.

The equations of motion following from (32) induce a δ-function in the Bianchi identity of the $8-p$-form field strength. In general, an elegant solution is difficult to find, but in the eight-brane case the situation simplifies.

QUANTIZATION OF MASS AND COSMOLOGICAL CONSTANT

Consider the eight-brane case only. The equation of motion of the nine-form is modified by the brane & plane actions such that the solution for $G^{(0)}$ is given by

$$G^{(0)} = \alpha \frac{n-8}{2\pi\ell_s} \varepsilon(x^9)\,. \qquad (33)$$

Thus we may identify the mass parameter of Type IIA supergravity as follows:

$$m = \begin{cases} \alpha \dfrac{n-8}{2\pi\ell_s}\,, & x^9 > 0\,, \\[2ex] -\alpha \dfrac{n-8}{2\pi\ell_s}\,, & x^9 < 0\,. \end{cases} \qquad (34)$$

The mass is quantized in string units and it is proportional to $n-8$ where there are $2n$ and $2(16-n)$ D8-branes at each O8-plane. The mass vanishes only in the special case $n=8$ when the contribution from the D8-branes cancels exactly the contribution from the O8-planes. In general, the mass takes only the restricted values

$$2\pi\ell_s|m| = 0,\ 1,\ 2,\ 3,\ 4,\ 5,\ 6,\ 7,\ 8. \qquad (35)$$

This is a quantization of our mass parameter, and for the cosmological constant it follows that

$$m^2 = (G^{(0)})^2 = \left(\frac{n-8}{2\pi\ell_s}\right)^2. \qquad (36)$$

Thus the mass parameter and the cosmological constant are quantized in the units of the string length in terms of the integers $n-8$.

The quantization of the mass and of the cosmological constant in $D=10$ was discussed before in [2, 4, 22] as well as in [5, 23]. In the latter two references, two independent derivations of the quantization condition were given. In [5], the T-duality between a 7-brane & 8-brane solution was investigated. Here it was pointed out that, in the presence of a cosmological constant, the relation between the $D = 10$ IIB R-R scalar $C^{(0)}$ and the one reduced to $D = 9$, $c^{(0)}$, is given via a generalized Scherk–Schwarz prescription:

$$C^{(0)} = c^{(0)}(x^9) + mx^8\,. \tag{37}$$

Here (x^8, x^9) parametrize the 2-dimensional space transverse to the 7-brane. x^9 is a radial coordinate whereas x^8 is periodically identified (it corresponds to a U(1) Killing vector field):

$$x^8 \sim x^8 + 1\,. \tag{38}$$

Furthermore, due to the $SL(2,\mathbb{Z})$ U-duality, the R-R scalar $C^{(0)}$ is also periodically identified:

$$C^{(0)} \sim C^{(0)} + 1\,. \tag{39}$$

Combining the two identifications with the reduction rule for $C^{(0)}$ leads to a quantization condition for m of the form

$$m \sim \frac{n}{\ell_s}\,, \qquad n \text{ integer}. \tag{40}$$

The same result was obtained by a different method in [23].

We are able to give a new, and independent, derivation of the quantization condition for the mass and cosmological constant. The conditions given in (34), (36) follow straightforwardly from our construction of the bulk & brane & plane action.

Note that the Scherk–Schwarz reduction in (37) and the quantization of $SL(2,\mathbb{R})$ were essential in deriving the quantization of m. In the new dual formulation we can derive a similar T-duality relation between the 7-brane and the 8-brane, including the source terms. However, in this case the T-duality relation does not imply a quantization condition for m since we do not know how to realize the $SL(2,\mathbb{R})$ symmetry in the dual formulation. Another noteworthy feature is that the derivation of the T-duality rules in the dual formulation does not require a Scherk-Schwarz reduction. This is possible due to the fact that the R-R scalar only appears after solving the equations of motion.

CONCLUSIONS

We have constructed new formulations of Type II $D = 10$ supergravity. For both Type IIA and IIB theories, we constructed democratic bulk theories with a unified treatment of all R-R potentials. Due to the doubling of R-R degrees of freedom one had to impose extra duality constraints and thus a proper action was not possible. A so-called pseudo-action, containing kinetic terms for all R-R potentials but without Chern-Simons terms, was discussed. Furthermore, we have broken the self-duality explicitly in the IIA case, allowing for a proper action. Instead of all R-R potentials only half of the $C^{(p)}$'s occur in these theories. Both the standard ($p = 1, 3$) as well as the dual ($p = 5, 7, 9$) formulations were discussed. Using these actions all bulk & brane systems can be described.

A notable difference of our scenario from the HW [25, 26] scenario is that the walls are the O8 and D8 objects which exist in string theory. The main goal of the HW theory was to present a scenario for appearance of chiral fermions starting with $D = 11$ supersymmetric theory with non-chiral fermions. Our O8-D8 construction may reach this precise goal in an interesting and controllable way due to stringy nature of this construction and due to the complete control over supersymmetries in the bulk & on the walls. We remark that the strong coupling limit of Type I$'$ string theory is equal to the HW theory. Using the results of this paper, it would be interesting to investigate whether and how in this limit the O8-D8 objects can be related to the HW branes.

ACKNOWLEDGMENTS

The work described in this talk is based upon the work of [11]. Two of the authors (E.B. and A.V.P.) would like to thank the organizing committee of the Karpacz School for the hospitality and stimulating atmosphere provided to us.

This work was supported by the European Commission RTN program HPRN-CT-2000-00131, in which E.B. is associated with Utrecht University. The work of R.K. was supported by NSF grant PHY-9870115. The work of T.O. has been supported in part by the Spanish grant FPA2000-1584. T.O. wouldlike to thank the C.E.R.N. TH Division and the I.T.P. of the University of Groningen for their financial support and warm hospitality. E.B. would like to thank the I.F.T.-U.A.M./C.S.I.C. for its hospitality.

REFERENCES

1. Polchinski, J., *String Theory*. Cambridge University Press, 1999.
2. Polchinski, J., *Phys. Rev. Lett.* **75**, 4724–4727 (1995), hep-th/9510017.
3. Romans, L. J., *Phys. Lett.*, **169B**, 374 (1986).
4. Polchinski, J., and Witten, E., *Nucl. Phys.*, **B460**, 525–540 (1996), hep-th/9510169.
5. Bergshoeff, E., de Roo, M., Green, M. B., Papadopoulos, G., and Townsend, P. K., *Nucl. Phys.*, **B470**, 113–135 (1996), hep-th/9601150.
6. Duff, M. J., and van Nieuwenhuizen, P., *Phys. Lett.*, **B94**, 179 (1980).
7. Aurilia, A., Nicolai, H., and Townsend, P. K., *Nucl. Phys.*, **B176**, 509 (1980).
8. Alonso-Alberca, N., Meessen, P., and Ortín, T., *Phys. Lett.*, **B482**, 400–408 (2000), hep-th/0003248.
9. Randall, L., and Sundrum, R., *Phys. Rev. Lett.*, **83**, 3370–3373 (1999), hep-ph/9905221. *Phys. Rev. Lett.*, **83**, 4690 (1999), hep-th/9906064.
10. Bergshoeff, E., Kallosh, R., and Van Proeyen, A., JHEP **10**, 033 (2000), hep-th/0007044.
11. Bergshoeff, E., Kallosh, R., Ortín, T., Roest, D., and Van Proeyen, A., hep-th/0103233.
12. Huq, M., and Namazie, M. A., *Class. Quant. Grav.*, **2**, 293 (1985).
13. Giani, F., and Pernici, M., *Phys. Rev.*, **D30**, 325–333 (1984).
14. Campbell, I. C. G., and West, P. C., *Nucl. Phys.*, **B243**, 112 (1984).
15. Schwarz, J. H. *Nucl. Phys.*, **B226**, 269 (1983).
16. Howe, P. S., and West, P. C. *Nucl. Phys.*, **B238**, 181 (1984).
17. Fukuma, M., Oota, T., and Tanaka, H., *Prog. Theor. Phys.*, **103**, 425–446 (2000), hep-th/9907132.
18. Bergshoeff, E., Boonstra, H. J., and Ortín, T., *Phys. Rev.*, **D53**, 7206–7212 (1996), hep-th/9508091.
19. Bergshoeff, E. de Roo, M., Janssen, B., and Ortín, T., *Nucl. Phys.*, **B550**, 289–302 (1999), hep-th/9901055.

20. Bergshoeff, E., Eyras, E., Halbersma, R., van der Schaar, J. P., Hull, C. M., and Lozano, Y., *Nucl. Phys.*, **B564**, 29–59 (2000), hep-th/9812224.
21. Meessen, P., and Ortin, *Nucl. Phys.*, **B541**, 195 (1999), hep-th/9806120.
22. Polchinski, J., and Strominger, A., *Phys. Lett.*, **B388**, 736–742 (1996), hep-th/9510227.
23. Green, M.B., Hull, C.M., and Townsend, P.K., *Phys. Lett.*, **B382**, 65–72 (1996), hep-th/9604119.
24. Dabholkar, A., hep-th/9804208. in 'High energy physics and cosmology 1997', Proceedings Trieste 1997, The ICTP Series in Theoretical Physics, Vol. 14, Eds. E. Gava et al., Singapore, World Scientific, 1998, p. 128.
25. Hořava, P., and Witten, E., *Nucl. Phys.*, **B475**, 94–114 (1996), hep-th/9603142.
26. Lukas, A., Ovrut, B.A., Stelle, K.S., and Waldram, D., *Phys. Rev.*, **D59**, 086001 (1999), hep-th/9803235.
Lukas, A., Ovrut, B.A., Stelle, K.S., and Waldram, D., *Nucl. Phys.*, **B552**, 246 (1999), hep-th/9806051;
Stelle, K.S., *Domain walls and the universe,* in 'Nonperturbative aspects of strings, branes and supersymmetry', Proc. Trieste 1998, Eds. M. Duff et al., World Scientific, 1999, p. 396, hep-th/9812086.

Superspace Methods in String Theory, Supergravity and Gauge Theory

M. Cederwall

Institute for Theoretical Physics, Göteborg University and Chalmers University of Technology, SE-412 96 Göteborg, Sweden, email: martin.cederwall@fy.chalmers.se

Abstract. In these two lectures, delivered at the XXXVII Karpacz Winter School, February 2001, I review some applications of superspace in various topics related to string theory and M-theory. The first lecture is mainly devoted to descriptions of brane dynamics formulated in supergravity backgrounds. The second lecture concerns the use of superspace techniques for determining consistent interactions in supersymmetric gauge theory and supergravity, *e.g.* α'-corrections from string/M-theory.

BRANES, SUPERGRAVITY AND SUPERSPACE

p-Branes: $D=11$, $p=2$

Branes play important rôles in string theory and M-theory. They are non-perturbative objects that may be described as solitons of the low-energy effective supergravity theories (see refs. [1] and [2] for extensive reviews). Here, I will concentrate on the dynamics of branes, as described by their actions [3, 4, 5, 6, 7]. There are a number of different branes in string theory and M-theory, most conveniently characterised by their field content when seen as a field theory on the world-volume. The simplest ones, the so-called p-branes, have a scalar multiplet on the world-volume. D-branes contain a vector multiplet, coupling to string endpoints [8], and the M5-brane has a self-dual tensor.

As a model for the simplest branes I will treat the membrane in eleven dimensions [4]. The action for a brane typically consists of two parts, a kinetic term proportional to the invariant volume, and a Wess–Zumino term specifying the minimal coupling to a $(p+1)$-form potential, under which the brane carries charge:

$$S=-T\int d^3\xi\sqrt{-|g|}+T\int C\,, \qquad (1)$$

T is the membrane tension. The action (1) looks like an action describing bosonic degrees of freedom contained in its transverse fluctuations. How do we describe a supersymmetric brane, containing an equal number of fermionic degrees of freedom? One simple and very efficient way, in many aspects much simpler than a component approach, is to consider the dynamics to be described by the same formal action, but

CP589, *Fundamental Interaction Theories*, edited by J. Lukierski and J. Rembieliński

where the bosonic world-volume is embedded in superspace[1]. The target superspace has coordinates $Z^M = (X^m, \theta^\mu)$ (the corresponding inertial indices are $A = (a, \alpha)$, a Lorentz vector and some spinor), and the background fields entering the action (1) are pullbacks from superspace to the world-volume, $g_{ij} = E_i{}^a E_j{}^b \eta_{ab}$, $C_{ijk} = E_k{}^C E_j{}^B E_i{}^a C_{ABC}$, with $E_i{}^A = \partial_i Z^M E_M{}^A$, $E_M{}^A$ being the target space super-vielbein.

Let us now investigate what this means for the supermembrane. The eight transverse bosonic oscillations must be matched in number by eight fermionic degrees of freedom if the action is to be supersymmetric. A (Majorana) spinor in $D = 11$ has 32 components. The number is reduced by half by the equations of motion, as usual, but it is clear that an additional local symmetry is required in order to get eight physical spinor degrees of freedom. This is the so called κ-symmetry, parametrised by a half spinor. κ-symmetry is a local (in terms of the location on the brane) translation of the brane in a fermionic direction in superspace. As such, it is generated by a superspace vector field pointing in fermionic directions only: $\kappa = \kappa^M \partial_M = \kappa^\alpha E_\alpha{}^M \partial_M$, and the transformation of the coordinates is $\delta_\kappa Z^M = \kappa^M$. Pullbacks of superspace forms are transformed by the Lie derivative, $\delta_\kappa f^* \Omega = f^* \mathcal{L}_\kappa \Omega = f^*(i_\kappa d + d i_\kappa)\Omega$, in the following)

$$\begin{aligned} \delta_\kappa C &= i_\kappa H + d i_\kappa C \,, & (2)\\ \delta_\kappa E^A &= D\kappa^A + i_\kappa T^A \,, & (3)\\ \delta_\kappa g_{ij} &= 2E_{(i}{}^a E_{j)}^B \kappa^\alpha T_{\alpha B}{}^b \eta_{ab} \,, & (4) \end{aligned}$$

where $H = dC$ is the background tensor superfield strength and $T^A = DE^A$ the superspace torsion.

To determine how the action transforms (modulo boundary terms), we only need $i_\kappa H$ and $i_\kappa T^a$. In $D = 11$ supergravity [9, 10] this is particularly simple, the only non-vanishing components of H and T^a with at least one spinorial form-index are the dimension 0 ones, $T_{\alpha\beta}{}^a = 2\Gamma^a_{\alpha\beta}$, $H_{ab\alpha\beta} = 2(\Gamma_{ab})_{\alpha\beta}$. A short calculation yields

$$\delta_\kappa S = -\int d^3\xi 2\sqrt{-g} E_i{}^\alpha (\Gamma^i - \tfrac{1}{2\sqrt{-g}} \varepsilon^{ijk} \Gamma_{jk})_{\alpha\beta} \kappa^\beta \,, \tag{5}$$

with the obvious notation for pullbacks of Γ-matrices. The combination of Γ-matrices in the last term may be written as

$$\Gamma^i - \tfrac{1}{2\sqrt{-g}} \varepsilon^{ijk} \Gamma_{jk} = \Gamma^i (\mathbb{1} - \tfrac{1}{6\sqrt{-g}} \varepsilon^{ijk} \Gamma_{ijk}) = \Gamma^i (\mathbb{1} - \Gamma) \,, \tag{6}$$

and is seen to provide a projection on κ, since $\Pi_\pm = \frac{1}{2}(\mathbb{1} \pm \Gamma)$, due to the identities $\Gamma^2 = \mathbb{1}$ and $\mathrm{tr}\,\Gamma = 0$, are projection matrices splitting a 32-component spinor in two halves. The only chance that this variation vanishes is thus that $\Pi_- \kappa = 0$. This is indeed the half spinor of local fermionic symmetry that was needed for the matching of bosonic and fermionic degrees of freedom. Since setting the dimension 0 torsion to a Γ-matrix

[1] There exists a framework, the so called embedding formalism, where both target space and the world-volume are superspaces [6]. I will not consider it here.

puts the background on shell [11], the supermembrane has κ-symmetry in any on-shell background of $D = 11$ supergravity.

Analogous calculations hold for other p-branes in other supergravities, and show that for general on-shell backgrounds, the actions are κ-symmetric. κ-symmetry is related to the fact that the branes are BPS-saturated configurations—the supersymmetry algebra generating the multiplets on the branes (in the present case a scalar multiplet) contains half the number of fermionic generators compared to the target space supersymmetry, and half of the target space supersymmetry is broken (the world-volume fields are Goldstone fields corresponding to broken symmetries of the background). The projection matrices are related to (target space) supersymmetry algebras with "central" tensorial charges, that get projected out by a half-rank projection $\Pi_\pm$.

We may also note that the formalism presented here, with the brane embedded in an arbitrary target superspace background, actually is as simple as in a flat superspace. Working with explicit fermionic coordinates becomes complicated, since the expression for a tensor potential is complicated, while the (gauge invariant) field strength is simple.

As presented here, the branes are viewed as infinitely thin objects moving in superspace. They may also be seen as solitons in the low-energy effective supergravity theories [12]. All fields on branes arise as Goldstone modes corresponding to broken symmetries of the background theory. Scalars and fermions correspond to broken translational symmetries and supersymmetry, while vectors on D-branes and tensors on M5-branes arise as Goldstone modes for large gauge symmetries of target space tensors, *i.e.*, gauge transformations that take different values "on the brane" and in the asymptotic region [13].

$D = 11$ Supergravity

The Γ-matrix constraint on the dimension 0 torsion puts the theory on shell [11]. The tensor field arises naturally from the superspace geometry, and it is not necessary to separately require the existence of a closed 4-form on superspace. The Bianchi identity for H at dimension 0 becomes $0 = (dH)_{a\alpha\beta\gamma\delta} = 6T_{(\alpha\beta}{}^{b}H_{|ba|\gamma\delta)} = 24\Gamma^{b}_{(\alpha\beta}\Gamma_{|ba|\gamma\delta)}$, and at dimension 1, the non-vanishing torsion is

$$T_{a\alpha}{}^{\beta} = \tfrac{1}{36}\Gamma^{bcd}{}_{\alpha}{}^{\beta}H_{abcd} + \tfrac{1}{288}\Gamma_{a}{}^{bcde}{}_{\alpha}{}^{\beta}H_{bcde}\,. \tag{7}$$

Actually, the superspace Bianchi identities also leave room for a spinor ω_α at dimension $\frac{1}{2}$ and a vector ω_a at dimension 1 in T, the Bianchi identities further require that these be integrable to $\omega_A = D_A\phi$, and the "conformal compensator" ϕ can then be removed by a conventional constraint, or alternatively by the enlargement of the structure group to include Weyl rescalings[2].

---

[2] There is a disagreement on this point. The view presented here is that of refs. [11, 14], while the authors of ref. [15] claim that the conformal compensator has to play a rôle in a (yet unknown) supersymmetric off-shell formulation of eleven-dimensional supergravity.

D-Branes, Type II Supergravity

Type II superstring theories, and their low-energy effective theories, type IIA and IIB supergravity, contain tensor fields in the Ramond-Ramond sector. For type IIA the potentials have odd rank, $C = C_{(1)} \oplus C_{(3)} \oplus C_{(5)} \oplus \ldots$, and for type IIB even, $C = C_{(0)} \oplus C_{(2)} \oplus C_{(4)} \oplus \ldots$. The corresponding field strengths are required to be self-dual, so in principle we have a redundant set of potentials, which is useful when considering brane actions. A five-brane, *e.g.*, couples minimally to a 6-form potential, whose 7-form field strength is dual to the 3-form. In addition there is the the NS-NS 2-form B.

D-branes are exactly the non-perturbative objects carrying charge under the RR fields. They act as hypersurfaces where fundamental strings are allowed to end, and contain vector degrees of freedom, coupling minimally to the world-lines of the string ends [8]. This picture resulted in an effective action for D-branes [16, 17]:

$$\begin{aligned} S &= -\int d^{p+1}\xi e^{-\phi}\sqrt{-|g_s+F|}+\int e^F C \\ &= -\int d^{p+1}\xi e^{\frac{p-3}{4}\phi}\sqrt{-|g_E+e^{-\phi/2}F|}+\int e^F C\,. \end{aligned} \tag{8}$$

There are some things to explain in this expression. The field ϕ is the dilaton field, and the factor $\exp(-\phi)$ means that the D-brane tension in the "string frame" is proportional to g^{-1}, where $g = \exp(\phi)$ is the string coupling. In the second line, the action has been rewritten in terms of the Einstein metric $g_E = \exp(\phi/2)g_s$, which is sometimes convenient, especially when I later want to consider SL(2;$\mathbb{Z}$) duality symmetry. The second, Wess–Zumino, term in the action is evaluated with wedge products, and the $(p+1)$-form is extracted, so that for the D3-brane, *e.g.*, it reads $\int(C_{(4)}+F\wedge C_{(2)}+\frac{1}{2}F\wedge F\wedge C_{(0)})$. There is a U(1) vector field A on the world-volume, and the field strength F contains the NS-NS 2-form potential B trough $F = dA - B$. The gauge transformations of B, $\delta_\lambda B = d\lambda$, also act on A as $\delta_\lambda A = \lambda$, so that F is invariant. This means that an expectation value for F can be traded for a background B field[3]. Apart from the dilaton factor, the first, kinetic, term is of Dirac–Born–Infeld type.

The RR tensors have "modified" field strengths $R = e^B d(e^{-B}C) = dC - H\wedge C$, and their Bianchi identities read $dR + H\wedge R = 0$. Gauge transformations $\delta_\Lambda C = e^B d\Lambda$ leave the WZ term invariant up to a total derivative.

As was done for the membrane in the previous section, the D-brane actions (8) are promoted to actions for supersymmetric D-branes by letting the the embedding be in a target superspace of type IIA or IIB. Superspace formulations of the type IIA and type IIB supergravities are given in refs. [19] and [20].

The essential check is again κ-symmetry. The calculations are somewhat more complicated than for a brane with a scalar multiplet, so I refer to ref. [7] for more details. In type IIB, the two spinor coordinates have the same chirality, and instead of introducing

---

[3] This implies that such configurations do not break supersymmetry, which can be seen in the supergravity solutions corresponding to D-branes with constant F, or equivalently, in a background B field [18]. What instead happens is that the part of supersymmetry remaining unbroken is a different projection than for $F = 0$, due to the F-dependence of the projector (13).

explicit indices, I include this in a IIB spinor index $\alpha = 1, \ldots, 32$, and introduce the basis of 2×2-matrices $\{\mathbb{1}, I, J, K\}$ with $I = i\sigma_2$, $J = \sigma_1$, $K = \sigma_3$ (they can be seen as a basis for the split quaternions). The the relevant fields at dimension 0 are

$$T_{\alpha\beta}{}^{a} = 2\Gamma^{a}_{\alpha\beta}\,, \tag{9}$$

$$H_{a\alpha\beta} = -2e^{\frac{\phi}{2}}(\Gamma_a K)_{\alpha\beta}\,, \tag{10}$$

$$R_{a_1\ldots a_{n-2}\alpha\beta} = 2e^{\frac{n-5}{4}\phi}(\Gamma_{a_1\ldots a_{n-2}}K^{\frac{n-1}{2}}I)_{\alpha\beta}\,. \tag{11}$$

Since the ten-dimensional supergravities have spinors, $\lambda_\alpha = D_\alpha\phi$ of dimension $\frac{1}{2}$, these will also occur in the dimensions $\frac{1}{2}$ components of H, R and T, which I do not list here.

The variation of the action requires that we specify the transformation of the vector A. In order to get something gauge invariant we must take $\delta_\kappa A = i_\kappa B$, which implies that $\delta_\kappa F = -i_\kappa H$. Then the variation of the WZ term becomes $\delta_\kappa(e^F C) = e^F i_\kappa R$ modulo boundary terms. Going through the procedure of inserting the variations of the fields in the lagrangian yields an expression

$$\delta_\kappa L \propto E_i{}^{\alpha}\Gamma^i(\mathbb{1} - \Gamma)\kappa^\alpha\,, \tag{12}$$

where Γ is a more complicated expression than eq. (6), containing different powers of the field strength F, and thus providing field-dependent half-rank projections of a spinor. For the D3-brane, it takes the form

$$\Gamma = \frac{\varepsilon^{ijkl}}{\sqrt{-|g + e^{-\phi/2}F|}}\left(\tfrac{1}{24}\Gamma_{ijkl}I - \tfrac{1}{4}e^{-\frac{\phi}{2}}F_{ij}\Gamma_{kl}J + \tfrac{1}{8}e^{-\phi}F_{ij}F_{kl}I\right)\,, \tag{13}$$

and similar expressions hold for other D-branes. This shows κ-symmetry of the D-branes in an arbitrary on-shell supergravity background.

SL(2;ℤ), Tensor Democracy

Type IIB supergravity has an SL(2;ℝ) symmetry, which at the quantum level is broken to the SL(2;ℤ) S-duality group. Since S-duality is non-perturbative, the representations under SL(2;ℤ) contain perturbative and non-perturbative states, and can not be manifested in perturbative string theory. Nevertheless, it can be manifested in the effective supergravity, and it is meaningful to ask whether it is possible to treat all branes with NS-NS and RR charges, including the fundamental string, on an equal footing, thus manifesting the S-duality symmetry.

The scalars, the dilaton and axion, belong to the coset SL(2;ℝ)/U(1). This is a combination of NS-NS and RR fields, since the axion is identified with $C_{(0)}$. The NS-NS and RR 2-forms $B_{(2)}$ and $C_{(2)}$ combine into an SL(2) doublet, and the 4-form (with selfdual field strength) is an SL(2) singlet. Higher rank tensors are dual to those already mentioned. The representations of branes reflect those of the tensor fields: The strings and five-branes come with charges that form an SL(2;ℤ) doublet (p,q), while the D3-brane forms a singlet. I will not examine higher-dimensional branes here.

The scalars of type IIB supergravity are describe as a complex doublet $\mathcal{U}^r$, $r = 1,2$, subject to the constraint $\frac{i}{2}\varepsilon_{rs}\mathcal{U}^r\bar{\mathcal{U}}^s = 1$ (which is the condition that $\mathcal{U}$ has unit determinant when seen as a real 2×2 matrix). The coset is obtained from gauging the U(1) acting as $\mathcal{U} \to e^{i\theta}\mathcal{U}$. One forms the left-invariant Maurer–Cartan forms

$$Q = \tfrac{1}{2}\varepsilon_{rs} d\mathcal{U}^r \bar{\mathcal{U}}^s , \tag{14}$$
$$P = \tfrac{1}{2}\varepsilon_{rs} d\mathcal{U}^r \mathcal{U}^s , \tag{15}$$

with Bianchi identities (Maurer–Cartan equations)

$$dQ - iP\wedge\bar{P} = 0 , \tag{16}$$
$$DP \equiv dP - 2iP\wedge Q = 0 . \tag{17}$$

The scalars act as a bridge between objects that are SL(2) doublets and real and objects that are SL(2) singlets but carry U(1) charge. If we write the 3-form doublet of field strengths as $H_{(3)r} = dC_{(2)r}$, the SL(2) singlet field strength is $\mathcal{H}_{(3)} = \mathcal{U}^r H_{(3)r}$. Notice that this is necessary when writing a kinetic term as proportional to $\mathcal{H}\cdot\bar{\mathcal{H}}$. The Bianchi identity is $D\mathcal{H} + i\bar{\mathcal{H}}\wedge P = 0$ (recall that $\mathcal{H}$ has U(1) charge 1 while P has charge 2). The singlet 5-form is constructed as $H_{(5)} = dC_{(4)} + \mathrm{Im}(\mathcal{C}_{(2)}\wedge\bar{\mathcal{H}}_{(3)})$, with Bianchi identity $dH_{(5)} - i\mathcal{H}_{(3)}\wedge\bar{\mathcal{H}}_{(3)} = 0$.

We now come to the crucial point in describing brane dynamics SL(2)-covariantly. It is not sufficient to introduce one vector field on the brane. Remember that the field strength was $F = dA - B$, where B was the NS-NS 2-form. It is clear that another vector, combining with the RR 2-form is needed, so that they form a doublet. One should thus have $F_r = dA_r - C_{(2)r}$, reflecting the fact that strings of different charges (p,q) can end on a brane. Once this step has been taken, it is equally natural to introduce a form of rank p for each background tensor fields of rank $p+1$, reflecting the fact that a p-brane can end on the brane we describe, and coupling minimally to its boundary. For this reason, such a formulation, with complete "tensor democracy" on the branes, should most naturally encode the coupling of branes to background fields. Gauge invariance (in target space and on the brane) demands that also the tensors on the brane have modified Bianchi identities. The 2-form and 4-form on any brane are

$$\mathcal{F}_{(2)} = \mathcal{U}^r dA_{(1)r} - \mathcal{C}_{(2)} , \tag{18}$$
$$F_{(4)} = dA_{(3)} - C_{(4)} + \mathrm{Im}(\mathcal{A}_{(1)}\wedge\bar{\mathcal{H}}_{(3)}) , \tag{19}$$

with Bianchi identities

$$D\mathcal{F}_{(2)} + i\bar{\mathcal{F}}_{(2)}\wedge P = -\mathcal{H}_{(3)} , \tag{20}$$
$$dF_{(4)} = -H_{(5)} - \mathrm{Im}(\mathcal{F}_{(2)}\wedge\bar{\mathcal{H}}_{(3)}) . \tag{21}$$

In a generic situation, the procedure seems to give too many bosonic fields, and there must be ways to reduce the number in order to recover an SL(2)-covariant description of brane dynamics. The key is selfduality, and I will sketch how it works for different branes. I refer the readers to refs. [21, 22] for details. All cases described may be shown to be κ-symmetric, along similar lines as in the previous sections. The actions do not divide into Born–Infeld plus Wess–Zumino, since this presumes a division into NS-NS and RR fields.

The (p,q) strings. The vectors A_r have no local degrees of freedom on the two-dimensional world-sheet, so we do not have to worry about removing degrees of freedom. The only degrees of freedom of vectors is a quantised electric flux (see ref. [23] for one vector), so the description gives rise to a pair of integers (p,q), which are the charges of string. In this way, the whole spectrum of (p,q) strings is described within one single action [21]. That the description is correct is checked by κ-symmetry and by the fact that the correct tensions [24] are produced.

The 3-brane. Having two vector potentials gives too many degrees of freedom, and one of them has effectively to be removed. This is obtained by imposing a selfduality relation on the complex field strength $\mathcal{F}$: $\mathcal{F} = i\star\mathcal{F}$ + higher order terms. It turns out that not any non-linear selfduality relation is allowed. Its exact form is dictated by consistency with the coupling to the background fields, and also, independently, by κ-symmetry, and it encodes in a manifest way the earlier observed Poincaré selfduality of the 3-brane. A formulation of the dynamics of the type IIB 3-brane is obtained [22] that naturally encodes in a most symmetric way all couplings to background fields, and thereby the possibilities for the 3-brane to host brane boundaries [22].

The (p,q) 5-branes. This case is not constructed in detail, but the general scheme is described in ref. [22]. There is a duality relation between the 4-form and the 2-forms. The fact that the corresponding supergravity solution could be described analytically [25] makes it reasonable to believe that the dynamics can be described covariantly, in spite of problems with dualisation in six dimensions [26].

The M5-brane and type IIA. The formalism is not restricted to type IIB. It was successfully applied to write down a "quasi-action" (the equations of motion follows, but not the selfduality, which however is uniquely determined by consistency with background couplings and by κ-symmetry) for the M5-brane [27]. It is also applicable to type IIA branes, and will also there encode the background interactions in the most natural way.

Summary

I have described brane dynamics by embedding in superspace, given a detailed account of the mechanisms behind κ-symmetry and focussed on the couplings of branes to fields in the background effective supergravity.

It is known that the effective supergravity theories following from string theory or M-theory receive corrections to higher order in α' than the lowest order ones used in this talk. Some α'-corrections to the brane actions themselves are also known [28, 29]. What happens to the brane dynamics when α' corrections are turned in in target space? It is clear that a superspace formulation is desirable in order to answer such questions. In the following lecture I will describe some recent progress in the superspace formulations of $D=11$ supergravity and $D=10$ super-Yang–Mills theory, both relevant for string/M-theory.

STRING/M-THEORY CORRECTIONS TO SUPERGRAVITY AND SUPER-YANG–MILLS

$D=11$ Supergravity Cont'd

We will now continue the discussion of the superspace formulation of eleven-dimensional supergravity in superspace [10, 11, 14]. The vielbeins are $E^A = dZ^M E_M{}^A$, and the resulting torsion 2-form is $T^A = DE^A = dE^A + E^B \wedge \omega_B{}^A$, where the structure group is the Lorentz group, *i.e.*, the spin connection satisfies $\omega_\alpha{}^\beta = \frac{1}{4}(\Gamma^a{}_b)_\alpha{}^\beta \omega_a{}^b$. The curvature is $R_A{}^B = d\omega_A{}^B + \omega_A{}^C \wedge \omega_C{}^B$. The Bianchi identities for torsion and curvature are $DT^A = E^B \wedge R_B{}^A$ and $DR_A{}^B = 0$. Of these, one needs only to use the first one.

As long as torsion and curvature are constructed from vielbeins and spin connections, the Bianchi identities are automatically fulfilled. In order to reduce the enormous amount of fields contained in these, one has however to impose "conventional constraints" connecting the different components. Then the Bianchi identities become integrability conditions that have to be checked, and which imply the equations of motion (this is true for the maximally supersymmetric theories I deal with in this lecture).

The conventional constraints are of two types. The first one uses the freedom in the definition of the torsion to shift it into the spin connection when possible. These constraints do not eliminate all of the torsion (as it does in bosonic gravity), but have the effect of determining the spin connection if terms of the vielbein, which is desirable. The second type uses a redefinition of the tangent bundle, $E^A \to E^B M_B{}^A$, while keeping the spin connection, and thus the curvature, invariant (although their components vary due to the change of basis). We want to use this freedom to the extent that it enables us to express all vielbein components in terms of the dimension $-\frac{1}{2}$ one, $E_\mu{}^a$.

Let us now examine the lowest-dimensional torsion components, $T_{\alpha\beta}{}^a$ at dimension 0. I already mentioned that putting it equal to a Γ-matrix takes the theory on-shell, so in order to incorporate corrections to the ordinary supergravity this constraint (which is not a conventional constraint) has to be modified. A general expansion yields, since the torsion is symmetric in the spinor indices,

$$T_{\alpha\beta}{}^c = 2(\Gamma^d_{\alpha\beta} X_d{}^c + \tfrac{1}{2}\Gamma^{d_1d_2}_{\alpha\beta} X_{d_1d_2}{}^c + \tfrac{1}{5!}\Gamma^{d_1\ldots d_5}_{\alpha\beta} X_{d_1\ldots d_5}{}^c)\,. \tag{22}$$

Decomposing into irreducible representations of the Lorentz group, we find that the three "X-tensors" contain $((20000)\oplus(01000)\oplus(00000))\oplus((11000)\oplus(00100)\oplus(10000))\oplus((10002)\oplus(00002)\oplus(00010))$. where standard Dynkin labels for highest weights are used.

If this representation content is compared to the one in the dimension-0 matrices for redefining the vielbein, $M_a{}^b$ and $M_\alpha{}^\beta$, which is $((20000)\oplus(01000)\oplus(00000))\oplus((00002)\oplus(00010)\oplus(00100)\oplus(01000)\oplus(10000)\oplus(00000))$, we see that the only remaining components are $X_{d_1d_2}{}^c|_{(11000)}$ and $X_{d_1\ldots d_5}{}^c|_{(10002)}$, *i.e.*, the "irreducible hooks" [30, 14]. These superfields should encode which the corrections to the supergravity are, and the equations of motion for any version of $D=11$ supergravity should follow from the solution of the Bianchi identities with a suitable choice of these tensors. This will be even clearer when we consider spinorial cohomology in a little while.

Solving the Bianchi identities turns out to be quite complicated, and we have not succeeded in doing it in full generality. In ref. [14], we were able to show that the gravitino equation of motion received a correction, by solving the Bianchi identities up to dimension $\frac{3}{2}$, encountering on the way some remarkable numerical coincidences. We found no contribution to the Weyl curvatures up to this level, which means that the elimination of the conformal compensator by a conventional constraint is still valid.

$D=10$ Super-Yang–Mills

The study of the general superspace formulation of $D=10$ super-Yang–Mills is motivated by its connection to string theory and the relevance for finding non-abelian analogues of the Born–Infeld action [31, 32]. An advantage with the system is that it is much easier to analyse than $D=11$ supergravity, so we hoped that it would be more manageable.

I now work in flat superspace, with $T_{\alpha\beta}{}^a=2\Gamma^a_{\alpha\beta}$ and the rest of the torsion vanishing. The gauge potential is a superspace 1-form with components $A_A=(A_a,A_\alpha)$, and the field strength is $F=dA+A\wedge A$ with Bianchi identity $DF=0$. In components, the Bianchi identity reads:

$$
\begin{aligned}
\text{dim. } \tfrac{3}{2}: &\qquad D_{(\alpha}F_{\beta\gamma)}+2\Gamma^c_{(\alpha\beta}F_{|c|\gamma)}=0\,, &(23)\\
2: &\qquad 2D_{(\alpha}F_{\beta)c}+D_cF_{\alpha\beta}+2\Gamma^d_{\alpha\beta}F_{dc}=0\,, &(24)\\
\tfrac{5}{2}: &\qquad D_\alpha F_{bc}+2D_{[b}F_{c]\alpha}=0\,, &(25)\\
3: &\qquad D_{[a}F_{bc]}=0\,. &(26)
\end{aligned}
$$

Taking $F_{\alpha\beta}=0$ puts the theory on-shell [33], and it must be relaxed if we want to incorporate corrections. The general expansion is

$$
F_{\alpha\beta}=\Gamma^a_{\alpha\beta}J_a+\tfrac{1}{5!}\Gamma^{a_1\ldots a_5}_{\alpha\beta}J_{a_1\ldots a_5}\,. \qquad (27)
$$

The vector can always be set to zero as a conventional constraint, to eliminate the "extra" vector potential occurring at θ-level in A_α. We then have $A_a=-\frac{1}{32}\Gamma_a^{\alpha\beta}D_\alpha A_\beta$ (in the abelian case). The relevant deformation lies in the five-form, which is automatically (anti-)selfdual, due to the chirality of the spinors. In reference [34] we were able to solve the Bianchi identities completely for arbitrary J, whose components act as a super-current multiplet, and obtain the equations of motion,

$$
\begin{aligned}
0 &= D^bF_{ab}-\lambda\Gamma_a\lambda-8D^bK_{ab}+36w_a-\tfrac{4}{3}\{\lambda,\tilde{J}_a\}-2\tilde{J}_b\Gamma_a\tilde{J}^b\\
&\quad+\tfrac{1}{140\cdot 3!}\tilde{J}_{bcd}\Gamma_a\tilde{J}^{bcd}+\tfrac{1}{42}[K_{bcde},J_a{}^{bcde}]+\tfrac{1}{42\cdot 4!}[D^fJ_{fbcde},J_a{}^{bcde}]\,, \qquad (28)\\
0 &= \not{D}\lambda-30\psi+\tfrac{4}{3}D^a\tilde{J}_a+\tfrac{5}{126\cdot 5!}\Gamma abcde[\lambda,J_{abcde}]\,.
\end{aligned}
$$

Apart from F_{ab} and λ^α (which appears in the field strength as $\lambda^\alpha = \frac{1}{10}\Gamma^{a\alpha\beta}F_{a\beta}$), the quantities appearing in these equations all arise, as explained in ref. [34], in the θ expansion of J_{abcde}; $\tilde{J}'s$ at first, $K's$ at second, ψ at third, and ω at fourth order in θ. Explicitly, their precise relations to J_{abcde} are given by

$$\begin{aligned}
\tilde{J}_a &= \tfrac{1}{1680}\Gamma^{bcde}DJ_{bcdea}\,, & (29)\\
\tilde{J}_{abc} &= -\tfrac{1}{12}\Gamma^{de}DJ_{deabc} - \tfrac{1}{224}\Gamma_{[ab}\Gamma^{defg}DJ_{|defg|c]}\,, & (30)\\
\tilde{J}_{abcde} &= DJ_{abcde} + \tfrac{5}{6}\Gamma_{[ab}\Gamma^{fg}DJ_{|fg|cde]} + \tfrac{1}{24}\Gamma_{[abcd}\Gamma^{fghi}DJ_{|fghi|e]}\,, & (31)
\end{aligned}$$

$$\begin{aligned}
K_{ab} &= \tfrac{1}{5376}(D\Gamma^{cde}D)J_{cdeab}\,, & (32)\\
K_{abcd} &= \tfrac{1}{480}(D\Gamma_{[a}{}^{fg}D)J_{|fg|bcd]}\,, & (33)
\end{aligned}$$

$$\psi_\alpha = -\frac{1}{840\cdot 3!\cdot 5!}\Gamma_{abc}{}^{\beta\gamma}\Gamma_{de\alpha}{}^{\delta}D_{[\beta}D_\gamma D_{\delta]}J^{abcde}\,, \qquad (34)$$

and finally

$$w_a = \frac{1}{4032\cdot 4!\cdot 5!}\Gamma^{[\alpha\beta}_{abc}\Gamma^{\gamma\delta]}_{def}D_\alpha D_\beta D_\gamma D_\delta J^{bcdef}\,. \qquad (35)$$

I will soon show how one may use this formalism to deduce possible forms of α'-corrections allowed by supersymmetry. The idea is thus to take advantage of the fact (normally considered as a drawback) that the superspace formulation takes the theory on-shell.

Fields and Deformations from Spinorial Cohomology

Before becoming more specific about string-related corrections to super-Yang–Mills theory, I would like to digress on an amusing mathematical structure that has something to tell about maximally supersymmetric theories.

The basic idea is that the theories we consider are gauge theories, and that, in a superspace formulation, where all potentials and field strengths are forms on superspace, all components except the purely spinorial ones are redundant. Since all physical fields are contained in the objects carrying spinorial form indices only, it is interesting to examine the structure arising from these. Our complexes are of the form

$$r_0 \xrightarrow{\Delta_0} r_1 \xrightarrow{\Delta_1} r_2 \xrightarrow{\Delta_2} \ldots \xrightarrow{\Delta_{n-1}} r_n \xrightarrow{\Delta_n} \ldots\,, \qquad (36)$$

where r_p, for some $p \geq 0$, is the representation carried by a gauge transformation, r_{p+1} that of a potential and r_{p+2} that of a field strength. I will refer to the representations r_n as n-forms, a notation not to be confused with that of a tensor antisymmetric in vector indices. The exact definitions are given, both for gauge theory and supergravity, in the following sections, where it will also be clear why Δ is a nilpotent operator. The rôle of r_{p+3} is as a Bianchi identity.

Let me describe in more detail how the complexes work, with the super-Yang–Mills theory as an example. We have already seen that A_α contains the fields of the theory.

The relevant part of the field strength, as argued above, lies in (00020)[4], and does not contain A_a. We also note [33, 34] that part of the dimension-$\frac{3}{2}$ Bianchi identity states the vanishing of the (00030) component of $D_\alpha F_{\beta\gamma}$. These observations make it natural to consider, not the sequence of completely symmetric representations in spinor indices, but a restriction of it, namely the sequence of Spin(1,9) representations $r_n \equiv (000n0)$. They are the part of the totally symmetric product of n chiral spinors that has vanishing Γ-trace, and may be represented tensorially as $C_{\alpha_1\ldots\alpha_n} = C_{(\alpha_1\ldots\alpha_n)}$, $\Gamma_a{}^{\alpha_1\alpha_2}C_{\alpha_1\alpha_2\alpha_3\ldots\alpha_n} = 0$. For $n=2$, C is an anti-selfdual five-form, for $n=3$ a Γ-traceless anti-selfdual five-form spinor, etc.

The operator Δ_n: $r_n \longrightarrow r_{n+1}$ can schematically be written as $\Delta_n C_n = \Pi(r_{n+1})DC_n$, where D is the exterior covariant derivative $D = d\theta^\alpha D_\alpha$ and $\Pi(r_n)$ is the algebraic projection from $\otimes_s^n(00010)$ to $(000n0)$. It is straightforward to write an explicit tensorial form for Δ by subtracting Γ-traces from DC, but it will not be used here. It is also straightforward to show that, for an abelian gauge group and standard flat superspace, the sequence (36) forms a complex, *i.e.*, that $\Delta^2 = 0$. This follows simply from the fact that while $\{D_\alpha, D_\beta\} = -T_{\alpha\beta}{}^c D_c$, the torsion only has a component $2\Gamma_{\alpha\beta}{}^c$ which is projected out by $\Pi(r_n)$. This means that for non-abelian gauge theory the complex should be considered in a flat background, and the deformations yielded are infinitesimal.

We would now like to calculate the cohomology $\mathcal{H}^n = \mathrm{Ker}\Delta_n/\mathrm{Im}\Delta_{n-1}$ of the complex associated with $D=10$ super-Yang–Mills. This can be done by considering the decomposition into irreducible representations of the representation sitting at level ℓ in r_n, $r_n^\ell \equiv \wedge^\ell S \otimes r_n$. This is easily done, *e.g.* with the help of the program LiE [35]. One then follows each of the irreducible representations at a given dimension through the subcomplex

$$r_0^\ell \rightarrow r_1^{\ell-1} \rightarrow r_2^{\ell-2} \rightarrow \ldots \rightarrow r_{\ell-1}^1 \rightarrow r_\ell \,.$$

Let me illustrate the calculation by examining the field content. We then look into the spinor potential of dimension $\frac{1}{2}$, which contains all fields in the vector multiplet, so we should examine the first cohomology. The vector (dimension 1) sits at $\ell = \frac{1}{2}$ and the spinor (dimension $\frac{3}{2}$) at $\ell = 1$. The subcomplexes under consideration are $r_0^2 \rightarrow r_1^1 \rightarrow r_2$ and $r_0^3 \rightarrow r_1^2 \rightarrow r_2^1 \rightarrow r_3$. Checking the multiplicities of the relevant representations, (10000) and (00001), in these, we obtain the sequences $0 \rightarrow 1 \rightarrow 0$ and $0 \rightarrow 1 \rightarrow 0 \rightarrow 0$. The components of the cohomology in these representations and dimensions clearly contain the physical fields. This can be understood in a traditional framework as removing degrees of freedom in a superfield gauge transformation (removing the image from the left) and imposing the vanishing of the field strength $F_{\alpha\beta}$ (removing the complement of the kernel from the right). Analogous considerations tell us that the second cohomology contains a spinor of dimension $\frac{5}{2}$ and a vector of dimension 3. These are interpreted as belonging to a current supermultiplet, *i.e.*, fields entering the right hand sides of the equations of motion. This goes well together with the observation that modifications of the theory are introduced by deforming the constraint $F_{\alpha\beta} = 0$ [33, 36, 34, 37]. The relevance of the cohomology is explained by the facts that deformations introduced by relaxing $F_{\alpha\beta} = 0$ have to fulfill the Bianchi identity (removing the complement of the kernel from the right), and that

[4] I use standard Dynkin labels for SO(1,9)

relevant deformations are counted modulo field redefinitions (removing the image from the left). See also the following section for a fuller discussion.

A complete calculation of the cohomology requires that one considers all irreducible representations occurring at arbitrary levels. This quickly becomes untractable to do by hand. The method for calculating cohomologies is by using the program LiE [35]. The method will be presented in detail in a forthcoming publication [38]. The complete cohomology consists of

$$\mathcal{H}^0 = (00000)_0 \qquad \text{(gauge transformations)} \tag{37}$$

$$\mathcal{H}^1 = (10000)_1 \oplus (00001)_{3/2} \qquad \text{(fields)} \tag{38}$$

$$\mathcal{H}^2 = (00010)_{5/2} \oplus (10000)_3 \qquad \text{(deformations)} \tag{39}$$

$$\mathcal{H}^3 = (00000)_4 \qquad \text{(?)} \tag{40}$$

where the subscript indicates dimension.

Similar cohomologies may be calculated for the $D = 11$ supergravity [38], and they confirm in a nice way the conclusions presented earlier in this lecture. An interesting observation is that one can choose either to consider the vielbein or the 3-form, and in either case are all the fields and deformations of the supergravity contained. It looks as though a superspace 3-form potential automatically contains gravitational degrees of freedom, although it is difficult to envisage how the dynamics should be formulated without reference to geometry.

F^4 Terms

I would like to sketch how the superspace methods already described are used to derive α'-corrections to $D = 10$ super-Yang–Mills. The method for $D = 11$ supergravity is in principle analogous, but much more complicated. So far, the corrections allowed by supersymmetry have been determined up to order α'^2 [37], and although the level of technical complexity is high, it seems reasonable to continue one or two levels.

We need to specify what J_{abcde} is in terms of the fundamental superfields F and λ. We first observe that there are no corrections at order α'. For dimensional reasons, $F_{\alpha\beta}$ has to be proportional to λ^2, which does not contain the representation (00020). Then, starting at order α'^2, there are two types of possible terms, modulo the lowest order field equations ($A, B, \ldots$ are adjoint gauge group indices, not to be confused with $A = (a, \alpha)$ used earlier):

$$\begin{aligned} J^A_{abcde} &= -\tfrac{1}{2}\alpha'^2 M^A{}_{BCD}(\lambda^B\Gamma^f\Gamma_{abcde}\Gamma^g\lambda^C)F^D_{fg} \\ &\quad +\tfrac{1}{6}\alpha'^2 N^A{}_{BC}\left(D_{[a}\lambda^B\Gamma_{bcd}D_{e]}\lambda^C - \text{dual}\right) . \end{aligned} \tag{41}$$

These satisfy the (00030) constraint at linear order, which is easily seen by acting with a spinor derivative and perform tensor multiplication of the representations of the fields. Here, M and N are some invariant tensors carrying adjoint indices of the gauge group.

Not all deformations in (00020) are relevant, as explained in the previous section. Those that are in the image of Δ_1 correspond to field redefinitions of A_α and are trivial.

A careful examination of field redefinitions shows that only the first term in eq. (41) is relevant, and the other can be discarded. In addition, M_{ABCD} can be taken to be completely symmetric in adjoint indices.

A lengthy calculation gives the deformed equations of motion at order α'^2 by acting with spinor derivatives on J_{abcde}, and inserting in eq. (28). These may subsequently be integrated to a component action, which reads

$$\begin{aligned}\Lambda \;=\; & -\tfrac{1}{4}G^{Aij}G^{A}_{ij}+\tfrac{1}{2}\chi^{A}\not{D}\chi^{A} \\ & -6\alpha'^2 M_{ABCD}\Big[\mathrm{tr}\,G^{A}G^{B}G^{C}G^{D}-\tfrac{1}{4}(\mathrm{tr}\,G^{A}G^{B})(\mathrm{tr}\,G^{C}G^{D}) \\ & \quad -2G^{Ai}{}_{k}G^{Bjk}(\chi^{C}\Gamma_{i}D_{j}\chi^{D})+\tfrac{1}{2}G^{Ail}D_{l}G^{Bjk}(\chi^{C}\Gamma_{ijk}\chi^{D}) \\ & \quad +\tfrac{1}{180}(\chi^{A}\Gamma^{ijk}\chi^{B})(D_{l}\chi^{C}\Gamma_{ijk}D^{l}\chi^{D})+\tfrac{3}{10}(\chi^{A}\Gamma^{ijk}\chi^{B})(D_{i}\chi^{C}\Gamma_{j}D_{k}\chi^{D}) \\ & \quad +\tfrac{7}{60}f^{D}{}_{EF}G^{Aij}(\chi^{B}\Gamma_{ijk}\chi^{C})(\chi^{E}\Gamma^{k}\chi^{F}) \\ & \quad -\tfrac{1}{360}f^{D}{}_{EF}G^{Aij}(\chi^{B}\Gamma^{klm}\chi^{C})(\chi^{E}\Gamma_{ijklm}\chi^{F})\Big]+O(\alpha'^3)\,. \end{aligned} \tag{42}$$

The spinor λ has been replaced by χ and F by G, since there is a field redefinition involved in reaching this final form. It agrees with previous work [39] on previously known terms (up to quadratic in fermions).

With only a minor further restriction on M, the action has a second non-linearly realised supersymmetry when the gauge group has a U(1) factor, as is the case when one considers field theory on multiple 9-branes. The "symmetrised trace prescription" of Tseytlin [31] is consistent with our results, but supersymmetry does not completely specify it, even at the F^4 level. It will of course be interesting to continue the analysis to higher orders. The (00030) Bianchi identity will necessarily lead to corrections at order α'^4 and higher, and a complete action will be non-polynomial. It it is not clear whether any closed, Born–Infeld-like form exists. It is even not known if new "invariants" arise that start at higher orders, or if everything follows uniquely once the α'^2 correction is determined.

Branes? Conclusions

The properties of the maximally supersymmetric field theories we have considered have been turned into a tool for studying restrictions imposed by supersymmetry on self-interactions. Much more is to be done, both for super-Yang–Mills and supergravity, but it will be necessary to use computer programs, *e.g.* LiE [35] and the Mathematica package GAMMA [40], to a higher degree.

A question which so-far remains unaddressed is what happens to branes moving in backgrounds with α'-corrections from string/M-theory. To investigate this one will need more informations about α'-corrected supergravity. Will the actions still be formally the same, and the dynamics only change through the coupling to background fields? I would tend to answer in the positive, although nothing is known. One difficulty immediately presents itself, namely that the tensor field strengths will take non-zero values even for the components of negative dimension [14]. Since κ-symmetry relies on cancellations of contributions from the kinetic and WZ terms, the resulting variations would have no contribution from the torsion to cancel against. One possibility is that also the condition that κ is purely spinorial is modified. One preliminary investigation would consist of checking κ-symmetry for supersymmetric Wilson loops [41] in a deformed super-Yang–Mills background.

ACKNOWLEDGMENTS

The author would like to thank the organisers of the XXXVII Karpacz Winter School for two very pleasant and inspiring weeks.

REFERENCES

1. Stelle, K.S., "BPS branes in supergravity", hep-th/9803116.
2. Duff, M.J., "TASI lectures on branes, black holes and anti-de Sitter space", hep-th/9912164.
3. Achucarro, A., Evans, J.M., Townsend, P.K., and Wiltshire, D.L., *Phys. Lett.*, **198**, 441 1987.
4. Bergshoeff, E., Sezgin, E., and Townsend, P.K., *Phys. Lett.*, **189**, 75 (1987); *Ann. Phys.*, **185**, 330 1988.
5. Duff, M.J., and Lu, J.X., *Nucl. Phys.*, **B390**, 276 1993; hep-th/9207060.
6. Howe, P.S., and Sezgin, E., *Phys. Lett.*, **390**, 133 1997; hep-th/9607227.
7. Cederwall, M., von Gussich, A., Nilsson, B.E.W., and Westerberg, A., *Nucl. Phys.*, **B490**, 163 1997 hep-th/9610148;
   Aganagić, M., Popescu, C., and Schwarz, J.H., *Phys. Lett.* **393**, 311 1997, hep-th/9610249;
   Cederwall, M., von Gussich, A., Nilsson, B.E.W., Sundell, P., and Westerberg, A., *Nucl. Phys.*, **B490**, 179 1997, hep-th/9611159;
   Bergshoeff, E., and Townsend, P.K., *Nucl. Phys.*, **B490**, 145 1997, hep-th/9611173.
8. Polchinski, J., *Phys. Rev. Lett.*, **75**, 4724 1995, hep-th/9510017.
9. Cremmer, E., Julia, B., and Sherk, J., *Phys. Lett.*, 76, 409 1978.
10. Brink, L., and Howe, P., *Phys. Lett.*, **91**, 384 1980; Cremmer, E., and Ferrara, S., *Phys. Lett.*, **91**, 61 1980.
11. Howe, P., *Phys. Lett.*, **415**, 149 1997, hep-th/9707184.
12. Duff, J.M., Khuri, R.R., and Lu, J.X., *Phys. Rept.*, **259**, 213 1995, hep-th/9412184.
13. Adawi, T., Cederwall, M., Gran, U., Nilsson, B.E.W., and Razaznejad, B., *J. High Energy Phys.*, **9902**, 0011999, hep-th/9811145.
14. Cederwall, M., Gran, U.. Nielsen, M., and Nilsson, B.E.W., *J. High Energy Phys.*, **0010**, 041 2000, hep-th/0007035; hep-th/0010042.
15. Gates, S.J. Jr. and Nishino, H., "Deliberations on 11D superspace for the M-theory effective action", hep-th/0001037.
16. Leigh, R.G., *Mod. Phys. Lett.*, **A4**, 2767 (1989);
    Callan, C.G., Lovelace, C., Nappi, C.R., and Yost, S.A., *Nucl. Phys.*, **B288**, 525 (1987).
17. Douglas, M., "Branes within branes", hep-th/9512077;
    Green, M.B., Hull, C.M., and Townsend, P.K., *Phys. Lett.*, **B382**, 65 (1996), hep-th/9604119.
18. Izquierdo,J.M., Lambert, N.D., Papadopoulos, G., and Townsend, P.K., *Nucl. Phys.*, **B460**, 560 (1996), hep-th/9508177;
    Russo, J.G. and Tseytlin, A.A., *Nucl. Phys.*, **B490**, 121 (1997), hep-th/9611047;
    Cederwall, M., Gran, U., Holm, M., and Nilsson, B.E.W., JHEP **9902**, 003 (1999), hep-th/9812144;
    Cederwall, M., Gran, U., Nielsen, M., and Nilsson, B.E.W., JHEP **0001**, 037 (2000), hep-th/9912106.
19. Carr, J.L., Gates, S.J.Jr., and Oerter, R.N., *Phys. Lett.*, **189**, 68 1987.
20. Howe, P.S., and West, P.C., *Nucl. Phys.*, **B238**, 181 1984.

21. Cederwall, M., and Townsend, P.K., *J. High Energy Phys.*, **9709**, 003 1997, hep-th/9709002
22. Cederwall, M., and Westerberg, A., *J. High Energy Phys.*, **9802**, 004 1998, hep-th/9710007.
23. Witten, E., *Nucl. Phys.*, **B460**, 335 1996, hep-th/9510135.
24. Schwarz, J.H., *Phys. Lett.*, **B360**, 13 (1995), hep-th/9508143; Erratum: ibid. **B364**, 252 (1995.
25. Cederwall, M., Gran, U., Nielsen, M., and Nilsson, B.E.W., *J. High Energy Phys.*, **0001**, 037 (2000), hep-th/9912106.
26. Aganagic, M., Park, J., Popescu, C., and Schwarz, J.H., *Nucl. Phys.*, **B496**, 215 1997, hep-th/9702133.
27. Cederwall, M., Nilsson, B.E.W., and Sundell, P., *J. High Energy Phys.*, **04**, 007 1998, hep-th/9712059.
28. Bachas, C.P., Bain, P., and Green, M.B., *J. High Energy Phys.*, **05**, 11 1999, hep-th/9903210.
29. Wyllard, N., "Derivative corrections to D-brane actions with constant background fields", hep-th/0008125.
30. Nilsson, B.E.W., "A supersymmetric approach to branes and supergravity", in "Theory of elementary particles", Proc. of the 31st international symposium Ahrenshoop, September 2-6, 1997, Buckow, Eds H. Dorn et al. (Wiley-VCH 1998), Göteborg-ITP-98-09 hep-th/0007017.
31. Tseytlin, A.A., "Born–Infeld action, supersymmetry and string theory", in the Yuri Golfand memorial volume, Ed. M. Shifman, World Scientific (2000) hep-th/9908105; *Nucl. Phys.*, **B501**, 41 1997, hep-th/9701125.
32. See also the talks by Ivanov, Krivonos and Bergshoeff at this school.
33. Nilsson, B.E.W., "Off-shell fields for the 10-dimensional supersymmetric Yang–Mills theory", Göteborg-ITP-81-6; *Class. Quantum Grav.*, **3**, L 41 1986.
34. Cederwall, M., Nilsson, B.E.W., and Tsimpis, D., "The structure of maximally supersymmetric gauge theories: constraining higher order interactions", hep-th/0102009.
35. Cohen, A.M., van Leeuwen, M., and Lisser, B., LiE v. 2.2 (1998), http://wallis.univ-poitiers.fr/˜maavl/LiE/
36. Gates, S.J. Jr., and Vashakidze, Sh., *Nucl. Phys.*, **B291**, 172 1987.
37. Cederwall, M., Nilsson, B.E.W., and Tsimpis, D., "$D = 10$ super-Yang–Mills at $O(\alpha'^2)$", hep-th/0104236.
38. Cederwall, M., Nilsson, B.E.W., and Tsimpis, D., in preparation.
39. Bergshoeff, E., Rakowski, M., and Sezgin, E., *Phys. Lett.*, **185**, 371 1987.
40. Gran, U., "GAMMA: A Mathematica package for performing gamma-matrix algebra and Fierz transformations in arbitrary dimensions", hep-th/0105086.
41. Ooguri, H., Rahmfeld, J., Robins, H., and Tannenhauser, J., "Holography in superspace", *J. High Energy Phys.*, **0007**, 045 2000, hep-th/0007104.

Supersymmetric Born-Infeld Theories from Nonlinear Realizations

E. Ivanov

BLTP, JINR, 141980 Dubna, Moscow Region, Russia, email: eivanov@thsun1.jinr.ru

Abstract. We discuss how supersymmetric Born-Infeld theories can be derived from the concept of partial breaking of global supersymmetry (PBGS).

INTRODUCTION

One of the approaches to the description of superbranes is based on the concept of partial breaking of global supersymmetry (PBGS) [2], [3]-[11]. In such a description the objects accommodating the physical worldvolume superbrane multiplets are Goldstone superfields. The worldvolume supersymmetry is realized on them by linear transformations and so is manifest. The rest of the full target supersymmetry is realized nonlinearly. After passing to the components in the invariant Goldstone superfield action, and, in general, after eliminating auxiliary fields, one is left with the action which coincides with a "static gauge" of the corresponding Green-Schwarz-type action.

As is well known (see [12] and refs. therein), the worldvolume supermultiplets of Dp-branes are vector ones with the Born-Infeld dynamics for gauge fields. So the corresponding PBGS actions should form a subclass of manifestly supersymmetric extensions of the Born-Infeld (BI) action [13], [14], [15]. It is characterized by the presence of hidden and nonlinearly realized second supersymmetry. In other words, the PBGS approach can be used as an efficient tool of deducing such superextensions of the BI action. Until now, only superextensions of abelian BI theory were derived in this way [4], [6], [10], [11]. However, there are some indications that this approach can be workable in the nonabelian case too.

In this lecture I explain, on a few instructive examples, how the PBGS approach augmented with the general methods of the theory of nonlinear realizations [1] leads to supersymmetric extensions of the BI theory. What can be directly derived from the nonlinear realizations formalism in most cases, is the Goldstone superfield equations of motion which give a superextension of the equations of motion of BI theory. The construction of the off-shell superfield actions (if they exist for the relevant supersymmetric Maxwell theory) is more tricky and requires constructing a *linear* realization of the corresponding PBGS pattern. The superfield BI Lagrangian density is identified with the proper component of a linear supermultiplet of the full underlying supersymmetry after imposing on this multiplet covariant constraints which express the whole multiplet in terms of the Goldstone vector multiplet of the unbroken supersymmetry. The precise form of these constraints can be found using the general relationship between linear and

CP589, *Fundamental Interaction Theories*, edited by J. Lukierski and J. Rembieliński

nonlinear realizations of supersymmetries [16] adapted to the PBGS case [17], [18].

SPACE-FILLING D2- AND D3-BRANES

D2-Brane

Let us first consider the case of the "space-filling" D2-brane having $N = 1, d = 3$ vector multiplet as the worldvolume one [6].

The full supersymmetry in this case is $N = 2, d = 3$ supersymmetry without central charge

$$\{Q_a, Q_b\} = \{S_a, S_b\} = P_{ab},\ \{Q_a, S_b\} = 0\,. \tag{1}$$

To describe its 1/2 partial breaking we should construct its nonlinear realization with the linearly realized $N = 1$ supersymmetry. We associate the $N = 1\,, d = 3$ superspace coordinates $\{\theta^a, x^{ab}\}$ with the unbroken supersymmetry generators Q_a, P_{ab} and associate a Goldstone fermionic superfield $\psi^a \equiv \psi^a(x,\theta)$ with the broken generator S_a. A coset element g is defined by

$$g = e^{x^{ab}P_{ab}} e^{\theta^a Q_a} e^{\psi^a S_a}\,. \tag{2}$$

Defining, in the standard way, Cartan 1-forms, one can construct the appropriate covariant derivatives

$$\mathcal{D}_{ab} = (E^{-1})_{ab}^{cd}\partial_{cd}, \qquad \mathcal{D}_a = D_a + \frac{1}{2}\psi^b D_a \psi^c\, \mathcal{D}_{bc}\,, \tag{3}$$

where

$$D_a = \frac{\partial}{\partial\theta^a} + \frac{1}{2}\theta^b\partial_{ab}, \qquad \{D_a, D_b\} = \partial_{ab}\,, \tag{4}$$

$$E_{ab}^{cd} = \frac{1}{2}(\delta_a^c\delta_b^d + \delta_a^d\delta_b^c) + \frac{1}{4}(\psi^c\partial_{ab}\psi^d + \psi^d\partial_{ab}\psi^c)\,.$$

Now, like in the $N = 2 \to N = 1, d = 4$ PBGS case considered in [4], we wish to view the Goldstone fermion superfield $\psi^a(x,\theta)$ as a nonlinear generalization of the standard linear $N = 1, d = 3$ Maxwell superfield strength $\mu^a(x,\theta)$ and to see what is the corresponding generalization of the free $N = 1, d = 3$ Maxwell theory.

The superfield strength μ_a satisfies the following Bianchi identity (which is off-shell) and the equation of motion [19]:

$$\text{(a)}\ \ D^a\mu_a = 0\,, \quad \text{(b)}\ \ D^2\mu_a = 0\,. \tag{5}$$

As was argued in [9], the covariantization of these equations is achieved by simply replacing the flat derivatives by the covariant ones (3):

$$\text{(a)}\ \ \mathcal{D}^a\psi_a = 0\,, \quad \text{(b)}\ \ \mathcal{D}^2\psi_a = 0\,. \tag{6}$$

In order to see which system these equations describe, let us consider their bosonic sector. It amounts to the following equation for the vector $V_{ab} \equiv \mathcal{D}_a\psi_b|_{\theta=0}$:

$$(\partial_{ac} + V_a^m V_c^n \partial_{mn})V_b^c = 0\,. \tag{7}$$

Rewrite the antisymmetric and symmetric parts of this equation as follows:

$$\text{(a)}\ \partial_{ab}\left(\frac{V^{ab}}{2-V^2}\right)=0\,,\qquad \text{(b)}\ \partial_{ac}\left(\frac{V^c_b}{2+V^2}\right)+\partial_{bc}\left(\frac{V^c_a}{2+V^2}\right)=0\,, \tag{8}$$

where $V^2 \equiv V^{mn}V_{mn}$. After passing to the "genuine" field strength

$$F^{ab}=\frac{2V^{ab}}{2-V^2}\Rightarrow \partial_{ab}F^{ab}=0\,. \tag{9}$$

Eq. (8b) is recognized as the equation of motion of the $d=3$ BI theory,

$$\partial_{ac}\left(\frac{F^c_b}{\sqrt{1+2F^2}}\right)+\partial_{bc}\left(\frac{F^c_a}{\sqrt{1+2F^2}}\right)=0\,, \tag{10}$$

while (8a) as the Bianchi identity for the $d=3$ abelian gauge field strength. One can show that the full superfield equations (6) are equivalent to the worldvolume superfield equation following from the off-shell superfield D2-brane action of [6].

D3-Brane

As another example we consider the space-filling D3-brane in $d=4$. This system amounts to the PBGS pattern $N=2\to N=1$ in $d=4$, with a nonlinear generalization of $N=1, d=4$ vector multiplet as the Goldstone multiplet [4], [5]. The off-shell superfield action for this system and the related equations of motion are known [4]. We wish to re-derive the latter directly from the coset approach, like in the previous example.

Our starting point is the $N=2$, $d=4$ Poincaré superalgebra without central charges:

$$\left\{Q_\alpha,\bar{Q}_{\dot\alpha}\right\}=2P_{\alpha\dot\alpha}\,,\ \left\{S_\alpha,\bar{S}_{\dot\alpha}\right\}=2P_{\alpha\dot\alpha}\,. \tag{11}$$

Assuming the $S_\alpha, \bar{S}_{\dot\alpha}$ supersymmetries to be spontaneously broken, we introduce the Goldstone superfields $\psi^\alpha(x,\theta,\bar\theta)$, $\bar\psi^{\dot\alpha}(x,\theta,\bar\theta)$ as the corresponding parameters in the following coset

$$g=e^{ix^{\alpha\dot\alpha}P_{\alpha\dot\alpha}}e^{i\theta^\alpha Q_\alpha+i\bar\theta_{\dot\alpha}\bar{Q}^{\dot\alpha}}e^{i\psi^\alpha S_\alpha+i\bar\psi_{\dot\alpha}\bar{S}^{\dot\alpha}}\,. \tag{12}$$

With the help of the Cartan forms one can define the covariant derivatives

$$\mathcal{D}_\alpha=D_\alpha-i\left(\bar\psi^{\dot\beta}D_\alpha\psi^\beta+\psi^\beta D_\alpha\bar\psi^{\dot\beta}\right)\mathcal{D}_{\beta\dot\beta},\quad \mathcal{D}_{\alpha\dot\alpha}=\left(E^{-1}\right)^{\beta\dot\beta}_{\alpha\dot\alpha}\partial_{\beta\dot\beta}\,, \tag{13}$$

where

$$E^{\beta\dot\beta}_{\alpha\dot\alpha}=\delta^\beta_\alpha\delta^{\dot\beta}_{\dot\alpha}-i\psi^\beta\partial_{\alpha\dot\alpha}\bar\psi^{\dot\beta}-i\bar\psi^{\dot\beta}\partial_{\alpha\dot\alpha}\psi^\beta\,,$$
$$D_\alpha=\frac{\partial}{\partial\theta^\alpha}-i\bar\theta^{\dot\alpha}\partial_{\alpha\dot\alpha}\,,\ \bar{D}_{\dot\alpha}=-\frac{\partial}{\partial\bar\theta^{\dot\alpha}}+i\theta^\alpha\partial_{\alpha\dot\alpha}\,. \tag{14}$$

Now we can write the covariant version of the constraints on ψ^α, $\bar\psi^{\dot\alpha}$ which define the superbrane generalization of $N=1, d=4$ vector multiplet, together with the covariant equations of motion for this system. They are a direct covariantization of the free $N=1, d=4$ Maxwell superfield strength constraints and equation of motion.

It was shown in [4] that the chirality constraints can be directly covariantized

$$\overline{\mathcal{D}}_{\dot\alpha}\psi_\alpha = 0\,, \quad \mathcal{D}_\alpha \bar\psi_{\dot\alpha} = 0\,. \tag{15}$$

These conditions are compatible with the algebra of the covariant derivatives (13).

The second off-shell constraint defining $N=1, d=3$ Maxwell superfield strength, the Bianchi identity, is non-trivial to covariantize. Its "naive" generalization

$$\mathcal{D}^\alpha\psi_\alpha + \overline{\mathcal{D}}_{\dot\alpha}\bar\psi^{\dot\alpha} = 0\,, \tag{16}$$

was demonstrated in [4] to be contradictory. It turns out, however, that the simultaneous straightforward covariantization of *both* the flat Bianchi identity and equation of motion yields the self-consistent on-shell system [9]

$$\mathcal{D}^\alpha\psi_\alpha = 0\,, \quad \overline{\mathcal{D}}_{\dot\alpha}\bar\psi^{\dot\alpha} = 0\,. \tag{17}$$

This can be shown by applying the covariant spinor derivatives to (17), using their algebra and the covariant chirality condition (15).

Thus the full set of equations describing the dynamics of the D3-brane supposedly consists of the generalized chirality constraint (15) and the equations (17). To prove its equivalence to the $N=1$ superfield description of D3-brane proposed in [4], recall that the latter is the $N=1$ supersymmetrization [13] of the $d=4$ BI action with one extra nonlinearly realized $N=1$ supersymmetry. So, let us consider the bosonic part of the above set of equations. Our superfields $\psi, \bar\psi$ contain the following bosonic components:

$$V^{\alpha\beta} \equiv \mathcal{D}^\alpha\psi^\beta|_{\theta=0}\,, \quad \bar V^{\dot\alpha\dot\beta} =\equiv \overline{\mathcal{D}}^{\dot\alpha}\bar\psi^{\dot\beta}|_{\theta=0}\,, \tag{18}$$

which, as a consequence of (17), satisfies the following simple equations

$$\partial_{\alpha\dot\alpha}V^{\alpha\beta} - V^\gamma_\alpha \bar V^{\dot\gamma}_{\dot\alpha}\partial_{\gamma\dot\gamma}V^{\alpha\beta} = 0\,, \quad \partial_{\alpha\dot\alpha}\bar V^{\dot\alpha\dot\beta} - V^\gamma_\alpha \bar V^{\dot\gamma}_{\dot\alpha}\partial_{\gamma\dot\gamma}\bar V^{\dot\alpha\dot\beta} = 0\,. \tag{19}$$

One can put them into the following suggestive form

$$\partial_{\beta\dot\alpha}\left(fV^\beta_\alpha\right) - \partial_{\alpha\dot\beta}\left(\bar f\bar V^{\dot\beta}_{\dot\alpha}\right) = 0\,, \qquad \partial_{\beta\dot\alpha}\left(gV^\beta_\alpha\right) + \partial_{\alpha\dot\beta}\left(\bar g\bar V^{\dot\beta}_{\dot\alpha}\right) = 0\,, \tag{20}$$

where

$$f = \frac{\bar V^2 - 2}{1 - \frac{1}{4}V^2\bar V^2}\,, \quad g = \frac{\bar V^2 + 2}{1 - \frac{1}{4}V^2\bar V^2}\,. \tag{21}$$

In terms of the "genuine" field strengths

$$F^\beta_\alpha \equiv \frac{1}{2\sqrt{2}} fV^\beta_\alpha\,, \quad \bar F^{\dot\beta}_{\dot\alpha} \equiv \frac{1}{2\sqrt{2}}\bar f\bar V^{\dot\beta}_{\dot\alpha}\,, \tag{22}$$

first of Eqs. (20) is recognized as the Bianchi identity

$$\partial_{\beta\dot{\alpha}} F^{\beta}_{\alpha} - \partial_{\alpha\dot{\beta}} \bar{F}^{\dot{\beta}}_{\dot{\alpha}} = 0 \,, \tag{23}$$

while the second one as the BI equation

$$\partial_{\beta\dot{\alpha}} \left(\frac{1+A}{B} F^{\beta}_{\alpha} \right) + \partial_{\alpha\dot{\beta}} \left(\frac{1-A}{B} \bar{F}^{\dot{\beta}}_{\dot{\alpha}} \right) = 0 \,, \tag{24}$$

where

$$A = F^2 - \bar{F}^2 \,, \qquad B = \sqrt{(F^2 - \bar{F}^2)^2 - 2(F^2 + \bar{F}^2) + 1} \,. \tag{25}$$

Indeed, (24) follows from the $d = 4$ BI action:

$$S = \int d^4x \sqrt{(F^2 - \bar{F}^2)^2 - 2(F^2 + \bar{F}^2) + 1} \,. \tag{26}$$

Note that at the full superfield level the redefinition (22) should correspond to passing from the Goldstone fermions ψ_α, $\bar{\psi}_{\dot{\alpha}}$ which have the simple transformation properties in the nonlinear realization of $N = 1, d = 4$ supersymmetry but obey the nonlinear irreducibility constraints, to the standard Maxwell superfield strength. The nonlinear action in [4] was written just in terms of this latter object. The equivalent form (17) of the equations of motion and Bianchi identity is advantageous in that it manifests the second (hidden) supersymmetry, being constructed out of the covariant objects.

N=2 BI THEORY WITH PARTIALLY BROKEN N=4 SUPERSYMMETRY

$N = 4 \rightarrow N = 2$ BI Equations

In this case no closed expression is known for the full off-shell BI action with second nonlinearly realized $N = 2$ supersymmetry. At present it exists merely as an iterative series in the $N = 2$ Maxwell superfield strength and its derivatives. The origin of this difficulty is the presence of a complex scalar field in the vector $N = 2$ Goldstone multiplet. The corresponding D3-brane is not space-filling: it is a D3-brane in $D = 6$, and the scalar field just mentioned parametrizes two its transverse directions. As was argued in [12] and [10], [11], the manifest worldvolume supersymmetry is not compatible with having simultaneously the manifest Nambu-Goto and Born-Infeld dynamics for the scalar and vector fields. For one of them one always gets the equations of motion, or the action, in a disguised form which can be brought into the standard one only after a complicated nonlinear field redefinition.

Leaving the problem of the off-shell $N = 2$ BI action aside for a moment, let us point out that the superfield equations for this theory can be still derived in a closed form by applying the approach exemplified in the preceding Section.

First, we recall that the basic object of $N=2$ Maxwell theory is a complex scalar $N=2$ off-shell superfield strength $\mathcal{W}$ which is chiral and satisfies one additional Bianchi identity:

$$\text{(a) } \bar{D}_{\dot{\alpha}i}\mathcal{W}=0\,,\ D^i_\alpha\bar{\mathcal{W}}=0\,,\quad \text{(b) } D^{ik}\mathcal{W}=\bar{D}^{ik}\bar{\mathcal{W}}\,. \tag{27}$$

Here,

$$D^i_\alpha=\frac{\partial}{\partial\theta^\alpha_i}+i\bar{\theta}^{\dot{\alpha}i}\partial_{\alpha\dot{\alpha}}\,,\quad \bar{D}_{\dot{\alpha}i}=-\frac{\partial}{\partial\bar{\theta}^{\dot{\alpha}i}}-i\theta^\alpha_i\partial_{\alpha\dot{\alpha}}\,, \tag{28}$$

$$D^{ij}\equiv D^{\alpha i}D^j_\alpha\,,\quad \bar{D}^{ij}\equiv\bar{D}^i_{\dot{\alpha}}\bar{D}^{\dot{\alpha}j}\,,\qquad (i,j=1,2)\,. \tag{29}$$

The superfield equation of motion for $\mathcal{W}$ reads

$$D^{ik}\mathcal{W}+\bar{D}^{ik}\bar{\mathcal{W}}=0 \tag{30}$$

and, together with (27b), amounts to

$$D^{ik}\mathcal{W}=\bar{D}^{ik}\bar{\mathcal{W}}=0\,. \tag{31}$$

It is a crucial difference from the previous two cases that the basic object of the $N=2$ theory is scalar. Hence, in order to incorporate an appropriate generalization of $\mathcal{W}$ into the nonlinear realization framework as the Goldstone superfield, we need the proper bosonic generator in the algebra. The following central extension of $N=4, d=4$ Poincaré superalgebra meets this demand

$$\begin{aligned}&\{Q^i_\alpha,\bar{Q}_{\dot{\alpha}j}\}=\{S^i_\alpha,\bar{S}_{\dot{\alpha}j}\}=2\delta^i_jP_{\alpha\dot{\alpha}}\,,\\ &\left\{Q^i_\alpha,S^j_\beta\right\}=2\varepsilon^{ij}\varepsilon_{\alpha\beta}Z\,,\quad \left\{\bar{Q}_{\dot{\alpha}i},\bar{S}_{\dot{\beta}j}\right\}=-2\varepsilon^{ij}\varepsilon_{\dot{\alpha}\dot{\beta}}\bar{Z}\,,\end{aligned} \tag{32}$$

with all other (anti)commutators vanishing. Note an important feature that the complex central charge Z appears in the crossing anticommutator, while the generators $(Q,\bar{Q})$ and $(S,\bar{S})$ on their own form two $N=2$ superalgebras without central charges. The superalgebra (32) is a specific $d=4$ form of $N=(2,0)$ (or $N=(0,2)$) Poincaré superalgebra in $D=6$.

Let us split the set of generators of the superalgebra (32) into the unbroken $\{Q^i_\alpha,\bar{Q}_{\dot{\alpha}j},P_{\alpha\dot{\alpha}}\}$ and broken $\{S^i_\alpha,\bar{S}_{\dot{\alpha}j},Z,\bar{Z}\}$ parts and define a coset element g as:

$$g=\exp i\left(-x^{\alpha\dot{\alpha}}P_{\alpha\dot{\alpha}}+\theta^\alpha_iQ^i_\alpha+\bar{\theta}^i_{\dot{\alpha}}\bar{Q}^{\dot{\alpha}}_i\right)\exp i\left(\psi^\alpha_iS^i_\alpha+\bar{\psi}^i_{\dot{\alpha}}\bar{S}^{\dot{\alpha}}_i\right)\exp i\left(WZ+\bar{W}\bar{Z}\right)\,. \tag{33}$$

By this element on can construct left-invariant Cartan 1-forms. We shall quote only the expression for the 1-form associated with the generator Z as the most essential for our discussion:

$$\omega_Z\;=\;dW-2id\theta^\alpha_i\psi^i_\alpha\,. \tag{34}$$

The covariant derivatives are defined in the standard way (like in the previous examples)

$$\begin{aligned}&\nabla_{\alpha\dot{\alpha}}=\left(E^{-1}\right)^{\beta\dot{\beta}}_{\alpha\dot{\alpha}}\partial_{\beta\dot{\beta}}\,,\quad \mathcal{D}^i_\alpha=D^i_\alpha+i\left(\psi^\beta_jD^i_\alpha\bar{\psi}^{\dot{\beta}j}+\bar{\psi}^{\dot{\beta}j}D^i_\alpha\psi^\beta_j\right)\nabla_{\beta\dot{\beta}}\,,\\ &E^{\beta\dot{\beta}}_{\alpha\dot{\alpha}}\equiv\delta^\beta_\alpha\delta^{\dot{\beta}}_{\dot{\alpha}}+i\psi^\beta_i\partial_{\alpha\dot{\alpha}}\bar{\psi}^{\dot{\beta}i}+i\bar{\psi}^{\dot{\beta}i}\partial_{\alpha\dot{\alpha}}\psi^\beta_i\,.\end{aligned} \tag{35}$$

The Goldstone fermionic superfields ψ^i_α, $\bar{\psi}_{\dot{\alpha}i}$ can be covariantly expressed in terms of the central-charge Goldstone superfields $\mathcal{W}$, $\bar{\mathcal{W}}$ by imposing the inverse Higgs constraints [20] on the above central-charge Cartan 1-form:

$$\omega_Z|_{d\theta,d\bar{\theta}} = \bar{\omega}_Z|_{d\theta,d\bar{\theta}} = 0\,, \tag{36}$$

where $|$ means the covariant projections on the differentials of the spinor coordinates. These constraints amount to the sought expressions for the fermionic Goldstone superfields

$$\psi^i_\alpha = -\frac{i}{2}\mathcal{D}^i_\alpha W\,, \quad \bar{\psi}_{\dot{\alpha}i} = -\frac{i}{2}\bar{\mathcal{D}}_{\dot{\alpha}i}\bar{W}\,, \tag{37}$$

and, simultaneously, to the covariantization of the chirality conditions (27a)

$$\bar{\mathcal{D}}_{\dot{\alpha}i}W = 0\,, \quad \mathcal{D}^i_\alpha \bar{W} = 0\,. \tag{38}$$

It is also straightforward to write the covariant generalization of the dynamical equations of the $N = 2$ abelian vector multiplet (27), (31)

$$\mathcal{D}^{\alpha(i}\mathcal{D}^{j)}_\alpha W = 0\,, \quad \bar{\mathcal{D}}^{(i}_{\dot{\alpha}}\bar{\mathcal{D}}^{\dot{\alpha}j)}\bar{W} = 0\,. \tag{39}$$

The equations (38), (39) with the superfield Goldstone fermions eliminated by (37) constitute a manifestly covariant form of the superfield equations of motion of $N = 2$ BI theory with the second hidden nonlinearly realized $N = 2$ supersymmetry. It was shown in [10] that these equations leave in W just the field content corresponding to the on-shell vector $N = 2$ multiplet. In the limit when all other fields decouple, for the nonlinear vector field strength $F_{\alpha\beta} \sim \mathcal{D}^i_{(\alpha}\psi_{\beta)i}|_{\theta=\bar{\theta}=0}$ one again has the disguised form of the $d = 4$ BI equations (plus Bianchi identity), while for the scalar field $w = W|_{\theta=\bar{\theta}=0}$ one gets the equation of motion corresponding to the standard static-gauge Nambu-Goto action for 3-brane in $D = 6$:

$$S = \int d^4x \left(\sqrt{(1-B)^2 - C} - 1 \right)\,, \quad B = (\partial w \cdot \partial \bar{w})\,,\, C = (\partial w)^2 (\partial \bar{w})^2\,. \tag{40}$$

Thus, the system of the superfield equations (37) (39) is self consistent and gives a $N = 2$ superextension of both the equations of $D = 4$ BI theory and those of the static-gauge 3-brane in $D = 6$, with the nonlinearly realized second $N = 2$ supersymmetry. Similarly to the previous examples, the nonlinear realization approach yields the BI equations in a disguised form, with the Bianchi identities and dynamical equations mixed in a tricky way. At the same time, for the scalars we get the familiar static-gauge Nambu-Goto-type equations. This is in agreement with the fact that $W, \bar{W}$ undergo pure shifts under the action of the central charge generators $Z, \bar{Z}$, suggesting the interpretation of these superfields as the transverse brane coordinates conjugated to $Z, \bar{Z}$.

Off-Shell $N = 4 \rightarrow N = 2$ BI Action

As was mentioned in Introduction, the only way of constructing superfield actions for the BI systems with the nonlinearly realized half of the full supersymmetry available at present is to proceed from a linear off-shell representation of the latter and to reproduce the nonlinear realization by imposing proper covariant constraints on this representation (still leaving the theory off shell). The BI superfield Lagrangian density turns out to be one of the components of the linear representation one starts with. The actions for the space-filling D2- and D3-brane can be constructed using the $N=2, d=3$ and $N=2, d=4$ vector multiplets, respectively, as such linear realizations of the relevant PBGS patterns. A direct generalization of this construction to the case at hand does not work: the standard $N=4, d=4$ vector multiplet is essentially on-shell, and therefore one cannot hope to employ it to set up an off-shell $N=4 \to N=2$ BI action. Besides, the relevant $N=4$ superalgebra (32) includes the complex central charge as the necessary ingredient, while the standard $N=4$ gauge multiplet corresponds to the $N=4$ superalgebra without central charges.

A way out was proposed in a recent preprint [11], and its basic idea is to employ an infinite-dimensional linear multiplet of (32), with a non-trivial realization of the central charges Z, $\bar{Z}$. Without going into details, one embeds the $N=2$ Maxwell superfield strength $\mathcal{W}$ defined by the off-shell conditions (27) into the infinite-dimensional $N=4$ multiplet

$$\mathcal{W},\ \bar{\mathcal{W}},\ \mathcal{A}_n,\ \bar{\mathcal{A}}_n, \qquad (n=0,1,2,\ldots)\,, \tag{41}$$

where $\mathcal{A}_n$ are chiral (otherwise unconstrained) $N=2$ superfields,

$$\bar{D}_{\dot{\alpha}i}\mathcal{A}_n = 0\,, \quad D^i_\alpha \bar{\mathcal{A}}_n = 0\,. \tag{42}$$

The following transformations

$$\delta \mathcal{W} = f - \frac{1}{2}\bar{D}^4(f\bar{\mathcal{A}}_0) + \frac{1}{4}\Box(\bar{f}\mathcal{A}_0) + \frac{1}{4i}\bar{D}^{i\dot{\alpha}}\bar{f}D^\alpha_i \partial_{\alpha\dot{\alpha}}\mathcal{A}_0\,, \tag{43}$$

$$\delta \mathcal{A}_0 = 2f\mathcal{W} + \frac{1}{4}\bar{f}\Box\mathcal{A}_1 + \frac{1}{4i}\bar{D}^{i\dot{\alpha}}\bar{f}D^\alpha_i \partial_{\alpha\dot{\alpha}}\mathcal{A}_1, \tag{44}$$

$$\delta \mathcal{A}_1 = 2f\mathcal{A}_0 + \frac{1}{4}\bar{f}\Box\mathcal{A}_2 + \frac{1}{4i}\bar{D}^{i\dot{\alpha}}\bar{f}D^\alpha_i \partial_{\alpha\dot{\alpha}}\mathcal{A}_2\,,$$

....

$$\delta \mathcal{A}_n = 2f\mathcal{A}_{n-1} + \frac{1}{4}\bar{f}\Box\mathcal{A}_{n+1} + \frac{1}{4i}\bar{D}^{i\dot{\alpha}}\bar{f}D^\alpha_i \partial_{\alpha\dot{\alpha}}\mathcal{A}_{n+1}, \quad (n \geq 1) \tag{45}$$

$$D^4 \equiv \frac{1}{48}D^{\alpha i}D^j_\alpha D^\beta_i D_{\beta j}\,, \quad \bar{D}^4 \equiv \overline{(D^4)}\,, \quad \Box = \partial^{\alpha\dot{\alpha}}\partial_{\alpha\dot{\alpha}}\,,$$

where

$$f = c + 2i\eta^{i\alpha}\theta_{i\alpha}\,, \tag{46}$$

and $c, \eta^{i\alpha}$ are the infinitesimal parameters associated with the Z, S generators, close off shell both among themselves and with those of the manifest $N=2$ supersymmetry just according to the superalgebra (32). It is seen that the central charge $(c, \bar{c})$ transformations non-trivially act on this infinite tower of $N=2$ superfields.

A good candidate for the chiral $N=2$ Lagrangian density is the superfield $\mathcal{A}_0$. Indeed, the "action"

$$S = \int d^4x d^4\theta \mathcal{A}_0 + \int d^4x d^4\bar{\theta}\bar{\mathcal{A}}_0 \tag{47}$$

is invariant with respect to the transformation (44) up to surface terms.

It remains to define covariant constraints which would express $\mathcal{A}_0$, $\bar{\mathcal{A}}_0$ in terms of $\mathcal{W}$, $\bar{\mathcal{W}}$, with preserving the linear representation structure (43), (44), (45). In view of an infinite number of $N=2$ superfields $\mathcal{A}_n$, there should exist an infinite set of constraints expressing these superfields through the basic Goldstone ones $\mathcal{W}$, $\bar{\mathcal{W}}$. The procedure of deducing this set of constraints is described in [11]. The first two ones read

$$\begin{aligned}
\phi_0 &= \mathcal{A}_0\left(1-\frac{1}{2}\bar{D}^4\bar{\mathcal{A}}_0\right) - \mathcal{W}^2 - \sum_{k=1} \frac{(-1)^k}{2\cdot 8^k}\mathcal{A}_k\Box^k\bar{D}^4\bar{\mathcal{A}}_k = 0\,, \\
\phi_1 &= \Box\mathcal{A}_1 + 2\left(\mathcal{A}_0\Box\mathcal{W} - \mathcal{W}\Box\mathcal{A}_0\right) \\
&- \sum_{k=0} \frac{(-1)^k}{2\cdot 8^k}\left(\Box\mathcal{A}_{k+1}\Box^k\bar{D}^4\bar{\mathcal{A}}_k - \mathcal{A}_{k+1}\Box^{k+1}\bar{D}^4\bar{\mathcal{A}}_k\right) = 0\,, \qquad (48)
\end{aligned}$$

and so on.

At present we do not know how to explicitly solve this set of constraints and to find a closed expression for the Lagrangian densities $\mathcal{A}_0$, $\bar{\mathcal{A}}_0$. We can only restore the general solution by iterations. E.g., in order to restore the action up to the 8th order, we have to know the following orders in $\mathcal{A}_k$:

$$\begin{aligned}
&\mathcal{A}_0 = \mathcal{W}^2 + \mathcal{A}_0^{(4)} + \mathcal{A}_0^{(6)} + \mathcal{A}_0^{(8)} + \dots\,, \qquad \mathcal{A}_1 = \mathcal{A}_1^{(3)} + \mathcal{A}_1^{(5)} + \mathcal{A}_1^{(7)} + \dots\,, \\
&\mathcal{A}_2 = \mathcal{A}_2^{(4)} + \mathcal{A}_2^{(6)} + \dots\,, \qquad \mathcal{A}_3 = \mathcal{A}_3^{(5)} + \dots\,. \qquad (49)
\end{aligned}$$

These terms can be explicitly found, e.g.,

$$\begin{aligned}
&\mathcal{A}_0^{(4)} = \frac{1}{2}\mathcal{W}^2\bar{D}^4\bar{\mathcal{W}}^2\,,\ \mathcal{A}_1^{(3)} = \frac{2}{3}\mathcal{W}^3\,, \quad \mathcal{A}_1^{(5)} = \frac{2}{3}\mathcal{W}^3\bar{D}^4\bar{\mathcal{W}}^2\,, \\
&\mathcal{A}_2^{(4)} = \frac{1}{3}\mathcal{W}^4\,, \quad \mathcal{A}_2^{(6)} = \frac{1}{2}\mathcal{W}^4\bar{D}^4\bar{\mathcal{W}}^2\,, \quad \mathcal{A}_3^{(5)} = \frac{2}{15}\mathcal{W}^5\,. \qquad (50)
\end{aligned}$$

The action up to the 10th order in $\mathcal{W}$, $\bar{\mathcal{W}}$ was found in [11]. It turned out to coincide, at least up to the 8th order, with the action restored by Kuzenko and Theisen [15] from the requirements of self-duality and invariance under nonlinear shifts of $\mathcal{W}$, $\bar{\mathcal{W}}$ (the $c, \bar{c}$ transformations in our notation). This is an indication that the full $N=4 \to N=2$ BI action can be also self-dual like its $N=2 \to N=1$ prototype [4]. Just to give a feeling how the action looks, I present it up to the 6th order:

$$\begin{aligned}
S &= \left(\int d^4x d^4\theta\,\mathcal{W}^2 + \text{c.c.}\right) + \int dZ\left\{\mathcal{W}^2\bar{\mathcal{W}}^2\left[1+\frac{1}{2}\left(D^4\mathcal{W}^2 + \bar{D}^4\bar{\mathcal{W}}^2\right)\right]\right. \\
&\left. -\frac{1}{18}\mathcal{W}^3\Box\bar{\mathcal{W}}^3\right\}\,. \qquad (51)
\end{aligned}$$

CONCLUSIONS

In this lecture I outlined the systematic approach to deducing the supersymmetric BI theories from the nonlinear realization formalism as theories of the 1/2 partial breaking of supersymmetry, with the vector multiplets of the unbroken supersymmetry as the Goldstone multiplets. It still remains to fully understand the procedure of constructing superfield actions in this approach and its relationship to other geometric approaches to the superbranes, in particular, to the superembedding approach [21]. The off-shell PBGS actions of the space-filling D3-brane [4] and D2-brane [6] were recovered from this general approach in recent papers [22] and [23].

ACKNOWLEDGMENTS

I am grateful to the Organizers of XXXVII Winter School in Karpacz, especially to Jurek Lukierski and Ziemek Popowicz, for inviting me to give this lecture and for the warm hospitality in Wrocław and Karpacz. I thank S. Bellucci and S. Krivonos in collaboration with whom most of the results reviewed here was obtained. This work was supported in part by the grants RFBR-CNRS 98-02-22034, RFBR-DFG-99-02-04022, RFBR 99-02-18417 and NATO Grant PST.CLG 974874.

REFERENCES

1. Coleman, S., Wess, J., Zumino, B., *Phys. Rev.*, **177**, 2239 (1969); Callan, C., Coleman, S., Wess, J., Zumino, B., *ibid*, **177**, 2247 (1969); Volkov, D.V., *Sov. J. Part. Nucl.*, **4**, 3 (1973); Ogievetsky, V.I., In: *Proceedings of Xth Winter School of Theoretical Physics in Karpacz*, Acta Universitatis Wratislaviensis, Vol **207**, Wroclaw, 1974, p.227.
2. Bagger, J., Wess, J., *Phys. Lett*,**B 138**, 105 (1984).
3. Hughes, J., Liu, J., Polchinski, J., *Phys. Lett.*, **B 180**, 370 (1986); Hughes, J., Polchinski, J., *Nucl.Phys.*, **B 278**, 147 (1986).
4. Bagger, J., Galperin, A., *Phys. Rev.*, **D 55**, 1091 (1997).
5. Roček, M., Tseytlin, A., *Phys. Rev.*, **D 59**, 106001 (1999).
6. Ivanov, E., Krivonos, S., *Phys. Lett.*, **B 453**, 237 (1999).
7. Bellucci, S., Ivanov, E., Krivonos, S., *Phys. Lett.*, **B 460**, 348 (1999).
8. Bellucci, S., Ivanov, E., Krivonos, S., In: *Proceedings of 32nd International Symposium Ahrenshoop*, Septemeber 1 - 5, 1998, Buckow, Germany, *Fortsch. Phys.*, **48**, 19 (2000).
9. Bellucci, S., Ivanov, E., Krivonos, S., *Phys. Lett.* **B 482**, 233 (2000).
10. Bellucci, S., Ivanov, E., Krivonos, S., *Phys. Lett.*, **B 502**, 279 (2001).
11. Bellucci, S., Ivanov, E., Krivonos, S., "Towards the complete N=2 superfield Born-Infeld action with partially broken N=4 supersymmetry", hep-th/0101195.
12. Tseytlin, A.A., "Born-Infeld Action, Supersymmetry and String Theory", hep-th/9908105.
13. Cecotti, S., Ferrara, S., *Phys. Lett.*, **B 187**, 335 (1987).
14. Ketov, S., *Mod. Phys. Lett.*, **A 14**, 501 (1999); *Class. Quant. Grav.*, **17**, L91 (2000).
15. Kuzenko, S.M., Theisen, S., "Nonlinear Self-Duality and Supersymmetry", LMU-TPW-00-19, July 2000, hep-th/0007231; *JHEP*, **0003**, 034 (2000).
16. Ivanov, E., Kapustnikov, A., *J. Phys.*, **A 11**, 2375 (1978).
17. Delduc, F., Ivanov, E., Krivonos, S., *Nucl.Phys*, **B 576**, 196 (2000).
18. Ivanov, E., Krivonos, S., Lechtenfeld, O., Zupnik, B., "Partial spontaneous breaking of two-dimensional supersymmetry", hep-th/0012199 (*Nucl. Phys.* **B**, in press).
19. Gates, S.J., Grisaru, M.T., Roček, M., Siegel, W., *Superspace*, Benjamin/ Cummings, Reading, Massachusetts, 1983.
20. Ivanov, E.A., Ogievetsky, V.I., *Teor. Mat. Fiz.*, **25**, 164 (1975).
21. Sorokin, D., *Phys. Reports*, **329**, 1 (2000).

22. Bandos, I., Pasti, P., Pokotilov, A., Sorokin, D., Tonin, M., "Space Filling Dirichlet 3-Brane in $N=2, D=4$ Superspace", hep-th/0103152.
23. Drummond, J.M., Howe, P.S., "Codimension zero superembeddings", hep-th/0103191.

Finite GUTs Within Conventional N=1 Gauge Theories

T. Kobayashi*, J. Kubo†, M. Mondragón** and G. Zoupanos‡

*Dept. of Phys, Kyoto Univ., Kyoto 606-8502, Japan
†Dept. of Physics, Kanazawa Univ., Kanazawa 920-1192, Japan
**Inst. de Física, UNAM, Apdo. Postal 20-364, México 01000 D.F., México
‡Physics Dept., Nat. Technical Univ., GR-157 80 Zografou, Athens, Greece, email: george.zoupanos@cern.ch

Abstract. Finite Unified Theories (FUTs) are conventional N=1 supersymmetric Grand Unified Theories, which can be made all-loop finite, both in the dimensionless (gauge and Yukawa couplings) and dimensionful (soft supersymmetry breaking terms) sectors. This remarkable property leads to a drastic reduction in the number of free parameters, which in turn leads to an accurate prediction of the top quark mass in the dimensionless sector, and predictions for the Higgs boson mass and the s-spectrum in the dimensionful sector. Among the characteristics of the latter ones are: 1) The lightest Higgs boson mass is predicted to be in the window 120-130 GeV, in case the LSP is neutralino, while in case the LSP is the $\tilde{\tau}$ (which can be consistently accommodated in presence of bilinear R-parity violating terms) it can be as light as 111 GeV. 2) The s-spectrum starts above several hundreds of GeV.

INTRODUCTION

In recent years new frameworks have been developed aiming to provide a unified description of all interactions including gravity. Theories based on superstrings, noncommutative geometry and quantum groups, although at a different stage of development in each area, have common unification targets and share similar hopes for exhibiting improved renormalization properties in the ultraviolet as compared to ordinary field theories. Moreover, recent progress shows that all above theoretical endeavours could be related and thus they might be understood in a unified manner too. In particular gauge theories on noncommutative spaces have recently attracted much attention for their connection to string and matrix theories [1, 2, 3]. Supersymmetric gauge theories arise as the low energy description of open strings ending on D-branes in the presence of a constant B-field which gives rise to space-noncommutativity. However inspite the importance of having frameworks to discuss quantum gravity in a self consistent way, the main goal expected from a unified description of interactions by the particle physics community is to understand the present day free parameters of the Standard Model (SM) in terms of a few fundamental ones, or in other words to achieve *reduction of couplings* at a more fundamental level. Unfortunately all the above theoretical frameworks did not offer anything in the understanding of the free parameters of the SM, and in the best case they have managed to accommodate earlier tools such as supersymmetry and ideas like Grand Unified Theories (GUTs) but without providing any further predictive power

CP589, *Fundamental Interaction Theories*, edited by J. Lukierski and J. Rembieliński

in these constructions. On the contrary the noncommutative gauge theories, instead of being UV finite, according to the hopes of the early motivations [4, 5], they exhibit a curious mixing between the short and long distance modes in their loop expansion. For the time being this mixing, called UV/IR mixing, looks like doubling the problem of divergences in field theories, instead of solving it.

In our recent studies [6, 7, 8, 9, 10, 11, 12] we have developed a complementary strategy in searching for a more fundamental theory possibly at the Planck scale, whose basic ingredients are GUTs and supersymmetry, but its consequences certainly go beyond the known ones. Our method consists of hunting for renormalization group invariant (RGI) relations holding below the Planck scale, which in turn are preserved down to the GUT scale. This programme, called Gauge–Yukawa unification scheme, applied in the dimensionless couplings of supersymmetric GUTs, such as gauge and Yukawa couplings, had already noticable successes by predicting correctly, among others, the top quark mass in the finite and in the minimal $N = 1$ supersymmetric SU(5) GUTs. An impressive aspect of the RGI relations is that one can guarantee their validity to all-orders in perturbation theory by studying the uniqueness of the resulting relations at one-loop, as was proven in the early days of the programme of *reduction of couplings* [13]. Even more remarkable is the fact that it is possible to find RGI relations among couplings that guarantee finiteness to all-orders in perturbation theory [14, 15].

Although supersymmetry seems to be an essential feature for a successful realization of the above programme, its breaking has to be understood too, since it has the ambition to supply the SM with predictions for several of its free parameters. Indeed, the search for RGI relations has been extended to the soft supersymmetry breaking sector (SSB) of these theories [9, 16], which involves parameters of dimension one and two. More recently a very interesting progress has been made [17, 24, 18, 19, 20, 21, 22, 23] concerning the renormalization properties of the SSB parameters based conceptually and technically on the work of ref. [25]. In this work the powerful supergraph method [26] for studying supersymmetric theories has been applied to the softly broken ones by using the "spurion" external space-time independent superfields [27]. In the latter method a softly broken supersymmetric gauge theory is considered as a supersymmetric one in which the various parameters such as couplings and masses have been promoted to external superfields that acquire "vacuum expectation values". Based on this method the relations among the soft term renormalization and that of an unbroken supersymmetric theory have been derived. In particular the β-functions of the parameters of the softly broken theory are expressed in terms of partial differential operators involving the dimensionless parameters of the unbroken theory. The key point in the strategy of refs. [20, 21, 22, 23, 24] in solving the set of coupled differential equations so as to be able to express all parameters in a RGI way, was to transform the partial differential operators involved to total derivative operators. This is indeed possible to be done on the RGI surface which is defined by the solution of the reduction equations.

On the phenomenological side there exist some serious developments too. Previously an appealing "universal" set of soft scalar masses was asummed in the SSB sector of supersymmetric theories, given that apart from economy and simplicity (1) they are part of the constraints that preserve finiteness up to two-loops [28, 29], (2) they are RGI up to two-loops in more general supersymmetric gauge theories, subject to the condition known as $P = 1/3\ Q$ [16] and (3) they appear in the attractive dilaton dominated supersymmetry breaking superstring scenarios [30]. However, further studies have exhibited a number of problems all due to the restrictive nature of the "universality" assumption for the soft scalar masses. For instance (a) in finite unified theories the universality predicts that the lightest supersymmetric particle is a charged particle, namely the superpartner of the τ lepton $\tilde{\tau}$ (b) the MSSM with universal soft scalar masses is inconsistent with the attractive radiative electroweak symmetry breaking [31] and (c) which is the worst of all, the universal soft scalar masses lead to charge and/or colour breaking minima deeper than the standard vacuum [32]. Therefore, there have been attempts to relax this constraint without loosing its attractive features. First an interesting observation was made that in $N = 1$ Gauge–Yukawa unified theories there exists a RGI sum rule for the soft scalar masses at lower orders; at one-loop for the non-finite case [10] and at two-loops for the finite case [11]. The sum rule manages to overcome the above unpleasant phenomenological consequences. Moreover it was proven [24] that the sum rule for the soft scalar massses is RGI to all-orders for both the general as well as for the finite case. Finally the exact β-function for the soft scalar masses in the Novikov-Shifman-Vainstein-Zakharov (NSVZ) scheme [33] for the softly broken supersymmetric QCD has been obtained [24]. Armed with the above tools and results we are in a position to study the spectrum of the full finite and minimal supersymmetric SU(5) models in terms of few free parameters with emphasis on the predictions for the masses of the lightest Higgs and LSP and on the constraints imposed by having a large $\tan\beta$.

REDUCTION OF COUPLINGS AND FINITENESS IN $N = 1$ SUSY GAUGE THEORIES

A RGI relation among couplings, $\Phi(g_1, \cdots, g_N) = 0$, has to satisfy the partial differential equation (PDE) $\mu d\Phi/d\mu = \sum_{i=1}^{N} \beta_i \partial\Phi/\partial g_i = 0$, where β_i is the β-function of g_i. There exist $(N-1)$ independent Φ's, and finding the complete set of these solutions is equivalent to solve the so-called reduction equations (REs), $\beta_g (dg_i/dg) = \beta_i$, $i = 1, \cdots, N$, where g and β_g are the primary coupling and its β-function. Using all the $(N-1)\,\Phi$'s to impose RGI relations, one can in principle express all the couplings in terms of a single coupling g. The complete reduction, which formally preserves perturbative renormalizability, can be achieved by demanding a power series solution, whose uniqueness can be investigated at the one-loop level. The completely reduced theory contains only one independent coupling with the corresponding β-function. This possibility of coupling unification is attractive, but it can be too restrictive and hence unrealistic. In practice one may use fewer Φ's as RGI constraints.

It is clear by examining specific examples, that the various couplings in supersymmetric theories have easily the same asymptotic behaviour. Therefore searching for a power series solution to the REs is justified. This is not the case in non-supersymmetric theories.

Let us then consider a chiral, anomaly free, $N = 1$ globally supersymmetric gauge theory based on a group G with gauge coupling constant g. The superpotential of the theory is given by

$$W = \frac{1}{2} m^{ij} \Phi_i \Phi_j + \frac{1}{6} C^{ijk} \Phi_i \Phi_j \Phi_k \ , \tag{1}$$

where m^{ij} and C^{ijk} are gauge invariant tensors and the matter field Φ_i transforms according to the irreducible representation R_i of the gauge group G.

The one-loop β-function of the gauge coupling g is given by

$$\beta_g^{(1)} = \frac{dg}{dt} = \frac{g^3}{16\pi^2}[\sum_i l(R_i) - 3C_2(G)] \ , \tag{2}$$

where $l(R_i)$ is the Dynkin index of R_i and $C_2(G)$ is the quadratic Casimir of the adjoint representation of the gauge group G. The β-functions of C^{ijk}, by virtue of the non-renormalization theorem, are related to the anomalous dimension matrix γ_i^j of the matter fields Φ_i as

$$\beta_C^{ijk} = \frac{d}{dt} C^{ijk} = C^{ijp} \sum_{n=1} \frac{1}{(16\pi^2)^n} \gamma_p^{k(n)} + (k \leftrightarrow i) + (k \leftrightarrow j) \ . \tag{3}$$

At one-loop level the γ_i^j are given by

$$\gamma_i^{j(1)} = \frac{1}{2} C_{ipq} C^{jpq} - 2g^2 C_2(R_i) \delta_i^j \ , \tag{4}$$

where $C_2(R_i)$ is the quadratic Casimir of the representation R_i, and $C^{ijk} = C_{ijk}^*$.

As one can see from Eqs. (2) and (4) all the one-loop β-functions of the theory vanish if $\beta_g^{(1)}$ and $\gamma_i^{j(1)}$ vanish, i.e.

$$\sum_i \ell(R_i) = 3C_2(G) \ , \qquad \frac{1}{2} C_{ipq} C^{jpq} = 2\delta_i^j g^2 C_2(R_i) \ . \tag{5}$$

A very interesting result is that the conditions (5) are necessary and sufficient for finiteness at the two-loop level.

The one- and two-loop finiteness conditions (5) restrict considerably the possible choices of the irreps. R_i for a given group G as well as the Yukawa couplings in the superpotential (1). Note in particular that the finiteness conditions cannot be applied to the supersymmetric standard model (SSM), since the presence of a $U(1)$ gauge group is incompatible with the condition (5), due to $C_2[U(1)] = 0$. This naturally leads to the expectation that finiteness should be attained at the grand unified level only, the SSM being just the corresponding, low-energy, effective theory.

A natural question to ask is what happens at higher loop orders. There exists a very interesting theorem [14] which guarantees the vanishing of the β-functions to all orders in perturbation theory, if we demand reduction of couplings, and that all the one-loop anomalous dimensions of the matter field in the completely and uniquely reduced theory vanish identically.

SOFT SUPERSYMMETRY BREAKING-SUM RULE OF SOFT SCALAR MASSES

The above described method of reducing the dimensionless couplings has been extended [9] to the soft supersymmetry breaking (SSB) dimensionful parameters of $N=1$ supersymmetric theories. In addition it was found [10] that RGI SSB scalar masses in Gauge-Yukawa unified models satisfy a universal sum rule. Here we will describe first how the use of the available two-loop RG functions and the requirement of finiteness of the SSB parameters up to this order leads to the soft scalar-mass sum rule [11].

Consider the superpotential given by (1) along with the Lagrangian for SSB terms

$$-\mathcal{L}_{\rm SB} = \frac{1}{6}h^{ijk}\phi_i\phi_j\phi_k + \frac{1}{2}b^{ij}\phi_i\phi_j + \frac{1}{2}(m^2)^j_i\phi^{*i}\phi_j + \frac{1}{2}M\lambda\lambda + \text{h.c.}, \qquad (6)$$

where the ϕ_i are the scalar parts of the chiral superfields Φ_i, λ are the gauginos and M their unified mass. Since we would like to consider only finite theories here, we assume that the gauge group is a simple group and the one-loop β-function of the gauge coupling g vanishes. We also assume that the reduction equations admit power series solutions of the form

$$C^{ijk} = g\sum_{n=0}\rho^{ijk}_{(n)}g^{2n}\,. \qquad (7)$$

According to the finiteness theorem of ref. [14], the theory is then finite to all orders in perturbation theory, if, among others, the one-loop anomalous dimensions $\gamma_i^{j(1)}$ vanish. The one- and two-loop finiteness for h^{ijk} can be achieved by

$$h^{ijk} = -MC^{ijk} + \cdots = -M\rho^{ijk}_{(0)}\,g + O(g^5)\,. \qquad (8)$$

With the above assumptions (and a couple of minor ones [11]) we find the following soft scalar-mass sum rule

$$(\,m_i^2 + m_j^2 + m_k^2\,)/MM^{\dagger} = 1 + \frac{g^2}{16\pi^2}\Delta^{(1)} + O(g^4) \qquad (9)$$

for i, j, k with $\rho^{ijk}_{(0)} \neq 0$, where $\Delta^{(1)}$ is the two-loop correction

$$\Delta^{(1)} = -2\sum_l[(m_l^2/MM^{\dagger}) - (1/3)]\,T(R_l), \qquad (10)$$

which vanishes for the universal choice in accordance with the previous findings of ref. [29].

If we know higher loop β-functions explicitly, we can follow the same procedure and find higher loop RGI relations among SSB terms. However, the β-functions of the soft scalar masses are explicitly known only up to two loops. In order to obtain higher loop results, we need something else instead of knowledge of explicit β-functions, e.g. some relations among β-functions.

The recent progress made using the spurion technique [26, 27] leads to the following all-loop relations among SSB β-functions [17, 21, 18, 19, 20],

$$\beta_M = 2\mathcal{O}\left(\frac{\beta_g}{g}\right), \tag{11}$$

$$\begin{aligned}\beta_h^{ijk} &= \gamma^i{}_l h^{ljk} + \gamma^j{}_l h^{ilk} + \gamma^k{}_l h^{ijl} \\ &\quad -2\gamma^i_1{}_l C^{ljk} - 2\gamma^j_1{}_l C^{ilk} - 2\gamma^k_1{}_l C^{ijl}, \end{aligned}\tag{12}$$

$$(\beta_{m^2})^i{}_j = \left[\Delta + X\frac{\partial}{\partial g}\right]\gamma^i{}_j, \tag{13}$$

$$\mathcal{O} = \left(Mg^2\frac{\partial}{\partial g^2} - h^{lmn}\frac{\partial}{\partial C^{lmn}}\right), \tag{14}$$

$$\Delta = 2\mathcal{O}\mathcal{O}^* + 2|M|^2 g^2\frac{\partial}{\partial g^2} + \tilde{C}_{lmn}\frac{\partial}{\partial C_{lmn}} + \tilde{C}^{lmn}\frac{\partial}{\partial C^{lmn}}, \tag{15}$$

where $(\gamma_1)^i{}_j = \mathcal{O}\gamma^i{}_j$, $C_{lmn} = (C^{lmn})^*$, and

$$\tilde{C}^{ijk} = (m^2)^i{}_l C^{ljk} + (m^2)^j{}_l C^{ilk} + (m^2)^k{}_l C^{ijl}. \tag{16}$$

The X term in (13) is explicitly known only in the lowest order [21, 22]

$$X^{(2)} = -\frac{Sg^3}{8\pi^2}, \quad S\delta_{AB} = (M^2)^k_l (R_A R_B)^l_k - |M|^2 C(G)\delta_{AB}. \tag{17}$$

In the NSVZ scheme it is given by [23]

$$X^{NSVZ} = -4\frac{\alpha^2}{16\pi^2}\frac{S}{[1-2\alpha C(G)(16\pi^2)^{-1})]}. \tag{18}$$

Assuming more generally than the finite case described above, (a) the existence of a RGI surface on which $C = C(g)$, or equivalently that

$$\frac{dC^{ijk}}{dg} = \frac{\beta_C^{ijk}}{\beta_g} \tag{19}$$

holds, and (b) the existence of a RGI surface on which the following relation holds too

$$h^{ijk} = -M(C^{ijk})' \equiv -M\frac{dC^{ijk}(g)}{d\ln g}, \tag{20}$$

and using the all-loop gauge β-function of Novikov *et al.* [33] given by

$$\beta_g^{\rm NSVZ} = \frac{g^3}{16\pi^2}\left[\frac{\sum_l T(R_l)(1-\gamma_l/2) - 3C(G)}{1 - g^2 C(G)/8\pi^2}\right], \tag{21}$$

it was found the all-loop RGI sum rule [24],

$$
\begin{aligned}
m_i^2+m_j^2+m_k^2 &= |M|^2\left\{\frac{1}{1-g^2C(G)/(8\pi^2)}\frac{d\ln C^{ijk}}{d\ln g}+\frac{1}{2}\frac{d^2\ln C^{ijk}}{d(\ln g)^2}\right\} \\
&+\sum_l\frac{m_l^2T(R_l)}{C(G)-8\pi^2/g^2}\frac{d\ln C^{ijk}}{d\ln g}\,. \qquad (22)
\end{aligned}
$$

In addition the exact β-function for m^2 in the NSVZ scheme has been obtained [24] for the first time and is given by

$$
\begin{aligned}
\beta_{m_i^2}^{\rm NSVZ} &= \left[|M|^2\left\{\frac{1}{1-g^2C(G)/(8\pi^2)}\frac{d}{d\ln g}+\frac{1}{2}\frac{d^2}{d(\ln g)^2}\right\}\right. \\
&\left.+\sum_l\frac{m_l^2T(R_l)}{C(G)-8\pi^2/g^2}\frac{d}{d\ln g}\right]\gamma_i^{\rm NSVZ}\,. \qquad (23)
\end{aligned}
$$

FINITE UNIFIED THEORIES

In this section we examine two concrete $SU(5)$ finite models, where the reduction of couplings in the dimensionless and dimensionful sector has been achieved. A predictive Gauge-Yukawa unified $SU(5)$ model which is finite to all orders, in addition to the requirements mentioned already, should also have the following properties:

1. One-loop anomalous dimensions are diagonal, i.e., $\gamma_i^{(1)\,j}\propto\delta_i^j$.
2. Three fermion generations, in the irreps $\overline{\mathbf{5}}_i,\mathbf{10}_i$ $(i=1,2,3)$, which obviously should not couple to the adjoint **24**.
3. The two Higgs doublets of the MSSM should mostly be made out of a pair of Higgs quintet and anti-quintet, which couple to the third generation.

In the following we discuss two versions of the all-order finite model. The model of ref. [6], which will be labeled **A**, and a slight variation of this model (labeled **B**) , which can also be obtained from the class of the models suggested by Kazakov *et al.* [19] with a modification to suppress non-diagonal anomalous dimensions.

The superpotential which describes the two models takes the form [6, 11]

$$
\begin{aligned}
W &= \sum_{i=1}^{3}\left[\frac{1}{2}g_i^u\,\mathbf{10}_i\mathbf{10}_iH_i+g_i^d\,\mathbf{10}_i\overline{\mathbf{5}}_i\overline{H}_i\right]+g_{23}^u\,\mathbf{10}_2\mathbf{10}_3H_4 \qquad (24)\\
&+g_{23}^d\,\mathbf{10}_2\overline{\mathbf{5}}_3\overline{H}_4+g_{32}^d\,\mathbf{10}_3\overline{\mathbf{5}}_2\overline{H}_4+\sum_{a=1}^{4}g_a^f\,H_a\mathbf{24}\overline{H}_a+\frac{g^\lambda}{3}(\mathbf{24})^3\,,
\end{aligned}
$$

where H_a and $\overline{H}_a$ $(a=1,\ldots,4)$ stand for the Higgs quintets and anti-quintets.

The non-degenerate and isolated solutions to $\gamma_i^{(1)} = 0$ for the models $\{\mathbf{A}\ ,\ \mathbf{B}\}$ are:

$$
\begin{aligned}
(g_1^u)^2 &= \{\frac{8}{5},\frac{8}{5}\}g^2\ ,\ (g_1^d)^2 = \{\frac{6}{5},\frac{6}{5}\}g^2\ ,\ (g_2^u)^2 = (g_3^u)^2 = \{\frac{8}{5},\frac{4}{5}\}g^2\ , \qquad (25)\\
(g_2^d)^2 &= (g_3^d)^2 = \{\frac{6}{5},\frac{3}{5}\}g^2\ ,\ (g_{23}^u)^2 = \{0,\frac{4}{5}\}g^2\ ,\ (g_{23}^d)^2 = (g_{32}^d)^2 = \{0,\frac{3}{5}\}g^2\ ,\\
(g^\lambda)^2 &= \frac{15}{7}g^2\ ,\ (g_2^f)^2 = (g_3^f)^2 = \{0,\frac{1}{2}\}g^2\ ,\ (g_1^f)^2 = 0\ ,\ (g_4^f)^2 = \{1,0\}g^2\ .
\end{aligned}
$$

According to the theorem of ref. [14] these models are finite to all orders. After the reduction of couplings the symmetry of W is enhanced [6, 11].

The main difference of the models **A** and **B** is that three pairs of Higgs quintets and anti-quintets couple to the **24** for **B** so that it is not necessary to mix them with H_4 and $\overline{H}_4$ in order to achieve the triplet-doublet splitting after the symmetry breaking of $SU(5)$.

In the dimensionful sector, the sum rule gives us the following boundary conditions at the GUT scale [11]:

$$
\begin{aligned}
m_{H_u}^2 + 2m_{\mathbf{10}}^2 &= m_{H_d}^2 + m_{\overline{\mathbf{5}}}^2 + m_{\mathbf{10}}^2 = M^2 \ \text{ for } \mathbf{A}\ , \qquad (26)\\
m_{H_u}^2 + 2m_{\mathbf{10}}^2 &= M^2\ ,\ m_{H_d}^2 - 2m_{\mathbf{10}}^2 = -\frac{M^2}{3}\ ,\\
m_{\overline{\mathbf{5}}}^2 + 3m_{\mathbf{10}}^2 &= \frac{4M^2}{3} \quad \text{for } \mathbf{B}, \qquad (27)
\end{aligned}
$$

where we use as free parameters $m_{\overline{\mathbf{5}}} \equiv m_{\overline{\mathbf{5}}_3}$ and $m_{\mathbf{10}} \equiv m_{\mathbf{10}_3}$ for the model **A**, and $m_{\mathbf{10}}$ for **B**, in addition to M.

PREDICTIONS OF LOW ENERGY PARAMETERS

Since the gauge symmetry is spontaneously broken below $M_{\rm GUT}$, the finiteness and Gauge-Yukawa unification conditions do not restrict the renormalization property at low energies, and all it remains are boundary conditions on the gauge and Yukawa couplings (25), the $h = -MC$ relation (8) and the soft scalar-mass sum rule (9) at $M_{\rm GUT}$, as applied in the various models. So we examine the evolution of these parameters according to their renormalization group equations at two-loop for dimensionless parameters and at one-loop for dimensionful ones with the relevant boundary conditions. Below $M_{\rm GUT}$ their evolution is assumed to be governed by the MSSM. We further assume a unique supersymmetry breaking scale M_s so that below M_s the SM is the correct effective theory.

The predictions for the top quark mass M_t are ~ 183 and ~ 174 GeV in models **A** and **B** respectively. Comparing these predictions with the most recent experimental value $M_t = (173.8 \pm 5.2)$ GeV, and recalling that the theoretical values for M_t may suffer from a correction of less than $\sim 4\%$ [12], we see that they are consistent with the experimental data. In addition the value of $\tan\beta$ is obtained as $\tan\beta = 54$ and 48 for models **A** and **B** respectively.

In the SSB sector, besides the constraints imposed by reduction of couplings and finiteness, we also look for solutions which are compatible with radiative electroweak

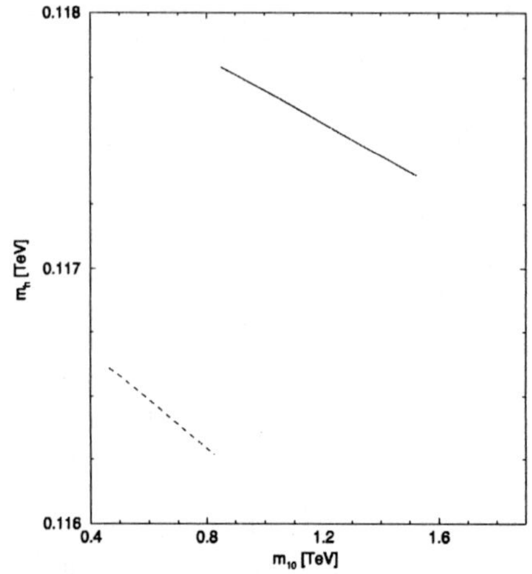

FIGURE 1. m_h as function of $m_{\mathbf{10}}$ for $M = 0.8$ (dashed) 1.0 (solid) TeV for the finite model **B**.

symmetry breaking. Thus, we look for the parameter space in which the lighter $\tilde{\tau}$ mass squared $m^2_{\tilde{\tau}}$ is larger than the lightest neutralino mass squared m^2_{χ} (which is the LSP). Notice that in the case where all the soft scalar masses are universal at the unfication scale, there is no region of $M_s = M$ below O(few) TeV in which $m^2_{\tilde{\tau}} > m^2_{\chi}$ is satisfied. But once the universality condition is relaxed this problem can be solved naturally (provided the sum rule). More specifically, using the sum rule (9) and imposing the conditions a) successful radiative electroweak symmetry breaking b) $m_{\tilde{\tau}^2} > 0$ and c) $m_{\tilde{\tau}^2} > m_{\chi^2}$, we find a comfortable parameter space for both models (although model **B** requires large $M \sim 1$ TeV).

As a final constraint, we calculate $BR(b \to s\gamma)$ [34], whose experimental value is $1 \times 10^{-4} < BR(b \to s\gamma) < 4 \times 10^{-4}$. The SM predicts $BR(b \to s\gamma) = 3.1 \times 10^{-4}$. This imposes a further restriction in our parameter space, namely $M \sim 1$ TeV if $\mu < 0$ for both models. This restriction is less strong in the case that $\mu > 0$. For example, the minimal model with $M = 1$ TeV leads to $BR(b \to s\gamma) = 3.8 \times 10^{-4}$ for $\mu < 0$. The prediction for the Higgs mass for the models is

$$m_h = 120 - 130 \;\; GeV \tag{28}$$

where the uncertainty comes from variations of the gaugino mass M and the soft scalar masses, and from finite (i.e. not logarithmically divergent) corrections in changing renormalization scheme.

In Tables 1 and 2 we present representative examples of the values obtained for the sparticle spectra in each of the models. The value of the lightest Higgs physical mass m_h has already the one-loop radiative corrections included, evaluated at the appropriate scale [35].

TABLE 1. A representative example of the predictions for the s-spectrum for the finite model **A** with $M = 1.0$ TeV, $m_{\bar{5}} = 0.8$ TeV and $m_{\mathbf{10}} = 0.6$ TeV.

$m_\chi = m_{\chi_1}$ (TeV)	0.45	$m_{\tilde{b}_2}$ (TeV)	1.76
m_{χ_2} (TeV)	0.84	$m_{\tilde{\tau}} = m_{\tilde{\tau}_1}$ (TeV)	0.63
m_{χ_3} (TeV)	1.49	$m_{\tilde{\tau}_2}$ (TeV)	0.85
m_{χ_4} (TeV)	1.49	$m_{\tilde{\nu}_1}$ (TeV)	0.88
$m_{\chi_1^\pm}$ (TeV)	0.84	m_A (TeV)	0.64
$m_{\chi_2^\pm}$ (TeV)	1.49	$m_{H^\pm}$ (TeV)	0.65
$m_{\tilde{t}_1}$ (TeV)	1.57	m_H (TeV)	0.65
$m_{\tilde{t}_2}$ (TeV)	1.77	m_h (TeV)	0.122
$m_{\tilde{b}_1}$ (TeV)	1.54		

TABLE 2. A representative example of the predictions of the s-spectrum for the finite model **B** with $M = 1$ TeV and $m_{\mathbf{10}} = 0.65$ TeV.

$m_\chi = m_{\chi_1}$ (TeV)	0.45	$m_{\tilde{b}_2}$ (TeV)	1.70
m_{χ_2} (TeV)	0.84	$m_{\tilde{\tau}} = m_{\tilde{\tau}_1}$ (TeV)	0.47
m_{χ_3} (TeV)	1.30	$m_{\tilde{\tau}_2}$ (TeV)	0.67
m_{χ_4} (TeV)	1.31	$m_{\tilde{\nu}_1}$ (TeV)	0.88
$m_{\chi_1^\pm}$ (TeV)	0.84	m_A (TeV)	0.73
$m_{\chi_2^\pm}$ (TeV)	1.31	$m_{H^\pm}$ (TeV)	0.73
$m_{\tilde{t}_1}$ (TeV)	1.51	m_H (TeV)	0.73
$m_{\tilde{t}_2}$ (TeV)	1.73	m_h (TeV)	0.118
$m_{\tilde{b}_1}$ (TeV)	1.56		

CONCLUSIONS

The programme of searching for exact RGI relations among dimensionless couplings in supersymmetric GUTs, started few years ago, has now supplemented with the derivation of similar relations involving dimensionful parameters in the SSB sector of these theories. In the earlier attempts it was possible to derive RGI relations among gauge and Yukawa couplings of supersymmetric GUTs, which could lead even to all-loop finiteness under certain conditions. These theoretically attractive theories have been shown not only to be realistic but also to lead to a successful prediction of the top quark mass. The new theoretical developments include the existence of a RGI sum rule for the soft scalar masses in the SSB sector of $N = 1$ supersymmetric gauge theories exhibiting gauge-Yukawa unification. The all-loop sum rule substitutes now the universal soft scalar masses and overcomes its phenomenological problems. Of particular theoretical interest is the fact that the finite unified theories, which could be made all-loop finite in the supersymmetric sector can now be made completely *finite*. In addition it is interesting to note that the sum rule coincides with that of a certain class of string models in which the massive string modes are organized into $N = 4$ supermultiplets. Last but not least in ref. [24], the exact β-function for the soft scalar masses in the NSVZ scheme was obtained for the first time. On the other hand the above theories have a remarkable predictive power leading to testable predictions of their spectrum in terms of very few parameters. In addition to the prediction of the top quark mass, which holds unchanged, the characteristic features that will judge the viability of these models in the future are 1) the lightest Higgs mass is found to be around 120 GeV and the s-spectrum starts beyond several hundreds of GeV. Therefore the next important test of Gauge-Yukawa and Finite Unified theories will be given with the measurement of the Higgs mass, for which these models show an appreciable stability, which is alarmingly close to the IR quasi fixed point prediction of the MSSM for large tan β [36]. Our preliminary search in the available parameter space of the above models shows that in case we relax the requirement that the mass of the s-tau should be smaller than the neutralinos masses, we obtain a wider window in the prediction of the lightest Higgs mass starting from 111 GeV. This possibility has no obvious problem in case we introduce bilinear R-parity violating terms that preserve finiteness. Actually, the introduction of such terms might be unavoidable given that it is a necessary ingredient of the only known mechanism to introduce neutrino masses in these models [37].

ACKNOWLEDGMENTS

It is a pleasure to thank the Organizing Committee for the very warm hospitality offered to one of us (G.Z.). Supported by the projects PAPIIT-125298 and ERBFM-RXCT960090.

REFERENCES

1. Connes, A., Douglas, M. R. and Schwarz,A., *JHEP*, **9802**, 003 (1998), hep-th/9711162.
2. Douglas, M. R. and Hull, C. *JHEP*, **9802**, 008 (1998), hep-th/9711165.
3. Seiberg, N. and Witten, E., *JHEP*, **9909**, 032 (1999), hep-th/9908142.
4. Connes, A., *Noncommutative Geometry*, Academic Press, U.S.A., 1994.
5. J.Madore, *An Introduction to Noncommutative Differential Geometry and its Physical Applications*, Cambridge Univ. Press, Cambridge, England, 1995.
6. Kapetanakis D., Mondragón, M. and Zoupanos, G., (1993) *Zeit. f. Phys.*, **C60** 181; Mondragón, M. and Zoupanos, G. (1995) *Nucl. Phys. B* (Proc. Suppl.) **37C**) 98.
7. Kubo, J., Mondragón, M. and Zoupanos, G., *Nucl. Phys.* **B424**, 291 (1994).
8. Kubo, J., Mondragón, M., Tracas, N.D. and Zoupanos, G. *Phys. Lett.* **B342**, 155 (1995); Kubo, J., Mondragón, M., Shoda, S. and Zoupanos, G., *Nucl. Phys.* **B469**, 3 (1996); Kubo, J., Mondragón, M., Olechowski, M. and Zoupanos, G., *Nucl. Phys.* **B479**, 25 (1996).
9. Kubo, J., Mondragón, M. and Zoupanos, G., *Phys. Lett.* **B389**, 523 (1996).
10. Kawamura, T., Kobayashi, T. and Kubo, J., *Phys. Lett.* **B405**, 64 (1997).
11. Kobayashi, T., Kubo, J., Mondragón, M. and Zoupanos, G., *Nucl. Phys.* **B511**, 45 (1998); in Proc. of ICHEP 1998, vol. 2, p. 1597 (Vancouver 1998); *Acta Phys. Polon.* **B30**, 2013 (1999); in Proc. of HEP99, p. 804 (Tampere 1999).

12. For an extended discussion and a complete list of references see: Kubo, J., Mondragón, M. and Zoupanos, G. *Acta Phys. Polon.* **B27**, 3911 (1997).
13. Zimmermann, W., *Com. Math. Phys.* **97**, 211 (1985); Oehme, R. and Zimmermann, W. *Com. Math. Phys.* **97**, 569 (1985).
14. Lucchesi, C., Piguet, O. and Sibold, K., *Helv. Phys. Acta* **61**, 321 (1988); Piguet, O. and Sibold, K. *Intr. J. Mod. Phys.* **A1**, 913 (1986); *Phys. Lett.* **B177**, 373 (1986); see also Lucchesi, C. and Zoupanos, G., *Fortsch. Phys.* **45**, 129 (1997).
15. Ermushev, A.Z., Kazakov, D.I. and Tarasov, O.V., *Nucl. Phys.* **281**, 72 (1987); Kazakov, D.I. *Mod. Phys. Lett.* **A9**, 663 (1987).
16. Jack, I. and Jones, D.R.T., *Phys. Lett.* **B349**, 294 (1995).
17. Hisano, J. and Shifman, M., *Phys. Rev.* **D56**, 5475 (1997).
18. Jack, I. and Jones, D.R.T. *Phys. Lett.* **B415**, 383 (1997).
19. Avdeev, L.V., Kazakov, D.I. and Kondrashuk, I.N., *Nucl. Phys.* **B510**, 289 (1998); Kazakov, D.I., *Phys.Lett.* **B449**, 201 (1999).
20. Kazakov, D.I. *Phys. Lett.* **B412**, 21 (1998).
21. Jack, I., Jones, D.R.T. and Pickering, A., *Phys. Lett.* **B426**, 73 (1998).
22. Jack, I., Jones, D.R.T., Martin, S.P., Vaughn, M.T., and Yamada, Y., *Phys. Rev.* **D50**, R5481 (1994).
23. Jack, I., Jones, D.R., and Pickering, A., *Phys. Lett.* **B435**, 61 (1998), hep-ph/9805482.
24. Kobayashi, T., Kubo, J., and Zoupanos, G., *Phys. Lett.* **B427**, 291 (1998); Kobayashi, T., et al, in Proc. of *"Supersymmetry, Supergravity and Superstrings"*, pp. 242-268, Seoul, 1999.
25. Yamada, Y., *Phys.Rev.* **D50**, 3537 (1994).
26. Delbourgo, R., *Nuovo Cim* **25A**, 646 (1975); Salam, A. and Strathdee, J., *Nucl. Phys.* **B86**, 142 (1975); Fujikawa, K. and Lang, W., *Nucl. Phys.* **B88**, 61 (1975); Grisaru, M.T., Rocek, M. and Siegel, W., *Nucl. Phys.* **B59**, 429 (1979).
27. Girardello, L. and Grisaru, M.T., *Nucl. Phys.* **B194**, 65 (1982); Helayel-Neto, J.A. *Phys. Lett.* **B135**, 78 (1984); Feruglio, F., Helayel-Neto, J.A., and Legovini, F., *Nucl. Phys.* **B249**, 533 (1985); Scholl, M. *Zeit. f. Phys.* **C28**, 545 (1985).
28. Jones, D.R.T., Mezincescu, L. and Yao, Y.-P., *Phys. Lett.* **B148**, 317 (1984).
29. Jack, I., and Jones, D.R.T., *Phys. Lett.* **B333**, 372 (1994).
30. Ibáñez, L.E. and Lüst, D., *Nucl. Phys.* **B382**, 305 (1992); Kaplunovsky, V.S. and Louis, J., *Phys. Lett.* **B306**, 269 (1993); Brignole, A., Ibañez, L.E. and Muñoz, C., *Nucl. Phys.* **B422**, 125 (1994), (Erratum: **B436**, 747 (1995)).
31. Brignole, A., Ibáñez, L.E. and Muñoz, C., *Phys. Lett.* **B387**, 305 (1996).
32. Casas, J.A., Lleyda, A. and Muñoz, C., *Phys. Lett.* **B380**, 59 (1996).
33. Novikov, V., Shifman, M., Vainstein, A., and Zakharov, V., *Nucl. Phys.***B229**, 381 (1983); *Phys. Lett.* **B166**, 329 (1986); Shifman, M., *Int.J. Mod. Phys.* **A11**, 5761 (1986) and references therein.
34. Bertolini, S., Borzumati, F., Masiero, A. and Ridolfi, G., *Nucl. Phys.* **B353**, 591 (1991).
35. Gladyshev, A.V., Kazakov, D.I., de Boer, W., Burkart, G. and Ehret, R., *Nucl. Phys* **B498**, 3 (1997); Carena, M. et. al., *Phys. Lett.* **B355**, 209 (1995).
36. Jurčišin, M., and Kazakov, D.I., *Mod. Phys. Lett.* **A14**, 671 (1999).
37. Hirsch,M. et.al., *Phys. Rev.* **D62**, 113008 (2000).

Supersymmetry in Five-Dimensional Brane Worlds

Z. Lalak

Theory Division, CERN, CH-1211 Geneva 23
Institute of Theoretical Physics, University of Warsaw, email: Zygmunt.Lalak@fuw.edu.pl

Abstract. We construct the explicit form of the four-dimensional effective supergravity action, which describes low-energy physics of the Randall–Sundrum model with moduli fields in the bulk and charged chiral matter living on the branes. The low-energy action is derived from the compactification of a locally supersymmetric model in five dimensions. We deduce a generalization of the effective 4d action to the case of a general, not necessarily exponential, warp factor. The mechanism of supersymmetry breaking mediation is described, which relies on the non-trivial configuration of the Z_2-odd bulk fields. Broken supersymmetry leads to stabilization of the interbrane distance.

INTRODUCTION

This talk is based on results obtained in collaboration with Adam Falkowski and Stefan Pokorski, published in refs. [1, 2, 3, 4].

Brane worlds with warped geometries offer new perspectives in understanding the hierarchy of mass scales in field theory models [5, 6, 7]. The initial hope was that the mere presence of extra dimensions would be a natural tool to control mass scales in gauge theories coupled to gravity. However, the realization of these simple ideas in terms of consistent models eventually called for quite sophisticated constructions, such as the brane–bulk supersymmetry that we are going to discuss in this talk.
The basic five-dimensional setup of brane world models is that of four-dimensional hypersurfaces (branes) hosting familiar gauge and charged matter fields, which are embedded in a five-dimensional ambient space, the bulk, populated by gravitational and gauge-neutral fields. The bulk degrees of freedom couple to the fields living on branes through various types of interactions. Some of these interactions are analogues of an interaction between electromagnetic potential and charge density located on branes – this is the case of the fields that are Z_2-even on an S^1/Z_2 orbifold forming the fifth dimension; some of them are rather analogues of the derivative coupling of the potential to the electric dipole moment density located on branes, i.e. analogues of the interactions of the Z_2-odd fields on S^1/Z_2. These interactions lead to the formation of nontrivial vacuum configurations in the brane system. In particular, the solutions of Einstein equations of the form $ds^2 = a^2(x^5)ds_4^2 + b^2(x^5)(dx^5)^2$ are usually allowed, as in the original Randall–Sundrum (RS) models, where ds_4^2 is the Minkowski, anti-deSitter or deSitter metric in 4d. The two basic observations pertaining to the hierarchy problem are the following.

CP589, *Fundamental Interaction Theories*, edited by J. Lukierski and J. Rembieliński

First, on the brane located at a position x^5, all the fundamental mass scales defining the 5d Lagrangian become down-scaled by the factor $a(x^5)$: $m \to ma(x^5)$ when written down in the frame canonical with respect to the 4d line element ds_4^2. Thus, if the warp factor falls down exponentially, as in the RS model, one is given a natural exponential mass hierarchy between branes which is directly related to their separation. In fact, the effective mass measuring the interaction of a test body with the gravitational zero-mode is modulated by the warp factor, $m_{\text{eff}} = ma(x^5)$. In addition, the heavy Kaluza-Klein modes of the metric tensor couple to the brane matter energy momentum tensor at x^5 with the strength $\Lambda^{-1}(x^5)$ where $\Lambda(x^5) = M_P a(x^5)$, thus implying, at least naively, an UV cut-off of that scale on perturbative physics on the brane. Second, as pointed out long ago by Rubakov and Shaposhnikov [8], the gradient energy associated with the variation of the warp factor in the direction transverse to the brane can cancel the contribution of the brane physics to the effective 4d cosmological constant. However, the stumbling observation is that whenever one finds a flat 4d foliation as the solution of higher-dimensional Einstein equations, which seems to be necessary for the existence of a realistic 4d effective theory, it is accompanied by a special choice of various parameters in the higher dimensional Lagrangian (see [9]). The fine-tuning seems to be even worse in 5d than in 4d, since typically one must correlate parameters living on spatially separated branes. Then there appears immediately the problem of stabilizing these special relations against quantum corrections. This situation has prompted the proposal [10, 11, 1], that it is a version of brane-bulk supersymmetry that may be able to explain apparent fine-tunings and stabilize hierarchies against quantum corrections. Indeed, the brane-bulk supersymmetry turns out to correlate in the right way the brane tensions and bulk cosmological constant in the supersymmetric Randall–Sundrum model. Moreover, local supersymmetry is likely to be necessary to embed brane worlds in string theory. Hence, the quest for consistent supersymmetric versions of brane worlds goes on, see [10, 11, 1, 12, 2, 14, 15, 16, 17, 18, 19, 20, 22, 23, 24, 25, 26, 3, 27]. First explicit supersymmetric models with delta-type (thin) branes were constructed in [10, 1, 2, 14]. The distinguishing feature of the pure supergravity Lagrangians proposed in [1] is imposing the Z_2 symmetry, such that gravitino masses are Z_2-odd. An elegant formulation of the model is given in Ref. [14] where an additional non propagating fields are introduced to independently supersymmetrize the branes and the bulk. In the on-shell picture for these fields the models of Ref. [14] and refs. [1, 2] are the same. On the other hand, in refs. [1, 2] it has been noted, that supersymmetric Randall–Sundrum-type models can be generalized to include the universal hypermultiplet and gauge fields and matter on the branes. The Lagrangian of such a construction has been given in [1, 2, 3, 13]. This has allowed us [1, 3] to study issues such as supersymmetry breaking and its transmission through the bulk. Finally, in [3] the effective low energy theory was formulated, which describes properly the physics of the warped five dimensional models with gauge sectors on the branes.

We have shown that in the class of models without nontrivial gauge sectors in the bulk, unbroken $N = 1$ local supersymmetry (classical solutions with four unbroken supercharges) implies vanishing of the effective cosmological constant. We have demonstrated the link between vanishing of the 4d cosmological constant, minimization of effective potentials in 5d and 4d, and moduli stabilization. We have also described supersymmetry breaking due to a global obstruction against the extension of bulk Killing spinors to the branes, which is a phenomenon observed earlier in the Horava–Witten model in 11d and 5d.

First steps towards the 4d effective theory were made in [25],[26] (where the Kähler function for the radion field was identified). In the set-up we consider in this paper supersymmetry in 5d is first broken from eight down to four supercharges by the BPS vacuum wall, and then again broken spontanously down to $N = 0$ due to switching on expectation values of sources living on the branes. The general strategy follows the one [28, 29, 30, 31] that led to the complete and accurate description of the low-energy supersymmetry breakdown in the Horava–Witten models, see [28, 31, 32]. We were able to deduce the Kähler potential, superpotential and gauge kinetic functions describing physics of corresponding vacua in four dimensions. It turns out that the warped background modifies in an interesting way the kinetic terms for matter fields and the gauge kinetic function on the warped wall. There also appears a potential for the radion superfield, its origin being a modulus-dependent prefactor multiplying the superpotential on the warped wall in the expression for the 4d effective superpotential.

SUPERSYMMETRY IN THE BRANE-WORLD SCENARIOS

Let us begin with a brief review of the original RS model. The action is that of 5d gravity on $M_4 \times S_1/Z_2$, with negative cosmological constant:

$$\begin{aligned} S &= M^3 \int d^5x\sqrt{-g}(\frac{1}{2}R + 6k^2) \\ &\quad + \int d^5x\sqrt{-g_i}(-\lambda_1\delta(x^5) - \lambda_2\delta(x^5 - \pi\rho)) . \end{aligned} \tag{1}$$

Three-branes of non-zero tension are located at Z_2 fixed points. The ansatz for vacuum solution preserving 4d Poincaré invariance has the warped product form:

$$ds^2 = a^2(x^5)\eta_{\mu\nu}dx^\mu dx^\nu + R_0^2(dx^5)^2 . \tag{2}$$

The breathing mode of the fifth dimension is parametrized by R_0. The solution for the warp factor $a(x^5)$ is:

$$a(x^5) = \exp(-R_0 k|x^5|) . \tag{3}$$

It has an exponential form, which can generate large hierarchy of scales between the branes. Matching delta functions in the equations of motion requires fine-tuning of the brane tensions:

$$\lambda_1 = -\lambda_2 = 6k . \tag{4}$$

With the choice (4) the matching conditions are satisfied for arbitrary R_0, so the fifth dimension is not stabilized in the original RS model. Thus R_0 enters the 4d effective theory as a massless scalar (radion), which couples to gravity in the manner of a Brans–Dicke scalar. This is at odds with the precision tests of general relativity, so any realistic model should contain a potential for the radion field.

Relaxing the condition (4) we are still able to find a solution in the maximally symmetric form, but only if we allow for non-zero 4d curvature (adS_4 or dS_4) [33].

In such a case the radion is stabilized and its vacuum expectation value is determined by the brane tensions and the bulk cosmological constant.

The Randall–Sundrum model can be extended to a locally supersymmetric model [10, 11, 1]. The basic set-up consists of 5d $N=2$ gauged supergravity [34, 35], which includes the gravity multiplet $(e^m_\alpha, \psi^A_\alpha, \mathcal{A}_\alpha)$, that is the metric (vielbein), a pair of symplectic Majorana gravitinos, and a vector field called the graviphoton. The 5d SUGRA action is

$$\begin{aligned} S &= \int d^5x e_5 \frac{1}{\kappa^2}\Big(\frac{1}{2}R - \frac{1}{2}\overline{\psi_\alpha}^A\gamma^{\alpha\beta\gamma}D_\beta\psi_{A\gamma} \\ &\quad - \frac{3}{4}\mathcal{F}_{\alpha\beta}\mathcal{F}^{\alpha\beta} + \dots\Big) \end{aligned} \tag{5}$$

and the supersymmetry transformations are given by

$$\begin{aligned} \delta e^m_\alpha &= \frac{1}{2}\bar{\varepsilon}^A\gamma^m\psi_{A\alpha}, \\ \delta\psi^A_\alpha &= D_\alpha\varepsilon^A - \frac{i}{4\sqrt{2}}(\gamma_\alpha^{\beta\gamma} - 4\delta^\beta_\alpha\gamma^\gamma)\mathcal{F}_{\beta\gamma}\varepsilon^A, \\ \delta\mathcal{A}_\alpha &= -\frac{i}{2\sqrt{2}}\overline{\psi_\alpha}^A\varepsilon_A. \end{aligned}$$

Let us now add the brane tension at the brane located at $x^5=0$, $S_1=\int d^5xe_4(-6k)\delta(x^5)$, and perform the supersymmetry transformation on the determinant of the induced vierbein. This produces a delta-type variation in the action: $\delta e \Rightarrow 3\delta(x^5)e_4k(\bar{\psi^1}_\mu\gamma^\mu\varepsilon^1 + (1\leftrightarrow 2))$. It is straightforward to notice that this can be cancelled through the variation of the term $\bar{\psi^A}_\mu\gamma^{\mu 5\rho}D_5\psi^A_\rho$ upon introducing new terms in the transformations of gravitini: $\delta\psi^1_\alpha = +\frac{k}{2}\varepsilon(x^5)\gamma_\alpha\varepsilon^1$, $\delta\psi^2_\alpha = -\frac{k}{2}\varepsilon(x^5)\gamma_\alpha\varepsilon^2$. These corrections introduce further variations in the bulk Lagrangian, which require further new terms in the bulk Lagrangian: $\mathcal{L}_{\psi^2} = +\frac{3e_5}{4}k\varepsilon(x^5)(\bar{\psi^1}_\alpha\gamma^{\alpha\beta}\psi^1_\beta - \bar{\psi^2}_\alpha\gamma^{\alpha\beta}\psi^2_\beta)$ and $\mathcal{L}_{cc} = 6e_5k^2$, which is precisely the bulk potential needed in the RS model. The continuation through $x^5=\pi\rho$ gives on the second brane the tension term $+\delta(x^5-\pi\rho)e_4 6k$. The resulting locally supersymmetric Lagrangian is in fact that of a gauged supergravity. The symmetry that is gauged is the $U(1)$ subgroup of the R-symmetry, $\psi^A_\alpha \to e^{i\phi}\psi^A_\alpha$, the gauge field being $\mathcal{A}^R_\alpha = -\frac{1}{2\sqrt{2}}\mathcal{A}_\alpha$. Gauging of the $U(1)_R$ symmetry means that we replace the derivative acting on the gravitino with the $U(1)_R$ covariant derivative:

$$D_\alpha\psi^A_\beta \to D_\alpha\psi^A_\beta - \frac{3}{\sqrt{2}}(\sigma^3)^A_B k\varepsilon(x^5)\mathcal{A}_\alpha\psi^B_\beta,$$

where D_α denotes the ordinary space-time covariant derivative. The coefficient of the coupling $\mathcal{A}_\alpha\psi^B_\beta$ defines the prepotential $\mathcal{P} = \frac{i}{4}\sigma^3$ and the Z_2-odd gauge coupling $g = \frac{6k\varepsilon(x^5)}{\sqrt{2}}$. The warp factor turns out to be $a(x^5) = e^{-kR_0|x^5|}$, precisely the one of the original RS model. Thus the fine-tuning present in the original RS model can be explained by the requirement of local supersymmetry [1].

New bosonic and fermionic fields do not affect the vacuum solution, so that the equations of motion for the warp factor are the same as in the original, non-supersymmetric RS model. The RS solution satisfies the BPS conditions and preserves one half of the supercharges, which corresponds to unbroken $N=1$ supersymmetry in four dimensions. In the supersymmetric version the brane tensions are fixed. In consequence, the exponential solution (3) is the only maximally symmetric solution and the radion is still not stabilized.

We now turn to studying the supersymmetric RS model coupled to matter fields. We want to investigate how general the features present in the minimal supersymmetric RS model are. We find that unbroken local supersymmetry implies flat 4d space-time in a wider class of 5d supergravities coupled to hyper- or vector multiplets, in which the scalar potential is generated by gauging a subgroup of R-symmetry.

Let us add neutral matter in the bulk (moduli) in the form of a universal hypermultiplet, $(\lambda^a, V, \sigma, \xi, \bar{\xi})$. The quaternionic metric h_{uv} of the scalar manifold can be read from the Kähler potential: $K = -\ln(S+\bar{S}-2\xi\bar{\xi}),\ \ S = V+\xi\bar{\xi}+i\sigma$.
New kinetic terms in the bulk Lagrangian contain

$$S = \int d^5x e_5 \frac{1}{\kappa^2}\Big(-\frac{1}{4V^2}(\partial_\alpha V \partial^\alpha V + \partial_\alpha \sigma \partial^\alpha \sigma) - \frac{1}{V}\partial_\alpha \xi \partial^\alpha \bar{\xi} - \frac{1}{2}\bar{\lambda}^a \gamma^\alpha D_\alpha \lambda_a + \dots \Big) . \tag{6}$$

Under Z_2, the bosonic fields (V,σ) are even, and (ξ) is odd. To arrive at supersymmetric Lagrangian in the presence of these additional fields it is necessary to gauge, in addition to $U(1)_R$, the isometry of the hypermultiplet scalar manifold: $\xi \to e^{i\theta}\xi$. Components of the Killing vector corresponding to this isometry are $k^\xi = i\xi,\ k^{\bar{\xi}} = -i\bar{\xi},\ \ k^x = -y,\ k^y = x$, where $\xi = x+iy$. This results in the prepotential

$$\mathcal{P} = \left(\tfrac{1}{4} - \tfrac{x^2+y^2}{4V}\right) i\sigma^3 - \tfrac{x}{2V^{1/2}} i\sigma^1 + \tfrac{y}{2V^{1/2}} i\sigma^2 \tag{7}$$

and scalar potential:

$$V = -k^2\left(6 + \tfrac{3}{V}|\xi|^2 - \tfrac{3}{V^2}|\xi|^4\right). \tag{8}$$

Finally, supersymmetry requires new terms localized on the branes: $\mathcal{L}_B = -\frac{e_4}{\kappa^2} 6k(1 - \frac{|\xi|^2}{V})(\delta(x^5) - \delta(x^5 - \pi\rho))$ and modifications of supersymmetry transformation laws that were given in [3]. To discuss both unbroken and broken $N=1$ supersymmetry, let us use from now on a more general ansatz for the metric allowing adS_4 geometry in four dimensions (the case of dS_4 is similar): $g_{\alpha\beta} = diag\,(-a^2(x^5)e^{2Lx^3}, a^2(x^5)e^{2Lx^3}, a^2(x^5)e^{2Lx^3}, a^2(x^5), R_0^2)$, where R_0 is the radion, and for later convenience we define $f = \exp(Lx^3)$. Let us note that $R^{(4)} = -12L^2$ and $\Lambda_{cc}^{(4)} = \langle V_{\rm eff}\rangle / M_P^2 = -6L^2$. The conditions for unbroken supersymmetry in the bulk (BPS conditions) $\delta\psi_\alpha^A = \delta\lambda^a = 0$ correlate scalars, the warp factor, Killing vector and prepotential:

$$\frac{a'}{a} = -4kR_0\sqrt{\vec{P}^2},\ h_{uw}\partial_5 q^w = 6kR_0 \partial_u \sqrt{\vec{P}^2}$$
$$h_{uv}\partial_5 q^u \partial_5 q^v = 9R_0^2 k^2 h_{uv} k^u k^v . \tag{9}$$

After substituting these relations into the expression for the 4d action and integrating over x^5 one obtains 4d effective action of the form

$$\begin{aligned} S_4 &= M_{PL}^2 \int d^4x(\frac{1}{2}\bar{R} - V_{\text{eff}}), \\ M_{PL}^2 V_{\text{eff}} &= -M^3 a^4(0)(24k\sqrt{\vec{P}^2}(0) - \lambda_1) \\ &\quad + M^3 a^4(\pi\rho)(-24k\sqrt{\vec{P}^2}(\pi\rho) - \lambda_2). \end{aligned} \tag{10}$$

Unbroken supersymmetry corresponds to $\sqrt{\vec{P}^2} = P_3$, which immediately implies $V_{\text{eff}} = \langle R^{(4)} \rangle = 0$. This is a general result, which holds for any hypermultiplet manifold, not only for the universal one, which implies that unbroken SUSY excludes solutions with non-zero 4d space-time curvature. Only Minkowski foliations correspond to unbroken $N = 1$ supersymmetry and the converse is also true in the class of models which are discussed in Ref. [3]. A further consequence is that stabilization of the radion field is impossible without breaking of residual four supercharges. It should be noted that in 4d supergravities one can a'priori obtain an anti-deSitter solution and preserve supersymmetry at the same time. Hence, compactifications of the supersymmetric RS scenarios yield a special subclass of 4d supergravities.

SUPERSYMMETRY BREAKING MEDIATED BY Z_2-ODD FIELDS AND RADION STABILIZATION

In this section we investigate an alternative mechanism of supersymmetry breaking, similar to that studied in M-theoretical scenarios [31]. It is triggered by brane sources coupled to the scalar fields in the bulk, which are odd with respect to the Z_2 parity. One way to see that supersymmetry is broken is to notice that the Killing spinor cannot be defined globally. The odd fields are the agents that transmit supersymmetry breaking between the hidden and visible branes. We present below a general description of our mechanism and then apply it to a specific model of 5d gauged supergravity with the universal hypermultiplet.

To see how to break the remaining supersymmetry (corresponding to $N = 1$ in 4d) down to nothing ($N = 0$ in 4d) one should inspect the parts of supersymmetry transformations of bulk fermions that contain the Z_2-odd fields ξ and their transverse derivatives:

$$\begin{aligned} \delta\psi_5^1 &= -\frac{1}{\sqrt{V}}\partial_5\xi\varepsilon^2 + 2k\varepsilon(x^5)(-\frac{Re(\xi)}{2\sqrt{V}}(\sigma^1)^{1B} \\ &\quad + \frac{Im(\xi)}{2\sqrt{V}}(\sigma^2)^{1B})\gamma_5\varepsilon_B, \\ \delta\lambda^1 &= +\frac{i}{\sqrt{2V}}\gamma^5\partial_5\xi\varepsilon^2 + 3k\varepsilon(x^5)V_u^{A1}k^u\varepsilon_A. \end{aligned} \tag{11}$$

It is clear that vacuum expectation values of ξ and $\partial_5\xi$ are parameters of the supersymmetry breakdown and the field ξ is the agent of local mediation of supersymmetry breakdown. Another way to see that ξ, $\partial_5\xi \neq 0$ imply completely broken SUSY is to notice that this means that the components $\mathcal{P}^1$ and $\mathcal{P}^2$ of the prepotential are non-zero as well. The BPS conditions give:

$$\varepsilon^2 = \frac{\sqrt{\vec{P}^2}\gamma_5\varepsilon^1 - P_3\varepsilon^1}{P_1 - iP_2}. \tag{12}$$

As an immediate consequence we find that if P_1 or $P_2 \neq 0$, then the Killing spinor is non-chiral, and is a mixture of the even and the odd components of the SUSY parameter ε. But only the even component survives the Z_2 projection at the orbifold fixed points, hence the Killing spinor cannot be defined globally and supersymmetry is broken because of the 'misalignment' between the bulk and the brane supersymmetries. In other words, the projection of the general bulk Killing spinor onto the brane would contain insufficient number of degrees of freedom to generate the minimal supersymmetry on the brane.

To excite ξ, $\partial_5\xi$ one needs to couple them to the branes. Details are given in [3]. Forgetting for a while about coupling to gaugino condensates on branes, the relevant part of the brane–bulk coupling is

$$\begin{aligned} &-\int d^5x\, e_5 \tfrac{2}{\sqrt{g_{55}}}(\delta(x^5)W_1(\partial_5\xi + 2\delta(x^5)\bar{W}_1) \\ &+\delta(x^5-\pi\rho)W_2(\partial_5\xi + 2\delta(x^5-\pi\rho)\bar{W}_2) + h.c.). \end{aligned} \tag{13}$$

We solve equations of motion perturbatively to order $(\frac{W}{M^3})^2$ with adS_4 foliation and ansatz $\xi = \xi(x^5) = \varepsilon(x^5)\zeta(|x^5|)$, $V = V(x^5)$, $\sigma = const$, $\mathcal{A}_\mu = 0$, $\mathcal{A}_5 = const$. Matching the δ' in EOMs yields the boundary conditions: $\zeta(0) = -\frac{\bar{W}_1}{M^3}$, $\zeta(\pi\rho) = \frac{\bar{W}_2}{M_3}$. As a consequence:

$$C = -\frac{\bar{W}_1}{M^3}, \quad Ce^{k(R_0 - 3\sqrt{2}i\mathcal{A}_5)\pi\rho} = \frac{\bar{W}_2}{M^3}. \tag{14}$$

One can see that as long as supersymmetry is unbroken, moduli R_0 and $\mathcal{A}_5$ are arbitrary. When sources are switched on for odd fields, then the expectation value of the radion is determined by the boundary sources W_i, assuming that V gets frozen. Perturbative $(o(|W^2|))$ solution to EOMs is

$$\begin{aligned} &\xi = C\varepsilon(x^5)e^{k(R_0-3\sqrt{2}ik\mathcal{A}_5)|y|}, \quad C = -\frac{\bar{W}_1}{M^3}, \quad Ce^{k(R_0-3\sqrt{2}ik\mathcal{A}_5)\pi\rho} = \frac{\bar{W}_2}{M^3} \\ &a = e^{kR_0|y|} + \frac{|C|^2}{2V}e^{-kR_0|y|}, \quad L^2 = \frac{16}{6}\frac{k^2|C|^2}{V} \\ &V = V_0 - |C|^2e^{2kR_0|y|}, \quad \sigma = \sigma_0, \end{aligned} \tag{15}$$

except that boundary conditions for V are fullfilled at the zeroth order only; hence it is better to say that one assumes V to be 'frozen'. The four-dimensional curvature is $R^{(4)} = -32\frac{k^2|C|^2}{V}$. This means that in our family of models broken supersymmetry implies negative 4d curvature, and that curvature vanishes only if the SUSY-breaking sources are switched off.

Solving the Einstein equations in the 4d effective theory yields $\bar{R} = -32\frac{k^2|C|^2}{V_0}$, which is consistent with the value of $L^2 \equiv -\frac{1}{12}\bar{R}$ in (15). We also see that V_0 enters the denominator of the effective potential, which explains its runaway behaviour commented on earlier.

Before closing the discussion of the 5d classical solutions and supersymmetry breakdown, let us comment on proposals [20, 21, 36] to solve the cosmological constant problem by virtue of supersymmetry of the bulk-brane system. To put the issue into perspective, let us note that the Einstein equation with indices (55) does not contain second derivatives of fields, hence it acts as a sort of constraint on the solutions of the remaining equations. This becomes clearer in the Hamiltonian approach towards the flow along the fifth dimension, where this equation arises as the Hamiltonian constraint $\mathcal{H} = 0$, and is usually used to illustrate the way the conservation of the 4d curvature L^2 is achieved through the compensation between gradient and potential terms along the classical flow. However, this classical conservation hinges upon fulfilling certain consistency conditions between brane sources, or between boundary conditions induced by them, as illustrated by the model above. When one perturbs the boundary terms on one wall, then to stay within the family of maximally symmetric foliations one of two things must happen. Either the distance between branes must change, or the source at the distant brane must be retuned. In the class of models which we constructed, if the 4d curvature is present then supersymmetry is broken, and doesn't take care of such a retuning. Moreover, even if retuning takes place, the size of 4d curvature, i.e. of the effective cosmological constant, does change as well, moreover, the magnitude of the effective cosmological constant has quadratic dependence on the boundary terms which induce supersymmetry breakdown. Hence, any perturbation of the boundary, instead of being screened by the bulk physics, contributes quadratically to the effective cosmological constant. Of course, we are talking about perturbations which can be considered quasi-classical on the brane. Thus we do not see here any special new effect of the extra dimension in the the cancellation of the cosmological constant. The positive aspect of supersymmetry is exactly the one which we know from the 4d physics. Supersymmetry, even the broken one, limits the size of the brane terms inducing supersymmetry breakdown, and limits in this way the magnitude of the 4d cosmological constant, since the two effects are strictly related to each other.

EFFECTIVE LOW-ENERGY THEORY IN FOUR DIMENSIONS

In this section we give the form of the effective four-dimensional supergravity describing zero-mode fluctuations in the models presented in the previous sections. Since in the 5d set-up supersymmetry is broken spontaneously, it is safe to assume that in 4d this supersymmetry breakdown can be considered as a spontaneous breakdown in a 4d supergravity Lagrangian described with the help of a Kähler potential K, superpotential W and gauge kinetic functions H. The goal is to identify these functions reliably starting from the maximally symmetric approximate solutions (15) that we have described in the previous section. Our procedure is perturbative in the supersymmetry breaking parameter $\frac{W}{M^3}$. Fortunately, it is sufficient to identify the functions we are looking for from the terms that can be reliably read at the order $(\frac{W_1}{M^3})^1$. Such terms include the gravitino mass term. In addition, we have at our disposal the complete kinetic terms for moduli, gauge and matter fields, which are of order $(W_1)^0$ and are sufficient to read off the Kähler potential for moduli and matter fields. The complete procedure has been given in Ref. [3]. Here we summarize the results and discuss the basic features of the warped 4d supergravities.

The Kähler function for moduli fields $S = V_0 + i\sigma_0$ and $T = k\pi\rho(R_0 + i\sqrt{2}\mathcal{A}_5)$, and for the charged fields Φ_1 and Φ_2 living on the Planck brane and warped brane respectively is

$$K(S,\bar{S};T,\bar{T};\Phi,\bar{\Phi}) = -M_P^2 \log(S+\bar{S}) - 3M_P^2 \log(f(T+\bar{T} - \frac{k}{3M^3}|\Phi_2|^2) - \frac{\beta k}{3M^3}|\Phi_1|^2) \tag{16}$$

where we defined $M_P^2 = \frac{M^3}{k}(1 - e^{-2k\pi\rho\langle R_0\rangle})$, $f = \beta(1 - e^{-(T+\bar{T})})$ with $\beta = \frac{M^3}{kM_P^2}$. The effective 4d superpotential is

$$W = 2\sqrt{2}(W_1 + e^{-3T} W_2) \tag{17}$$

and the gauge kinetic functions are

$$H_{\mathrm{warped}}(S,T) = S + 2b_0 T,\; H_{\mathrm{Planck}}(S) = S. \tag{18}$$

One can study the 4d effective scalar potential derived from K, W, H. The standard formulation gives: $\mathcal{V} = e_4 e^G (G_i G^{i\bar{j}} G_{\bar{j}} - 3)$ with $G = K + \log|W|^2$. Explicitly:

$$\begin{aligned}\mathcal{V} &= e_4 \frac{4}{V_0 M_P^2 \beta^3 (1 - e^{-2k\pi\rho R_0})^3} (|W_1|^2 (3e^{-2k\pi\rho R_0} - 2) + |W_2|^2 (3e^{-4k\pi\rho R_0} - 2e^{-6k\pi\rho R_0}) \\ &\quad + W_1 \bar{W}_2 e^{-3k\pi\rho R_0 - i\sqrt{2}\mathcal{A}_5} + W_2 \bar{W}_1 e^{-3k\pi\rho R_0 + i\sqrt{2}\mathcal{A}_5}).\end{aligned} \tag{19}$$

Minimizing the above scalar potential wrt $\mathcal{A}_5$ yields

$$\begin{aligned}Arg(W_2) - Arg(W_1) - 3\sqrt{2}k\pi\rho\mathcal{A}_5 &= \pi n, \\ n &= 0, \pm 1, \pm 2, \ldots .\end{aligned} \tag{20}$$

Taking $n = 1$ and minimizing wrt R_0 one obtains

$$e^{-k\pi\rho R_0} = \frac{|W_1|}{|W_2|}, \tag{21}$$

consistently with the 5d picture. Let us summarize the basic features of our model. The F-terms take at the minimum the expectation values

$$\begin{aligned}|F^S|^2 &= 8e^K (S+\bar{S})^2 a^2(\pi\rho)|W_2|^2 \\ &\times (1 - a^2(\pi\rho))^2 \neq 0 \\ |F^T|^2 &= 0,\end{aligned} \tag{22}$$

which means that supersymmetry is broken along the dilaton direction. The potential energy at this vacuum is negative:

$$V_{vac} = -\frac{8|W_2|^2}{V_0(M^3/kM_P^2)^3 M_P^2} \frac{a^2(\pi\rho)}{1-a^2(\pi\rho)}. \tag{23}$$

The mass of the canonically normalized radion is

$$m_R^2 = \frac{24}{V_0(M^3/kM_P^2)^3} \frac{a^2(\pi\rho)|W_2|^2}{(1-a^2(\pi\rho))} \frac{1}{M_P^4} \tag{24}$$

and the gravitino mass term is given by the expression

$$m_{3/2} = \frac{2}{\sqrt{V_0}(M^3/kM_P^2)^{3/2}} \frac{a(\pi\rho)}{(1-a^2(\pi\rho))^{3/2}} \frac{|W_2|}{M_P^2}. \tag{25}$$

It is interesting to compare these features to those of the models, which are low-energy limits of weakly and strongly coupled heterotic string theories. As is well known, in the leading, tree-level, approximation, heterotic string gives the four-dimensional supergravity which enjoys the no-scale structure [37]. The gauge kinetic functions in all, visible and hidden, sectors, are universal and depend only on the 4d dilaton superfield S, $H = S$, $\partial_T H = 0$. In addition, at the perturbative level there is no superpotential for moduli superfields S and T. As a consequence, the effective potential is positive semi-definite and takes the form $V = K_{S\bar{S}}|F^S|^2 \geq 0$. Hence, the vacuum configuration corresponds to $F^S = 0$, but F^T, which doesn't enter the potential, is allowed to take a non-zero value, so that supersymmetry is broken along the T direction in moduli space. The problem with this is that the scale of supersymmetry breakdown is arbitrary, and it can only be hoped that it becomes fixed at a proper value, after taking into account various perturbative and/or nonperturbative corrections to the Lagrangian. In the supergravity model which is the low energy limit of weakly coupled heterotic string with one-loop corrections, and at the same time that of a strongly coupled heterotic string where these corrections arise in classical expansion taking into account the presence of an extra dimension, see [38], the situation changes. When the interplay of sources of supersymmetry breakdown, such as condensates in various sectors and expectation value of the superpotential, is taken into account, it is possible to arrange for unbroken supersymmetry in the anti-deSitter background, or for vacua where both F^S and F^T are sizable - see [39, 40] for details. In all these considerations the effects of the nonperturbative warping of an extra dimension, like the one in the RS models, were not taken into account. The model which was constructed in [3] finally allows to discuss the impact of the rapidly changing warp-factor on the low-energy physics. In the model with just expectation values of the superpotentials serving as a supersymmetry breaking source, it is crucial that the effective superpotential acquires the exponential dependence on the modulus T. This leads to vanishing of the F^T upon using the equation of motion for T. As a result, this time it is F^S which is non-zero, although its value remains undetermined (as long as the corrections to the Lagrangian are not taken into account). The vacuum energy takes the form $V = -\frac{2}{M_P^2} e^K |W|^2 \leq 0$, which describes, at tree-level, an unstable background with negative energy density, in agreement with five-dimensional considerations.

GENERALIZATION TO AN ARBITRARY WARP FACTOR

Let us assume, that as a result of solving the 5d Einstein equations with the Ansatz

$$ds^2 = a^2(kR_0|y|)ds_4^2 + R_0^2(dx^5)^2 \tag{26}$$

we obtain a specific function $a(kR_0|y|)$, corresponding to the BPS vacuum of the 5d supergravity, preserving four unbroken supercharges. To obtain the effective theory, one needs to substitue the solution for the warp factor into the action, and integrate over x^5. The most relevant term from the point of view of the radion Lagrangian is the 5d Einstein–Hilbert term. Upon the integration over x^5 one obtains some derivative terms for the radion, which are a part of its kinetic terms, and the four-dimensional Einstein–Hilbert term with the radion-dependent coefficient

$$\frac{1}{2}M_P^2 f(R_0(x))R^{(4)}, \tag{27}$$

where $f = 2\beta\int_0^{k\pi R_0\rho} dy\, a^2(y)$ with $\beta = \frac{M^3}{kM_P^2}$. Now, following closely the steps given in [3], one can deduce the Kähler potential for the radion,

$$K = -3\log f(T+\bar{T}), \tag{28}$$

where $f(T+\bar{T})$ is obtainded by substituting $2k\pi\rho R_0 = T+\bar{T}$ everywhere in the function f. In this, lowest order, approximmation to the complete theory the whole information about the bulk physics is encoded in f through the R_0-dependence of the warp factor. A comment is in order about the way the second Z_2-even bulk scalar V_0, which is the real part of the 4d dilaton S, enters the Kähler potential. Firstly, in consequence of the fact, that it is borne as a member of the universal hypermultiplet, the Kähler potential for S must contain a piece $-\log(S+\bar{S})$. Secondly, in general, the solution for the warp factor contains not only R_0, but also V_0. An example is the warp factor in the 5d Horava–Witten model: $a(x^5) = (1+\alpha\sqrt{2}\frac{R_0}{V_0}(|x^5|-\pi\rho/2))^{1/6}$. This will necessarily lead to a mixing between S and T in the complete Kähler function. We cannot control this mixing without committing ourselves to a specific model. The attitude which we shall assume in what follows is that we freeze S in the effective Lagrangian, assuming it forms a background for the evolution of T and possible matter fields, and shall discuss the stabilization of the radion in such a background. Finally, it is straightforward to extend the model by the matter on the unwarped brane and by gauge fields, see [4], to deduce a complete 4d supergravity model given by

$$K(\Phi_1,\Phi_2) = -M_P^2\log(S+\bar{S}) - 3M_P^2\log\left(f(T+\bar{T}-\gamma|\Phi_2|^2) - \beta\gamma|\Phi_1|^2\right) \tag{29}$$

where $\gamma = \frac{k}{3M^3}$,

$$W = 2\sqrt{2}(W_1 + a^3(T)W_2), \tag{30}$$

and gauge kinetic functions are

$$H_1(S) = S, \quad H_2(S) = S - 2b_0\log a(T). \tag{31}$$

There are terms of which we know that they are missing here. In addition to the kinetic S–T mixing, they include mixing between S and matter fields and further T-dependent threshold corrections to the gauge kinetic functions (in Horava-Witten model these are of the form $\pm\alpha T$), but at least the last two types of corrections one may assume to be of the higher order in some perturbative expansion (an example being again the H–W model, see [1]). The supergravity potential generated by the above choices of the Kähler function and the superpotential lead generically to stabilization of the value of the real part of T, hence of R_0, in terms of the expectation values of the brane superpotentials W_1 and W_2. The basic reason for that is that due to the warp factor the effective superpotential for T appears in the effective 4d supergravity. Interestingly enough this is not always sufficient to stabilize the interbrane distance, and, again, the exapmle is the 5d Horava–Witten model.

It turns out that one can enhance the models presented here in such a way, that it is possible to have cosmological constant zero while stabilizing all moduli and breaking supersymmetry. The enhancement which does the job consists in adding a Polonyi superfield on the Planck brane and a racetrack–type superpotential for S on the warped brane. The price which has to be paid is assigning an active role to the new Polonyi field. In fact, the 4d goldstino turns out to be a mixture of the dilatino and the Polonyi fermion; the contribution of the T-modulino is suppressed by the factor $a(T)$. For details see [4].

SCALES IN SUPERSYMMETRIC RANDALL–SUNDRUM MODEL

The RS2 model with two branes is often said to offer a solution to the perturbative hierarchy problem. The basic observation is that on the warped brane all the mass scales initially present in the Lagrangian become universally scaled down by the factor $a(\pi\rho)$: $m \to a(\pi\rho)m$, so that on that brane the local reference scale becomes $\mu = Ma(\pi\rho)$ rather than the 5d gravitational scale M. Moreover, the masses of the Kaluza–Klein towers of the bulk fields start with $m_{KK} = \mu$ and the coupling of heavy KK modes to matter localized on the warped brane is set by μ^{-1}. This means, that at energies E somewhat larger than μ the KK modes of gravitons and bulk moduli couple strongly to the brane fields, and the whole mixture needs to be treated as a single model, unfortunately – strongly coupled in the (super)gravity description. One can argue on the basis of the conjectured gravity–CFT duality that the upper scale of perturbative physics is slightly higher than μ, $\Lambda_{\text{cut-off}} = \mu(M/k)$, where $c = (M/k)^3$ has the interpretation of the central charge of the conjectured dual CFT, so that there is a region where the 5d weakly coupled bulk-brane supergravity is a good description of physics.

Let us review then the mass scale patterns that are possible in the weakly couplead brane-bulk supergravity regime. First of all, the gravitational mass splittings in the multiplets living on the branes are set by $m_{3/2} \sim \frac{|W_1|}{M_P^2} = \frac{a(\pi\rho)|W_2|}{M_P^2}$. Now, we would call $|W_2|$ natural on the warped brane, if $|W_2| \leq \mu^3$. If we insist that $\mu = 1\ TeV$, we $m_{3/2} \leq 10^{-45}\ TeV$. This cannot be completely excluded, but to obtain a realistic mass split in the multiplets one would need to invoke some additional mechanism of supersymmetry breakdown mediation, like gauge mediation, which makes the whole scheme complicated. The other possibility is to say that $m_{3/2} \approx \mu = 1\ TeV$, and to find out that $|W_2| = M_P^3 \gg \mu^3$. It is easy to see that in such a case $|W_1| = (10^{13}\ GeV)^3$. Hence, on this case one has a set of three scales in the perturbative sector, $M_2 = 1\ TeV$, $M_1 = 10^{13}\ TeV$ and M_P, two of which are beyond the naive range of perturbativity. A different option is possible. Let us request that $m_{3/2} = 1\ TeV$ and W_2 be natural on the warped brane. Then we obtain the needed value of the warp factor $a(\pi\rho) \approx 10^{-4}$. One finds that $M_1 \approx 10^{14}\ GeV$ and $M_2 \approx 10^{-1} M_1$. The scales of the brane physics are within the perturbative regime, and the scale μ is very close to the usual GUT scale. This might suggest, that actualy it is more natural in the supersymmetric Randall–Sundrum to think of the warped brane as of a M_{GUT}-brane, rather than a TeV brane; with supersymmetry required anyway, there is simply no obvious preference for the TeV version.

SUMMARY

The main result of [3, 4] summarized in this talk is the four-dimensional effective supergravity action, which describes low-energy physics of the Randall–Sundrum-type models with moduli fields in the bulk and charged chiral matter living on the branes.

The low-energy action has been read off from a compactification of a locally supersymmetric model in five dimensions. The exponential warp factor has interesting consequences for the form of the effective 4d supergravity. The asymmetry between the warped and unwarped walls is visible in the Kähler function, in the gauge kinetic functions and in the superpotential. Roughly speaking the contributions to these functions which come from the warped wall are suppressed by an exponential factor containing the radion superfield. This is the way the warp factor and (and RS brane tensions) is encoded in the low-energy Lagrangian.

We have described the mechanism of supersymmetry breaking mediation which relies on a non-trivial configuration of the Z_2 odd fields in the bulk. We point out, that the odd Z_2 parity fields can be an important ingredient of 5d supersymmetric models. They play a crucial role in communication between spatially separated branes.

After freezing the dilaton, it is possible to stabilize the radion field in the backgrounds with broken supersymmetry and excited odd-parity fields.

We have shown that in the class of brane world models without charged matter and gauge fields in the bulk, unbroken $N = 1$ supersymmetry implies vanishing cosmological constant. In the case where the sources that induce supersymmetry breakdown are represented simply by constant superpotentials on the branes, broken supersymmetry gives rise to anti-deSitter-type geometry in four dimensions. However, in more sophisticated models, for instance in a model with a Polonyi field on the Planck brane and a racetrack-type superpotential for S on the warped brane, one is able to find solutions with broken supersymmetry, stabilized moduli and vanishing vacuum energy.

We believe that the class of models we have constructed in [3] provides a useful, explicit, setup to study low-energy phenomenology of the supersymmetric brane models with warped vacua.

ACKNOWLEDGMENTS

This work has been supported by RTN programs HPRN-CT-2000-00152 and HPRN-CT-2000-00148 and by the Polish Committee for Scientific Research grant 5 P03B 119 20 (2001-2002).

REFERENCES

1. Falkowski, A., Lalak, Z., and Pokorski, S., *Phys. Lett.*, **B491**, 172 (2000).
2. Falkowski, A., Lalak, Z., and Pokorski, S., *Five dimensional supergravities with universal hypermultiplet and warped brane worlds, Phys. Lett.*, **B** in press, hep-th/0009167.
3. Falkowski, A., Lalak, Z., and Pokorski, S., *Four-dimensional supergravities from five-dimensional brane worlds*, hep-th/0102145.
4. Falkowski, A., Lalak, Z., and Pokorski, S., *Phenomenology of warped brane worlds, to appear.*
5. Randall, L., and Sundrum, R., *Phys. Rev. Lett.*, **83**, 3370 (1999).
6. Randall, L, and Sundrum, R., *Phys. Rev. Lett.*, **83**, 4690 (1999).
7. Lykken, J., and Randall, L., *JHEP*, **0006**, 014 (2000).
8. Rubakov, V.A., and Shaposhnikov, M.E., *Phys. Lett.*, **B125**, 139 (1983).
9. Förste, S., Lalak, Z., Lavignac, S., and Nilles, H., *JHEP*, **0009**, 034 (2000); *Phys. Lett.*, **B481**, 360 (2000).
10. Altendorfer, R., Bagger, J., and Nemeschansky, D., *Supersymmetric Randall–Sundrum scenario*, hep-th/0003117.
11. Gherghetta, T., and Pomarol, A., *Bulk fields and supersymmetry in a slice of AdS*, hep-th/0003129.
12. Alonso-Alberca, N., Meesen, P., and Ortin, T., *Supersymmetric Brane-Worlds*, hep-th/0003248.
13. Falkowski, A., *M.Sc.Thesis*: info.fuw.edu.pl/~afalkows.
14. Bergshoeff, E., Kallosh, R., and Van Proeyen, A., *Supersymmetry in singular spaces*, hep-th/0007044.
15. Behrndt, K., and Cvetić, M., *Phys. Rev.*, **D61**, 101901 (2000).
16. Behrndt, K., and Cvetić, M., *Phys. Lett.*, **B475**, 253 (2000).
17. Kallosh, R., and Linde, A., *JHEP*, **0002**, 0005 (2000) and references therein.
18. Kallosh, R., Linde, A., and Shmakova, M., *JHEP*, **9911**, 010 (1999).
19. Behrndt, K., *Nonsingular infrared flow from D=5 gauged supergravity*, hep-th/0005185.
20. Mayr, P., *Stringy world branes and exponential hierarchies*, hep-th/0006204.
21. Verlinde, H., *Supersymmetry at large distance scales*, hep-th/0004003.
22. Duff, M.J., Liu, J.T., and Stelle, K.S., *A supersymmetric Type IIB Randall–Sundrum realization*, hep-th/0007120.
23. Behrndt, K., Herrmann, C., Louis, J., and Thomas, S., *Domain walls in five dimensional supergravity with non-trivial hypermultiplets*, hep-th/0008112.
24. Zucker, M., *Supersymmetric brane world scenarios from off-shell supergravity*, hep-th/0009083.
25. Bagger, J., Nemeschansky, D., and Zhang, R., *Supersymmetric radion in the Randall–Sundrum scenario*, hep-th/0012163.
26. Luty, M., and Sundrum, R., *Hierarchy stabilization in warped supersymmetry*, hep-th/0012158.
27. Cvetic, M., Lu, H., and Pope, C., *Localized gravity in the singular domain wall backgrounds?*, hep-th/0002054.
28. Mirabelli, E., and Peskin, M., *Phys. Rev.*, **D58**, 065002 (1998).
29. Lukas, A., Ovrut, B.A., Stelle, K.S., and Waldram, D., *Phys. Rev.*, **D59**, 086001 (1999).
30. Lukas, A., Ovrut, B.A., Stelle, K.S., and Waldram, D., *Nucl. Phys.*, **B552**, 246 (1999).
31. Ellis, J., Lalak, Z., Pokorski, S., and Pokorski, W., *Nucl. Phys.*, **B540**, 149 (1999).
32. Ellis, J., Lalak, Z., and Pokorski, W., *Nucl. Phys.*, **B559**, 71 (1999).
33. Karch, A., and Randall, L., *Locally localized gravity*, hep-th/0011156.

Introduction to the Superembedding Description of Superbranes

D. Sorokin

INFN, Sezione di Padova, via F. Marzolo 8, 35131, Padova, Italia, email: sorokin@pd.infn.it and Institute for Theoretical Physics, National Science Center, KIPT, Kharkov, Ukraine,

Abstract. Basics of the geometrical formulation of the dynamics of supersymmetric objects are considered and its relation to conventional formulations of superbranes is discussed. In particular, we demonstrate how the kappa–symmetry of the Green–Schwarz formulation shows up from local worldvolume supersymmetry, and briefly discuss applications of the superembedding approach.

INTRODUCTION

Superembedding is an elegant and geometrically profound approach which is based on a supersymmetric extension of the classical surface theory to the description of superbrane dynamics by means of embedding worldvolume supersurfaces into target superspaces (see [1] for a detailed review).

The superembedding approach arose as a proposal to solve the problem of the covariant quantization of the Green–Schwarz superstring by combining in a more general formulation main properties of both the Neveu–Schwarz–Ramond (NSR) and the Green–Schwarz (GS) formulation, and by now it has developed into a generic geometrical method for formulating the theory of superbranes.

Being manifestly doubly supersymmetric (on the worldvolume and in target superspace) the superembedding approach has explained the origin and the nature of the fermionic κ–symmetry of the GS formulation as a manifestation of the conventional local supersymmetry of the worldvolume, and thus solved the problem of infinite reducibility of the κ–symmetry by realizing it as an irreducible extended worldvolume supersymmetry [2]. This stimulated progress in the covariant quantization of the Green–Schwarz superstring mainly due to persistent work of N. Berkovits [3].

The superembedding approach established or clarified a classical relationship between various formulations of the dynamics of superparticles and superstrings, such as the NSR and the GS formulation [4], the twistor [5] and harmonic [6] descriptions.

This approach has proved to be a universal and powerful method applicable to the description of all known supersymmetric branes, in particular, to those of them for which standard methods encountered problems because of their specific structure, such as the 5–brane of M–theory. The superembedding methods allowed, for the first time, to derive the complete set of covariant equations of motion of the M5–brane [7]. And only later these equations were obtained [8] from the M5–brane action [9] based on a different technique adapted to deal with self–dual fields.

CP589, *Fundamental Interaction Theories*, edited by J. Lukierski and J. Rembieliński

The superembedding formulation has also proved to be useful for studying the "brany" mechanism of partial supersymmetry breaking [10] by giving a geometrical recipe [11, 12, 13] for constructing covariant worldvolume supersymmetric actions for superbranes. Upon gauge fixing worldvolume superdiffeomorphisms these actions become those of effective field theories with non–linearly realized spontaneously broken supersymmetries. This has demonstrated an intrinsic link of the superembedding approach and the method of nonlinear realizations developed in application to supersymmetric theories in [14] (see [15] for a review and references).

In this contribution I would like to make an introduction into the superembedding formalism and describe its basic properties.

DOUBLE SUPERSYMMETRY

As it has been mentioned in the Introduction the superembedding description of superbrane dynamics is formulated in such a way that it possesses manifest supersymmetry on the brane worldvolume and in target superspace, and thus has properties of both the NSR and the GS formulation.

The Neveu–Schwarz–Ramond Formulation

In the NSR formulation, which describes spinning particles [16] and spinning strings [17][1], the worldline or worldsheet of the spinning object is a supersurface $\mathcal{M}_w$ parametrized by bosonic coordinates ξ and fermionic coordinates η which we will collectively call $z^M = (\xi, \eta)$. Depending on the model considered $\mathcal{M}_w$ may have a various number of fermionic directions η. For simplicity we here take only one η. The dynamics of the spinning object is described by embedding $\mathcal{M}_w$ into a bosonic target space–time M_T parametrized by coordinates $X^{\underline{m}}$ ($\underline{m} = 0, 1, \cdots, D-1$). In the classical problems the number of space–time dimensions can be arbitrary, but for quantum consistency the spinning string, whose worldsheet has one or two fermionic directions, must live in a ten–dimensional target space.

The motion of the spinning object is described by the image of $\mathcal{M}_w$ in M_T

$$X^{\underline{m}}(z^M) = x^{\underline{m}}(\xi) + i\eta\chi^{\underline{m}}(\xi), \tag{1}$$

where $x^{\underline{m}}(\xi)$ is associated with the bosonic degrees of freedom of the spinning particle or string in M_T, and the Grassmann–odd vector $\chi^{\underline{m}}(\xi)$ is associated with its spin degrees of freedom.

The NSR formulation is invariant under worldsheet superdiffeomorphisms $z^M \to z'^M(z^M)$, which include $\mathcal{M}_w$ bosonic reparametrizations

$$\delta\xi = a(\xi), \tag{2}$$

[1] No spinning branes have been consistently constructed so far [18].

and local worldsheet supersymmetry

$$\delta\eta = \kappa(\xi), \qquad \delta\xi = i\kappa(\xi)\eta. \tag{3}$$

The presence of the local symmetries (2) and (3) implies that the dynamics of the spinning objects is subject to bosonic and fermionic first–class constraints, respectively, which I will *schematically* write down in the form

$$(\partial x^{\underline{m}})^2 = 0, \qquad \partial x^{\underline{m}}\chi_{\underline{m}} = 0. \tag{4}$$

The bosonic constraint in (4) is the mass shell or the Virasoro conditions, and the fermionic constraint produces, upon quantization, the Dirac equation for the spin wave functions of the dynamical system.

The NSR formulation does not have target–space supersymmetry. The latter appears, in the case of the spinning string, only at the quantum level upon imposing the Gliozzi–Scherk–Olive projection. The merit of this formulation is that it is covariantly quantizable.

The Green–Schwarz Formulation

This formulation is applicable to all known superparticles, superstrings and superbranes[2].

Now the worldvolume $\mathcal{M}_w$ is a (p+1)–dimesnional bosonic surface parametrized by the coordinates ξ^m $(m = 0, 1, \cdots, p)$ and the target space, into which $\mathcal{M}_w$ is embedded, is a superspace M_{TS} parametrized by bosonic coordinates $x^{\underline{m}}$ $(\underline{m} = 0, 1, \cdots, D-1)$ and by an appropriate number of fermionic coordinates $\theta^{\underline{\alpha}}$ $(\underline{\alpha} = 1, \cdots, 2n)$

$$Z^{\underline{M}}_{TS}(\xi) = (x^{\underline{m}}(\xi), \;\; \theta^{\underline{\alpha}}(\xi)). \tag{5}$$

The GS formulation is manifestly invariant under bosonic $\mathcal{M}_w$ reparametrizations $\xi \to \xi'(\xi)$, and target space superdiffeomorphisms $Z^{\underline{M}} \to Z'^{\underline{M}}(Z^{\underline{M}})$, which in the case of flat target superspace reduce to the translations along $x^{\underline{m}}$ and to the global target–space supersymmetry transformations

$$\delta\theta^{\underline{\alpha}} = \varepsilon^{\underline{\alpha}}, \qquad \delta x^{\underline{m}} = i\bar{\theta}\Gamma^{\underline{m}}\delta\theta. \tag{6}$$

There is also another (non–manifest) local worldvolume fermionic symmetry, so–called κ–symmetry, inherent to the GS formulation. This is an important symmetry which implies and reflects the existence of supersymmetric BPS brane–like solutions of corresponding supergravity theories. It is thus responsible for the 'brane scan' (i.e. which brane lives in which target superspace). The κ–symmetry was first observed in the case

---

[2] The name "Green–Schwarz" used for this formulation should be regarded as cumulative, since for different extended objects it has been developed by different people. This also concerns the "NSR" fromulation. A detailed list of references the reader may find in [1].

of superparticles [19] and has the following generic form of transformations (in flat target superspace):

$$\delta\theta^{\underline{\alpha}} = \Pi^{\underline{\alpha}}{}_{\underline{\beta}}\kappa^{\underline{\beta}}(\xi), \qquad \delta x^{\underline{m}} = -i\bar{\theta}\Gamma^{\underline{m}}\delta\theta, \tag{7}$$

where $\Pi^{\underline{\alpha}}{}_{\underline{\beta}}$ is a projector matrix ($\det\Pi^{\underline{\alpha}}{}_{\underline{\beta}} = 0$), whose form depends on the object under consideration.

Because of the presence of the projector in the κ–transformations they are infinite reducible, so only half of $\kappa^{\underline{\alpha}}$, i.e. n of the $2n$ Grassmann spinor components, effectively contribute to the variation of the worldvolume fields. This is a cause of the problem of the covariant quantization of the GS formulation.

To solve the infinite reducibility problem of the κ–symmetry it is natural to try to find its irreducible realization which is covariant in target superspace. A natural assumption is that this should be an extended local worldvolume supersymmetry (3) with the number of independent parameters equal to the number of independent (irreducible) κ–symmetries [2]. This reasoning brings us to

The Doubly–Supersymmetric Formulation

In this formulation the dynamics of superbranes is described by embedding a worldvolume *super*surface $\mathcal{M}_{sw}$ parametrized by the coordinates $z^M = (\xi^m, \eta^\alpha)$ ($m = 0, 1, \cdots, p$), ($\alpha = 1, \cdots, n$) into a target *super*space M_{TS} parametrized by the coordinates $Z^{\underline{M}} = (X^{\underline{m}}, \Theta^{\underline{\alpha}})$ ($\underline{m} = 0, 1, \cdots, D-1$), ($\underline{\alpha} = 1, ..., 2n$). Note that the number of the Grassmann directions of $\mathcal{M}_{sw}$ is half the number of the Grassmann directions of M_{TS}. Such a choice of the supermanifolds for superembedding is caused by our desire to identify n local supersymmetries on $\mathcal{M}_{sw}$ with n independent κ–symmetries of the GS formulation.

Thus in the doubly supersymmetric formulation the degrees of freedom of the superbranes are described by worldvolume superfields

$$X^{\underline{m}}(z^M) = x^{\underline{m}}(\xi) + i\eta^\alpha \chi_\alpha^{\underline{m}}(\xi) + \cdots, \quad \Theta^{\underline{\alpha}}(z^M) = \theta^{\underline{\alpha}}(\xi) + \eta^\alpha \lambda_\alpha^{\underline{\alpha}}(\xi) + \cdots, \tag{8}$$

where $\cdots$ stand for the terms of higher order in η^α. These terms contain auxiliary fields and in addition, for example in the case of the D–branes and the M5–brane, include the gauge fields propagating on the worldvolumes of these branes.

From Eq. (8) we see that in the doubly supersymmetric construction the number of degrees of freedom of the superbrane roughly speaking doubles. We now have $x^{\underline{m}}(\xi)$ describing the bosonic oscillations of the brane, the Grassmann 'spin'–vectors $\chi_\alpha^{\underline{m}}(\xi)$ as in the NSR formulation, the Green–Schwarz fermionic spinor degrees of freedom $\theta^{\underline{\alpha}}(\xi)$ and their bosonic counterparts $\lambda_\alpha^{\underline{\alpha}}(\xi)$. So if all these worldvolume fields are independent the corresponding models will not describe conventional superbranes. In this case one will get, for instance, so called spinning superparticles and spinning superstrings [20] which have more degrees of freedom then the conventional NSR and GS dynamical systems. They also have infinite reducible κ–symmetry as a fermionic symmetry independent of the local worldvolume supersymmetry.

To reach our goal of interpreting κ–symmetry as a manifestation of the local worldvolume supersymmetry we should find an appropriate doubly–supersymmetric description of the *conventional* superbranes. For this we should impose constraints on the superfields (8) which relate their components in such a way that the independent physical degrees of freedom described by these superfields will correspond to the standard GS formulation. The geometrical meaning of these constraints is that they cause the worldvolume supersurface $\mathcal{M}_{sw}$ to be imbedded into the target superspace M_{TS} in a specific way.

THE SUPEREMBEDDING CONDITION

Before discussing in more detail the geometrical meaning of the superembedding condition let us consider its dynamical consequences in the simplest case of a superparticle propagating in a flat $N=1, D=3$ target superspace. Then the supersurface $\mathcal{M}_{sw}$ of the previous subsection is associated with the superparticle "worldline" having one bosonic (time) and one fermionic coordinate (ξ,η), and the target superspace is parametrized by bosonic three–vector coordinates $X^{\underline{m}}$ $(m=0,1,2)$ and Grassmann Majorana two–spinor coordinates $\Theta^{\underline{\alpha}}$ $(\underline{\alpha}=1,2)$. The superembedding condition relates the worldline superfields $X^{\underline{m}}(z^M)$ and $\Theta^{\underline{\alpha}}(z^M)$ in the following way

$$DX^{\underline{m}} - iD\bar{\Theta}\Gamma^{\underline{m}}\Theta = 0, \tag{9}$$

where D is a Grassmann covariant derivative on $\mathcal{M}_{sw}$ which in the case of the superparticles can be chosen to be flat

$$D = \frac{\partial}{\partial\eta} + i\eta\frac{\partial}{\partial\xi}, \qquad \{D,D\} = 2i\partial_\xi. \tag{10}$$

Using the η–expansion (8) which in the case under consideration does not contain the "..."–terms we obtain the following relation between the components of the superfields $X^{\underline{m}}(z^M)$ and $\Theta^{\underline{\alpha}}(z^M)$

$$\partial_\xi x^{\underline{m}} - i\partial_\xi\bar{\theta}\Gamma^{\underline{m}}\theta = \bar{\lambda}\Gamma^{\underline{m}}\lambda, \tag{11}$$

$$\chi^{\underline{m}} = \bar{\theta}\Gamma^{\underline{m}}\lambda. \tag{12}$$

From Eq. (11) we see that $\lambda^{\underline{\alpha}}(\xi)$ are not independent fields and are expressed in terms of the derivatives of $x^{\underline{m}}$ and $\theta^{\underline{\alpha}}$. Moreover in the l.h.s. of (11) one can recognize the canonical momentum of the superparticle $\Pi^{\underline{m}} = \partial_\xi x^{\underline{m}} - i\partial_\xi\bar{\theta}\Gamma^{\underline{m}}\theta$ whose square is identically zero

$$\Pi^{\underline{m}}\Pi_{\underline{m}} = 0, \tag{13}$$

because of its so called Cartan–Penrose (or twistor) representation as a bilinear combination of commuting spinor components and Γ–matrix identities. We conclude that the superparticle is massless.

In the case of the superstrings the superembedding condition will produce in a similar way the Virasoro constraints, and it will produce the corresponding constraints for the other superbranes.

Eq. (12) implies the relation between the Grassmann vector and the Grassmann spinor variables, so that only one or another can be regarded as describing independent fermionic degrees of freedom. This is a basic relation which allows one to establish a classical correspondence between the NSR and the GS formulation of supersymmetric particles and strings [4].

Local Worldvolume Supersymmetry Versus κ–Symmetry

Let us now demonstrate how κ–symmetry appears in the superembedding formulation as a weird realization of the local worldvolume supersymmetry.

The components (7) of the superfields $X^{\underline{m}}(z^M)$ and $\Theta^{\underline{\alpha}}(z^M)$ transform under the local worldline supersymmetry (3) in the standard way

$$\delta\theta^{\underline{\alpha}} = -\lambda^{\underline{\alpha}}\kappa(\xi)\,, \qquad \delta\lambda^{\underline{\alpha}} = i\partial_\xi\theta^{\underline{\alpha}}\kappa(\xi)\,, \tag{14}$$

$$\delta x^{\underline{m}} = i\chi^{\underline{m}}\kappa(\xi)\,, \qquad \delta\chi^{\underline{m}} = -\partial_\xi x^{\underline{m}}\kappa(\xi)\,. \tag{15}$$

We now substitute into the first equation of (15) the solution (12) of the superembedding condition (9) and observe that, due to the form of the θ–variation (14), the variation of $x^{\underline{m}}$ can be rewritten as follows

$$\delta x^{\underline{m}} = i\bar{\theta}\Gamma^{\underline{m}}\lambda\kappa(\xi) = -i\bar{\theta}\Gamma^{\underline{m}}\delta\theta\,. \tag{16}$$

The next step is to replace the Grassmann scalar parameter of the local supersymmetry by the scalar product of the spinor $\lambda_{\underline{\beta}}$ with a Grassmann spinor parameter $\kappa^{\underline{\beta}}(\xi)$, which is always possible,

$$\kappa(\xi) = 2\lambda_{\underline{\beta}}\kappa^{\underline{\beta}}(\xi)\,, \tag{17}$$

and to substitute (17) into the θ–variation (14). We thus get

$$\delta\theta^{\underline{\alpha}} = -2\lambda^{\underline{\alpha}}\lambda_{\underline{\beta}}\kappa^{\underline{\beta}}(\xi)\,. \tag{18}$$

We now note that, due to the superembedding condition (11), the bilinear combination of λ in (18) is nothing but

$$\Pi^{\underline{\alpha}}{}_{\underline{\beta}} = -2\lambda^{\underline{\alpha}}\lambda_{\underline{\beta}} = (\partial_\xi x^{\underline{m}} - i\partial_\xi\bar{\theta}\Gamma^{\underline{m}}\theta)(\Gamma_{\underline{m}})^{\underline{\alpha}}{}_{\underline{\beta}}\,, \tag{19}$$

which is the projector matrix in the κ–symmetry variation of θ (7). Hence, the local supersymmetry variations (16) and (18) reduce to the κ–variations (7).

We have thus demonstrated how, in virtue of the superembedding condition (9), the κ–symmetry of the GS formulation of the superbranes shows up from the irreducible local worldvolume supersymmetry.

One might have already noticed the difference in sign in the target–space supersymmetry variations of $x^{\underline{m}}$ (6) and in the worldvolume supersymmetry variations (16) and corresponding κ–variations (7). The target–space supersymmetry and the local worldvolume supersymmetry (or κ–symmetry) can be therefore regarded as, respectively, 'left' and 'right' supertranslations of $x^{\underline{m}}$.

The Geometrical Meaning of the Superembedding Condition

To understand the superembedding condition from the geometrical point of view let us note that the left hand side of (9) is the Grassmann component of the pull–back onto the superworldline of the target–space vector supervielbein one–form $E^{\underline{a}} = dX^{\underline{a}} - id\bar{\Theta}\Gamma^{\underline{a}}\Theta$

$$E^{\underline{a}}|_{\mathcal{M}_{sw}} = (d\xi + i\eta d\eta)(\partial_\xi X^{\underline{m}} - i\partial_\xi\bar{\Theta}\Gamma^{\underline{m}}\Theta) + d\eta(DX^{\underline{m}} - iD\bar{\Theta}\Gamma^{\underline{m}}\Theta), \tag{20}$$

where $d\xi + i\eta d\eta$ and $d\eta$ form a basis of the supercovariant one–forms (supervielbeins) on the superworldline.

We see that the superembedding condition (9) requires that the pullback of the target–space *vector* supervielbein $E^{\underline{a}}$ vanishes along the *Grassmann* directions of the superworldvolume.

This is the generic requirement for the superembedding to be appropriate to the description of the dynamics of the superbranes.

In general, if we take a supersurface $\mathcal{M}_{sw}$, whose geometry is described by supervielbein one–forms $e^a(z^M)$ $(a = 0, 1, \cdots, p)$ and $e^\alpha(z^M)$ $(\alpha = 1, \cdots n)$, and a curved target superspace M_{TS}, whose geometry is described by supervielbein one–forms $E^{\underline{a}}(Z^{\underline{M}})$ $(\underline{a} = 0, 1, \cdots, D-1)$ and $E^{\underline{\alpha}}(Z^{\underline{M}})$ $(\underline{\alpha} = 1, \cdots 2n)$[3], and consider the embedding of $\mathcal{M}_{sw}$ into M_{TS}, then for the superembedding to describe a super–p–brane propagating in M_{TS} the pull–back of E^a along the Grassmann directions of $\mathcal{M}_{sw}$ must vanish, i.e. in

$$E^{\underline{a}}|_{\mathcal{M}_{sw}} = e^{\underline{a}}E_a{}^{\underline{a}} + e^{\underline{\alpha}}E_\alpha{}^{\underline{a}} \tag{21}$$

the Grassmann components are zero

$$E_\alpha{}^{\underline{a}}(Z(z)) = 0. \tag{22}$$

For the most of the superbranes (with some subtleties for the space filling and codimension one branes [21, 22, 12, 13]), the superembedding condition (22), accompanied by the M_{TS} and/or $\mathcal{M}_{sw}$ supergravity constraints, implies that

- the geometry of the superworldvolume $\mathcal{M}_{sw}$ is induced by its imbedding into M_{TS}, i.e. the $\mathcal{M}_{sw}$ supergravity on the brane is not propagative;
- the dynamics of the superbrane is subject to the standard constraints of the Green–Schwarz formulation, such as the Virasoro constraints and their fermionic counterparts;
- κ–symmetry is a particular form of worldvolume superdiffeomorphisms;
- the consistency of the superembedding condition results in the same 'brane scan' as that of the GS formulation.

In addition, when the number of the Grassmann directions of $\mathcal{M}_{sw}$ is 16 or higher, the integrability of the superembedding condition requires the worldvolume superfields to

---

[3] Note that the supergeometries of $\mathcal{M}_{sw}$ and M_{TS} should be that of corresponding supergravities, which implies that the torsions and curvatures of $\mathcal{M}_{sw}$ and M_{TS} are subject to appropriate supergravity constraints.

satisfy the dynamical equations of motion of the superbrane [23] [4]. It is in this way the covariant equations of motion of the M5–brane were obtained for the first time [7].

SUPEREMBEDDING ACTIONS

In the cases when the superembedding condition does not imply dynamical equations but only the kinematic constraints, one can construct (doubly supersymmetric) worldvolume superfield actions for corresponding superbranes. Several related methods have been proposed so far to construct the superembedding actions [2, 24, 25, 26, 27, 11, 12, 13].

For the massless superparticles the action is simply the integral over the superworldline of the product of the left hand side of the superembedding condition (22) with a Lagrange multiplier $P^{\alpha}_{\underline{a}}(z)$

$$S = \int d\xi d^n\eta P^{\alpha}_{\underline{a}}\, E^{\underline{a}}_{\alpha}(Z(z))\,. \tag{23}$$

In the case of the massive superparticles and superbranes to the action (23) one must add a second term which governs the dynamics of the superbrane and produces both the Nambu–Goto (or Dirac–Born–Infeld) term and the Wess–Zumino term of the GS formulation.

Let me explain the general structure of this second term with the example of a space filling D3–brane in $N=2$, $D=4$ superspace. In this case, the geometry of the worldvolume $\mathcal{M}_{sw}$ is that of chiral $N=1, D=4$ supergravity [12, 13].

The D3–brane couples to supergravity fields via the worldvolume pull–back of the Wess–Zumino four–form

$$\hat{C} = C_4 + F_2 \wedge C_2 + \frac{1}{2} F_2 \wedge F_2 C_0\,, \tag{24}$$

where F_2 is the field strength of the BI gauge field propagating on the D3–brane and C_p (p=0,2,4) are 'Ramond-Ramond' p–form fields.

To construct the D3–brane action as an integral over the chiral $N=1$, $D=4$ superspace $\mathcal{M}_{sw}$ parametrized by $z^M_L = (\xi^m_L, \eta^{\alpha})$ $(\alpha = 1,2)$ we take a component of the pull–back of (24) onto $\mathcal{M}_{sw}$ which has the appropriate dimensionality. This is $\hat{C}_{\dot{\alpha}\dot{\beta}ab}$. We then contract $\hat{C}_{\dot{\alpha}\dot{\beta}ab}$ with the antisymmetric product of the Pauli matrices $(\sigma^{ab})^{\dot{\alpha}\dot{\beta}}$. The D3–brane action (accompanied by the superembedding term (23)) is

$$S = \int d^4\xi_L d^2\eta\; \mathcal{E}_L (\sigma^{ab})^{\dot{\alpha}\dot{\beta}} \hat{C}_{\dot{\alpha}\dot{\beta}ab} \;+\; h.c.\,, \tag{25}$$

where $\mathcal{E}_L$ is a chiral superspace integration measure (see [12]).

Upon solving for the superembedding condition, integrating over η and eliminating auxiliary fields one can get from this simply–looking superembedding action the D3–brane action of the GS formulation. It is still a problem for future study to demonstrate

---

[4] This is similar to the case of, say, higher dimensional super–Yang–Mills and supergravity theories whose superfield constraints produce the dynamical equations of motion.

how, in the static gauge, the action (25) is related to the super field DBI action discussed in [29, 14, 15].

As a final remark to this section we note that when the superembedding condition (22) produces dynamical equations of motion and the worldvolume superfield actions cannot be constructed there exist geometrically well–grounded recipes [28, 27] of how to construct Green–Schwarz–type actions for corresponding superbranes.

CONCLUSION

We have tried to describe in simple terms basic features of the superembedding approach, which has not only allowed one to explain and clarify various classical and quantum properties of superstring and superbrane theory, but has also found applications in the construction and description of new superbrane models, and of field theories with partially broken supersymmetry, as well as for solving practical problems. For instance, recently it has been used for calculating vertex operators in (M2–M5)–brane systems [30].

One may expect the superembedding methods to be also useful for other purposes, such as a unified (S–duality) description of fundamental and solitonic extended objects [31], and, in particular, for a covariant description of N coincident Dp-branes and corresponding supersymmetric non–Abelian Dirac–Born–Infeld theories.

ACKNOWLEDGMENTS

I am grateful to the Organizers of the XXXVII Karpacz Winter School for their warm hospitality, and would like to thank I. Bars, E. Bergshoeff, M. Cederwall, J. de Azcarraga, E. Ivanov, S. Krivonos, J. Lukierski, A. Pashnev, P. Pasti and M. Tonin for interest to this work and discussions. This work was partially supported by the European Commission RTN Programme HPRN-CT-2000-00131 and by the Grant N 2.51.1/52-F5/1795-98 of the Ukrainian Ministry of Science and Technology.

REFERENCES

1. Sorokin, D., *Physics Reports*, **329**, 1–101 (2000).
2. Sorokin, D., Tkach, V. and Volkov, D., *Mod. Phys. Lett.* **A4**, 901 (1989).
3. Berkovits, N., *Nucl. Phys.* **B358**, 169 (1991); *Ibid* **B379**, 96 (1992); *Ibid* **B420**, 332 (1994); Covariant Quantization of the Superstring, hep-th/0008145; Relating the RNS and Pure Spinor Formalisms for the Superstring, hep-th/0104247.
4. Volkov, D. V. and Zheltukhin, A., *Lett. Math. Phys.* **17**, 141 (1989); *Nulc. Phys.* **B335**, 723 (1990).
   Sorokin, D., Tkach, V., Volkov, D. V. and Zheltukhin, A., *Phys. Lett.* **B216**, 302 (1989).
   Aoyama, S., Pasti, P. and Tonin, M., *Phys. Lett.* **B283**, 213 (1992);
   Uvarov, D., Covariant κ–Symmetry Gauge Fixing and the Classical Relation Between Physical Variables of the NSR String and the Type II GS Superstring, hep-th/0104235.
5. Ferber, A., *Nucl. Phys.* **132**, 55 (1977); Shirafuji, T., *Progr. Theor. Phys.* **70**, 18 (1983);
   Bengtsson, A. K. H., Bengtsson, I., Cederwall, M. and Linden, N., *Phys. Rev.* **D36**, 1766 (1987);

Eisenberg, Y. and Solomon, S., *Nucl. Phys.* **B309**, 709 (1988);
Plyushchay, M. S., *Phys. Lett.* **B240**, 133 (1990).
6. Nissimov, E., Pacheva, S. and Solomon, S., *Nucl. Phys.* **B296**, 469 (1988);
Bandos, I. A., *Sov. J. Nucl. Phys.* **51**, 906 (1990); *JETP. Lett.* **52**, 205 (1990).
7. Howe, P. S. and Sezgin, E., *Phys. Lett.* **B394**, 62 (1997); Howe, P. S., Sezgin, E. and West, P. C., *Phys. Lett.* **B399**, 49 (1997).
8. Bandos, I., Lechner, K., Nurmagambetov, A., Pasti, P., Sorokin, D. and Tonin, M., *Phys. Lett.* **B408**, 135 (1997).
9. Pasti, P., Sorokin, D. and Tonin, M., *Phys. Lett.* **B398**, 41 (1997);
Bandos, I., Lechner, K., Nurmagambetov, A., Pasti, P., Sorokin, D. and Tonin, M., *Phys. Rev. Lett.* **78**, 4332 (1997);
Aganagic, M., Park, J., Popescu, C. and Schwarz, J. H., *Nucl. Phys.* **B496**, 191 (1997).
10. Hughes, J., Liu, J. and Polchinski, J., *Phys. Lett.* **B180**, 370 (1986); Hughes, J. and Polchinski, J., *Nucl. Phys.* **B278**, 147 (1986);
Achucarro, A., Gauntlett, J., Itoh, K., and Townsend, P. K., *Nucl. Phys.* **B314**, 129 (1989).
11. Pasti, P., Sorokin, D. and Tonin, M., *Nucl. Phys.* **B591**, 109 (2000); Geometrical aspects of superbrane dynamics, hep-th/0011020.
12. Bandos, I., Pasti, P., Pokotilov, A., Sorokin, D. and Tonin, M., The space filling Dirichlet 3-brane in N = 2, d = 4 superspace, hep-th/0103152.
13. Drummond, J. M. and Howe, P. S., Codimension zero superembeddings, hep-th/0103191.
14. Bagger, J. and Galperin, A., *Phys. Rev.* **D55**, 1091 (1997).
Roček, M. and Tseytlin, A., *Phys. Rev.* **D59**, 106001 (1999).
Ivanov, E., and Krivonos, S., *Phys. Lett.* **B453**, 237 (1999).
Kapustnikov, A. and Shcherbakov, A., Linear and nonlinear realizations of superbranes, hep-th/0104196.
15. Bellucci, S., Ivanov, E. and Krivonos, S., *Phys. Lett.* **B482**, 233 (2000). Superbranes and Super Born-Infeld Theories from Nonlinear Realizations, hep-th/0103136.
16. Brink, L., Deser, S., Zumino, B., Di Vecchia, P. and Howe, P., *Phys. Lett.* **B64**, 435 (1976);
Gershun, V. D. and Tkach, V. I., *JETP Letters* **29**, 320 (1979);
Howe, P. S., Penati, P., Pernici, M. and Townsend, P., *Phys. Lett.* **B215**, 555 (1988).
17. Neveu, A. and Schwarz, J. H., *Nucl. Phys.* **B31**, 86 (1971); Ramond, P., *Phys. Rev.* **D3**, 2415 (1971).
18. Howe, P. S. and Tucker, R. W., *J. Phys.* **A10**, L155 (1977).
19. De Azcarraga, J. and Lukierski, J., *Phys. Lett.* **B113**, 170 (1982);
Siegel, W., *Phys. Lett.* **B128**, 397 (1983).
20. Gates Jr., S. J. and Nishino, H., *Class. Quantum Grav.* **3**, 391 (1986).
Kowalski-Glikman, J., van Holten, J. W., Aoyama, S. and Lukierski, J. *Phys. Lett.* **B201**, 487 (1987).
Kavalov A., and Mkrtchyan, R. L., Spinning superparticles. Preprint Yer.PhI 1068(31)-88, Yerevan, 1988 (unpublished).
21. Howe, P. S., Kaya, A., Sezgin, E. and Sundell, P., *Nucl. Phys.* **B587**, 481 (2000).
22. Akulov, V., Bandos, I., Kummer W., and Zima, V., *Nucl. Phys.* **B527**, 61 (1998).
23. Galperin, A. and Sokatchev, E., *Phys. Rev.* **D48**, 4810 (1993);
Bandos, I., Pasti, P., Sorokin, D., Tonin, M. and Volkov, D., *Nucl. Phys.* **B446**, 79 (1995);
Howe, P. S. and Sezgin, E., *Phys. Lett.* **B390**, 133 (1997).
24. Tonin, M., *Phys. Lett.* **B266**, 312 (1991); *Int. J. Mod. Phys.* **A7**, 6013 (1992).
25. Ivanov, E. A. and Kapustnikov, A. A. *Phys. Lett.* **B267**, 175 (1991);
Delduc, F. and Sokatchev, E., *Class. Quantum Grav.* **9**, 361 (1992);
Pashnev, A. and Sorokin, D., *Class. Quant. Grav.* **10**, 625 (1993);
Pashnev, A. and Chikalov, V., *Mod. Phys. Lett.* **A8**, 285 (1993); *Phys. Rev.* **D50**, 7450 (1994).
26. Galperin, A. and Sokatchev, E., *Phys. Rev.* **D46**, 714 (1992); Delduc, F., Galperin, A., Howe, P. S., and Sokatchev, E., *Phys. Rev.* **D47**, 587 (1992).
27. Howe, P. S., Raetzel, O. and Sezgin, E., *JHEP* **9808**, 011 (1998).
28. Bandos, D., Sorokin, D. and Volkov, D., *Phys. Lett.* **B352**, 269 (1995). n
29. Cecotti, S. and Ferrara, S., *Phys. Lett.* **B187**, 335 (1987).
30. Moore, G., Peradze, G. and Saulina, N., Instabilities in heterotic M-theory induced by open membrane instantons, hep-th/0012104.
31. Bandos, I., *Nucl. Phys.* **B599**, 197 (2001).

Revisiting Supergravity and Super Yang-Mills Renormalization

K.S. Stelle

Department of Physics, Imperial College, London SW7 2BW, UK,
email: k.stelle@ic.ac.uk

Abstract. Standard superspace Feynman diagram rules give one estimate of the onset of ultraviolet divergences in supergravity and super Yang-Mills theories. Newer techniques motivated by string theory but which also make essential use of unitarity cutting rules give another in certain cases. We trace the difference to the treatment of higher-dimensional gauge invariance in supersymmetric theories that can be dimensionally oxidized to pure supersymmetric gauge theories.

NON-RENORMALIZATION THEOREMS

The problem we shall consider in this article[1] is to reconcile the predictions for the onset of ultraviolet divergences in supergravity and super Yang-Mills theories [1]made from traditional superspace Feynman diagram analysis and those made on the basis of unitarity and factorization implied by embedding in string theory [10]. The question of ultraviolet divergences is one of the oldest concerns in the subject of quantum gravity. Although the problem of order-by-order perturbative finiteness has been solved in superstring theory, the question remains of interest because superstring theory acts as a physical regulator for its limiting supergravity theory, so the orders at which various operators become relevant to supergravity divergences gives information about the presence of the same operator with a finite coefficient in superstring theory.

Many problems of current interest involve the *scaling* behavior of quantum corrections, which is determined by how strongly they are renormalized. And the verification of duality conjectures makes detailed use of supergravity radiative corrections. And a number of mysteries have arisen: some correlation functions appear to be mysteriously "protected" for reasons that are not yet understood. So, in this context it is important to get to the bottom of any disagreements between different viewpoints on the ultraviolet problem.

One good way to approach the study of non-renormalization theorems is through the background-field method. A classic example of a non-renormalization theorem is the Adler-Bardeen theorem [2]. This may be understood from the standpoint of the background field method [3] as follows. Consider the Yang-Mills Chern-Simons operator

[1] This article is based upon work done in collaboration with P.S. Howe and M. Petrini.

CP589, *Fundamental Interaction Theories*, edited by J. Lukierski and J. Rembieliński

under a background-quantum split $A^r_a = \mathcal{B}^r_a + Q^r_a$:

$$K^a = 4g^2\varepsilon^{abcd}[A^r_b(\partial_c A^r_d + \tfrac{1}{3}gf^{rst}A^s_cA^t_d)]\,. \tag{1}$$

This operator has the property that its divergence gives the Pontryagin density: $\partial_a K^a = 2g^2 F^r_{ab}{}^*F^{rab}$, but it is not itself gauge invariant. It's renormalization properties, however, are important in the context of anomalies. When one expands it in powers of the quantum field Q^r_a, the lowest term is just the expression (1) with $\mathcal{B}^r_a$ substituted for A^r_a, so it is no more background gauge invariant than (1). Nor is the term linear in Q^r_a invariant under background gauge transformations. But the term quadratic in Q^r_a can be written $4g^2\varepsilon^{abcd}Q^r_aD_b(\mathcal{B})Q^r_c$, which is fully background gauge invariant. Similarly, the term cubic in Q^r_a is background invariant, since Q^r_a transforms as a tensor under the background transformations.

The above structure for K^a has the striking consequence that although K^a itself is not gauge invariant, the terms actually used in calculating one-particle-irreducible (1PI) Feynman diagrams with operator insertions of K^a are actually background gauge invariant. Accordingly, the renormalization of this operator can only be through gauge invariant operators, of which class it is not itself a member. Consequently, neither K^a itself nor does any gauge-invariant operator O mix with it under renormalization, and so for the anomalous dimensions one has $\gamma_{KK} = \gamma_{OK}=0$.

The Adler-Bardeen theorem then arises as follows. In the presence of axial anomalies, the axial current j^5_a is not conserved; one has the famous relation $\partial_a J^{5a} = 2g^2cF^r_{ab}{}^*F^{rab}$, where c is the anomaly coefficient, first occurring at the one-loop level. However, there is a non-gauge-invariant current $j'^5_a = J^5_a - cK_a$ that is conserved. Since conserved operators are not renormalized, one has the renormalization group equation $\mu\frac{\partial}{\partial\mu}j'^5_a = 0$. In consequence, and taking into account the K^a anomalous dimensions from above, $\gamma_{KK} = \gamma_{jK} = 0$ together with independence of the operators j^5_a and K_a, one finds

$$\gamma_{jj} - c\gamma_{Kj} = 0 \tag{2}$$

$$\mu\frac{\partial c}{\partial\mu} = 0\,; \tag{3}$$

and (3) is just the Adler-Bardeen theorem.

We may draw from this brief review the moral that "vestigial" consequences of gauge invariance, such as those revealed by the above use of the background field method, can give rise to important non-renormalization properties, such as the Adler-Bardeen theorem.

Now let us review the basic non-renormalization theorem [4] of $N=1$ supersymmetry, also from the point of view of the background field method. We consider the basic Wess-Zumino model, based upon a chiral superfield ϕ; $\bar{D}_{\dot\alpha}\phi = 0$. In the background field method, one again makes a background-quantum split, $\phi = \varphi + Q$. The background superfield φ then appears only on the external legs of 1PI Feynman diagrams used in calculating the effective action $\Gamma(\varphi)$, while the quantum field Q occurs on the inside lines of diagrams.

The derivation of the Feynman rules for the Q lines may be carried out either taking particular care with the functional delta functions arising from variational derivatives with respect to chiral superfields, or one may solve the chirality constraint for Q and work henceforth with unconstrained general scalar superfields. Adopting the second approach, one writes $Q = \bar{D}^2 X$, where $\bar{D}^2 = \bar{D}^{\dot{\alpha}}\bar{D}_{\dot{\alpha}}$. Expanding then the action into background and quantum fields, one finds that all terms except for the $O(X^0)$ terms can now be re-written as integrals over the full superspace. Thus, *e.g.*, a mass term decomposes as $m\int d^4xd^2\theta\phi^2 \longrightarrow m\int d^4xd^2\theta\varphi^2 + 2m\int d^4xd^4\theta\varphi X + \int d^4xd^4\theta X\bar{D}^2X$.

The vertices and propagators for the quantum X fields used in 1PI diagrams are derived from terms containing 2 or more Q fields, expanded as above and written as full superspace integrals. On the other hand, the background φ superfields continue to appear always as constrained, chiral, superfields. Hence the non-renormalization theorem for $D = 4$, $N = 1$ supersymmetry arises: perfectly supersymmetric invariants such as chiral superspace integrals over prepotentials may occur in the original action of the theory, but they may not occur as counterterms. From the structure of the Feynman rules, all quantum corrections to the effective action Γ must be written as full superspace integrals. Thus, chiral superspace integrals like $\int d^4xd^2\theta f(\phi)$ are not renormalized. Note that one cannot try to frustrate the strictures of this non-renormalization theorem by writing chiral superspace integrals in forms like $\int d^4xd^4\theta\phi\Box^{-1}D^2\phi$, because the operator $\Box^{-1}$ is spatially non-local, and this then is not in accord with the requirements of Weinberg's theorem [5] on the locality of counterterms.

For extended supersymmetries that can be given a complete linear realization in superspace, similar results are found. In particular, for matter fields such as the $N = 2$ hypermultiplet, one directly obtains a similar result: hypermultiplet self-interactions are not renormalized [6].

For gauge multiplets and supergravity multiplets, there is a further subtlety, which affects however only the one-loop diagrams. As in the case of the $N = 1$ chiral multiplet, one introduces prepotentials for the quantum fields that appear on internal lines of diagrams, while background fields appear through constrained superfields, the original superspace gauge connections. The procedure of gauge fixing and introduction of the ghost actions, however, introduces some complications. Superspace gauge symmetry parameters are described by constrained superfields, *e.g.* the chiral gauge superfield Λ for $N = 1$, $D = 4$ super Yang-Mills theory, which is constrained in the background field method by a background-covariant constraint $\bar{\mathcal{D}}_{\dot{\alpha}}(\mathcal{B})\Lambda = 0$. In the construction of the corresponding ghost action for a chiral ghost superfield written with the chirality constraint solved *via* the introduction of a prepotential, *e.g.* $\bar{\mathcal{D}}^2 S$, one finds that S has its own gauge invariance $\delta S = \bar{\mathcal{D}}^{\dot{\alpha}}\bar{\Lambda}_{\dot{\alpha}}$. This requires a new gauge fixing and ghost action, which in turn has a new gauge invariance with chiral parameter $\bar{\Lambda}_{\dot{\alpha}\dot{\beta}}$. This process continues indefinitely, with further gauge parameters $\bar{\Lambda}_{\dot{\alpha}\dot{\beta}\dot{\gamma}}$, *etc*, and these will all couple to the background fields unless one cuts off the interactions with the background at some order. This necessitates the introduction of a background gauge field prepotential, thus breaking the non-renormalization theorem, but only for the one-loop diagrams.

DIVERGENCE ESTIMATES FOR EXTENDED SUPERSYMMETRY

Theories with extended supersymmetry, and in particular the maximal supergravity and super Yang-Mills theories, face another problem in estimating the onset of ultraviolet divergences: not all of a given theory's supersymmetry can be linearly realized "off-shell," *i.e.* with an algebra that closes without use of the equations of motion. This is an old problem in supersymmetric theories, whose full implications are still to this day not clear. In analyzing the possibilities for infinities, one can choose between several alternatives for the "next best thing." For example, one can see what is the maximal degree of supersymmetry that can be linearly realized off-shell and then use that, keeping at the same time full Lorentz invariance [7]. Or, alternatively, one can keep the full automorphism symmetry that appears in the on-shell formalism without auxiliary fields but sacrifice Lorentz invariance, *e.g.* through use of a light-cone formalism [8]. Similar conclusions are obtained in either formalism; however, the Lorentz-covariant approach has been used to study more cases. The upshot is that for $N = 4 \leftrightarrow 3, 2, 1$ super Yang-Mills one can use formalisms with $M = 2, 2, 1$ off-shell supersymmetry. For $N = 8 \leftrightarrow 7, 6, 5, 4, 3, 2, 1$ supergravity theories one can use formalisms with $M = 4, 3, 3, 2, 2, 2, 1$ off-shell supersymmetry.

Putting all of this together, one has a general supersymmetric non-renormalization theorem: *For gauge and supergravity multiplets at loops $\ell \geq 2$, and for matter multiplets at all loop orders, counterterms must be written as full superspace integrals for the maximal off-shell linearly realizable supersymmetry, with background gauge invariant integrands written without using prepotentials for the background fields.*

Although the full power of a given theory's supersymmetry cannot always be used to determine the structure of counterterms from supersymmetric Ward identities – only the linearly realized supersymmetry gives useable Ward identities in general – the full supersymmetry can still play a rôle. Things simplify greatly if one considers the counterterms subject to the classical field equations. For the first non-vanishing one of these, the Ward identities for the full (nonlinear) supersymmetry requires full supersymmetry invariance, subject to the imposition of the classical field equations [9]. There is no requirement, however, on being able to write the counterterm as a full superspace integral for the full nonlinear supersymmetry. Nonetheless, this requirement does serve to rule out some counterterms that would by themselves be acceptable according to the above theorem, but whose coefficients become linked *via* the full supersymmetry to the coefficients of terms disallowed by the above theorem.

Of course, the above considerations cannot say exactly when the first divergences can occur in a given theory – the most they can say is up to which loop orders the divergences cannot. Nonetheless, in the study of divergences one has become used to the maxim that if something isn't ruled out, it will occur. So any further cancellations would appear "miraculous" from the standpoint of the above analysis. Here is a table of the expectations for divergences in maximal (*i.e.* $N = 1$ in $D = 10$ or $N = 4$ in $D = 4$) super Yang-Mills theory:

and in maximal (*i.e.* $N = 1$ in $D = 11$ or $N = 8$ in $D = 4$) supergravity:

TABLE 1. Super Yang-Mills divergence expectations, standard Feynman rules.

Dimension	10	8	7	6	5	4
loop L	1	1	2	3	4	∞
gen. form	$\partial^2 F^4$	F^4	$\partial^2 F^4$	$\partial^2 F^4$	F^4	finite

TABLE 2. Supergravity divergence expectations, standard Feynman rules.

Dimension	11	10	8	7	6	5	4
loop L	2	1	1	2	2	2	3
gen. form	$\partial^6 R^4$	$\partial^2 R^4$	R^4	$\partial^4 R^4$	$\partial^2 R^4$	R^4	R^4

STRING-INSPIRED CALCULATIONS VIA CUTTING RULES

Traditional Feynman diagram techniques quickly become unmanageable. For example, a 5-loop diagram in $N = 8$, $D = 4$ supergravity may have on the order of 10^{30} terms in its algebraic expression. So a better procedure is clearly desirable.

Just such a procedure (let us call it the BD^{2}PR procedure [10]) has been offered in recent years, using factorization properties of supergravity amplitudes inherited from string theory, coupled with a procedure for extracting the ultraviolet divergences from the absorptive parts of amplitudes, using the cutting rules. The key elements of this procedure are:

1) the optical theorem
2) dimensional regularization

Specifically, from the optical theorem, one determines the imaginary part of a one-loop amplitude in terms of a product of tree amplitudes, for example. One might worry that by this procedure one looses information about the real parts of amplitudes. But this is avoided by a careful use of dimensional regularization. The optical theorem works best for amplitudes containing logarithms, where one can clearly see the relation between real and imaginary parts, *e.g.* $\ln(s) = \ln(|s|) + i\pi\theta(s)$. This can be enhanced by dimensional regularization, which produces many additional logarithms, *e.g.* $s^{-2\varepsilon} = 1 - 2\varepsilon\ln(s) + \ldots$, where $\varepsilon = D - D_{\text{physical}}$; these help to determine amplitudes providing one calculates diagrams factorized using the cutting rules *to all powers in* ε. The remaining essentials in the BD^{2}PR procedure are

3) field-theory amplitudes result from string amplitudes in the $\alpha' \to 0$ limit
4) the KLT relations [11] between closed and open string tree-level S-matrices.

As a combined example of the two last points, one has a relation between supergravity and super Yang-Mills 4-point amplitudes, arising from the $\alpha' \to 0$ limit of superstring theory, where M_4^{tree} is a supergravity amplitude and A_4^{tree} is a SYM amplitude:

$$M_4^{\text{tree}}(1,2,3,4) = -is_{12}A_4^{\text{tree}}(1,2,3,4)A_4^{\text{tree}}(1,2,3,4)\,. \tag{4}$$

Using the KLT relations, one may organize supergravity amplitudes and super Yang-Mills amplitudes to display a squaring relationship. Moreover, amplitude calculations with irreducible supermultiplets reduce to scalar amplitudes multiplied by index-bearing factors. This is illustrated in Fig. 1.

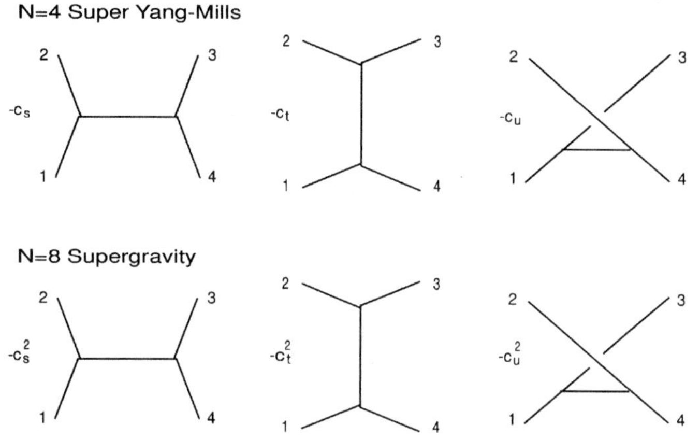

FIGURE 1. Tree-level color-ordered $N = 4$ SYM and $N = 8$ SG amplitudes expressed in terms of ϕ^3 diagrams. Supergravity coefficients are squares of the SYM coefficients.

The relations between supergravity and super Yang-Mills amplitudes continues on to higher loops, enabling SG and SYM amplitudes to be calculated in terms of scalare diagrams. A one-loop example is illustrated in Fig. 2

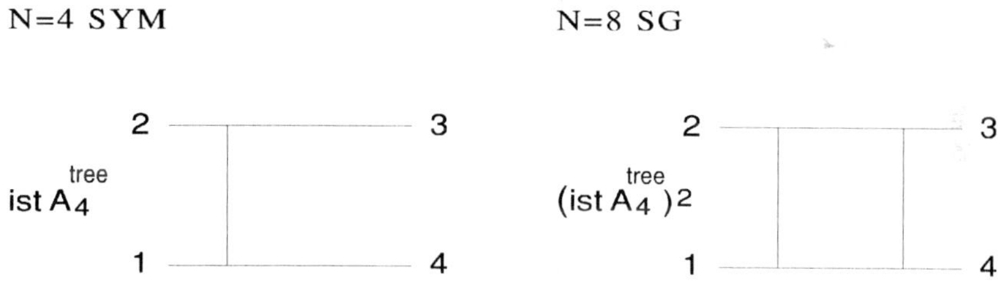

FIGURE 2. Relation between $N = 4$ SYM theory and $N = 8$ SG amplitudes at the one-loop level. Tree amplitudes multiply scalar field loop diagrams.

Using these techniques, BD^{2}PR obtain the following predictions for the first on-shell divergences in the maximal super Yang Mills theories:

TABLE 3. Maximal super Yang-Mills divergence expectations from cutting rules.

Dimension	10	8	7	6	5	4
loop L	1	1	2	3	6	∞
gen. form	$\partial^2 F^4$	F^4	$\partial^2 F^4$	$\partial^2 F^4$	$\partial^2 F^4$	finite

and in maximal supergravity they obtain:

TABLE 4. Maximal supergravity divergence expectations from cutting rules.

Dimension	11	10	8	7	6	5	4
loop L	2	1	1	2	3	4	5
gen. form	$\partial^6 R^4$	$\partial^2 R^4$	R^4	$\partial^4 R^4$	$\partial^6 R^4$	$\partial^6 R^4$	$\partial^4 R^4$

So one can see that the BD^{2}PR results appear to be predicting a "miracle" in the onset of divergences in the maximal SYM and SG theories. One may summarize their results for the two maximal theories by the rule that the maximal SYM theories are finite for $D < \frac{6}{L} + 4$ ($L > 1$) and the maximal supergravity theories are finite for $D < \frac{10}{L} + 2$ ($L > 1$). A striking aspect of this improvement is that it applies only to the maximal supersymmetric theories. The cutting-rule analysis of $N \leq 6$ supergravity in $D = 4$ shows the onset of divergences to be exactly as predicted on the basis of standard Feynman diagram analysis, equal to those summarized for the maximal theories in tables 1 and 2.

Of course, there is strictly speaking no firm contradiction here, for the standard Feynman diagram analysis gave only a lower bound on the possible loop order for the first on shell surviving infinities. Nonetheless, the difference is striking. The BD^{2}PR results are in fact more than an estimate. They give an actual calculation of the coefficients of the various divergences, limited only in that for loops $L > 2$ only two-particle cuts have been considered; a full calculation would require general m particle cuts.

COUNTERTERM ANALYSIS

The study of counterterms for super Yang-Mills and supergravity theories has been going on since the discovery of these theories. Pure Einstein gravity diverges at the two-loop order with an R^3 counterterm, as found by explicit calculations [12]. In supergravity theories, on-shell supersymmetry rules out all on-shell surviving counterterms at the $L = 1$ and $L = 2$ loop orders. In particular, this dramatically causes the cancellation of the R^3 counterterm that is present in pure Einstein gravity.

The first dangerous supergravity counterterm occurs at $L = 3$ loops in $D = 4$, and has an R^4 generic structure. This counterterm involves the curvature through the square of the Bel-Robinson tensor

$$T_{\mu\nu\rho\sigma} = R_{\mu\alpha\nu\beta} R^{\alpha}{}_{\rho}{}^{\beta}{}_{\sigma} + {}^*R_{\mu\alpha\nu\beta}{}^*R^{\alpha}{}_{\rho}{}^{\beta}{}_{\sigma} \,. \tag{5}$$

This tensor is a kind of analogue of the stress tensor for gravity: it is totally symmetric and totally traceless (for pure gravity) and it is covariantly conserved on any index. In an analogy to the supersymmetric $T_{\mu\nu}T^{\mu\nu}$ counterterms that exist for matter systems in a gravitational background, pure supergravity has a $T_{\mu\nu\rho\sigma}T^{\mu\nu\rho\sigma}$ + fermionic counterterm [13], appropriate by standard power counting for the $L = 3$ order of perturbation theory. The existence of similar counterterms was also found for all $D = 4$ extended supergravities [14, 15, 16]. These same expressions occur as important corrections with finite coefficients to superstring theories; the question in that case is also at which loop order they first appear.

Consider now in some more detail the structure of the R^4 counterterm for the higher $N \geq 4$ supergravities in $D = 4$ [16]. The on-shell supergravity multiplet is described by a superfield $W_{[ijkl]}$ carrying automorphism group $SU(N)$ indices and satisfying the full on-shell superspace constraints

$$
\begin{aligned}
D^i_\alpha W_{jklm} &= \frac{-4}{N-3}\delta^i_{[j} D^n_\alpha W_{klm]n} && (6)\\
\bar{D}_{\dot{\alpha} i} W_{jklm} &= \bar{D}_{\dot{\alpha}[i} W_{jklm]}\,. && (7)
\end{aligned}
$$

At the linearized level, which all that we need concern ourselves with here, each component of W_{ijkl} is a field F with $2s$ symmetrized spinor indices of the same chirality. The constraints imply $\partial^{\alpha_1}_{\dot{\alpha}} F_{\alpha_1\cdots\alpha_{2s}} = \Box F_{\alpha_1\cdots\alpha_{2s}} = 0$. These on-shell component field strengths describe massless spin s for $s \leq 2$, including $N(N-1)(N-2)(N-3)/12$ $s = 0$ scalars for $N \leq 7$, and 70 scalars for the self-conjugate $N = 8$ theory. All of these scalars arise from dimensional reduction of the metric and the higher-dimensional gauge field $A_{\mu\nu\rho}$.

The R^3 counterterm is written for the $N = 8$ theory as an on-shell superspace integral over a subset of 16 out of the 32 fermionic coordinates. We recall that the non-renormalization theorem requires that this counterterm (which is the first one non-vanishing on shell) be fully supersymmetric under the full (in general nonlinear, although here we consider only the linearized level) $N = 8$ supersymmetry, but that it also be possible to be expressed in off-shell $N = 4$ superfields in which no prepotentials are introduced and the counterterm is written as a full superspace integral. The $\int d^{16}\theta$ structure of the R^4 counterterm allows this to happen. In linearized superspace, the structure of the on-shell counterterm is [16]

$$
\Delta\Gamma = \kappa^4 \int d^4x (d^{16}\theta)_{232848} (W^4)_{232848} \qquad (8)
$$

where the 232848 representation of $SU(8)$ is described by the 4×4 square Young tableau with four rows of four boxes each. One can also view this as a superinvariant built from the 232848 rep. extracted from the quadratic product of two $(W^2)_{1764}$ superfields, where the 1764 rep. is described by two columns of 4 boxes. This is the multiplet that contains the Bel-Robinson tensor, which in 2-component indices is $T_{\alpha\dot{\alpha},\beta\dot{\beta},\gamma\dot{\gamma},\delta\dot{\delta}} = C_{\alpha\beta\gamma\delta} C_{\dot{\alpha}\dot{\beta}\dot{\gamma}\dot{\delta}}$, where $C_{\alpha\beta\gamma\delta}$ is the Weyl tensor in two-component notation. Since this counterterm satisfies all the requirements of the non-renormalization theorem discussed earlier.

GAUGE INVARIANCE AND THE DISCREPANCY IN DIVERGENCE ESTIMATES

It is striking that the standard Feynman diagram analysis and the BD^{2}PR analysis agree in almost all cases, with the only disagreements coming for the maximal super Yang-Mills and supergravity theories. This provides a clue to what might be going on. Again, it seems to come down to a difference in the treatment of gauge invariance. Both of the theories for which there is a discrepancy have the property that they can be dimensionally oxidized to theories in which all the scalar fields have disappeared into components of higher-dimensional gauge fields, metrics, and so on. In particular, the $D = 4$, $N = 8$ supergravity theory oxidizes to $D = 11, N = 1$ supergravity, in which *all* fields are gauge fields, including all the spinors.

This observation implies that the origin of the difference in divergence estimates lies in the way gauge invariance is being treated. Indeed, if one adopts the rule that background fields must appear in the counterterm in a way that can also be oxidized to a gauge-invariant structure in the higher dimension, then the two divergence estimates agree [17]. This also works for the super Yang-Mills theories.

Another example that shows how the discrepancy between the two divergence estimates is removed by the revised rule for gauge invariance occurs in $D = 5$ maximal super Yang-Mills theory. This has 5 scalars, but all of them are absorbed upon oxidization of the theory to $D = 10, N = 1$ super Yang-Mills theory. Another example is $D = 5$, $N = 1$ SYM, which has a single scalar. This can, however, be absorbed into the gauge field upon oxidization to $D = 6, N = 1$ SYM theory.

This leaves us with a question, on which we are going to end this review of divergence problems and their attendant mysteries. *Are there circumstances in which higher-dimensional gauge symmetries may still effectively govern the structure of counterterms in lower-dimensional theories?* If so, then the standard superspace Feynman rules and the BD^2PR cutting-rule analysis give the same results. But how this comes about remains mysterious. There does not appear to be any direct way to preserve more of the higher-dimensional gauge symmetries than that which is manifest in the lower dimension upon dimensional reduction. The gauge transformations depending upon the compactified dimensions are the ones that would have to impart structure to the counterterms, but these are lost when one suppresses dependence upon the compactified coordinates, or sets to zero the corresponding loop momenta.

So we end with this question: do supersymmetry and the vestiges of superstring theory present in supersymmetric gauge theories conspire to keep alive the higher gauge symmetries? These are the most unified theories that we know, and it just might be that in whatever dimension one formulates them, they remember their higher-dimensional origins.

REFERENCES

1. Howe, P.S., and Stelle, K.S., *Int. J. Mod. Phys.* **A4**, 1871(1989).
2. Adler, S., and Bardeen, W.A., *Phys. Rev.* **182**, 1517 (1969).
3. Breitenlohner, P., Maison, D., and Stelle, K.S., *Phys. Lett.* **134B**, 63 (1984.
4. Grisaru, M.T., Siegel, W., and Roček, M., *Nucl. Phys.* **B159**, 429 (1979).
5. Weinberg, S., *Phys. Rev.* **118**, 838 (1959).
6. Howe, P.S., Stelle, K.S., and Townsend, P.K., *Nucl. Phys.* **B214**, 519 (1983).
7. Howe, P.S., Stelle, K.S., and Townsend, P.K., *Nucl. Phys.* **B236**, 125 (1984).
8. Mandelstam, S., *Nucl. Phys.* **B213**, 149 (1983);
   Brink, L., Lindgren, O., and Nilsson, B., *Phys. Lett.* **123B**, 328 (1983).
9. Howe, P.S., and Stelle, K.S., *Phys. Lett.* **137B**, 175 (1984).
10. Bern, Z., Dixon, L., Dunbar, D., Perelstein, M., and Rosowsky, J., *Nucl. Phys.* **B530**, 401 (1998).
11. Kawai, H., Lewellyn, D.C., and Tye, S.-H.H., *Nucl. Phys.* **B269**, 1 (1986).
12. Goroff, M.H., and Sagnotti, A., *Nucl. Phys.* **B266**, 709 (1986);
    van de Ven, A.E.M., *Nucl. Phys.* **B378**, 309 (1992).

13. Deser, S., Kay, J.H., and Stelle, K.S., *Phys. Rev. Lett.* **38**, 527 (1977).
14. Deser, S., and Kay, J.H., *Phys. Lett.* **76B**, 400 (1978).
15. Kallosh, R.E., *Phys. Lett.* **99B**, 122 (1981).
16. Howe, P.S., Stelle, K.S., and Townsend, P.K., *Nucl. Phys.* **B191**, 445 (1981).
17. Howe, P.S., Petrini, M., and Stelle, K.S., in preparation.

The Scalars of $N = 2, D = 5$ and Attractor Equations

A. Van Proeyen

Institute for Theoretical Physics K.U. Leuven Celestijnenlaan 200D, B-3001 Leuven, Belgium, email: Antoine.VanProeyen@fys.kuleuven.ac.be

Abstract. Theories in 5 dimensions with minimal supersymmetry are studied for domain-wall solutions and in the context of the AdS/CFT correspondence. The scalar manifold is a product of a very special real manifold and a quaternionic-Kähler manifold. Superconformal methods can clarify the structure of these manifolds, which are part of the family of special manifolds. BPS solutions depending on the scalars and a warp factor of the 5-dimensional metric with a flat 4-dimensional metric can interpolate between critical points determined by algebraic attractor equations. The mixing of vector and hypermultiplets is essential to obtain UV and IR critical points.

INTRODUCTION

Supersymmetric theories in 5 dimensions have got new interest in the context of the AdS/CFT correspondence and for a supersymmetrisation of the Randall–Sundrum (RS) scenario. In both cases one uses at the end a metric of the form

$$ds^2 = a(x^5)^2 dx^{\underline{\mu}} dx^{\underline{\nu}} \eta_{\underline{\mu\nu}} + (dx^5)^2, \tag{1}$$

where $\underline{\mu}, \underline{\nu} = 0,1,2,3$. We thus have a flat 4-dimensional space with a warp factor a that depends on the fifth direction x^5. We first want to draw the attention on two different concepts, which are called either 'smooth solutions' or 'singular sources'. With 'smooth solutions' we [1] mean that we consider the generic 5-dimensional supergravity theory [2], and we look for a solution where the warp factor has the required form as explained above. On the other hand, 'singular sources' means that 3-brane sources are inserted at specific places. The fifth dimension is in that case an orbifold $S^1/\mathbb{Z}_2$, and the sources sit at its fixed points. The 5-dimensional bulk action is supplemented by a brane term that involves delta functions $\delta(x^5 - x^5_{\text{fixed}})$ times a four-dimensional action. We [3] provided a simple general mechanism how to implement supersymmetry in such a scenario despite the singularities in spacetime due to the delta functions. It introduces a new 4-form field in the bulk supergravity, that appears also as 4-form for the brane Wess–Zumino term. This construction has been reviewed also in [4] and we will omit therefore this part of the talk from these proceedings. The mechanism inspired a generalization to 10 dimensions, leading to new formulations of type IIA (and also type IIB) supergravity in [5], as reviewed in the talk of Eric Bergshoeff in this school. This setup is a similar to the Hořava–Witten [6] theory, which was reduced to 5 dimensions in [7]. It has been considered by various groups [8, 9, 10, 11], and our bulk& brane solution was used for cosmology in [12, 13].

CP589, *Fundamental Interaction Theories*, edited by J. Lukierski and J. Rembieliński

We restrict ourselves here to 'smooth solutions'. The main part of this review will be devoted to the structure of the manifolds of scalars that appear in these theories. We explain how the superconformal methods clarify the structure, referring to new results of [14, 15], and pay special attention to definitions of quaternionic(-Kähler) manifolds. At the end, we consider the scalar-dependent solutions in the warped background (1), and show how the critical points are determined by algebraic attractor equations [1], generalizing earlier similar equations with vector multiplets in 4 dimensions [16, 17, 18] to include hypermultiplets. To analyse the properties of the critical points, a general formula on the scalar mass matrix [19] gives a lot of insight.

$N = 2, D = 5$ SUPERGRAVITY AND ITS SCALARS

Pure $N = 2$, $D = 5$ supergravity[1] contains a graviton, two gravitini and a graviphoton (spin 1). The theory can contain vector multiplets, each containing a vector, a doublet of spinors and a scalar. These scalars define a 'very special real manifold', as we will explain. Furthermore, there are hypermultiplets each containing 2 spinors and 4 scalars. The latter define a quaternionic-Kähler manifold. Note that we will not consider tensor multiplets. The antisymmetric tensors are dual to vectors in 5 dimensions. That duality is only for Abelian couplings, but that is all that we consider here. Remark, however, that for non-Abelian theories, antisymmetric tensor multiplets lead to more general possibilities, as discussed in detail in [20].

We first review the superconformal construction of matter couplings in supergravity. We then discuss the two scalar manifolds. Finally we discuss the consequences of gauging isometries of the quaternionic manifold using the vector fields.

Superconformal Tensor Calculus

Superconformal tensor calculus provides a way to construct matter-coupled super-Poincaré theories. The aim here is thus not to end up with a superconformal theory, but rather to provide a way to construct the general supergravity theories. Using conformal invariance facilitates the construction of the theory and leads to more insight in the structure of the theory. E.g. the structure of special Kähler manifolds was developed by using superconformal tensor calculus.

The method thus consists of first constructing a theory invariant under the superconformal group, and then fixing all the invariances that are not required for a super-Poincaré invariant theory. A superconformal group consists of the conformal group (with translations P_μ, Lorentz rotations $M_{\mu\nu}$, dilatations D, and special conformal generators K_μ), supersymmetry Q, and a special supersymmetry S, and finally some R-symmetry. The latter is a bosonic group that acts on the supersymmetries and appears in the an-

[1] $N = 2$ means 8 real supercharges: 4-component spinors in an $SU(2)$ doublet. It is minimal supersymmetry in 5-dimensional Minkowski space. Note that for other signatures (2 or 3 time directions) one can impose Majorana conditions such that only 4 of them survive ($N = 1$).

TABLE 1. *Superconformal algebras, with the two parts of the bosonic subalgebra: one that contains the conformal algebra and the other one is the R-symmetry. In the cases $D = 4$ and $D = 8$, the $U(1)$ factor in the R-symmetry group can be omitted for $N \neq 4$ and $N \neq 16$, respectively.*

D	supergroup	bosonic group	
3	$OSp(N\|4)$	$Sp(4) = SO(3,2)$	$SO(N)$
4	$SU(2,2\|N)$	$SU(2,2) = SO(4,2)$	$SU(N) \times U(1)$
5	$OSp(8^*\|N)$	$SO^*(8) \supset SO(5,2)$	$USp(N)$
	$F(4)$	$SO(5,2)$	$SU(2)$
6	$OSp(8^*\|N)$	$SO^*(8) = SO(6,2)$	$USp(N)$
7	$OSp(16^*\|N)$	$SO^*(16) \supset SO(7,2)$	$USp(N)$
8	$SU(8,8\|N)$	$SU(8\|8) \supset SO(8,2)$	$SU(N) \times U(1)$
9	$OSp(N\|32)$	$Sp(32) \supset SO(9,2)$	$SO(N)$
10	$OSp(N\|32)$	$Sp(32) \supset SO(10,2)$	$SO(N)$
11	$OSp(N\|64)$	$Sp(64) \supset SO(11,2)$	$SO(N)$

ticommutator of ordinary and special supersymmetry. In our case, this group is $SU(2)$, and the full superconformal algebra defines the supergroup $F^2(4)$ (where the index '2' indicates the particular real form of the superalgebra, see Table 5 in [21]). Its bosonic subgroup is $SO(5,2) \times SU(2)$, where the first factor is the conformal group and the second is the R-symmetry group.

Note that in this case the bosonic subgroup is a direct product of the conformal group and the R-symmetry group. This is the case in the superconformal algebras classified by Nahm [22]. It implies that bosonic symmetries that are not in the conformal algebra are spacetime scalars. This is not a necessity. Other examples have been considered first in 10 and 11 dimensions in [23]. Recently, a new classification has appeared in [24] of which we can extract Table 1 for dimensions from 3 to 11. The bosonic subgroup contains always two factors. One contains the conformal group. If that factor is really the conformal group, then the algebra appears in Nahm's classification. Note that 5 dimensions is a special case. There is a generic superconformal algebra for any extension. But for the case $N = 2$ there exists a smaller superconformal algebra that is in Nahm's list. So far, superconformal tensor calculus has only been based on algebras of Nahm's type. Note that for $D = 6$ or $D = 10$, where one can have chiral spinors, only the case that all supersymmetries have the same chirality has been included.

For the methods that are used in superconformal tensor calculus, we refer to existing reviews, as the one that appeared in the Karpacz proceedings of 1983 [25]. A useful example is its application in $N = 1$, $D = 4$ supergravity [26, 27], that has been written down in detail in [28].

The basic multiplet is the one that contains the gauge field of all the symmetries in the superconformal group. This is called the Weyl multiplet, and has recently been constructed for $N = 2$, $D = 5$ in [14, 15]. There are two versions, as it is the case in six dimensions [29]. Both versions have 32+32 components and are equivalent. In fact, there is a procedure to go from one to the other [14]. We will restrict ourselves to one version, which is the one used primarily also in 4 and 6 dimensions. Its content is given in Table 2. We indicate for each field the number of components in each dimension, the symmetry for which it is a gauge field, and possibly other gauge transformations that have been used to reduce its number of degrees of freedom in this counting.

TABLE 2. *Standard Weyl multiplet in 4, 5 and 6 dimensions.*

(gauge) field	$D=4$	$D=5$	$D=6$	gauge transf.	subtracted
$e_\mu{}^a$	5	9	14	P^a	M_{ab}, D
b_μ		compensating K^a		D	K^a
$\omega_\mu{}^{ab}$		composite		M^{ab}	
$f_\mu{}^a$		composite		K^a	
$V_{\mu i}{}^j$	9	12	15	$SU(2)$	
A_μ	3	–	–	$U(1)$	
$\psi_\mu{}^i$	16	24	32	Q^i	S^i
$\phi_\mu{}^i$		composite		S^i	
T_{ab}, T^-_{abc}	6	10	10		
D	1	1	1		
χ^i	8	8	8		
TOTAL	$24+24$	$32+32$	$40+40$		

Once one has this multiplet, one can add other multiplets, i.e. representations of the superconformal algebra. In order to satisfy this algebra, the transformation laws of the fields in these multiplets will involve the fields of the Weyl multiplet. Then one constructs a superconformal invariant action, and finally one has to fix the superfluous symmetries. Remark that we already used the K^a symmetry to put the gauge field of dilatations, b_μ, equal to zero. The remaining superfluous symmetries are therefore the dilatations D, the special supersymmetries S^i, and the R-symmetry $SU(2)$.

TABLE 3. *Multiplets and fields of the super-Poincaré theories*

spin	pure SG	vector mult.	hypermult.	field	indices
2	1			e_μ^a	$\mu, a = 0,\ldots,4$
$\frac{3}{2}$	2			ψ_μ^i	$i = 1,2$
1	1	n		A_μ^I	$I = 0,\ldots,n$
$\frac{1}{2}$		$2n$	$2r$	λ_i^x, ζ^A	$A = 1,\ldots,2r$
0		n	$4r$	ϕ^x, q^X	$x = 1,\ldots,n;\ X = 1,\ldots,4r$

As we mentioned in the beginning of this chapter, we consider general couplings with n vector multiplets and r scalar multiplets. Table 3 gives their content, the names that we use for the fields, and the corresponding range of indices. In the superconformal method, these are obtained in a different way. One starts with the Weyl multiplet, and adds vector multiplets and hypermultiplets in representations of the superconformal algebra. As well for the vector multiplets as for the hypermultiplets, one starts by adding

TABLE 4. *Multiplets and fields in the superconformal construction*

spin	Weyl	vector	hyper	gauge fix	auxiliary
2	e_μ^a				
$\frac{3}{2}$	ψ_μ^i				
	$V_{\mu i}{}^j, T_{ab}$				auxiliary
1		$n+1$			
$\frac{1}{2}$	χ^i	$2(n+1)$	$2(r+1)$	2: S	χ^i with 2 others
0	D	$n+1$	$4(r+1)$	1: dilatations, 3: $SU(2)$	D and 1 other

one more multiplet than appears in the final super-Poincaré theory. These 'compensating multiplets' contain the degrees of freedom that will be gauge-fixed. This is schematically represented in Table 4. It is indicated how the superfluous symmetries are fixed, and how some of the fields of the Weyl multiplet serve as Lagrange multipliers eliminating degrees of freedom of the spin $1/2$ and scalar fields. The field $V_{\mu i}{}^{j}$ will be eliminated by its field equation, and will play the role of $SU(2)$ curvature of the quaternionic manifold defined by the hyperscalars. The field T_{ab} will become a function of the field strengths of the vectors in the vector multiplet (dressed by the scalars), and plays the role of gauge field that enters in the gravitino transformation (related to the central charge).

Very Special Real and Quaternionic-Kähler Manifolds

The manifolds of supergravity–matter couplings in $D = 5$ are similar to those that are known from $N = 2$ in 4 dimensions. Table 3 would be nearly identical for 4 dimensions, except that each vector multiplet then contains two scalars. The supersymmetry defines a complex structure, and the manifold is Kählerian. In $N = 1$ supergravity, general Kähler manifolds are possible. In $N = 2$ they are restricted to a category that is called 'special Kähler manifolds' [30]. The quartets of scalars in hypermultiplets are connected by 3 complex structures and the manifold is quaternionic-Kähler [31]. Another recent review containing the fundamental facts of these manifolds is given in [32].

Very Special Real Manifolds

We first consider the vector multiplets [33]. In 5 dimensions, these have real scalars (one of the scalars of 4 dimensions sits in the $5d$-vector). We define 'very special real manifolds' [34] as those that appear in these couplings of vector multiplets to 5-dimensional supergravity. It is clear from the above, that they can be described in superconformal tensor calculus by starting with $n+1$ scalars, which we denote h^I, as in Table 4. Then we impose a dilatational gauge choice. This defines an n-dimensional hypersurface in the $(n+1)$-dimensional space.

The locally supersymmetric action of the vector multiplets in 5 dimensions contains always a Chern–Simons term of the form $C_{IJK}A^I \mathrm{d}A^J \mathrm{d}A^K$. In order for this to be gauge-invariant, the C_{IJK} have to be constant. This tensor is completely symmetric in its indices, and supersymmetry implies that the full action is determined by these constants (up to the choice of coordinates on the manifold). Thus the set of numbers C_{IJK} are all one needs to specify a very special real manifold [33]. For an arbitrary set, one still has to verify whether they allow a non-empty domain with positive-definite metric on the scalar manifold.

The dilatational gauge choice that is most appropriate is the condition

$$C_{IJK}h^I(\phi)h^J(\phi)h^K(\phi) = 1\,. \tag{2}$$

ϕ^x are coordinates on this manifold such that the embedding $h(\phi)$ satisfies the condition. The metric on the scalar manifold is then

$$g_{xy} = -3(\partial_x h^I)(\partial_y h^J) C_{IJK} h^K\,. \tag{3}$$

Quaternionic-Kähler Manifolds

Let us now look at the other side: the hypermultiplets. We first define quaternionic manifolds. We start with a $4r$-dimensional manifold with coordinates q^X. At each point there is a tangent space where the vectors are labelled with indices (iA) (see ranges in Table 3). These are connected by $4r\times 4r$ vielbeins f_X^{iA} or their inverses f_{iA}^X. We will here introduce the quaternionic manifolds starting from these vielbeins. Quaternionic manifolds entered physics in [31], and [35] contains a lot of interesting properties. There were two workshops on quaternionic geometry where mathematics and physics results were brought together [36, 37]. Other recent papers that review the properties of quaternionic manifolds are [32, 38].

For supersymmetry, starting from vielbeins is a convenient approach because these are the objects that one uses from the very beginning, i.e. in the supersymmetry transformations of the hyperscalars:

$$\delta(\varepsilon) q^X = f_{iA}^X \bar{\varepsilon}^i \zeta^A\,. \tag{4}$$

Almost quaternionic manifolds. We thus have

$$f_Y^{iA} f_{iA}^X = \delta_Y{}^X\,, \qquad f_X^{iA} f_{jB}^X = \delta_j{}^i \delta_B{}^A\,. \tag{5}$$

These vielbeins satisfy a reality condition defined by matrices $E_i{}^j$ and $\rho_A{}^B$ that satisfy

$$E E^* = -\mathbb{1}_2\,, \qquad \rho\rho^* = -\mathbb{1}_{2r}\,. \tag{6}$$

One may choose a standard antisymmetric form for ρ and identify E with ε by a choice of basis. The reality condition for the vielbeins are

$$(f_X^{iA})^* = f_X^{jB} E_j{}^i \rho_B{}^A\,. \tag{7}$$

The transformations on variables with an A index are by the reality condition restricted to $G\ell(r,Q) = SU^*(2n)\times U(1)$.

We define complex structures as ($r=1,2,3$ and using the three sigma matrices)

$$J_X{}^{Yr} \equiv -\mathrm{i} f_X^{iA} (\sigma^r)_i{}^j f_{jA}^Y\,, \qquad \Rightarrow \qquad J_X{}^Y{}_i{}^j \equiv \mathrm{i} J_X{}^{Yr} (\sigma^r)_i{}^j = 2 f_X^{jA} f_{iA}^Y - \delta_i{}^j \delta_X{}^Y\,. \tag{8}$$

We use the same transition between triplet and doublet notation below for other quantities. The complex structures satisfy, due to (5), the quaternion algebra

$$J^r J^s = -\mathbb{1}_{4r} \delta^{rs} + \varepsilon^{rst} J^t\,. \tag{9}$$

This defines the manifold to be 'almost quaternionic'.

Quaternionic manifolds. We now suppose that there is a torsionless connection $\Gamma^Z_{XY} = \Gamma^Z_{YX}$. Consider then

$$\Omega_{X\,jB}{}^{iA} \equiv f^Y_{jB}\left(\partial_X f^{iA}_Y - \Gamma^Z_{XY} f^{iA}_Z\right) = -\omega_{X\,j}{}^i \delta_B{}^A - \omega_{XB}{}^A \delta_j{}^i\,, \tag{10}$$

where $\omega_{X\,j}{}^i$ is traceless. If this $\Omega_{X\,jB}{}^{iA}$, for each X, would be a general $4r\times 4r$ matrix, then we would say that the holonomy is not restricted (or sits in $G\ell(4r)$). The splitting as in the right-hand side of this equation implies that the holonomy group is restricted to $SU(2)\times G\ell(r,Q)$. We can write (10) as the covariant constancy of the vielbein:

$$\partial_X f^{iA}_Y - \Gamma^Z_{XY} f^{iA}_Z + f^{jA}_Y \omega_{X\,j}{}^i + f^{iB}_Y \omega_{XB}{}^A = 0\,, \tag{11}$$

with composite gauge fields for $SU(2)$ and $G\ell(r,Q)$. These conditions promote the almost quaternionic structure to a quaternionic structure, and the manifolds is 'quaternionic'. If the $SU(2)$ connection is zero, they are called 'hypercomplex'.

The integrability condition of (11), (multiplied by a vielbein) is

$$R^Z{}_{WXY} = f^Z_{iA} f^{jA}_W \mathcal{R}_{XY\,j}{}^i + f^Z_{iA} f^{iB}_W \mathcal{R}_{XYB}{}^A = -J_W{}^{Zr}\mathcal{R}_{XY}{}^r + f^Z_{iA} f^{iB}_W \mathcal{R}_{XYB}{}^A\,, \tag{12}$$

where respectively the metric curvature $R^Z{}_{WXY} \equiv 2\partial_{[X}\Gamma^Z_{Y]W} + 2\Gamma^Z_{V[X}\Gamma^V_{Y]W}$, the $SU(2)$ curvature $\mathcal{R}_{XY}{}^r \equiv 2\partial_{[X}\omega_{Y]}{}^r + 2\omega_{[X}{}^s\omega_{Y]}{}^t\varepsilon^{rst}$, and the $G\ell(r,Q)$ curvature $\mathcal{R}_{WXB}{}^A$ appear.

Quaternionic-Kähler manifolds. Quaternionic-Kähler manifolds (including 'hyperkähler' for the case that the $SU(2)$ curvature vanishes) by definition have a metric. First define an Hermitian metric d_{AB} such that $C_{AB} \equiv \rho_A{}^C d_{CB} = -C_{BA}$. By redefinitions of the basis [39] one may diagonalize d while simultaneously bringing C to a canonical form

$$d = \begin{pmatrix} \mathbb{1}_{2p} & 0 \\ 0 & -\mathbb{1}_{2(r-p)} \end{pmatrix}, \qquad C = \begin{pmatrix} \begin{matrix} 0 & 1 \\ -1 & 0 \end{matrix} & & \\ & \begin{matrix} 0 & 1 \\ -1 & 0 \end{matrix} & \\ & & \ddots \end{pmatrix}. \tag{13}$$

The subgroup of $G\ell(r,Q)$ that preserves d is $USp(2p,2r-2p)$. We define then the metric of the manifold to be[2]

$$g_{XY} = f^{iA}_X C_{AB}\varepsilon_{ij} f^{jB}_Y\,. \tag{14}$$

We use ε_{ij} and C_{AB} to raise and lower indices according to the NW–SE convention

$$A_i = A^j\varepsilon_{ji}\,, \qquad A^i = \varepsilon^{ij}A_j\,, \qquad A_A = A^B C_{BA}\,, \qquad A^A = C^{AB}A_B\,. \tag{15}$$

One can check that this raising and lowering of indices, together with the usual raising and lowering of indices X by g_{XY} and its inverse is consistent with defining f^{iA}_X and f^X_{iA} as each others inverses.

---

[2] In principle one could also introduce a metric in the $SU(2)$ part, but as there is no choice for the signature in this sector, this is irrelevant, and we identify $E_i{}^j = \varepsilon_{ij}$ (with $\varepsilon_{ik}\varepsilon^{jk} = \delta_i{}^j$).

The choice of the signature in (13) is relevant for the reality conditions, which are then $(f_X^{iA})^* d_{AB} = f_{XiB}$. Our notations hide d from other places.

For $r > 1$ one can prove that these manifolds are Einstein, and that the $SU(2)$ curvatures are proportional to the complex structures:

$$R_{XY} = \frac{1}{4r} g_{XY} R, \qquad \mathcal{R}^r_{XY} = \tfrac{1}{2} \nu J^r_{XY}, \qquad \nu = \frac{1}{4r(r+2)} R. \tag{16}$$

(with $R_{XY} = R^Z{}_{XZY}$). For $r = 1$ this is part of the definition of quaternionic-Kähler manifolds. Hyperkähler manifolds are those where the $SU(2)$ curvature is zero, and these are thus also Ricci-flat.

Supergravity. In supergravity we find all these constraints from requiring a supersymmetric action. Moreover, we need for the invariance of the action that the last equation of (16) is satisfied with $\nu = -1$. This implies that the scalar curvature is $R = -4r(r+2)$. This excludes e.g. the compact symmetric spaces.

The Family of Special Manifolds

We now place these manifolds in the context of the manifolds that are obtained for supersymmetries with 8 real supercharges. Note that a higher number of supercharges would restrict the possibilities for the scalar manifolds to a discrete number of symmetric spaces. We first consider vector multiplets in 5 or 4 dimensions with $N = 2$. Vector multiplets in 6 dimensions do not contain scalars. When reducing to 3 dimensions, the vectors become dual to scalars (we can perform duality transformations as we are just considering kinetic terms here, and we can thus restrict to Abelian vectors). Therefore the multiplet in 3 dimensions is dual to a multiplet with only scalars: the hypermultiplet. For hypermultiplets, the spacetime dimension is not really relevant as there are no vectors, and thus the results for hypermultiplets are the same for any dimension. In the picture it is convenient to consider them in 3 dimensions because of the dimensional reduction that we just described. With real scalars in the vector multiplets in 5 dimensions, these geometries are real geometries, those in 4 dimensions are Kählerian, and the hypermultiplets lead to 3 complex structures. Furthermore, we can distinguish between theories that appear in rigid supersymmetry, and those in supergravity. This leads to the overview in the upper part of Table 5. The geometries that are related to rigid supersymmetry have been called 'affine' in the mathematics literature [40, 41], while those for supergravity are called 'projective' (and these are the default, in the sense that e.g. special Kähler refers to the geometry that is found in supergravity. The analogous manifolds with 3 complex structures got already a name in the literature.

The name 'projective' versus 'affine' can be understood from the construction of the manifolds in supergravity using superconformal tensor calculus. We saw already (see Table 4) how the real very special manifolds are obtained starting from $(n+1)$ vector multiplets. Before any gauge fixing, these are just real manifolds with a dilatational invariance. This manifold has therefore a cone structure, with $C_{IJK} h^I h^J h^K$ as the radial coordinate. The physical scalars of the supergravity theory are thus defined modulo this

TABLE 5. *Geometries from supersymmetric theories with 8 real supercharges, and the connections provided by the* **r**-*map and the* **c**-*map.*

	$d=5$ vector multiplets	$d=4$ vector multiplets	hypermultiplets
rigid (affine)	affine very special real	affine special Kähler	hyperkähler
local (projective)	(projective) very special real	(projective) special Kähler	quaternionic-Kähler

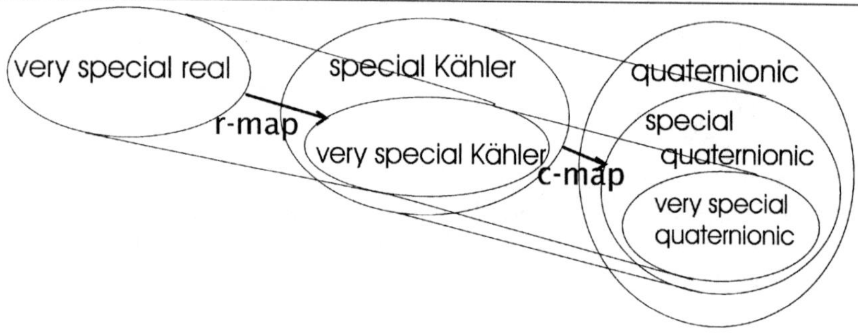

dilatational scaling. The manifolds that occurs in supergravity can thus be seen as a projective space of dimension n.

Similarly, to construct special Kähler geometry, one starts in 4 dimensions with the couplings as they occur in rigid supersymmetry, demanding the presence of a superconformal symmetry. Again, the manifold has a cone structure, and the dilatational gauge condition selects a submanifold at fixed radius. In this case, the superconformal group contains a $U(1)$ invariance and the manifold at fixed radius is a 'Sasakian manifold' of dimension $2n+1$, if this $U(1)$ is not gauged. In conformal supergravity the $U(1)$ is local and eliminates one more scalar. The gauge field of this $U(1)$, which is an auxiliary field in the superconformal tensor calculus (similar to $V_{\mu i}{}^{j}$ in Table 4), becomes by its field equation the $U(1)$ connection on the Kähler manifold. The final manifold in super-Poincaré has then non-trivial $U(1)$ curvature (and will be a Hodge-Kähler manifold).

The construction for quaternionic manifolds is similar, as has been demonstrated recently in 4 dimensions in [42]. One starts then from hyperkähler cones. The dilatational gauge choice leads to a tri-Sasakian manifold of dimensions $4r+3$ for ungauged $SU(2)$. The $SU(2)$ gauge fields of the Weyl multiplet get by their field equations the value $V_{\mu i}{}^{j} = \partial_\mu q^X \omega_{X i}{}^{j}$, using the $SU(2)$ connection that we had in the previous section. The $SU(2)$ curvature is thus non-zero as required by (16).

Dimensional reduction gives a mapping between these manifolds. These mappings have been called the **c**-map (from special Kähler to special quaternionic) [43], and the **r**-map (from very special real to very special Kähler) [44]. They are represented in the lower part of Table 5. Dimensional reduction of a manifold in 5 dimensions gives a 4-dimensional theory. But the 4-dimensional theories that can be obtained in this way, are only a subset of all 4-dimensional theories. The table shows the structure in the names given to various classes of manifolds. Very special Kähler manifolds are a subset of all special Kähler manifolds. The quaternionic manifolds that are in the image of the **c**-map are the special quaternionic manifolds, and those in the image of the **c**∘**r**-map are the very special quaternionic manifolds. It is remarkable that nearly all the homogeneous quaternionic manifolds are very special quaternionic manifolds [44] (The only non-special homogeneous quaternionic manifolds are the quaternionic projective spaces).

Gauging of Isometries and the Consequences

Having vectors in the vector multiplets (and one graviphoton), these can be used to gauge extra symmetries. The 'extra' refers here to the fact that these do not belong to the super-Poincaré or the superconformal algebra. However, at the end, in supergravity, we do not have a strict mathematical algebra of symmetries, but a soft algebra. This means that there are structure functions rather than structure constants. These functions depend on the fields in the theory. In this way the 'extra' gauge group can appear in the anticommutator of two supersymmetries with structure functions depending on the fields. When these have a non-zero expectation value, central charges appear. The generators of these gauge group may act also on the fields of the quaternionic manifold.

In supersymmetry, such a gauging has three consequences ([32] gives general properties of gauging in supergravity). The first is that new terms appear in the supersymmetry transformations of the fermions (proportional to a gauge coupling constant). Secondly, the scalar potential is completely determined by the gauging (this is true for theories with 8 supercharges and more). Finally, gauged R-symmetry leads to a cosmological constant. R-symmetry is the symmetry that rotates the supersymmetries and was mentioned in Table 1. In $N=2$, $D=5$ it is $SU(2)$. We obtain gauged R-symmetry if the extra gauge group contains an $SU(2)$ or a $U(1)$ subgroup thereof. This extra gauge group mixes with the R-symmetry. In the superconformal approach this is due to the gauge fixing of the $SU(2)$ of the superconformal algebra by fixing the scalars in the compensating hypermultiplet, see Table 4. The latter in general also transform under the gauge symmetry (see [45] for the structure of the symmetries in the hyperkähler cone). The remaining gauge group is a diagonal subgroup of the superconformal $SU(2)$ and the extra gauge group. The gauge fields of the extra gauge group then gauge the symmetry that acts the supersymmetries (and on the gravitini). In that case we use the terminology 'gauged supergravity'. We will do this explicitly below for a $U(1)\subset SU(2)$. The formulae can also be used for gauging the full $SU(2)$ as has been done in [46]. Gauged R-symmetry induces a cosmological constant[3], which is proportional to the square of the gauge coupling constant (the group theoretical argument was repeated in [4]).

The vectors are in the adjoint of the gauge group, and supersymmetry then implies the same for the scalars h^I in the $(n+1)$ dimensional space. If the constraint (2) is compatible with this, i.e. $C_{L(IJ}f_{K)M}^{L}=0$, then the group can be gauged. It is for these non-Abelian theories that tensor multiplets give extra possibilities [20, 2].

---

[3] To have really a 'constant', one still has to assume that there is a solution such that the scalars that determine the value of the potential are constant.

In order that these symmetries can act on the hypermultiplet, the quaternionic manifold should have isometries. Thus, we suppose that there are Killing vectors $K_\alpha^X(q)$ that determine transformations of the scalars q^X, and α denotes the different isometries. In general, only a subset of these can be gauged. We need for each one a gauge vector. Therefore, the appropriate index is I, labeling the vectors (see Table 3), and the gauged isometries are determined by $K_I^X(q)$. In quaternionic geometry, the isometries are determined by a triplet of prepotentials $P_I^r(q)$:

$$\mathcal{R}^r_{XY}K_I^Y = D_X P_I^r\,,\qquad D_X P_I^r \equiv \partial_X P_I^r + 2\varepsilon^{rst}\omega_X^s P_I^t\,. \tag{17}$$

These can be solved as well for the Killing vectors, or for the prepotentials:

$$K_I^Z = -\tfrac{4}{3}\mathcal{R}^{rZX}D_X P_I^r\,,\qquad P_I^r = \frac{1}{2r}\mathcal{R}^{rXY}D_X K_{IY}\,. \tag{18}$$

The latter equation [19] is obviously only true for $r \neq 0$. If there are no physical hypermultiplets ($r=0$), then P_I^r are just some constants. In the superconformal approach they determine the action of the symmetry on the compensating hypermultiplet. These are the analogues of the Fayet–Iliopoulos terms. For $r>0$ there is thus no Fayet–Iliopoulos term possible [1]. In rigid supersymmetry, the $SU(2)$ curvatures vanish, and P_I^r are again arbitrary constants or 'Fayet–Iliopoulos terms'.

The above quantities determine the modified supersymmetries. For that purpose one defines 'dressed' Killing vectors and 'dressed' prepotentials:

$$P^r \equiv h^I(\phi)P_I^r(q)\,,\qquad K^X \equiv h^I(\phi)K_I^X(q)\,. \tag{19}$$

These thus depend as well on the scalars of the vector multiplets as on those of the hypermultiplets, but in a well-structured way. The supersymmetry transformations of the fermions are then [2] (bosonic terms only)

$$\begin{aligned}
\delta_\varepsilon\psi_{\mu i} &= D_\mu(\omega)\varepsilon_i + \tfrac{1}{4\sqrt{6}}\mathrm{i}\left(\gamma_{\mu\nu\rho}-4g_{\mu\nu}\gamma_\rho\right)\varepsilon_i h_I F^{\nu\rho I} - \tfrac{1}{\sqrt{6}}\mathrm{i}g\gamma_\mu P_i{}^j\varepsilon_j\,,\\
\delta_\varepsilon\lambda_i^x &= -\tfrac{1}{2}\mathrm{i}(\not{D}\phi^x)\varepsilon_i + \tfrac{1}{4}h_I^x\gamma^{\mu\nu}\varepsilon_i F_{\mu\nu}^I - g\sqrt{\tfrac{3}{2}}\varepsilon^j\partial_x P_i{}^j\,,\\
\delta_\varepsilon\zeta^A &= f_X^{Ai}\left[\tfrac{1}{2}\mathrm{i}(\not{D}q^X)\varepsilon_i - g\tfrac{1}{4}\sqrt{6}\varepsilon_i K^X\right].
\end{aligned} \tag{20}$$

We used here notations of very special geometry:

$$h_I \equiv C_{IJK}h^J h^K\,,\qquad h_I^x = \sqrt{\tfrac{3}{2}}g^{xy}\partial_y h^I\,,\qquad h_x^I \equiv -\sqrt{\tfrac{3}{2}}\partial_x h^I\,. \tag{21}$$

Notice that the gauging produced an extra scalar-dependent term for each of the fermions, apart from covariantizations depending on the gauge vectors:

$$D_\mu\varepsilon_i = \ldots - gA_\mu^I P_{Ii}{}^j\varepsilon_j\,,\quad D_\mu\phi^x = \ldots + gA_\mu^I\sqrt{\tfrac{3}{2}}h_K f_{JI}^K h^{Jx}\,,\quad D_\mu q^X = \ldots + gA_\mu^I K_I^X\,. \tag{22}$$

Finally, as usual in supersymmetry, the scalar potential is a square of the transformation laws of the fermions. We have here:

$$V = g^2\left[-4P^rP^r + 3(\partial_x P^r)(\partial^x P^r) + \tfrac{3}{4}K^X K_X\right]. \tag{23}$$

A general form for scalar potentials has been put forward, guaranteeing stability [47, 48, 49]:

$$V = g^2\left(-6W^2 + \tfrac{9}{2}g^{\Lambda\Sigma}\partial_\Lambda W\partial_\Sigma W\right), \tag{24}$$

where Λ, Σ run over all the scalars, and W is some 'superpotential'. To make the transition from (23), we first split the dressed prepotential in a norm W, which is identified as superpotential, and a phase Q^r:

$$P^r = \sqrt{\tfrac{3}{2}}WQ^r, \qquad Q^rQ^r = 1. \tag{25}$$

The phase determines which $U(1)$ subgroup of the $SU(2)$ R-symmetry is gauged. Then it turns out [1] that we can write the potential as (24) if the derivative of the phase with respect to the scalars of the vector multiplets is zero:

$$\partial_x Q^r = 0. \tag{26}$$

An equivalent condition was found in [50]. This condition is automatic if there are no hypermultiplets (the prepotentials are then constants) or if there are no vector multiplets. However, if one has vector- and hypermultiplets then this is in general only satisfied on a submanifold of the total scalar manifold. We will see below that this condition is required also for BPS solutions of the theory.

SMOOTH SOLUTIONS: RS, FLOWS AND ATTRACTORS

BPS Conditions

We now look for bosonic solutions with vanishing vectors, a metric of the form (1), and preserving some amount of supersymmetry [1]. As the fermions are zero in such solutions, their transformation laws should vanish too. Therefore we investigate the vanishing of (20) in this background. The transformation of ψ_5 determines the x^5-dependence of the Killing spinors $\varepsilon^i(x)$, and we can neglect this further. In the transformation of the ψ_μ appears a contribution of the spin connection. It gives the same equation for each $\underline{\mu} = \bar{0}, 1, 2, 3$. This equation is still a triplet of equations that can be split in the norm and the phase. These equations are respectively (we use a prime to denote derivatives w.r.t. x^5)

$$gW = \pm\frac{a'}{a}, \qquad \mathrm{i}\gamma_5\varepsilon_i = \mp Q_i{}^j\varepsilon_j. \tag{27}$$

As we defined W to be positive (a norm of a 3-vector), the sign in the first equation depends on the sign of a'/a. The other signs then follow from this one. The second equation is a projection on the preserved supersymmetries. It implies that one half of the supersymmetries survives (4 real supercharges, i.e. $N = 1$ in 4 dimensions).

Note that we have considered only solutions that do not explicitly depend on the coordinates $x^{\underline{\mu}}$. On 'critical points' (see below), other solutions are possible, doubling the number of preserved supersymmetries (related to the S-supersymmetries in the dual conformal theory).

The transformations of the gauginos can also be split in their norm and phase as $SU(2)$ triplets. The phase gives again (26), as we announced already [1]. The equation of the norm and the equation for the hyperinos give a similar condition for all scalars ϕ^Λ [50]:

$$\phi^{\Lambda\prime} = \mp 3g\, g^{\Lambda\Sigma}\partial_\Sigma W\,. \tag{28}$$

Terminology of Renormalization Group Flow

In the duality between the 5-dimensional theories and 4-dimensional conformal theories, the scalars are dual to coupling constants. The dependence on x^5 is denoted as a 'flow'. The value of the warp factor a is dual to the energy scale. Therefore, β-functions, i.e. logarithmic derivatives of the coupling constants to the energy, are

$$\beta^\Lambda = a\frac{\partial}{\partial a}\phi^\Lambda = \frac{a}{a'}\phi^{\Lambda\prime} = -3\frac{\partial^\Lambda W}{W}\,, \tag{29}$$

where we used (27) and (28). Critical points ($\beta = 0$) correspond thus to extrema of the superpotential and at these points the scalars are constant. As well for a suitable RS scenario as for a renormalization group flow, the end points of the flow $x^5 = \pm\infty$ should be such critical points.

Whether a critical point is a UV or an IR critical point depends on whether the β-function decreases or increases while passing through its zero. In the first case, the value of the scalars is attracted to this point in the high-energy (large a) regime. In the second case they are attracted to this point in the low-energy (small a) regime. The type of critical point is thus determined by the matrix

$$U_\Sigma{}^\Lambda \equiv -\left.\frac{\partial\beta^\Lambda}{\partial\phi^\Sigma}\right|_{\beta=0} = 3\left.\frac{\partial_\Sigma\partial^\Lambda W}{W}\right|_{\partial W=0}\,. \tag{30}$$

Positive eigenvalues imply that the point is a UV attractor for flows in the direction of the corresponding eigenvector, while negative eigenvalues indicate that it can be an IR attractor. The eigenvalues u are the conformal dimensions in the dual theory and the scalar mass is $M^2 = u(u-4) \leq -4$, satisfying the Breitenlohner–Freedman bound [51].

One can prove [19] a general formula for U:

$$\mathcal{U} = \begin{pmatrix} 2\delta_x{}^y & -\frac{1}{W^2}(\partial_x K^Z)\mathcal{J}_Z{}^Y \\ \frac{1}{W^2}\mathcal{J}_{XZ}\partial^y K^Z & \frac{3}{2}\delta_X{}^Y - \frac{1}{W^2}\mathcal{J}_X{}^Z\mathcal{L}_Z{}^Y \end{pmatrix}, \tag{31}$$

where the first entries are for the vector multiplets and the second for the hypermultiplets, and $\mathcal{J}$ and $\mathcal{L}$ select respectively the $SU(2)$ and $USp(2r)$ part of the dressed gauged isometry, defined by

$$\mathcal{J}_{XY} \equiv 2P^r\mathcal{R}^r_{XY}\,, \qquad D_X K_Y = \mathcal{J}_{XY} + \mathcal{L}_{XY}\,. \tag{32}$$

Note that with only vector multiplets, we have only the upper-left entry of (31), and thus only UV attractors [52, 53]. The trace of $\mathcal{J}\mathcal{L}$ is zero, and thus the trace of the full matrix U is $2n+6r>0$. Therefore any critical point is UV in some directions.

The 'attractor equations', which determine the conditions for critical points, can be written as algebraic equations [1]

$$K^X \equiv h^I K_I^X = 0, \qquad P_I^r = h_I h^J P_J^r . \tag{33}$$

They have a group-theoretical meaning. The first one says that the 'dressed symmetry' at the critical point should be in the stability subgroup of the isometry group. The second has n components, as it is trivial when multiplied with h^I. It says that at the critical point the other n symmetries should have the same $SU(2)$ content as the dressed symmetry.

Also other equations can be understood in a group-theoretical way. The value of the cosmological constant at these points is $-6W^2 = -4P^rP^r$, i.e. determined by the part of the gauging in the $SU(2)$ direction. On the other hand, for an IR critical point, one needs that either the lower-right entry of (31) has to become negative, which means that $\mathcal{L}$ has to be large, corresponding to a gauging in the $USp(2r)$ part, or non-diagonal entries should be non-zero, which means that Killing vectors should have parts that are not in the isotropy group of that point.

These results allow to investigate the structure of flows for arbitrary vector- and hypermultiplets in $N=2$, $D=5$ supergravity. In [1] couplings of 1 vector and 1 hypermultiplet were considered, and a flow was found that generalizes the IR to UV flow of [54] with two arbitrary parameters. More general models can be investigated easily due to the algebraic nature of the attractor equations and their geometric significance.

ACKNOWLEDGMENTS

It was a pleasure to enjoy the friendly atmosphere of the Karpacz school. My understanding of several aspects of the problems mentioned in this text has grown by discussions with participants in this school, especially I. Bars, E. Bergshoeff, G. Gibbons, J. Lukierski, K. Stelle. This text summarizes work done in several collaborations. I thank D. Alekseevsky, E. Bergshoeff, A. Ceresole, V. Cortés, S. Cucu, G. Dall'Agata, M. Derix, C. Devchand, B. de Wit, T. de Wit, R. Halbersma and R. Kallosh for the fruitful collaborations. I also thank also J. Gheerardyn and S. Vandoren for discussions.

REFERENCES

1. Ceresole, A., Dall'Agata, G., Kallosh, R., and Van Proeyen, A., `hep-th/0104056`.
2. Ceresole, A., and Dall'Agata, G., *Nucl. Phys.*, **B585**, 143–170 (2000), `hep-th/0004111`.
3. Bergshoeff, E., Kallosh, R., and Van Proeyen, A., *JHEP*, **10**, 033 (2000), `hep-th/0007044`.
4. Bergshoeff, E., Kallosh, R., and Van Proeyen, A., `hep-th/0012110`, To be published in the proceedings of the NATO advanced research workshop *Noncommutative structures in mathematics and physics* in Kiev, Sep.2000, and in the proceedings of the EC-RTN workshop *The quantum structure of spacetime and the geometric nature of fundamental interactions* in Berlin, Okt. 2000.
5. Bergshoeff, E., Kallosh, R., Ortín, T., Roest, D., and Van Proeyen, A., `hep-th/0103233`.
6. Hořava, P., and Witten, E., *Nucl. Phys.*, 94-114 **B475** (1996), `hep-th/9603142`.

7. Lukas, A., Ovrut, B.A., Stelle, K.S., and Waldram, D., *Nucl. Phys.*, **B552**, 246–290 (1999), hep-th/9806051.
8. Gherghetta, T., and Pomarol, A., *Nucl. Phys.*, **B586**, 141–162 (2000), hep-ph/0003129.
9. Altendorfer, R., Bagger, J., and Nemeschansky, D., hep-th/0003117.
10. Falkowski, A., Lalak, Z., and Pokorski, S., *Phys. Lett.*, **B491**, 172–182 (2000), hep-th/0004093.
11. Zucker, M., *Nucl. Phys.*, **B570**, 267-283 (2000), hep-th/9907082.
12. Brax, P., and Davis, A.C., *Phys. Lett.*, **B497**, 289–295 (2001), hep-th/0011045.
13. Brax, P., and Davis, A.C., hep-th/0104023.
14. Bergshoeff, E., Cucu, S., Derix, M., de Wit, T., Halbersma, R., and Van Proeyen, A., hep-th/0104113.
15. Fujita, T., and Ohashi, K., hep-th/0104130.
16. Ferrara, S., Kallosh, R., and Strominger, A., *Phys. Rev.*, **D52**, 5412–5416 (1995), hep-th/9508072.
17. Strominger, A., *Phys. Lett.*, **B383**, 39–43 (1996), hep-th/9602111.
18. Ferrara, S., and Kallosh, R., *Phys. Rev.*, **D54**, 1514 and 1525 (1996), hep-th/9602136 and hep-th/9603090.
19. Alekseevsky, D., Cortés, V., Devchand, C., and Van Proeyen, A., in preparation.
20. Günaydin, M., and Zagermann, M., *Nucl. Phys.*, **B572**, 131–150 (2000), hep-th/9912027.
21. Van Proeyen, A., *Ann. Univ. Craiova, Physics AUC*, **9 (part I)**, 1 (1999), hep-th/9910030.
22. Nahm, W., *Nucl. Phys.*, **B135**, 149 (1978).
23. van Holten, J.W., and Van Proeyen, A., *J. Phys.*, **A15**, 3763 (1982).
24. D'Auria, R., Ferrara, S., Lledo, M.A., and Varadarajan, V.S., hep-th/0010124.
25. Van Proeyen, A., in *Supersymmetry and Supergravity 1983*, XIXth Winter School and Workshop of Theoretical Physics Karpacz, Poland, ed. B. Milewski (World Scientific, Singapore 1983).
26. Cremmer, E., Ferrara, S., Girardello, L., and Van Proeyen, A., *Nucl. Phys.*, **B212**, 413 (1983).
27. Kugo, T., and Uehara, S., *Nucl. Phys.*, **B222**, 125 (1983).
28. Kallosh, R., Kofman, L., Linde, A., and Van Proeyen, A., *Class. Quant. Grav.*, **17**, 4269–4338 (2000), hep-th/0006179.
29. Bergshoeff, E., Sezgin, E., and Van Proeyen, A., *Nucl. Phys.*, **B264**, 653 (1986).
30. de Wit, B., and Van Proeyen, A., *Nucl. Phys.*, **B245**, 89 (1984).
31. Bagger, J., and Witten, E., *Nucl. Phys.*, **B222**, 1 (1983).
32. Frè, P., hep-th/0102114. Lectures given at EC-RTN Workshop on Latest Development in M-Theory, Paris, France, Feb 2001.
33. Günaydin, M., Sierra, G., and Townsend, P.K., *Nucl. Phys.*, **B242**, 244 (1984).
34. de Wit, B., and Van Proeyen, A., *Phys. Lett.*, **B293**, 94–99 (1992), hep-th/9207091.
35. Galicki, K., *Commun. Math. Phys.*, **108**, 117 (1987).
36. Gentili, G., Marchiafava, S., and Pontecorvo, M., Eds., *Quaternionic Structures in Mathematics and Physics*. 1996. Proceedings of workshop in Trieste, September 1994; ILAS/FM-6/1996, available on http://www.emis.de/proceedings/QSMP94/.
37. Marchiafava, S., Piccinni, P., and Pontecorvo,M., Eds., *Quaternionic Structures in Mathematics and Physics*. Proceedings of workshop in Roma, September 1999; available on http://www.univie.ac.at/EMIS/proceedings/QSMP99/.
38. D'Auria, R., and Ferrara, S., hep-th/0103153.
39. de Wit, B., Lauwers, P.G., and Van Proeyen, A., *Nucl. Phys.*, **B255**, 569 (1985).
40. Freed, D.S. *Commun. Math. Phys.*, **203**, 31 (1999), hep-th/9712042.
41. Alekseevsky, D.V., Cortés, V., and Devchand, C., math.dg/9910091."
42. de Wit, B., Roček, M., and Vandoren, S., *JHEP*, **02**, 039 (2001), hep-th/0101161.
43. Cecotti, S., Ferrara, S., and Girardello, L., *Int. J. Mod. Phys.*, **A4**, 2475 (1989).
44. de Wit, B., and Van Proeyen, A., *Commun. Math. Phys.*, **149**, 307–334 (1992), hep-th/9112027.
45. de Wit, B., Roček, M., and Vandoren, S., hep-th/0104215.
46. Günaydin, M., and Zagermann, M., *Phys. Rev.*, **D63**, 064023 (2001), hep-th/0004117.
47. Boucher, W., *Nucl. Phys.*, **B242**, 282 (1984).
48. Townsend, P.K., *Phys. Lett.*, **B148**, 55 (1984).
49. Skenderis, K., and Townsend, P.K., *Phys. Lett.*, **B468**, 46–51 (1999), hep-th/9909070.
50. Behrndt, K., Herrmann, C., Louis, J., and Thomas, S., *JHEP*, **01**, 011 (2001), hep-th/0008112.
51. Breitenlohner, P., and Freedman, D.Z., *Phys. Lett.*, **B115**, 197 (1982).

52. Kallosh, R., and Linde, A., *JHEP*, **02**, 005 (2000), hep-th/0001071.
53. Behrndt, K., and Cvetič, M., *Phys. Rev.*, **D61**, 101901 (2000), hep-th/0001159.
54. Freedman, D.Z., Gubser, S.S., Pilch, K., and Warner, N.P., *Adv. Theor. Math. Phys.*, **3**, 363–417 (1999), hep-th/9904017.

NONCOMMUTATIVE GEOMETRY AND QUANTUM DEFORMATIONS

Status of Relativity with Observer-Independent Length and Velocity Scales

G. Amelino-Camelia

Dipartimento di Fisica, Universitá "La Sapienza", P.le Moro 2, I-00185 Roma, Italy
email: amelino@roma1.infn.it

Abstract. I have recently shown that it is possible to formulate the Relativity postulates in a way that does not lead to inconsistencies in the case of space-times whose structure is governed by observer-independent scales of both velocity and length. Here I give an update on the status of this proposal, including a brief review of some very recent developments. I also emphasize the role that one of the κ-Poincaré Hopf algebras could play in the realization of a particular example of the new type of postulates. I show that the new ideas on Relativity require us to extend the set of tools provided by κ-Poincaré and to revise our understanding of certain already available tools, such as the energy-momentum coproduct.

RELATIVITY AND OBSERVER-INDEPENDENT SCALES

In these notes I examine the status of my recent proposal [1, 2] attempting to identify consistent Relativity postulates that involve both an observer-independent velocity scale ($c \sim 3{\cdot}10^8 m/s$) and an observer-independent length scale ($L_p \sim 1.6{\cdot}10^{-35} m$). Readers already familiar with the proposal [1, 2] might find anyway useful my review for what concerns the results obtained in Refs. [3, 4, 5, 6] which were motivated by Refs. [1, 2] and provided important contributions to the programme.

It is often assumed that the Planck length L_p has a fundamental role in the short-distance structure of space-time; however, this appears to be conceptually troublesome for one of the cornerstones of Einstein's Special Relativity: FitzGerald-Lorentz length contraction. According to FitzGerald-Lorentz length contraction, different inertial observers would attribute different values to the same physical length. If the Planck length only has the role we presently attribute to it, which is basically the role of a coupling constant (an appropriately rescaled version of the coupling G), no problem arises for FitzGerald-Lorentz contraction, but if we try to promote L_p to the status of an intrinsic characteristic of space-time structure it is natural to find conflicts with FitzGerald-Lorentz contraction. For example, it is very hard (perhaps even impossible) to construct discretized versions or non-commutative versions of Minkowski space-time which enjoy ordinary Lorentz symmetry.[1] Therefore, unless the Relativity postulates are modified, it appears impossible to attribute to the Planck length a truly fundamental (observer-independent) intrinsic role in the microscopic structure of space-time.

---

[1] Pedagogical illustrative examples of this observation have been discussed, *e.g.*, in Ref. [9] for the case of discretization and in Refs. [10, 11, 12, 13] for the case of non-commutativity.

CP589, *Fundamental Interaction Theories*, edited by J. Lukierski and J. Rembieliński

We are of course not forced to introduce such a modification of the Relativity postulates. In fact, we do not (yet?) have any data that require us to attribute to L_p an observer-independent role in the microscopic structure of space-time (note, however, the intruiging indications emerging from the data analysed in Ref. [14] and references therein) and the theoretical arguments suggesting such a role are still rather debatable. On the other hand it is of course legitimate to explore this possibility.

In Refs. [1, 2] I set out to show that the Relativity Principle can coexist with some types of postulates stating that space-time structure is governed by observer-independent scales of both velocity (c) and length (L_p). The addition of an observer-independent length scale does not require major revisions of the physical interpretation of c. I shall just prudently avoid assuming a priori that it is legitimate to extrapolate from our long-wavelength data:

- (law1): The value of the fundamental velocity scale c can be measured by each inertial observer as the $\lambda/L_p \to \infty$ limit of the speed of light of wavelength λ.

While for c we can at least rely on long-wavelength data, we basically have no experimental information on the role (if any) of L_p in space-time structure. An illustrative example of new (L_p-dependent) Relativity postulates is introduced in the next Section, and is also the main focus of most of the remainder of these notes (and of Refs. [1, 2]).

AN ILLUSTRATIVE EXAMPLE OF NEW POSTULATES

In this Section I focus on one example of new Relativity postulates. In choosing an illustrative example of postulate attributing a role to L_p in space-time structure and kinematics, I found [1, 2] appropriate to give priority to ideas that would have significant phenomenological consequences and that can make some contact with preliminary indications of quantum-gravity theories. The hypothesis of an L_p-dependent dispersion relation finds some motivation in recent quantum-gravity studies and can lead to new effects that are small enough to be consistent with all presently-available data, while being large enough to be tested in the near future.

Motivated by these considerations, in Refs. [1, 2] I chose to use the following illustrative example of postulate attributing a role in space-time structure and kinematics to a length scale $\tilde{L}_p$:

- (law2): Each inertial observer can establish the value of $\tilde{L}_p$ (same value for all inertial observers) by determining the dispersion relation for photons, which takes the form $E^2 = c^2p^2 + f(E,p;\tilde{L}_p)$, where the function f has leading $\tilde{L}_p$ dependence given by: $f(E,p;L_p) \simeq \tilde{L}_p c E p^2$.

We could even contemplate the possibility that $\tilde{L}_p$ be completely unrelated to L_p, but it appears reasonable to explore in particular the possibility the quantity[2] setting the strength of the dispersion-relation deformation and the Planck length calculated *a la* Planck be identified up to a numerical coefficient not too different from 1 and a possible sign choice ($\tilde{L}_p \equiv \rho L_p$, with $\rho \in R$, $|\rho| \sim 1$).

Transformation Rules (One-Particle Case)

The logical consistency of the new postulates (law1) and (law2) requires that, in their analyses of photon data in leading order in $\tilde{L}_p$, all inertial observers agree on the dispersion relation $E^2 = c^2p^2 + \tilde{L}_p c p^2 E$, for fixed (observer-independent) values of c and $\tilde{L}_p$. The postulates do not explicitly concern massive particles. I tentatively adopt the simplest mass dependence $E^2 = c^4m^2 + c^2p^2 + \tilde{L}_p c p^2 E$, (possible generalizations are mentioned in Ref. [1]). I should therefore look for boost generators (generators of rotations clearly do not require modification) such that this dispersion relation is valid for all inertial observers. At this stage (see the wording adopted in (law2)) we shall be satisfied with checking logical consistency at leading order in $\tilde{L}_p$. For additional simplicity, here let me also limit my considerations to boosts along the z direction of particles with momentum only in the z direction. The Lorentz z-boost generator clearly requires a deformation. I make the ansatz $B_z^{\tilde{L}_p} = i[cp_z + \tilde{L}_p \Delta_E]\partial/\partial E + i[E/c + \tilde{L}_p \Delta_{p_z}]\partial/\partial p_z$, for which one easily finds that the sought invariance translates into the requirement $2E\Delta_E - 2p_z\Delta_{p_z} = -2E^2p_z - p_z^3$. The simplest solutions are of the type $2\Delta_E = 0\,,\;\; \Delta_{p_z} = E^2 + p_z^2/2$ and $\Delta_{p_z} = 0\,,\;\; \Delta_E = -Ep_z - p_z^3/(2E)$. Various arguments of simplicity [1] (including considerations involving combinations of boosts and rotations and the desire to have generators which would be well-behaved even off shell) lead me to adopt the first option, so the new z-boost generator takes the form

$$B_z^{\tilde{L}_p} = icp_z\frac{\partial}{\partial E} + i[E/c - \tilde{L}_p E^2/c^2 - \tilde{L}_p p_z^2/2]\frac{\partial}{\partial p_z}\,. \tag{1}$$

One important observation to be made at this point is that the generators of boosts (and rotations) constructed in the way I just described turn out to correspond to the leading-order-in-$\tilde{L}_p$ version of the Lorentz-sector generators of a well-known κ-Poincaré Hopf algebra [10, 11, 13], the example of κ-Poincaré Hopf algebra first introduced in Ref. [12]. In this sense just like the introduction of the Special Relativity postulates led to preexisting Lorentz-group mathematics, the example of new Relativity postulates I am analyzing leads to preexisting κ-Poincaré mathematics (note however that some of the observations reported in the following, concerning particle-production processes, do not fit in the κ-Poincaré mathematics, at least not in the way in which it is presently understood).

---

[2] As illustrated by the specific example (law2), the observer-independent length scale must not necessarily have the physical meaning of the length of something. For example, as indeed it happens in (law2), the role of L_p in space-time structure could be such that it provides a sort of reference scale for momenta (wavelengths).

Having obtained the new generators of boosts and rotations one immediately obtains infinitesimal transformations and then finite transformations are obtained by straightforward, but tedious, integration (explicit formulas in Ref. [1]).

Length Contraction

Concerning the space-time sector the fact that I was led to generators which had already emerged in preexisting κ-Poincaré mathematics suggests that one should consider κ-Minkowski space-time ($l,m = 1,2,3$):

$$[x_m,t] = i\tilde{L}_p x_m \ , \quad [x_m,x_l] = 0 \ , \tag{2}$$

which is known to be dual [12, 13] to κ-Poincaré energy-momentum space.

This fact that the space-time sector might be "quantum" invites one to be prudent [1] in making considerations on the space-time picture of the transformation rules. As I showed in Refs. [1, 2], we can however obtain some (partial) information on the nature of this space-time sector even just using structures obtained in energy-momentum space. One first observation concerns the possible emergence of a minimum wavelength (maximum momentum). In Refs. [1, 2] I showed (to leading order in $\tilde{L}_p$) that, for $\tilde{L}_p > 0$, the new energy-momentum transformation rules are such that there is indeed a maximum momentum.

Related findings emerge from the analysis of length contraction within the illustrative example of new Relativity theory on which I am focusing. This I showed [1, 2] by analysing a gedanken length-measurement procedure, also using the fact that the dispersion relation $E^2 \simeq c^2p^2 + \tilde{L}_p cEp^2$ corresponds[3] to the deformed speed-of-light law

$$v_\gamma = \frac{dE}{dp} = c(1 + \tilde{L}_p c^{-1} E) \ . \tag{3}$$

The analysis of the gedanken length-measurement procedure suggests [1, 2] that (again, for $\tilde{L}_p > 0$ in the new Relativity theory) when one inertial observer assigns to a length value greater than $\tilde{L}_p$ all other inertial observers also find that length to be greater than $\tilde{L}_p$.

Kinematical Conditions for Particle-Production Processes

As I emphasized in Ref. [1], an important requirement for the logical consistency of a Relativity theory is that all observers should agree on whether or not a certain particle-production process is allowed. This requirement is trivially satisfied in ordinary Special Relativity. For the new Relativity theory let me here consider the simple case of

---

[3] The careful reader will realize that by assuming that the relation $v = dE/dp$ is unmodified I am actually stating a (perhaps not very strong) property of the space-time sector. From the results obtained in Ref. [15] one can conclude that this property is enjoyed by the space-time of (2).

a scattering process $a+b \rightarrow c+d$ (collision processes with two incoming particles and two outgoing particles) in one space dimension (working again to leading order in $\tilde{L}_p$).

The fact that the energy-momentum transformation rules imposed by the postulates (law1),(law2) are non-linear (unlike special-relativistic transformation rules) provides room for various alternatives for the laws to be satisfied by particle-production processes [1]. Here I just want to mention two possibilities whose consistency with the postulates has been already verified. The first example is

- (cons): $a+b \rightarrow c+d$ collision processes must satisfy the requirements

$$E_a+E_b-\tilde{L}_p c p_a p_b - E_c - E_d + \tilde{L}_p c p_c p_d = 0 , \tag{4}$$

$$p_a + p_b - \tilde{L}_p (E_a p_b + E_b p_a)/c - p_c - p_d + \tilde{L}_p (E_c p_d + E_d p_c)/c = 0 . \tag{5}$$

The type of non-linearity encoded in the postulates (law1),(law2) is also consistent with a completely different alternative type of particle-production laws. Whereas (cons) is a "single-channel conservation law", just like its counter-part in Special Relativity, it is also possible to find consistent multi-channel laws for particle production. A significant example is

- (cons'): $a+b \rightarrow c+d$ collision processes must satisfy one of the requirements obtained by permutations of $p_a, p_b, \tilde{p}_c, \tilde{p}_d$ in the conditions

  $$E_a+E_b-E_c-E_d = 0 , \tag{6}$$

  $$p_a \dot{+} p_b \dot{+} \tilde{p}_c \dot{+} \tilde{p}_d = 0 , \tag{7}$$

  where the deformed sum $\dot{+}$ is defined by $k \dot{+} q \equiv k + q + \tilde{L}_p E_k q$ and $\tilde{k} \equiv -k - \tilde{L}_p E_k k$ (with E_k denoting the energy that corresponds to the momentum k).

It is somewhat more difficult to build some intuition for this scenario (cons'). However, if the expectation that the space-time sector is described by (2) is correct, the law (cons') is actually to be favoured. In fact, in the very recent Ref. [5] it was shown that the construction of a consistent theory on the space-time (2) leads to the law (cons').

The content of (cons') is not as shocking as it may seem at first sight. It says that, in the case of a $a+b \rightarrow c+d$ process, there are 24 channels available to the process (associated with the 24 possible permutations of the particle momenta $p_a, p_b, \tilde{p}_c, \tilde{p}_d$). The process will be allowed whenever one of the 24 cases is satisfied. Reassuringly in the limit in which the particles have energies (and momenta) much smaller than $1/|\tilde{L}_p|$ all 24 channels collapse into a single energy-momentum conservation condition, the one of Special Relativity.

Postulates Beyond Leading Order

The results described up to this point are the ones on which I based my proposal [1, 2] of Relativity postulates with more than one observer independent scale. The analysis was done in leading order in $\tilde{L}_p$, leaving for future studies the search of consistent choices of the all-order function $f(E, p; \tilde{L}_p)$ that appears in the postulate (law2). As mentioned above (and in Refs. [1, 2]) an important hint appears to come from the fact that my analyses led to leading-order expressions for the generators of boosts and rotations that are recognizable as the leading-order approximation of the generators in the Lorentz sector of the example of κ-Poincaré Hopf algebra proposed in Ref. [12]. The connection between the new Relativity postulates and this Hopf algebra was further explored, within an all-order analysis, by Kowalski-Glikman [3] who adopted as a natural candidate for the function $f(E,p;\tilde{L}_p)$ the one which is inferred from the all-order form of the relevant κ-Poincaré "M^2" casimir (*i.e.* in analogy with my leading-order proposal). Specifically, using the form of this casimir, Kowalski-Glikman provided an all-order generalization of one of the arguments which I used [2] in support of the emergence of a minimum wavelength (maximum momentum). Bruno, Kowalski-Glikman and I also showed, in a very recent study [6], that the use of the relevant κ-Poincaré "M^2" casimir in the postulates leads to consistent transformation rules between different inertial observers, again generalizing to all orders the leading-order results I reported in Refs. [1, 2]. An all-order formulation of the particle-production rules (cons') were reported by Arzano and myself in the very recent Ref. [5]. All these results appear to support my conjecture that the results reported in leading order in Refs. [1, 2] could be straightforwardly generalized to the level of an all-order analysis.

RELATION WITH QUANTUM SYMMETRIES

At least within the chosen illustrative example of new postulates, it appears that in the new Relativity theory the subject of "quantum groups" and "quantum algebras" should be in some way relevant [1]. In fact, as already observed above, the postulate (law2) involves a dispersion relation which corresponds to the leading-order-in-$\tilde{L}_p$ version of a casimir that has emerged [12, 13] in the quantum-algebra literature, and, upon imposing consistency with the Relativity Principle, I was led to boost (and rotation) generators which can also be recognized as the leading-order-in-$\tilde{L}_p$ version of the generators of the relevant quantum algebra.

While there is a wide-spread belief (see, *e.g.*, Ref. [11]) that κ-Poincaré quantum algebras do not have an associated group action (this action should only lead to a "quasi-group" in the sense of Batalin [16]), I have argued in Ref. [1] that the Lorentz sector of the specific κ-Poincaré algebra [12, 13] that appears to be relevant for my illustrative example of new Relativity postulates does reassuringly lead to ordinary group structure. This follows from the observation that the Lorentz sector of the relevant κ-Poincaré algebra does satisfy corresponding criteria derived by Batalin [16].

However, the analysis of particle-production processes reported above appears to require some new algebraic tools. In particular, at least according to the standard interpretation of the strictly mathematical language of analysis of quantum algebras, the mathematics literature would support the expectation [12, 13] that energy-momentum conservation in collision processes should involve a troubling lack of symmetry between pairs of identical particles. On the contrary, my analysis shows that consistency with the postulates does not require any such loss of symmetry under exchange of particles.

The very recent analysis reported in Ref. [5] appears to provide the tools for introducing these new concepts in κ-Poincaré, and provides solutions for some of the reasons of concern which have traditionally obstructed attempts to apply the κ-Poincaré formalism in physics. To these observations I devote the remainder of this Section.

Role of the κ-Poincaré Coproduct in Particle-Production Rules

One of the key obstacles for physics applications of the κ-Poincaré formalism is associated with the κ-Poincaré coproduct, which readers familiar with κ-Poincaré will recognize (of course, in leading order) in the "$\dot{+}$" operation. The fact that $p\dot{+}k$ is affected by a severe loss of p,k-exchange symmetry has motivated some skepticism toward the applicability of κ-Poincaré in physics. For example, in the κ-Poincaré literature it has been assumed that scattering processes involving two incoming and two outgoing particles should conserve total momentum in the sense that the coproduct sum of the incoming momenta should equal the coproduct sum of the outgoing momenta. This appears troubling since the lack of symmetry of the coproduct would imply, for example, that in the case of two identical particles colliding to produce two other identical particles one should choose which of the incoming momenta enters the coproduct from the left and a similar choice would have to be made for the outgoing particles. This problem is solved by the particle-production rule (cons'), which involves the coproduct in a way that does not force us to choose the ordering of the incoming and outgoing momenta. The price payed for this reassuring result is that (cons') does not admit interpretation as an ordinary rule of energy-momentum conservation: (cons') states that, *e.g.*, a process with two incoming and two outgoing particles can be realized through any one of 24 conservation rules, obtained by permutations of the four momenta involved in the process.

As a first example of application of the rule (cons') let me note here the explicit formulas that according to (cons') would describe the process in which a photon of energy E and a photon of energy ε, with $\varepsilon \ll E$, collide and produce an electron-positron pair. The analysis of (cons') is rather simple if we assume that the process is "at threshold" 'citegactp2 and we include only the leading-order corrections, of order $\tilde{L}_p E^2$. In this limit one easily finds that the 24 channels actually all give raise to the same (leading-order) conservation rule. While in conventional physics one would impose the relation $2p = E - \varepsilon$ on the common momentum of the produced pair, according to (cons') one should impose the condition $2p = E - \varepsilon + \tilde{L}_p E^2/4$. Combining this result with the structure of the dispersion relation one finds that there is no leading-order deformation of the threshold condition: the deformation of the dispersion relation is compensated by the deformation of momentum conservation, giving back the ordinary threshold condition $E\varepsilon = m_e^2$, with m_e the electron mass.

This cancellation of leading-order corrections to the threshold condition for processes described in a highly boosted frame (a frame which is highly boosted with respect to the center-of-mass frame) may at first appear reassuring. However, it is instead more correct to describe this result of cancellation of leading-order effect as disappointing. In fact, just for processes seen in a "LAB frame" which is highly boosted with respect to the center-of-mass frame there is growing evidence in support of an anomaly in the threshold conditions. This evidence emerges from astrophysical observations which I will discuss in the last Section, but the key point is that in order to explain these observations it would have been useful [14, 17] to encounter a leading-order correction to the threshold conditions.

The next question to ask is of course: Does the cancellation of leading-order corrections to the threshold conditions mean that the new Relativity theory cannot provide an explanation for these puzzling observations? A definite answer to this question still requires additional investigations. It seems to me that there are at least three promising avenues to seek a solution of the observational paradoxes within the new Relativity theory: (i) The peak of the cross section for the particle-physics processes [14, 17] relevant for the mentioned observational paradoxes is not exactly at threshold; it is somewhat above threshold. The fact that the leading-order corrections cancel each other out for the very special conditions required by threshold production does not necessarily imply that above threshold one should find a similar cancellation. This should be studied and compared with the observations. (ii) The structure of the new particle-production kinematical requirements of the new Relativity theory may actually combine in non-trivial way when a given process actually involves more than one microscopic process (*e.g.* a small cascade). Again, this may affect the comparison of the new theory with observations. (iii) As I shall emphasize in the next Subsection, it appears that the new Relativity theory requires careful handling of composite particles (particles composed by a few fundamental particles). The mentioned astrophysical observations are believed to involve [14, 17] photons, electrons, protons and pions. At least protons and pions cannot be treated as truly fundamental particles. In the next Subsection I suggest that collisions involving composite particles might behave quite differently from collisions among fundamental particles. This should be taken into account in seeking an explanation for the mentioned puzzling observations in astrophysics.

Differences Between Microscopic and Macroscopic Bodies

There is another, perhaps even more serious, obstacle for physics applications of the κ-Poincaré formalism: while deformed dispersion relations of the type $E^2 = c^2p^2 + \tilde{L}_p cEp^2$ are consistent with all available data on fundamental particles such a deformation is clearly unacceptable for macroscopic bodies. If κ-Poincaré should play a role in physics, the deformation of the dispersion relation must somehow be confined to systems of one or a few fundamental particles; it should not hold for macroscopic bodies.

In the way in which the κ-Poincaré formalism has been developed until now there is no room for such a separation between microscopic and macroscopic realms. In fact, it was assumed that the coproduct should characterize the total momentum of a multi-particle system, and this is found to lead to total momentum and total energy which transform just like the single-particle energy momentum, *i.e.* following the deformed dispersion relation.

The proposal of (cons') can be also used to motivate a solution of this problem, again inspired by the idea that some of the new Relativity theories of the general type proposed in Ref. [1, 2] might in some way involve the κ-Poincaré formalism. The point is that the concept of total momentum of a multi-particle system must be introduced to reflect an operatively-defined property of a physical system. A natural opportunity for attributing a physical meaning to the concept of total momentum is provided by collision processes: we will be able to give operative meaning to the concept of total momentum of two incoming particles if some combination of the energy-momenta of these incoming particles is conserved in the process. The rules (cons') do not admit this type of interpretation. According to (cons') it is in particular not legitimate to adopt the coproduct sum of two incoming momenta as the total momentum of that two-particle system. This is sufficient to provide the needed opportunity for a separation between microscopic and macroscopic realms: if the total momentum is not identified with the coproduct sum of the momenta it will not necessarily obey the same dispersion relation of the energy-momentum of a single particle.

Let me also observe, however, that in some weak sense it is possible to introduce some sort of total momentum. Using again (cons') it appears that we should describe the concept of total momentum only as some sort of average property of a macroscopic body. We will have a good definition of total momentum if we identify a characteristic momentum of a macroscopic body which is conserved in collisions between macroscopic bodies. (cons') does not allow to enforce such a condition exactly, but it does allow to introduce such a condition in an appropriate statistical sense. Let us consider the collision of two macroscopic bodies: such a collision at a fundamental level will actually involve a very large number of collisions between the fundamental particles that compose the macroscopic bodies. Each of these microscopic collisions will actually be characterized by a single one of the 24 channels (when the process is $2 \rightarrow 2$) but the collision between two macroscopic bodies will involve such a large number of these microscopic collisions that it will be characterized by the average of the 24 channels.

At least in leading order in $\tilde{L}_p$, it appears plausible that this authomatic averaging procedure would lead to the introduction of a (non-fundamental) concept of total momentum of a macroscopic body which, just as in conventional physics, is based on the ordinary sum of momenta. (This however might only be applicable to collisions between macroscopic bodies whose velocities are not very high, so that a small boost is sufficient to take the system to the center-of-mass frame.) In support for this possibility let me analyze the implications of (cons') for a center-of-mass collision of two identical particles with momenta p and $-p$ respectively, and of course same energy E, that produces two other identical particles just above threshold. This microscopic process would be one of the many microscopic processes that occur when two macroscopic bodies collide. Conventional physics would predict that the sum of the momenta of the outgoing particles should vanish: $p_1' + p_2' = p - p = 0$. From (cons') one easily finds that, in leading order, in this case the 24 channels that characterize (cons') split up into 8 channels with $p_1' + p_2' = 0$, 4 channels with $p_1' + p_2' + \tilde{L}_p(E/c)(p_1' - p_2')/2 = \tilde{L}_p E p/c$, 4 channels with $p_1' + p_2' + \tilde{L}_p(E/c)(p_1' - p_2')/2 = -\tilde{L}_p E p/c$, 4 channels with $p_1' + p_2' - \tilde{L}_p(E/c)(p_1' - p_2')/2 = \tilde{L}_p E p/c$, 4 channels with $p_1' + p_2' - \tilde{L}_p(E/c)(p_1' - p_2')/2 = -\tilde{L}_p E p/c$. A single process of this type would follow a single one of these options, but a collection of a large number of these processes would be primarily characterized by the average behaviour of the 24 channels, which is simply $< p_1' + p_2' >= 0$.

RELATIONS WITH OTHER QUANTUM-GRAVITY STUDIES

Before the proposal [1, 2] the only option for introducing the Planck length in the fundamental structure of space-time was provided by the present interpretation of L_p as a scale characteristic of the rules of dynamics, just a rescaled value of the gravitational coupling G. In that conventional picture the Planck length could enter space-time structure only when accompanied by an associated background [1, 18, 19, 20] (*e.g.* as a scale present in a/the vacuum solution of the equations of dynamics). The results I reported in Refs. [1, 2] show that in addition to this traditional scenario, it is also possible to follow another scenario for the introduction of the Planck length, in which there is no preferred class of inertial observers (no background) but a short-distance deformation of boosts and an associated modification of the Relativity postulates is required. My intuition that such a scenario should be explored found additional encouragement even after the announcement of Refs. [1, 2], especially through conversations in which I became aware of arguments put forward by other colleagues [21, 22] in support of the hypothesis that we might eventually encounter a deformation of boosts (of course, also those arguments were motivated [21, 22] by quantum-gravity issues, such as minimum length and the quantum mechanics of black holes).

I must also stress that, in light of my results [1, 2], it appears necessary for authors to be more careful in their description of certain popular quantum-gravity concepts, such a "minimum length" and "deformed dispersion relation". In many quantum-gravity approaches [23, 24, 25, 26, 27, 28, 29, 30, 31] one or another formulation of the concept of "minimum length" is discussed. However, these studies do not clarify how the presence of a minimum length could affect boosts. This appears to be a serious omission, since, as emphasized above, there are two options for introducing such concepts: either as a characteristic of quantum geometrodynamics (without any modification of the Special-Relativity postulates) or as a characteristic of the Relativity postulates. Similar issues arise in the analysis of approaches (see, *e.g.*, Refs. [18, 19]) predicting deformed dispersion relations. Clearly, assuming ordinary special-relativistic rules of transformation of energy and momentum, these dispersion relations would allow to select a preferred class of inertial frames, but I have shown that deformed dispersion relations can also be introduced as observer-independent laws.

Another class of studies which have emerged in more or less direct connection with quantum-gravity research and might be reanalyzed from the perspective advocated in my proposal [1, 2] is the one of deformations of various types of algebras motivated by the desire to implement the existence of concepts such as minimum length, minimum de Broglie wavelength or a maximum accelleration [32, 33, 34, 35, 36]. Again in these studies until now much emphasis has been placed on the algebraic tools, but the readers were left without any explicit remarks concerning the faith of the Special-Relativity postulates. It would be interesting to reanalyse the relevant proposals within the new relativistic conceptual framework here proposed, particularly working toward the identification of transformation rules such that the equations describing minimum length and/or minimum de Broglie wavelength and/or maximum accelleration acquire the status of being observer-independent (valid in every inertial frame). What are then the new relativistic transformation rules between observers? Are they physically acceptable? (For example, do the new Lorentz transformations form group, or just a quasigroup?)

CLOSING REMARKS

From the viewpoint advocated here and in Refs. [1, 2] the Relativity Principle is somewhat hostile to the introduction of observer-idependent physical scales. In that respect, Einstein's Relativity postulates, by introducing c, well deserve to be qualified as "special". My studies have shown that one can also consistently construct a "Doubly Special Relativity", in which the Relativity Principle coexists with observer-independent scales of both length (or momentum) and velocity.

While the motivation for my studies comes from the desire to eventually unify General Relativity and Quantum Mechanics, the approach is at present still only able to handle flat space-time. In working toward a general-relativistic generalization a useful intermediate step could be the one of applying the new postulates in contexts with a curved, but still fixed (non-dynamical), space-time, such as De Sitter or Schwarzschild.

Another interesting possibility is the one of describing space-time curvature as a requirement of non-commuting momenta. If this viewpoint turned out to be correct, one could perhaps obtain a general-relativistic generalization by an appropriate extension of the κ-Minkowski algebra (2) to some sort of "κ" x_i, t, p_i, E space.

Even before these preliminary steps are done, there are certain conceptual issues that must be analyzed. One key point is that at present the Planck length L_p is seen as a quantity which is derived from three fundamental constants c, G and $\hbar$. The fundamental constants c, G and $\hbar$ already have their own operative definitions. If L_p is operatively defined through the Relativity postulates then L_p is also promoted to the status of fundamental constant. The relation between L_p, c, G and $\hbar$ would accordingly acquire the very rare status of a relation between fundamental concepts, all with their own operative definition. This is very rare in physics, but it cannot be excluded since we have at least one example in which something like this happens: the Equivalence-Principle relation between inertial mass and the gravitational mass. My proposal [1, 2] might eventually force us to introduce a new principle somewhat analogous to the introduction of the Equivalence Principle. There is of course another conceptual alternative: somehow this formalism that provides an intrinsic operative definition for the Planck length might eventually lead to the understanding of one of the two scales not explicitly present in the new Relativity postulates, either G or $\hbar$, as a derived concept.

It is important to notice that, as shown by the illustrative example of new postulates on which I focused, new Relativity postulates can have significant phenomenological implications. These implications have been discussed in some detail in Ref. [1]. In particular, the deformation of the dispersion relation introduced in the postulate (law2) can be tested [18, 37, 38, 39] with forthcoming experiments.

The deformed rules for particle production could be most effectively tested in experiments sensitive to the structure of the threshold requirements for particle production. Interestingly, some of these experiments, observations of ultra-high-energy cosmic rays [40] and of Markarian501 photons [41], have recently obtained data that appear to be in conflict with conventional theories and appear to require [42, 43, 44, 45, 14, 17] a deformation of the kinematic conservation rules applied to collision processes. These observations clearly provide some encouragement for the idea of new Relativity postulates. As I emphasized above, the illustrative example of new postulates on which I focused appears to provide an avenue for explaining these paradoxical observations, but more work is needed in order to substantiate this hypothesis. Here and in Ref. [1] I discussed certain mechanisms for explaining the paradoxes within the new Relativity theory, but a detailed analysis of these mechanisms is postponed to future studies.

A third class of phenomenological studies that could be significantly affected by the new Relativity postulates is the one pertaining to cosmology and the early stages of evolution of the Universe. The interested reader can find brief remarks on this point in Ref. [2] and a more detailed (and preliminarily quantitative) study in Ref. [4].

While here and in Refs. [1, 2] I focused mostly on a specific illustrative example, my proposal of exploring the possibility of new Relativity postulates involving the Planck length could of course be followed investigating a large variety of classes of new postulates. In this respect a key point might emerge from the analysis of combinations of boosts. Whereas in the illustrative example pursued until now the new Lorentz transformations form group in the ordinary sense, it appears plausible that other choices of the new postulates would only lead to quasigroup structure [16], a rather undesireable feature.

Another potentially interesting possibility is the one of attempting to introduce even a third observer-independent scale in the postulates. Since my results showed that a logically consistent framework can emerge from Relativity postulates with a second observer-independent scale, it is now natural to wonder whether a third observer-independent scale could also be consistently introduced. Motivated by the studies I reported in Refs. [1, 2], Kowalski-Glikman has briefly presented in Ref. [3] some "aestethic arguments" (not guided by experimental input or by conceptual urgency, but by an intuition for the conceptual elegance of the fundamental laws of physics) in favour of Relativity with three observer-independent scales, but did not formulate any attempt to provide an operative definition of the third scale (the entire analysis reported in Ref. [3] relies on the type of deformation of the dispersion relation which I had introduced with (law2)). Consistently with the explorative spirit of my proposal [1, 2], I neither favour nor disfavour *a priori* any particular number of observer-independent scales. Examples of Relativity postulates with three observer-independent scales should be studied, and, if any class of such postulates turned out to be logically consistent, corresponding experimental tests are certainly well motivated. In this respect I should emphasize that, as discussed in Ref. [1], for each observer-independent scale the postulates should also provide an operative definition of that scale (this key point was omitted in Ref. [3]). Also important are some considerations related with the remarks I made above: by introducing an independent operative definition of the Planck length I am already forced to find a new conceptual understanding of the relation between L_p, c, G and $\hbar$; an even greater conceptual challenge emerges if a third operative definition is introduced in the Relativity postulates (*e.g.* we could contemplate the possibility of interpreting both G and $\hbar$ as derived scales!).

The illustrative example of new Relativity postulates on which I focused ended up making strong contact with preexisting mathematics of κ-Poincaré algebras. However, while preexisting Lorents mathematics really provided all the tools needed for analyses based on Special Relativity, my proposal was confronted [1, 2] with some missing pieces in the development of κ-Poincaré, particularly the lack of understanding of the role of the coproduct in the laws for particle production and the lack of the needed mechanism for confining the applicability of the κ-Poincaré dispersion relation to the microscopic realm. The new Relativity postulates led me to propose some solutions for these outstanding problems of κ-Poincaré. These results, besides playing a key role in my new type of Relativity theories, appear to have even wider significance, possibly of use in all contexts in which κ-Poincaré is being considered as a useful mathematical structure.

ACKNOWLEDGMENTS

I am indebted to several colleagues, starting of course with my collaborators Michele Arzano, Rossano Bruno, Jerzy Kowalski-Glikman and Tsvi Piran. I am also greatful to Jerzy Lukierski for conversations on the subject of "quasigroups", as described by Batalin [16]. Finally, I should thank all the colleagues who expressed interest in the proposal I put forward in Refs. [1, 2] and pointed to my attention some other formalisms and ideas which, like the κ-Poincaré formalism, may turn out to provide opportunities for the introduction of Relativity theories of the type I proposed; in particular, I must thank Dharam Ahluwalia for bringing to my attention the studies in Refs. [33, 34], Roberto Aloisio for bringing to my attention some remarks in Ref. [21] Michele Arzano for bringing to my attention some remarks in Ref. [32], Jerzy Kowalski-Glikman for bringing to my attention Ref. [22] and Giuseppe Marmo for bringing to my attention Refs. [35, 36].

REFERENCES

1. Amelino-Camelia, G., gr-qc/0012051, *Int. J. Mod. Phys.* D (in press).
2. Amelino-Camelia, G., hep-th/0012238, *Phys. Lett.* B (in press).
3. Kowalski-Glikman J., hep-th/0102098.
4. Alexander, S., and Magueijo, J., hep-th/0104093.
5. Amelino-Camelia, G., and Arzano, M., hep-th/0105120.
6. Bruno, R., Amelino-Camelia, G., and Kowalski-Glikman, J., in preparation.
7. Rovelli, C., gr-qc/0006061 (in Procedings of the 9th Marcel Grossmann Meeting on Recent Developments in Theoretical and Experimental General Relativity, Gravitation and Relativistic Field Theories, Rome, Italy, 2-9 Jul 2000).
8. Amelino-Camelia, G., *Nature* **408**, 661 (2000), gr-qc/0012049.
9. 't Hooft, G., *Class. Quant. Grav.* **13**, 1023 (1996).
10. Lukierski, J., A. Nowicki, H. Ruegg, and V.N. Tolstoy, *Phys. Lett.* **B264**, 331–338 (1991).
11. Lukierski, J., Ruegg, H., and Ruhl, W., *Phys. Lett.* **B313**, 357 (1993).
12. Majid, S., and Ruegg, H., *Phys. Lett.* **B334**, 348 (1994).
13. Lukierski, J., Ruegg, H., and Zakrzewski, W. J., *Ann. Phys.* **243**, 90 (1995).
14. Amelino-Camelia, G., and Piran, T., astro-ph/0008107, *Phys. Rev.* D (in press).
15. Amelino-Camelia, G., and Majid, S., *Int. J. Mod. Phys.* **A15**, 4301 (2000).
16. Batalin, I. A., *J. Math. Phys.* **22**, 1837 (1981).
17. Amelino-Camelia, G., and Piran, T., *Phys. Lett.* **B497**, 265 (2001).
18. Amelino-Camelia, G., Ellis, J., Mavromatos, N.E., Nanopoulos, D. V., and Sarkar, S., *Nature*, **393**, 763 (1998); astro-ph/9712103.
19. Gambini, R., and Pullin, J., *Phys. Rev.* **D59**, 124021 (1999).
20. Ellis, J., Mavromatos, N. E., and Nanopoulos, D. V., hep-th/0012216.
21. Garay, L. J., *Int. J. Mod. Phys.* **A10**, 145 (1065).
22. Susskind, L., *Phys. Rev.*, **D49**, 6606 (1994).

23. Mead, C. A., *Phys. Rev.* **135**, B849–B862 (1964).
24. Padmanabhan, T., *Class. Quantum Grav.* **4**, L107 (1987).
25. Ahluwalia, D. V., *Phys. Lett.* **B339**, 301 (1994) 301.
26. Ng, Y. J., and H. Van Dam, H., *Mod. Phys. Lett.* **A9**, 335 (1994).
27. Amelino-Camelia, G., *Nature* **398**, 216 (1999); *Mod. Phys. Lett.* **A9**, 3415 (1994).
28. Veneziano, G., *Europhys. Lett.* **2**, 199 (1986).
29. Gross, D. J., and Mende, P. F., *Nucl. Phys.* **B303**, 407 (1988).
30. Amati, D., Ciafaloni, M., and Veneziano, G., *Phys. Lett.* **B216**, 41 (1989).
31. Kabat, D., and Pouliot, P., *Phys. Rev. Lett.* **77**, 1004 (1996);
Douglas, M. R., Kabat, D., Pouliot, P., and Shenker, S. H., *Nucl. Phys.* **B485**, 85 (1997).
32. Maggiore, M., *Phys. Lett.* **B319**, 83 (1993).
33. Kempf, A., Mangano, G., and Mann, R. B., *Phys. Rev.* **D52**, 1108 (1995).
34. Ahluwalia, D. V., *Phys. Lett.* **A275**, 31 (2000).
35. Caianiello, E. R., *Riv. Nuovo Cim.* **15**, 1 (1992).
36. Caianiello, E. R., Gasperini, M., and Scarpetta, G., *Nuovo Cim.* **105B**, 259 (1990).
37. Schaefer, B. E., *Phys. Rev Lett.* **82**, 4964 (1999).
38. Biller, S. D., *et al*, *Phys. Rev. Lett.* **83**, 2108 (1999).
39. Norris, J. P., Bonnell, J. T., Marani, G. F., Scargle, J. D., `astro-ph/9912136`;
de Angelis, A., `astro-ph/0009271`.
40. Takeda, M., *et al*, *Phys. Rev. Lett.* **81**, 1163 (1998).
41. Aharonian, F. A., *et al*, *A&A* **349**, 11A (1999).
42. Coleman, S., Glashow, S. L., *Phys. Rev.* **D59**, 116008 (1999).
43. Kifune, T., *Astrophys. J. Lett.* **518**, L21 (1999).
44. Aloisio, R., Blasi, P., Ghia, P. L., and Grillo, A. F., *Phys. Rev.* **D62**, 053010 (2000).
45. Protheroe, R. J., and Meyer, H., *Phys. Lett.* **B493**, 1 (2000).

Renormalization Problems in Noncommutative Gauge Theories

L. Bonora and M. Salizzoni[1]

Scuola Internazionale Superiore di Studi Avanzati, Via Beirut 2-4, 34014 Trieste, Italy, and INFN, Sezione di Trieste, email: bonora@sissa.it, sali@sissa.it

Abstract. We discuss some renormalization problems that arise at one loop in noncommutative gauge theories in 4D. It is shown that the the noncommutative U(N) theory is one–loop renormalizable. When the analysis is then extended to the noncommutative orthogonal gauge theories we find that for one–loop amplitudes the situation is more problematic.

INTRODUCTION

When D–branes are in presence of a constant NSNS B-field the low energy effective action of the open strings attached to the branes can be represented by a Euclidean YM theory defined on a noncommutative spacetime endowed with a Moyal bracket. All this clearly holds at a semiclassical level (i.e. tree amplitudes computed in the field theory setting compare well with string theory in the $\alpha' \to 0$ limit). But it is natural to compare loop amplitudes calculated both in string theory and in the corresponding noncommutative field theory, in order to see how effective the noncommutative effective field theory is. Several calculations of this type have been carried out, see in particular [1, 2, 3]. Not everything is exactly parallel to ordinary gauge theories. For instance it would seem that more general * products appear in the effective actions. However this question can be settled with appropriate careful calculations, see [4, 5]. Other problems, such as the IR/UV mixing in renormalization are still an open problem.

It seems to be important therefore to know exactly what are the properties of a noncommutative YM theory we can rely on. One of the basic properties of a local field theory is renormalizability. In this report we study renormalizion of different noncommutative gauge field theories. We first consider the case of a noncommutative YM theory with $U(N)$ gauge group in 4D without matter and analyze its one–loop renormalizability properties. Since non–planar singularities are dumped by the noncommutative parameter θ, only planar one–loop contributions are relevant in the UV region. Actually one can prove, [1, 2, 3, 6], that noncommutative $U(N)$ gauge theories are one–loop renormalizable, exactly as the ordinary YM theories. Next we consider noncommutative gauge theories whose Lie algebra are determined by orthogonal or symplectic groups. One can show that these theories can indeed be defined and correspond to the field theory limit of

[1] This work was partially supported by the Italian MURST for the program "Fisica Teorica delle Interazioni Fondamentali"

CP589, *Fundamental Interaction Theories*, edited by J. Lukierski and J. Rembieliński

open string theories attached to D-branes at tree level in the presence of an orientifold, [7]. We show however, in a simple example, that the field theory limit of the one–loop string amplitudes disagree with the (naive) one–loop amplitudes calculated from the noncommutative field theory. A reformulation of the noncommutative gauge theory is needed in this case.

ONE–LOOP RENORMALIZATION OF NONCOMMUTATIVE $U(N)$ YM THEORIES

The noncommutative $U(N)$ gauge theories are specified by the action

$$S = \int d^4x \,\mathrm{Tr}\left(-\frac{1}{2}F_{\mu\nu}F^{\mu\nu} - \frac{1}{\alpha}(\partial_\mu A^\mu)^2 + i\bar{c} * \partial_\mu D^\mu c - i\partial_\mu D^\mu c * \bar{c}\right), \tag{1}$$

where

$$F_{\mu\nu} = \partial_\mu A_\nu - \partial_\nu A_\mu - ig(A_\mu * A_\nu - A_\nu * A_\mu) \tag{2}$$

and the Moyal product for functions in $\mathbb{R}^4$ is defined by

$$f * g(x) \equiv f(x) e^{\frac{i}{2}\theta^{\mu\nu}\overleftarrow{\partial}_\mu \overrightarrow{\partial}_\nu} g(x) . \tag{3}$$

The potential A_μ is valued in the Lie algebra $u(N)$, i.e. is an hermitian matrix, and we will choose the Feynman gauge $\alpha = 1$. Our conventions for the Lie algebra $u(N)$ tensors are as follows. Use is made of the basis t^a, $a = 1, \ldots, N^2 - 1$ of traceless hermitean matrices for the Lie algebra $su(N)$, with normalization

$$\mathrm{tr}(t^a t^b) = \frac{1}{2}\delta^{ab} \tag{4}$$

and structure constants f_{abc} defined by

$$[t^a, t^b] = i f_{abc} t^c . \tag{5}$$

The third order ad-invariant completely symmetric tensor d_{abc} is given by

$$\{t^a, t^b\} = \frac{1}{N}\delta_{ab} + d_{abc}t^c . \tag{6}$$

Next one passes to the Lie algebra $u(N)$ by introducing the additional generator $t^0 = \frac{1}{\sqrt{2N}}\mathbf{1}_N$. Corresponding to any index a for $su(N)$ one introduces the index $A = (0, a)$, so that A runs from 0 to $N^2 - 1$. So one has

$$[t^A, t^B] = i f_{ABC} t^C , \quad \{t^A, t^B\} = d_{ABC} t^C , \tag{7}$$

where f_{ABC} is completely antisymmetric, f_{abc} is the same as for $su(N)$ and $f_{0BC} = 0$, while d_{ABC} is completely symmetric; d_{abc} is the same as for $su(N)$, $d_{0BC} = \sqrt{\frac{2}{N}}\delta_{BC}$,

$d_{00c} = 0$ and $d_{000} = \sqrt{\frac{2}{N}}$. We have also

$$\mathrm{Tr}(t^A t^B) = \frac{1}{2}\delta^{AB} . \tag{8}$$

The Feynman rules for (1) are collected in the Appendix of [6]. Gluons carry Lorentz indices $\mu, \nu, \ldots$, color indices $A, B, \ldots$, and momenta $p, q, \ldots$. Ghosts carry only the last two type of labels. All the momenta are entering and we use the notation $p \times q = \frac{1}{2} p_\mu \theta^{\mu\nu} q_\nu$. Evaluating the one–loop contributions is lengthy but straightforward. One must evaluate the planar part of the 2–, 3–point and 4–point functions. Adopting the dimensional regularization ($\varepsilon = 4 - D$), we extract first the planar part (i.e. the part in which $\theta^{\mu\nu}$ is saturated only by external momenta) and, out of it, the divergent part. The relevant results are are as follows.

For the 2–point function we have two nonvanishing contribution to the UV divergent part:
– gluons circulating inside the loop:

$$i\frac{1}{(4\pi)^2}\frac{2}{\varepsilon}\delta_{AB}N\left[\frac{19}{12}g_{\mu\rho}p^2 - \frac{11}{6}p_\mu p_\nu\right] , \tag{9}$$

– ghosts circulating inside the loop:

$$i\frac{1}{(4\pi)^2}\frac{2}{\varepsilon}\delta_{AB}N\left[\frac{1}{12}g_{\mu\rho}p^2 + \frac{1}{6}p_\mu p_\nu\right] . \tag{10}$$

Their sum is:

$$i\frac{1}{(4\pi)^2}\frac{2}{\varepsilon}\delta_{AB}N\frac{5}{3}\left[g_{\mu\rho}p^2 - p_\mu p_\nu\right] , \tag{11}$$

which entails the usual renormalization constant $Z_3 = 1 + \frac{5}{3}g^2 N \frac{1}{(4\pi)^2}\frac{2}{\varepsilon}$.

Let us consider now the 3–point function. The external gluons carry labels (A, p, μ), (B, q, ν) and (C, k, λ) for the Lie algebra, momentum and Lorentz indices. They are ordered in anticlockwise sense. The triangle diagram gives

$$-\frac{13}{8}g^3 N \frac{1}{(4\pi)^2}\frac{2}{\varepsilon}(\cos(p\times q)f_{ABC} + \sin(p\times q)d_{ABC})$$
$$\left((p-q)_\lambda g_{\mu\nu} + (q-k)_\mu g_{\nu\lambda} + (k-p)_\nu g_{\mu\lambda}\right) . \tag{12}$$

The diagram with one three–gluon vertex and one four–gluon vertex gives:

$$\frac{9}{4}g^3 N \frac{1}{(4\pi)^2}\frac{2}{\varepsilon}(\cos(p\times q)f_{ABC} + \sin(p\times q)d_{ABC})$$
$$\left((p-q)_\lambda g_{\mu\nu} + (q-k)_\mu g_{\nu\lambda} + (k-p)_\nu g_{\mu\lambda}\right) . \tag{13}$$

The contribution of the ghost circulating diagram is:

$$\frac{1}{24}g^3 N \frac{1}{(4\pi)^2}\frac{2}{\varepsilon}(\cos(p\times q)f_{ABC} + \sin(p\times q)d_{ABC})$$
$$\left((p-q)_\lambda g_{\mu\nu} + (q-k)_\mu g_{\nu\lambda} + (k-p)_\nu g_{\mu\lambda}\right) . \tag{14}$$

The sum of the coefficients is

$$-\frac{13}{8}+\frac{9}{4}+\frac{1}{24}=\frac{2}{3}. \tag{15}$$

Therefore, as in the ordinary YM theory, the renormalization constant Z_1 is

$$Z_1 = 1+\frac{2}{3}g^2 N\frac{1}{(4\pi)^2}\frac{2}{\varepsilon}. \tag{16}$$

As for the four–point function, there are four distinct graphs contributing: the gluon box, the ghost box, the gluon triangle and the gluon 'candy' containing a loop with two internal lines. There are two main type of contributions, distinguished by their Lie algebra tensor structure. The first is characterized by Kronecker delta functions in the Lie algebra indices, while the second consists of d and f tensors. The first type contributions, which are potentially dangerous for renormalizability, luckily vanish graph by graph. The explicit calculations are reported in [6]. The result there entails that the four–A term in the action is renormalized with a Z_4 given by

$$Z_4 = 1-\frac{N}{3}g^2\frac{2}{\varepsilon}\frac{1}{(4\pi)^2}. \tag{17}$$

This is the same renormalization that occurs in ordinary $U(N)$ Yang–Mills theories. Therefore, *the noncommutative $U(N)$ Yang–Mills theories are one–loop renormalizable.*

It has been shown in [3, 6] that the same conclusion does not hold if one repeat the same calculations by restricting the Feynman rules to the $SU(N)$ indices. Consequently the theory (if it is a theory) defined by these rules is not renormalizable

NONCOMMUTATIVE ORTHOGONAL GAUGE THEORIES

At first sight trying to define a noncommutative gauge theory corresponding to subgroup of $U(n)$ and a string/brane theory configuration that corresponds to it, does not look very promising: the product of two gauge transformations valued in a Lie subalgebra of $u(N)$ is not valued in the same Lie subalgebra. However the question admit in some sense a positive answer, [7], see also [8]. It is in fact possible to find consistent noncommutative extension for gauge theories corresponding to certain subgroups of $U(n)$. The main point is that it is possible to define gauge transformations that close to form a subgroup of the group of $NCU(n)$ gauge transformations even though the corresponding gauge potentials and gauge transformations are not valued in a classical Lie subalgebra of the unitary Lie algebra $u(n)$.

We work in $\mathbb{R}^d$ and use the Moyal bracket (3). We define the algebra $\mathcal{A}_\theta$ as the vector space $\mathcal{A} \equiv C^\infty(\mathbb{R}^d, \mathbb{H})$ of functions endowed with this product. Although the product of two gauge transformations valued in a proper subalgebra of $u(n)$ is in general not valued in the same subalgebra, one remarks that in order to define a gauge theory one only needs to define a Lie algebra of (infinitesimal) gauge transformations. And for some subalgebras of $u(n)$ this is possible.

To start with we work in a setting in which θ is a parameter. Accordingly we consider $\mathcal{A}_\theta$ as an algebra of power series in θ. This algebra has an anti–automorphism r defined by

$$(.)^r : f(x,\theta) \mapsto f^r(x,\theta) \equiv f(x,-\theta)\,. \tag{18}$$

This map reduces to the identity on the generators x^μ and reverses the order in the product: $(x_1^\mu * \ldots * x_n^\mu)^r = (x_n^\mu)^r * \ldots * (x_1^\mu)^r$.

Next we consider the algebras $so(N)$ and $sp(N)$ as subalgebras of $u(N)$. In other words we keep the usual antihermiticity condition on the $u(n)$–valued connections A and gauge transformations λ, i.e.

$$\begin{aligned} A^*_{ij}(x,\theta) &= -A_{ji}(x,\theta)\,,\\ \lambda^*_{ij}(x,\theta) &= -\lambda_{ji}(x,\theta)\,. \end{aligned} \tag{19}$$

We use Greek letters for space-time indices and i and j for matrix (group) indices.

To define $NCSO(n)$ gauge theories we select connections and gauge transformations satisfying the following constraints:

$$\begin{aligned} A^r_{ij}(x,\theta) &= -A_{ji}(x,\theta)\,,\\ \lambda^r_{ij}(x,\theta) &= -\lambda_{ji}(x,\theta)\,. \end{aligned} \tag{20}$$

It is easy to see that they are preserved by gauge transformations. One can see it componentwise. Alternatively, rewrite (20) in the concise form $A = -(A^t)^r$ and $\lambda = -(\lambda^t)^r$, i.e. t is the matrix transposition. Define $((.)^t)^r \equiv (.)^{rt}$. The proof is now formally similar to the usual one for $U(n)$: $(\lambda * A - A * \lambda)^{rt} = A^{rt} * \lambda^{rt} - \lambda^{rt} * A^{rt} = -(\lambda * A - A * \lambda)$.

As anticipated above, under (20) connections and gauge parameters do not turn out to be $so(n)$–valued. Nevertheless (20) introduces restrictions on the matrix functions A_{ij}. To see what they are, let us write (20) more explicitly

$$\begin{aligned} A_{ij}(x,\theta) &= -A_{ji}(x,-\theta)\,,\\ \lambda_{ij}(x,\theta) &= -\lambda_{ji}(x,-\theta)\,. \end{aligned} \tag{21}$$

Inserting a power expansion in θ for A

$$A^\mu(x,\theta) = A_0^\mu(x) + i\theta_{\nu\rho}A_1^{\mu\nu\rho}(x) + \ldots\,, \tag{22}$$

we see that (20) implies that $A_0, A_2, \ldots$ are antisymmetric and $A_1, A_3 \ldots$ symmetric. The hermiticity condition (19) imposes that all the coefficients $A_0, A_1, \ldots$ be real. The same conclusions hold for the power expansion of λ.

Up to now, $A_0, A_1, \ldots$ are unrestricted, except for the just mentioned constraint. However, if we want to make connection with string theory, $A_1, A_2, \ldots$ are expected not to introduce new degrees of freedom, but to be functionally dependent on A_0. In practice they are thought to be given by the Seiberg–Witten map [9]:

$$A^\mu(A_0) = A_0^\mu - \frac{i}{4}\theta^{\nu\rho}\{A_{0\nu}, \partial_\rho A_0^\mu + F_{0\rho}{}^\mu\} + O(\theta^2)\,. \tag{23}$$

This is indeed consistent: the term linear in θ is symmetric if the first term is antisymmetric. In fact, one can also see that the next term is antisymmetric, and so on; so there is complete accord with (22).

To define a Yang–Mills $NCSO(n)$ theory, let $A = A(x,\theta)$ satisfy the constraint (20). The action is the usual one

$$S = -\frac{1}{4}\int d^d x F_{ij}^{\mu\nu} F_{ji\mu\nu}\,, \tag{24}$$

where F is defined as

$$F_{\mu\nu} = \partial_{[\mu}A_{\nu]} + A_\mu * A_\nu - A_\nu * A_\mu\,. \tag{25}$$

The action (24) is naturally gauge invariant under $NCSO(n)$ and positive. It reduces to the usual one for $SO(n)$ in the $\theta = 0$ case.

In a completely similar way one can define noncommutative $Sp(N)$ gauge theories.

In [7] it was shown that the above defined gauge theories can be regarded as $\alpha' \to 0$ limits of open string theories attached to D–branes collapsed over an orientifold in the presence of a background B field. In [7] it was also shown that the correlators of the vertex operators for the Yang–Mills fields living on the D–branes in the $\alpha' \to 0$ limit become the amplitudes of the field theory (24) at three level. Of course, like in the $U(N)$ noncommutative gauge theory case, the question arises of whether this correspondence holds also at one–loop order. A simple example will show that the situation here is less clearcut.

Let us consider the $NCSO(2)$ case. From the string theory point of view it can be argued ,[10], that the theory should not have UV divergences. The one–loop contributions to open string amplitudes with $SO(n)$ Chan–Paton factors are of three types: planar (P) with the world–sheet of the annulus, nonplanar (NP) with the same world–sheet and nonorientable (NO) with the world–sheet of the Moebius strip. Due to the structure of the string propagators on the annulus (P) and on the Moebius strip (NO), the contributions in the presence and in the absence of the B field differ only by noncommutative factors of type $\cos(p\times q)$ or $\sin(p\times q)$, where p and q are external momenta. It follows that the contributions which become divergent in the field theory limit are the same whether B is there or not. Now, in the ordinary $SO(N)$ case the divergent part comes from the planar contribution with a factor of N in front, and from the NO contribution with a factor of -2. So altogether the divergent field theory part is proportional to $N-2$, and therefore vanishes in the case $N = 2$. This is obvious from the ordinary field theory side, because the theory is free. However, as we noticed above, this conclusion holds also in the noncommutative case. Therefore the $NCSO(2)$ theory should be finite.

Now let us look at the one–loop order on the noncommutative field theory side. The Feynman rules are very simple in this case since only the four–point vertex is nonvanishing. If p,q,r,s and μ,ν,ρ,σ are the momenta and Lorentz indices of the four legs in clockwise order, the Feynman rule gives:

$$\begin{aligned} -2ig^2 \quad &\big[\cos(p\times r - q\times s)(g_{\mu\rho}g_{\nu\sigma} + g_{\mu\nu}g_{\rho\sigma} - 2g_{\mu\sigma}g_{\nu\rho}) \\ &+\cos(p\times s + q\times r)(g_{\mu\sigma}g_{\nu\rho} + g_{\mu\rho}g_{\nu\sigma} - 2g_{\mu\nu}g_{\rho\sigma}) \\ &+\cos(p\times s - q\times r)(g_{\mu\nu}g_{\rho\sigma} + g_{\mu\sigma}g_{\nu\rho} - 2g_{\mu\rho}g_{\nu\sigma})\big]\,. \end{aligned} \tag{26}$$

The one–loop correction is infinite. So the theory needs a renormalization. What is worse is that the divergent part is not of the form (26), but

$$\begin{aligned}\sim \frac{g^4}{\varepsilon} \quad & \left[\cos(p\times r - q\times s)(7g_{\mu\rho}g_{\nu\sigma} + 7g_{\mu\nu}g_{\rho\sigma} - 8g_{\mu\sigma}g_{\nu\rho}) \right. \\ & + \cos(p\times s + q\times r)(7g_{\mu\sigma}g_{\nu\rho} + 7g_{\mu\rho}g_{\nu\sigma} - 8g_{\mu\nu}g_{\rho\sigma}) \\ & \left. + \cos(p\times s - q\times r)(7g_{\mu\nu}g_{\rho\sigma} + 7g_{\mu\sigma}g_{\nu\rho} - 8g_{\mu\rho}g_{\nu\sigma})\right] .\end{aligned} \tag{27}$$

In order to eliminate this divergence we need a counterterm of the form

$$\sim (7A_\mu * A_\mu * A_\nu * A_\nu - 4A_\mu * A_\nu * A_\mu * A_\nu) . \tag{28}$$

Therefore not only the $NCSO(2)$ gauge field theory is not finite, but the divergent part breaks the gauge symmetry. It is apparent that either a careful adjustment of the gauge field theory at the one loop level is required.

ACKNOWLEDGMENTS

We would like to thank T. Krajewski, M. Schnabl, M. Sheikh–Jabbari, A. Tomasiello for useful discussions.

REFERENCES

1. Sheikh–Jabbari, M., *One loop renormalizability of supersymmetric Yang–Mills theories on noncommutative two–torus, JHEP* **9906**, 015 (1999), hep-th/9903107.
2. Krajewski, T., and Wulkenhaar, R., *Int.J.Mod.Phys.*, **A15**, 1011 (2000), hep-th/9903187.
3. Armoni, A., *Comments on Perturbative Dynamics of Non-Commutative Yang-Mills Theory*, hep-th/0005208.
4. Liu, H., and Michelson, J., **–Trek: The one–loop $\mathcal{N}=4$ noncommutative SYM action*, hep-th/0008205.
5. Zanon, D., *Noncommutative perturbation in superspace*, hep-th/0009196;
   Santambrogio, A., and Zanon, D., *One–loop four–point function in noncommutative $\mathcal{N}=4$ Yang–Mills theory*, hep-th/0010275.
6. Bonora, L., and Salizzoni, M., *Phys. Lett.*, **B504**, 80 (2001), hep-th/0011088.
7. Bonora, L., Schnabl, M., Sheikh–Jabbari, M.M., and Tomasiello, A., *Nucl.Phys.*, **B589**, 461 (2000), hep-th/0006091.
8. Jurco, B., Möller, L., Scharml, S., Schupp, P., and Wess, J., *Construction of nonabelian gauge theories onnoncommutative spaces*; hep-th/0104153.
9. Seiberg, N., and Witten, E., *String Theory and Noncommutative Geometry*, *JHEP* **09**, 032 (1999), hep-th/9908142.
10. Bonora, L., and Salizzoni, M., to appear.

The Explicit Construction of Irreducible Representations of the Quantum Algebras $U_q(sl(n))$

Č. Burdík*, R.C. King† and T.A. Welsh†**

*Department of Mathematics, Czech Technical University, Trojanova 13, Prague, Czech Republic, email: burdik@siduri.fjfi.cvut.cz

†Faculty of Mathematical Studies, University of Southampton, Southampton SO17 1BJ, England, email: rck@maths.soton.ac.uk

**Department of Mathematics and Statistics, University of Melbourne, Victoria 3010, Australia, email: trevor@ms.unimelb.edu.au

Abstract. The duality between the quantum algebra $U_q(sl(n))$ and the Hecke algebra $H_m(q^2)$ first pointed out by Jimbo is exploited to construct explicit irreducible representations of $U_q(sl(n))$. The method is based on the use of Young tableaux and involves the notion of q-dependent Young symmetrisers. A key role is played by q-dependent generalisations of the Garnir identities. The appropriate algorithm is first described and illustrated in the generic case for which q is not a root of unity. All matrix elements for the irreducible representations of $U_q(sl(3))$ are given. The complications that arise in the non-generic case for which q is a primitive p-th root of unity are then addressed. Explicit results on both irreducible and indecomposable representations are presented.

THE QUANTUM ALGEBRA $U_q(sl(n))$

For $q \neq 0, \pm 1$, the quantum algebra $U_q(sl(n))$, is the associative universal enveloping algebra generated by $\{e_i, f_i, k_i, k_i^{-1}; i = 1, 2, ..., n-1\}$ subject to the relations [1] :

$$\begin{array}{ll} k_i k_j = k_j k_i\,, & k_i k_i^{-1} = k_i^{-1} k_i = 1\,, \\ k_i e_j k_i^{-1} = q^{a_{ij}} e_j\,, & k_i f_j k_i^{-1} = q^{-a_{ij}} f_j\,, \\ e_i f_j - f_j e_i = \delta_{ij} \dfrac{k_i - k_i^{-1}}{q - q^{-1}}\,, & \\ e_i^2 e_{i\pm1} - (q+q^{-1}) e_i e_{i\pm1} e_i + e_{i\pm1} e_i^2 = 0\,, & \\ f_i^2 f_{i\pm1} - (q+q^{-1}) f_i f_{i\pm1} f_i + f_{i\pm1} f_i^2 = 0\,, & \\ e_i e_j = e_j e_i\,, \quad f_i f_j = f_j f_i & \text{for } |i-j| \geq 2\,. \end{array} \tag{1}$$

with co-product Δ defined by :

$$\Delta(k_i) = k_i \otimes k_i; \quad \Delta(e_i) = e_i \otimes 1 + k_i \otimes e_i; \quad \Delta(f_i) = f_i \otimes k_i^{-1} + 1 \otimes f_i. \tag{2}$$

It is not difficult to see that there exist 2^{n-1} inequivalent 1-dimensional irreducible representations of $U_q(sl(n))$ defined by $e_i \to (0)$, $f_i \to (0)$, and $k_i \to (\pm 1)$ for $i = 1, 2, ..., n-1$. The corresponding modules may be generated through tensor products

CP589, *Fundamental Interaction Theories*, edited by J. Lukierski and J. Rembieliński

from $V^{\omega_1}, V^{\omega_2}, \ldots, V^{\omega_{n-1}}$, where the 1-dimensional V^{ω_s} is defined by $e_i v = f_i v = 0$, $k_s v = -v$, $k_j v = v$, with $i, j = 1, 2, \ldots, n-1$ and $j \neq s$.

Let the action of $U_q(sl(n))$ on an n-dimensional linear vector space V spanned by $\{v_1, v_2, \ldots, v_n\}$ be defined by :

$$e_i v_j = \delta_{i,j-1} v_i; \quad f_i v_j = \delta_{i,j} v_{j+1}; \quad k_i^{\pm 1} v_j = q^{(\delta_{i,j} - \delta_{i,j-1})} v_j. \tag{3}$$

It may be verified that V is an irreducible $U_q(sl(n))$-module for all q. It is referred to as the defining module.

In the case $n = 2$, V has basis $\{v_1, v_2\}$ and the action (3) gives the representations:

$$k_1 \to \begin{pmatrix} q & 0 \\ 0 & q^{-1} \end{pmatrix}, \; e_1 \to \begin{pmatrix} 0 & 1 \\ 0 & 0 \end{pmatrix}, \; f_1 \to \begin{pmatrix} 0 & 0 \\ 1 & 0 \end{pmatrix}. \tag{4}$$

The tensor product module $V^{\otimes 2} = V \otimes V$ has basis $\{v_1 \otimes v_1, v_1 \otimes v_2, v_2 \otimes v_1, v_2 \otimes v_2\}$ and the co-product rules (2) combined with the action (3) immediately give the representation:

$$k_1 \to \begin{pmatrix} q^2 & 0 & 0 & 0 \\ 0 & 1 & 0 & 0 \\ 0 & 0 & 1 & 0 \\ 0 & 0 & 0 & q^{-2} \end{pmatrix}, \; e_1 \to \begin{pmatrix} 0 & q & 1 & 0 \\ 0 & 0 & 0 & 1 \\ 0 & 0 & 0 & q^{-1} \\ 0 & 0 & 0 & 0 \end{pmatrix}, \; f_1 \to \begin{pmatrix} 0 & 0 & 0 & 0 \\ 1 & 0 & 0 & 0 \\ q^{-1} & 0 & 0 & 0 \\ 0 & q & 1 & 0 \end{pmatrix}. \tag{5}$$

This is manifestly reducible, since a change of basis gives:

$$k_1 \to \begin{pmatrix} q^2 & 0 & 0 & 0 \\ 0 & 1 & 0 & 0 \\ 0 & 0 & q^{-2} & 0 \\ 0 & 0 & 0 & 1 \end{pmatrix}, \; e_1 \to \begin{pmatrix} 0 & q+q^{-1} & 0 & 0 \\ 0 & 0 & 1 & 0 \\ 0 & 0 & 0 & 0 \\ 0 & 0 & 0 & 0 \end{pmatrix}, \; f_1 \to \begin{pmatrix} 0 & 0 & 0 & 0 \\ q+q^{-1} & 0 & 0 & 0 \\ 0 & 1 & 0 & 0 \\ 0 & 0 & 0 & 0 \end{pmatrix}. \tag{6}$$

In fact, $V^{\otimes 2} = V^{(2)} \oplus V^{(0)}$, where $V^{(2)}$ and $V^{(0)}$ are of dimension 3 and 1, respectively, and have bases $\{v_1 \otimes v_1, v_1 \otimes v_2 + q^{-1} v_2 \otimes v_1, v_2 \otimes v_2\}$ and $\{v_1 \otimes v_2 - q v_2 \otimes v_1\}$. The reduction is therefore performed by taking certain q-symmetrised or antisymmetrised combinations of basis vectors. It is this process we wish to generalize.

It follows from (2) that the action of $U_q(sl(n))$ on $V^{\otimes m}$ is the linear extension of the action on the general tensor $v_{i_1 \ldots i_m} = v_{i_1} \otimes v_{i_2} \otimes \ldots \otimes v_{i_m}$ given by:

$$\begin{aligned} k_i^{\pm 1}(v_{i_1 \ldots i_m}) &= k_i^{\pm 1} v_{i_1} \otimes k_i^{\pm 1} v_{i_2} \otimes \ldots \otimes k_i^{\pm 1} v_{i_m}; \\ e_i(v_{i_1 \ldots i_m}) &= \textstyle\sum_{a=1}^m k_i v_{i_1} \otimes \ldots \otimes k_i v_{i_{a-1}} \otimes e_i v_{i_a} \otimes v_{i_{a+1}} \otimes \ldots \otimes v_{i_m}; \\ f_i(v_{i_1 \ldots i_m}) &= \textstyle\sum_{a=1}^m v_{i_1} \otimes \ldots \otimes v_{i_{a-1}} \otimes f_i v_{i_a} \otimes k_i^{-1} v_{i_{a+1}} \otimes \ldots \otimes k_i^{-1} v_{i_m}. \end{aligned} \tag{7}$$

In the generic case for which q is not a root of unity, decomposing the m-fold tensor product module $V^{\otimes m}$ into its irreducible constituents may be accomplished through the use of the Hecke algebra $H_m(q^2)$ which is a q-deformation of the symmetric group S_m.

THE HECKE ALGEBRA $H_m(q)$

The symmetric group S_m on m symbols is generated by the transpositions $s_i = (i, i+1)$ for $i = 1, 2, ..., m-1$. Each element $w \in S_m$ can be written in reduced form as a word $w = s_{i_1} s_{i_2} ... s_{i_l}$ of minimal length $l = l(w)$ in the generators. The Hecke algebra $H_m(q)$ has basis $\{h(w) : w \in S_m\}$, where $h(ww') = h(w)h(w')$ if $l(ww') = l(w) + l(w')$, and if we write $h(s_i) = h_i$ then:

$$h_i h_{i+1} h_i = h_{i+1} h_i h_{i+1}; \quad h_i h_j = h_j h_i \text{ if } |\, i-j \,| > 1; \; h_i^2 = (q-1)h_i + q. \tag{8}$$

In the case $m = 2$, $H_2(q^2)$ is generated by just h_1 whose action on $V^{\otimes 2}$ in the basis $\{v_1 \otimes v_1, v_1 \otimes v_2, v_2 \otimes v_1, v_2 \otimes v_2\}$ is defined so as to give the representation:

$$h_1 \to \begin{pmatrix} q^2 & 0 & 0 & 0 \\ 0 & q^2-1 & q & 0 \\ 0 & q & 0 & 0 \\ 0 & 0 & 0 & q^2 \end{pmatrix}. \tag{9}$$

It may be verified that this matrix commutes with those of (5) and that the minimal idempotents $1 + h_1$ and $1 - q^{-2} h_1$ of $H_2(q^2)$ serve to project out the irreducible $U_q(sl(2))$-modules $V^{(2)}$ and $V^{(0)}$ from $V^{\otimes 2}$.

More generally, the action of $H_m(q^2)$ on $V^{\otimes m}$ is defined to be the linear extension of:

$$h_a(v_{i_1} \otimes \cdots \otimes v_{i_a} \otimes v_{i_{a+1}} \otimes \cdots \otimes v_{i_m})$$
$$= \begin{cases} q^2(v_{i_1} \otimes \cdots \otimes v_{i_a} \otimes v_{i_{a+1}} \otimes \cdots \otimes v_{i_m}) & \text{if } i_a = i_{a+1}; \\ (q^2-1)(v_{i_1} \otimes \cdots \otimes v_{i_a} \otimes v_{i_{a+1}} \otimes \cdots \otimes v_{i_m}) & \\ \qquad +q(v_{i_1} \otimes \cdots \otimes v_{i_{a+1}} \otimes v_{i_a} \otimes \cdots \otimes v_{i_m}) & \text{if } i_a < i_{a+1}; \\ q(v_{i_1} \otimes \cdots \otimes v_{i_{a+1}} \otimes v_{i_a} \otimes \cdots \otimes v_{i_m}) & \text{if } i_a > i_{a+1}; \end{cases} \tag{10}$$

The actions (7) and (10) on $V^{\otimes m}$ commute with one another. In fact [2, 3], $U_q(sl(n))$ and $H_m(q^2)$ are mutual full centralising algebras in their actions on $V^{\otimes m}$, so that the minimal idempotents of $H_m(q^2)$ may be used to project out irreducible $U_q(sl(n))$-modules from $V^{\otimes m}$. This duality between $U_q(sl(n))$ and $H_m(q^2)$ is the analogue of the well-known Schur-Weyl duality between $sl(n)$ and S_m.

YOUNG SYMMETRISERS AND GARNIR ELEMENTS

In the generic case for which q is not a root of unity, $H_m(q)$ is isomorphic to S_m and much of the representation theory of S_m is directly applicable to $H_m(q)$. In particular, the irreducible representations S^λ of $H_m(q)$ are indexed by those partitions $\lambda = (\lambda_1, \lambda_2, ..., \lambda_r)$, with $\lambda_1 \geq \lambda_2 \geq ... \geq \lambda_r > 0$, for which $\sum_{i=1}^r \lambda_i = m$. Each such partition defines a Young diagram F^λ, consisting of m boxes arranged in r left-adjusted rows of lengths $\lambda_1, \lambda_2, ..., \lambda_r$. An S_m-standard Young tableau t^λ of shape λ is a numbering of F^λ with the entries $1, 2, ..., m$ placed one in each box without repetition, in such a way that the entries increase from left to right across each row, and from top to bottom down each column. The basis vectors of S^λ may themselves be indexed by the set of all S_m-standard Young tableaux of shape λ, and in such a basis it is possible to write down explicitly the corresponding representation matrix for each generator, h_a of $H_m(q)$.

In the present context, a more important role for S_m-standard Young tableaux is the part they play in the construction of idempotents of $H_m(q)$ [4]. To this end, let t_+^λ be the S_m-standard Young tableau of shape λ obtained by inserting the integers $1,2,...,m$ from left to right across each row in turn taken from top to bottom, and let t_-^λ be obtained similarly by inserting $1,2,...,m$ from top to bottom down each column taken in turn from left to right. For instance, if $\lambda = (3,2)$ then

$$t_+^{(3,2)} = \begin{matrix} 1\,2\,3 \\ 4\,5 \end{matrix}, \qquad t_-^{(3,2)} = \begin{matrix} 1\,3\,5 \\ 2\,4 \end{matrix}. \tag{11}$$

Let $w_\lambda \in S_m$ be such that $t_+^\lambda = w_\lambda t_-^\lambda$ where S_m acts directly on the numerals of the tableaux. For the above case, $w_{(3,2)} = s_3 s_2 s_4$. The Young subgroups W_+^λ and W_-^λ are defined as the subgroups of S_m which stabilise the rows of t_+^λ and the columns of t_-^λ, respectively. For example, $W_+^{(3,2)} = \langle s_1, s_2, s_4 \rangle$ and $W_-^{(3,2)} = \langle s_1, s_3 \rangle$. Now define

$$e_+^\lambda = \sum_{w \in W_+^\lambda} h(w), \quad e_-^\lambda = \sum_{w \in W_-^\lambda} (-q)^{-l(w)} h(w). \tag{12}$$

Then a q-dependent Young symmetriser is defined by

$$Y^\lambda(q) = h(w_\lambda^{-1}) e_+^\lambda h(w_\lambda) e_-^\lambda. \tag{13}$$

It is a minimal idempotent in $H_m(q)$ [4].

In what follows, particular use is made of certain elements of $H_m(q)$ which are annihilated by $Y^\lambda(q)$.

For each entry a of t_-^λ which is not at the bottom of any column, so that $a+1$ lies immediately below a in t_-^λ and $s_a \in W_-^\lambda$, let C_a^λ be the *column element* defined by $C_a^\lambda = 1 + h_a$. For example, for $\lambda = (3,2)$, there are two column elements, $C_1^{(3,2)} = 1 + h_1$ and $C_3^{(3,2)} = 1 + h_3$.

For each entry a of t_-^λ which is not at the right hand end of any row, so that some d lies immediately to the right of a in t_-^λ, let b lie at the bottom of the column containing a and let $c = b+1$ lie at the top of the column containing d. Let W_{xy} denote the subgroup of S_m generated by $\{s_x, s_{x+1}, ..., s_{y-1}\}$, and let $\mathcal{G}_a^\lambda$ be the set of right coset representatives of $W_{ab} \otimes W_{cd}$ in W_{ad} of minimal length, that is $\mathcal{G}_a^\lambda = \{w \in W_{ad} : l(uw) \geq l(w)$ for all $u \in W_{ab} \otimes W_{cd}\}$. Finally, let G_a^λ be the *Garnir element* defined by [5] :

$$G_a^\lambda = \sum_{w \in \mathcal{G}_a^\lambda} (-q)^{-l(w)} h(w). \tag{14}$$

For example for $\lambda = (3,2)$, there are three Garnir elements $G_1^{(3,2)} = 1 - q^{-1} h_2 + q^{-2} h_2 h_1$, $G_2^{(3,2)} = 1 - q^{-1} h_2 + q^{-2} h_2 h_3$, and $G_3^{(3,2)} = 1 - q^{-1} h_4 + q^{-2} h_4 h_3$.

The significance of these column and Garnir elements is that for each a for which they are defined:

$$Y^{\lambda}(q)C_a^{\lambda} = 0 \quad \text{and} \quad Y^{\lambda}(q)G_a^{\lambda} = 0. \tag{15}$$

In what follows, the column and Garnir identities, with q replaced by q^2, are exploited in the construction of explicit representations of $U_q(sl(n))$. These representations are irreducible in the generic case, but may be reducible if q is a root of unity.

THE WEYL MODULE

The Schur-Weyl duality between $sl(n)$ and S_m shows itself in the irreducibility of the Weyl modules $sl(n)$ which may be labelled by those partitions λ of m such that $\lambda = (\lambda_1, \lambda_2, ..., \lambda_r)$ with $r < n$. An $sl(n)$-standard Young tableau T^{λ} of shape λ is a numbering of the Young diagram F^{λ} with entries taken from the set $\{1, 2, ..., n\}$ placed one in each box, with repetitions allowed, in such a way that the entries weakly increase from left to right across each row and strictly increase from top to bottom down each column.

It is well known that the basis vectors of the irreducible Weyl modules of $sl(n)$ are indexed by $sl(n)$-standard Young tableaux. Remarkably, whether or not q is generic, these same standard Young tableaux T^{λ} of shape λ index basis vectors $v_{T^{\lambda}}$ of what we call the Weyl module, W^{λ}, of $U_q(sl(n))$. To be precise,

$$v_{T^{\lambda}} = Y^{\lambda}(q^2)(v_{i_1} \otimes v_{i_2} \otimes \cdots \otimes v_{i_m}), \tag{16}$$

where $i_1, i_2, ..., i_m$ are the entries of T^{λ} appearing in the positions of $1, 2, ..., m$ in t_{-}^{λ}. For example, for $\lambda = (3,2)$ we have:

$$T^{(3,2)} = \begin{matrix} i_1 & i_3 & i_5 \\ i_2 & i_4 & \end{matrix} \; . \tag{17}$$

In (16), $Y^{\lambda}(q^2)$ may be obtained from (13), and its action on $V^{\otimes m}$ from (10). Thanks to the fact that the actions of $U_q(sl(n))$ and $H_m(q^2)$ on $V^{\otimes m}$ commute, we then have, for any $g \in U_q(sl(n))$,

$$g(v_{T^{\lambda}}) = Y^{\lambda}(q^2)g(v_{i_1} \otimes v_{i_2} \otimes \cdots \otimes v_{i_m}). \tag{18}$$

The action of g on $V^{\otimes m}$ is defined by (7), from which it follows that the right hand side of (18) is a linear combination of terms of the form:

$$v_{X^{\lambda}} = Y^{\lambda}(q^2)(v_{j_1} \otimes v_{j_2} \otimes \cdots \otimes v_{j_m}), \tag{19}$$

where X^{λ} is a numbering of F^{λ} with entries $j_1, j_2, ..., j_m$ appearing in the positions of $1, 2, ..., m$ in t_{-}^{λ}. This numbering might not be $sl(n)$-standard, and in order to obtain the explicit matrix representing $g \in U_q(sl(n))$, the column and Garnir element identities of (15) may need to be exploited to express each term (19) as a linear combination of $sl(n)$-standard terms $v_{T^{\lambda}}$.

It follows from (10) and the column identities of (15) that firstly, if X^λ has two neighbouring entries in a column both equal to i, then $v_{X^\lambda} = 0$, and secondly, if X^λ has an entry j directly above a smaller entry i, then $v_{X^\lambda} = -qv_{Z^\lambda} = 0$, where Z^λ is obtained from X^λ by interchanging the pair i and j. Using these column identities, all v_{X^λ} may be expressed in terms of v_{Z^λ}, with Z^λ column strict in the sense that the entries of Z^λ are strictly increasing down each column.

It further follows from (10) and the Garnir identities of (15) that if Z^λ is column strict but has an entry j at position a of t^λ_- directly to the left of smaller entry i, then by using (14),

$$\sum_{w\in G_a^\lambda} (-q)^{-l(w)} v_{w\tilde{w}^{-1}Z^\lambda} = 0, \tag{20}$$

where $w, \tilde{w}^{-1} \in S_m$ act on Z^λ by permutation of positions as labelled by t^λ_-, and $\tilde{w}$ is the unique element of maximal length in G_a^λ. It may be shown that successive applications of column and Garnir identities enable $sl(n)$-standardisation to be effected. This provides a proof that the Weyl module of $U_q(sl(n))$ does indeed have as a basis the set of vectors v_{T^λ} indexed by $sl(n)$-standard Young tableaux T^λ.

Moreover, the use of (18) together with the standardisation algorithm allows the corresponding representations matrices to be constructed explicitly. For generic q, the Weyl module W^λ is irreducible [6, 7].

To illustrate the construction procedure, consider the 8-dimensional $U_q(sl(3))$-module $W^{(2,1)}$ for which the standard tableaux are :

$$\begin{array}{cccccccc} 23 & 13 & 22 & 12 & 11 & 13 & 12 & 11\,. \\ 3 & 3 & 3 & 3 & 3 & 2 & 2 & 2 \end{array} \tag{21}$$

Using (18), the application of the generator $e_1 \in U_q(sl(3))$ to each of the corresponding vectors gives:

$$e_1 v_{\substack{23\\3}} = v_{\substack{13\\3}}\,; \quad e_1 v_{\substack{13\\3}} = 0; \quad e_1 v_{\substack{12\\3}} = q v_{\substack{11\\3}}\,; \quad e_1 v_{\substack{11\\3}} = 0;$$

$$\begin{aligned} e_1 v_{\substack{22\\3}} &= v_{\substack{12\\3}} + q^{-1} v_{\substack{21\\3}} = v_{\substack{12\\3}} + q^{-1}\left(q^{-1} v_{\substack{12\\3}} - q^{-2} v_{\substack{13\\2}}\right) \\ &= (1+q^{-2}) v_{\substack{12\\3}} - q^{-3} v_{\substack{13\\2}}\,; \end{aligned} \tag{22}$$

$$e_1 v_{\substack{13\\2}} = q v_{\substack{13\\1}} = 0; \quad e_1 v_{\substack{12\\2}} = q v_{\substack{12\\1}} + v_{\substack{11\\2}} = v_{\substack{11\\2}}\,; \quad e_1 v_{\substack{11\\2}} = q v_{\substack{11\\1}} = 0.$$

Here a Garnir identity has only been used in the fifth calculation, whereas column identities are used in each of the last three. It follows that in the case of the $U_q(sl(3))$-

module $W^{(2,1)}$:

$$e_1 \to \begin{pmatrix} \cdot & \cdot & \cdot & \cdot & \cdot & \cdot & \cdot & \cdot \\ 1 & \cdot & \cdot & \cdot & \cdot & \cdot & \cdot & \cdot \\ \cdot & \cdot & \cdot & \cdot & \cdot & \cdot & \cdot & \cdot \\ \cdot & \cdot & 1+q^{-2} & \cdot & \cdot & \cdot & \cdot & \cdot \\ \cdot & \cdot & \cdot & q & \cdot & \cdot & \cdot & \cdot \\ \cdot & \cdot & -q^{-3} & \cdot & \cdot & \cdot & \cdot & \cdot \\ \cdot & \cdot & \cdot & \cdot & \cdot & \cdot & \cdot & \cdot \\ \cdot & \cdot & \cdot & \cdot & \cdot & \cdot & 1 & \cdot \end{pmatrix}. \tag{23}$$

In a similar way, the representation matrices may be found for the remaining generators of $U_q(sl(3))$. The matrices for k_i, with $i=1,2$, are diagonal. In fact, for general $U_q(sl(n))$, (3), (7) and (18) imply that $k_i v_{T^\lambda} = \tilde{k}_i v_{T^\lambda}$ where the eigenvalue $\tilde{k}_i$ of k_i is given by $\tilde{k}_i = q^{n_i - n_{i+1}}$, where n_i is the number of entries i in the tableau T^λ. We refer to $\tilde{k} = (\tilde{k}_1, \tilde{k}_2, ..., \tilde{k}_{n-1})$ as the q-weight of v_{T^λ}. The Weyl module W^λ is said to have highest weight λ, and the corresponding highest q-weight is $\tilde{k} = (q^{\lambda_1-\lambda_2}, q^{\lambda_2-\lambda_3}, \ldots, q^{\lambda_{n-1}-\lambda_n})$.

Returning to $U_q(sl(3))$, in the case of an arbitrary but fixed Weyl module $W^\lambda = W^{(\lambda_1,\lambda_2)}$ of highest q-weight $\tilde{k} = (q^{\lambda_1-\lambda_2}, q^{\lambda_2})$, it is convenient to introduce a Gelfand pattern notation whereby for each $sl(3)$-standard Young tableau $T^{(\lambda_1,\lambda_2)}$,

$$\left|\begin{matrix}\mu_1\ \mu_2\\ \nu_1\end{matrix}\right\rangle = v_{T^{(\lambda_1,\lambda_2)}},$$

where μ_1 and μ_2 are the numbers of entries ≤ 2 in the first and second rows,respectively, of $T^{(\lambda_1,\lambda_2)}$, and ν_1 is the number of 1s in the first row. The fact that $T^{(\lambda_1,\lambda_2)}$ is $sl(3)$-standard ensures that $\lambda_1 \geq \mu_1 \geq \lambda_2 \geq \mu_2 \geq 0$ and $\mu_1 \geq \nu_1 \geq \mu_2$. With this notation, the use of (18) and the standardization algorithm leads to the explicit representation:

$$e_1 \left|\begin{matrix}\mu_1\ \mu_2\\ \nu_1\end{matrix}\right\rangle = q^{2\nu_1-\mu_1-\mu_2+1}[\mu_1-\nu_1]\left|\begin{matrix}\mu_1\ \mu_2\\ \nu_1+1\end{matrix}\right\rangle - \alpha q^{2\nu_1-2\mu_1+1}[\mu_1-\lambda_2]\left|\begin{matrix}\mu_1-1\ \mu_2+1\\ \nu_1+1\end{matrix}\right\rangle,$$

$$f_1 \left|\begin{matrix}\mu_1\ \mu_2\\ \nu_1\end{matrix}\right\rangle = q^{\mu_1+\mu_2-2\nu_1+1}[\nu_1-\mu_2]\left|\begin{matrix}\mu_1\ \mu_2\\ \nu_1-1\end{matrix}\right\rangle - \beta q^{2\mu_2-2\nu_1+1}[\lambda_2-\mu_2]\left|\begin{matrix}\mu_1-1\ \mu_2+1\\ \nu_1-1\end{matrix}\right\rangle,$$

$$e_2 \left|\begin{matrix}\mu_1\ \mu_2\\ \nu_1\end{matrix}\right\rangle = q^{2\mu_2-\kappa_1+1}[\kappa_1-\mu_2]\left|\begin{matrix}\mu_1\ \mu_2+1\\ \nu_1\end{matrix}\right\rangle + q^{2\mu_1+2\mu_2-\nu_1-\lambda_1-\lambda_2+1}[\lambda_1-\mu_1]\left|\begin{matrix}\mu_1+1\ \mu_2\\ \nu_1\end{matrix}\right\rangle,$$

$$f_2 \left|\begin{matrix}\mu_1\ \mu_2\\ \nu_1\end{matrix}\right\rangle = q^{\lambda_1+\kappa_2-2\mu_1+1}[\mu_1-\kappa_2]\left|\begin{matrix}\mu_1-1\ \mu_2\\ \nu_1\end{matrix}\right\rangle + q^{\lambda_1+\lambda_2+\nu_1-2\mu_1-2\mu_2+1}[\mu_2]\left|\begin{matrix}\mu_1\ \mu_2-1\\ \nu_1\end{matrix}\right\rangle,$$

$$k_1 \left|\begin{matrix}\mu_1\ \mu_2\\ \nu_1\end{matrix}\right\rangle = q^{2\nu_1-\mu_1-\mu_2}\left|\begin{matrix}\mu_1\ \mu_2\\ \nu_1\end{matrix}\right\rangle, \qquad k_2 \left|\begin{matrix}\mu_1\ \mu_2\\ \nu_1\end{matrix}\right\rangle = q^{2\mu_1+2\mu_2-\nu_1-\lambda_1-\lambda_2}\left|\begin{matrix}\mu_1\ \mu_2\\ \nu_1\end{matrix}\right\rangle,$$

where $\alpha = 1$ if $\nu_1 < \lambda_2$ and 0 if $\nu_1 \geq \lambda_2$, $\beta = 1$ if $\nu_1 > \lambda_2$ and 0 if $\nu_1 \leq \lambda_2$, $\kappa_1 = \min(\lambda_2, \nu_1)$, $\kappa_2 = \max(\lambda_2, \nu_1)$ and $[x]$ signifies the usual q-number:

$$[x] = \frac{q^x - q^{-x}}{q - q^{-1}}. \tag{24}$$

It is to be noted that every matrix element is of the form $q^y[x]$ for some integers x and y. Initially it was conjectured on the basis of many examples that the matrix elements of every generator of $U_q(sl(n))$ in the Weyl module W^λ were all of the form $q^y[x]$ for some integers x and y. However, more recently we have found counter-examples to this conjecture. Nevertheless, it is certainly true that all matrix elements are a linear combination of terms of the form $q^y[x]$. The significance of this is that the representation is perfectly well defined whether or not q is generic. This is markedly different from the explicit representations of $U_q(sl(n))$ written down previously in the case of generic q [8]. These involve square roots of rational functions of q-numbers, which for q a root of unity are not, in general, well defined. A method has been devised of circumventing this difficulty quite generally [9], but here we wish to exploit our explicit construction of the Weyl module to construct irreducible highest weight modules in the root of unity case.

THE NON-GENERIC CASE

Henceforth, let q be a primitive Nth root of unity with $N > 2$, and let $p = N$ if N is odd, and $p = N/2$ if N is even. In the first case $q^p = 1$ and in the second $q^p = -1$, while in both cases p is the smallest integer such that $[p] = 0$. In such non-generic cases, the Weyl modules W^λ of $U_q(sl(n))$ may be either irreducible, or reducible but indecomposable, or even decomposable. The key problems are firstly, to determine the structure of each Weyl module W^λ, and secondly, to identify a complete set of inequivalent irreducible modules V^μ appearing as composition factors of Weyl modules.

Consider $U_q(sl(2))$, with generators e_1, f_1, k_1, k_1^{-1} in the case $N = 4$ so that $p = 2$ and $q^2 = -1$. The Weyl module $W^{(3)}$ has dimension 4 and basis vectors $v_{111}, v_{112}, v_{122}$ and v_{222} whose q-weights are given by the eigenvalues of k_1, namely $q^3 = -q, q, q^{-1} = -q$ and $q^{-3} = q$; respectively. The action of e_1 and f_1 on these basis vectors, as given by (7,18) and the use of (20), yields:

$$\begin{aligned}
e_1 v_{111} &= 0, & f_1 v_{111} &= (1+q^{-2}+q^{-4})v_{112} = v_{112},\\
e_1 v_{112} &= q^2 v_{111} = -v_{111}, & f_1 v_{112} &= q(1+q^{-2})v_{122} = 0,\\
e_1 v_{122} &= q(1+q^{-2})v_{112} = 0, & f_1 v_{122} &= q^2 v_{222} = -v_{222},\\
e_1 v_{222} &= (1+q^{-2}+q^{-4})v_{122} = v_{122}, & f_1 v_{222} &= 0.
\end{aligned} \tag{25}$$

It follows that the Weyl module $W^{(3)}$ is decomposable into two isomorphic 2-dimensional submodules with bases $\{v_{111}, v_{112}\}$ and $\{v_{122}, -v_{222}\}$, both of highest q-weight $(-q)$.

On the other hand, for the same Weyl module $W^{(3)}$ with $N = 3$ so that now $p = 3$ and $q^3 = 1$, we have:

$$\begin{aligned}
e_1 v_{111} &= 0, & f_1 v_{111} &= (1+q^{-2}+q^{-4})v_{112} = 0,\\
e_1 v_{112} &= q^2 v_{111}, & f_1 v_{112} &= q(1+q^{-2})v_{122} = -v_{122},\\
e_1 v_{122} &= q(1+q^{-2})v_{112} = -v_{112}, & f_1 v_{122} &= q^2 v_{222},\\
e_1 v_{222} &= (1+q^{-2}+q^{-4})v_{122} = 0, & f_1 v_{222} &= 0.
\end{aligned} \tag{26}$$

It follows this time that there exist two trivial 1-dimensional submodules spanned by v_{111} and v_{222} each of highest q-weight (1), since $q^3 = q^{-3} = 1$. The corresponding quotient module of dimension 2 has basis $\{v_{112}, v_{122}\}$, and is of highest q-weight (q). It is isomorphic to the defining module V of $U_q(sl(2))$.

More generally, for $U_q(sl(2))$ the Weyl module $W^{(m)}$ with $m = rp + l$ for $0 \le l < p$, contains a fully reducible submodule consisting of the direct sum of $r+1$ copies of $(V^{\omega})^r \otimes V^{(l)}$ each of dimension $l+1$ and highest q-weight $((\pm 1)^r q^l)$, and a corresponding fully reducible quotient module consisting of the direct sum of r copies of $(V^{\omega})^{r-1} \otimes V^{(p-l-2)}$ each of dimension $p-l-1$ and highest q-weight $((\pm 1)^{r-1} q^{p-l-2})$. Here V^{ω} signifies a 1-dimensional module of highest q-weight (± 1), where the $+$ sign is taken for N odd and the $-$ sign for N even. We have obtained by this constructive method all irreducible highest weight modules of $U_q(sl(2))$ for any q. The results are in accord with those obtained by other methods [10].

Turning to $U_q(sl(3))$, in the case $p = N = 3$ and $\lambda = (3)$ we can proceed in the same way. The action of the generators on the highest weight vector v_{111} of the Weyl module $W^{(3)}$ gives:

$$\begin{aligned} &e_1 v_{111} = 0, && f_1 v_{111} = (1 + q^{-2} + q^{-4}) v_{112} = 0, \\ &e_2 v_{111} = 0, && f_2 v_{111} = 0, \\ &k_1 v_{111} = q^3 v_{111} = v_{111}, && k_2 v_{111} = q^0 v_{111} = v_{111}; \end{aligned} \tag{27}$$

so that v_{111} spans a 1-dimensional submodule. Similarly, both v_{222} and v_{333} span 1-dimensional submodules. These three submodules have highest q-weight $\tilde{k} = (1,1)$ and are each isomorphic to the trivial module $V^{(0)} = W^{(0)}$. The quotient of $W^{(3)}$ with respect to these three submodules, obtained by setting $v_{111} = v_{222} = v_{333} = 0$, is an irreducible 7-dimensional module of highest q-weight $\tilde{k} = (q,q)$ spanned by $v_{112}, v_{122}, v_{113}, v_{123}, v_{223}, v_{133}, v_{223}$.

It is to be anticipated that the same irreducible 7-dimensional module should appear as a submodule of the 8-dimensional Weyl module $W^{(2,1)}$ whose highest weight vector $v_{\substack{11\\2}}$ also has q-weight $\tilde{k} = (q,q)$. That this is indeed the case can be seen by considering the action of the generators of $U_q(sl(3))$ as in (22). In particular it can be seen that:

$$\begin{aligned} e_1 v_{\substack{22\\3}} &= (1+q^{-2}) v_{\substack{12\\3}} - q^{-3} v_{\substack{13\\2}} = (-q^{-1}) v_{\substack{12\\3}} - v_{\substack{13\\2}}, \\ f_1 v_{\substack{11\\3}} &= (1+q^{-2}) v_{\substack{12\\3}} - q^{-3} v_{\substack{13\\2}} = (-q^{-1}) v_{\substack{12\\3}} - v_{\substack{13\\2}}, \\ e_2 v_{\substack{13\\3}} &= q^{-1} v_{\substack{12\\3}} + v_{\substack{13\\2}}, \quad f_2 v_{\substack{12\\2}} = q^{-1} v_{\substack{12\\3}} + v_{\substack{13\\2}}. \end{aligned} \tag{28}$$

Correspondingly, there exists an irreducible 7-dimensional submodule $V^{(2,1)}$ with basis vectors

$$v_{\substack{2\,3\\3}},\; v_{\substack{1\,3\\3}},\; v_{\substack{2\,2\\3}},\; q^{-1}v_{\substack{1\,2\\3}} + v_{\substack{1\,3\\2}},\; v_{\substack{1\,1\\3}},\; v_{\substack{1\,2\\2}},\; v_{\substack{1\,1\\2}},$$

so that here, the two-fold degeneracy of the basis states $v_{\substack{1\,2\\3}}$ and $v_{\substack{1\,3\\2}}$ has been removed.

It may be shown that this module $V^{(2,1)}$ is isomorphic to our previously identified 7-dimensional quotient module of $W^{(3)}$. Moreover, the quotient module of $W^{(2,1)}$ with respect to $V^{(2,1)}$ is isomorphic to $V^{(0)}$. To summarise, for $U_q(sl(3))$ with $p = 3$ we have

$$W^{(3)} = (V^{(0)} \oplus V^{(0)} \oplus V^{(0)}) \mathbin{\supset\!\!\!\!\!+} V^{(2,1)} \quad \text{and} \quad W^{(2,1)} = V^{(2,1)} \mathbin{\supset\!\!\!\!\!+} V^{(0)}.$$

A similar process of analysis has produced the data shown in Table 1.

Table 1. Dimensions of irreducible highest weight representations of $U_q(sl(3))$ at various roots of unity

			$p=2$	$p=3$	$p=4$	$p=5$	$p=6$
λ	$\tilde{k} = (\tilde{k}_1, \tilde{k}_2)$	$\dim W^{\lambda}$	$N=4$	$N=3,6$	$N=8$	$N=5,10$	$N=12$
$(0,0)$	$(1,1)$	1	1	1	1	1	1
$(1,0)$	$(q,1)$	3	3	3	3	3	3
$(1,1)$	$(1,q)$	3	3	3	3	3	3
$(2,1)$	(q,q)	8	8	7	8	8	8
$(2,0)$	$(q^2,1)$	6		6	6	6	6
$(2,2)$	$(1,q^2)$	6		6	6	6	6
$(3,1)$	(q^2,q)	15		15	12	15	15
$(3,2)$	(q,q^2)	15		15	12	15	15
$(4,2)$	(q^2,q^2)	27		27	26	19	27
$(3,0)$	$(q^3,1)$	10			10	10	10
$(3,3)$	$(1,q^3)$	10			10	10	10
$(4,1)$	(q^3,q)	24			24	18	24
$(4,3)$	(q,q^3)	24			24	18	24
$(5,2)$	(q^3,q^2)	42			42	39	27
$(5,3)$	(q^2,q^3)	42			42	39	27
$(6,3)$	(q^3,q^3)	64			64	63	56
⋮	⋮	⋮				⋮	⋮

For each λ, the table includes the highest q-weight $(q^{\lambda_1-\lambda_2}, q^{\lambda_2})$, and the dimensions of the corresponding Weyl module W^{λ} of $U_q(sl(3))$, followed by the dimensions $\dim V^{\lambda}$ of the irreducible submodules V^{λ} for successive values of p. For each p, the list is terminated after all irreducible modules, which are inequivalent and, in the case of even N, inequivalent up to tensor products with one of the four 1-dimensional modules.

For various values of p, the structure of the Weyl module W^λ of $U_q(sl(3))$ is:

$$W^{(\lambda_1,\lambda_2)} = \begin{cases} \ldots \supsetplus V^{(\lambda_1,\lambda_2)} & \text{if } 2 < p \leq \max(\lambda_1-\lambda_2,\lambda_2); \\ V^{(\lambda_1,\lambda_2)} & \text{if } p = \max(\lambda_1-\lambda_2+1,\lambda_2+1); \\ V^{(\lambda_1,\lambda_2)} \supsetplus V^{(2p-\lambda_1-4,p-\lambda_1+\lambda_2-2)} & \text{if } \max(\lambda_1-\lambda_2,\lambda_2) < p \leq \lambda_1+1; \\ V^{(\lambda_1,\lambda_2)} & \text{if } p > \lambda_1+1. \end{cases} \tag{29}$$

In the first case ... signifies a rich sub- and quotient-module structure which can be determined algorithmically for any given $W^{(\lambda_1,\lambda_2)}$. For example, for $p=6$ we have

$$W^{(10,3)} = ((V^{\omega_1}\otimes V^{(4,3)})\oplus(V^{\omega_2}\otimes V^{(4,3)})\oplus(V^{\omega_1}\otimes V^{\omega_2}\otimes V^{(4,3)}))\supsetplus V^{(8,5)}.$$

The three other cases are remarkably simple. They lead directly to the following formulae for the dimensions of the irreducible highest weight modules of $U_q(sl(3))$ in agreement with the results of Dobrev [11] :

$$\dim V^{(\lambda_1,\lambda_2)} = \begin{cases} \frac{1}{2}(\lambda_1+2)(\lambda_2+1)(\lambda_1-\lambda_2+1) \\ \qquad \text{if } p = \max(\lambda_1-\lambda_2+1,\lambda_2+1) \text{ or } p > \lambda_1+1, \\ \frac{1}{2}(\lambda_1+2)(\lambda_2+1)(\lambda_1-\lambda_2+1) \\ -\frac{1}{2}(2p-\lambda_1-2)(p-\lambda_1-\lambda_2+1)(p-\lambda_2-1) \\ \qquad \text{if } \max(\lambda_1-\lambda_2+1,\lambda_2+1) < p \leq \lambda_1+1. \end{cases} \tag{30}$$

A consideration of our explicit construction procedure for $U_q(sl(n))$ for $n \geq 2$ supports the following hypotheses:

1. Each Weyl module W^λ of $U_q(sl(n))$ contains an irreducible submodule V^λ of the same highest weight λ.
2. Each composition factor of any Weyl module W^λ of $U_q(sl(n))$ is isomorphic to an irreducible module V^μ, with highest q-weight $\tilde{k} = (q^{m_1}, q^{m_2}, \ldots, q^{m_{n-1}})$ such that $0 \leq m_i < N$ for $i = 1,2,\ldots,n-1$.
3. The Weyl module W^λ is irreducible if it is not possible to remove any continuous boundary strip of boxes of length p from the Young diagram F^λ and obtain thereby a Young diagram F^ν specified by any partition ν.
4. The dimension of each irreducible model V^λ is bounded by $p^{n(n-1)/2}$. This bound is attained in the case $\lambda = (p-1)\rho$ where $\rho = (n-1, n-2, \ldots, 1, 0)$. The corresponding Weyl module W^λ is irreducible, and has highest q-weight $\tilde{k} = (q^{p-1}, q^{p-1}, \ldots., q^{p-1})$.

All our data is consistent with the validity of these four hypotheses for all $U_q(sl(n))$. For example, the above results on the structure of the modules $W^{(3)}$ and $W^{(2,1)}$ of $U_q(sl(3))$ exemplify hypotheses 1. and 2., whereas the irreducibility of the module $W^{(6,3)}$ of dimension 64 in the case $p=4$ exemplifies both hypotheses 3. and 4.

EPILOGUE

Since the first presentation[1] of this work a number of developments have taken place. In the case of $U_q(sl(3))$ Dobrev and Truini [12, 13] have obtained an explicit description of the irreducible modules in the non-generic, root of unity case that have been identified here. More generally, some profound developments have taken place that apply to all $U_q(sl(n))$. These follow the seminal work of Lascoux, Leclerc and Thibon [14] who conjectured that the coefficients of the global lower crystal basis [16] of the basic representation $M(\Lambda_0)$ of the quantum affine algebra $U_v(\widehat{sl(p)})$ on a certain natural basis (indexed by partitions, or equivalently, semi-infinite wedges) provide the decomposition numbers for the Specht modules of the Hecke algebras $H_m(q)$ when q is a primitive pth root of unity. Moreover, [14] provides a fast algorithm for calculating these values. Their conjecture was proved by Ariki [15]. Leclerc and Thibon [17] were able to extend their construction to the (reducible) Fock space of $U_v(\widehat{sl(p)})$, and conjectured that the coefficients of a global lower crystal basis on the same natural basis, now provide decomposition numbers of the Weyl modules W^λ of $U_q(sl(n))$ for q^2 a primitive pth root of unity. This conjecture was proved by Varagnolo and Vasserot [18].

REFERENCES

1. De Concini, C. and Kac, V.G., *Progress in Math.* **92**, 471–506 (1990)
2. Jimbo, M., *Lett. Math. Phys.* **11**, 247–252 (1986).
3. Ram, A., *Invent. Math.* **106**, 461–488 (1991).
4. Gyoja, A., *Osaka J. Math.* **23**, 841–852 (1986).
5. King, R.C. and Wybourne, B.G., *J. Math. Phys.* **33**, 4–11 (1992).
6. Lusztig, G., *Adv. Math.* **70**, 237–249 (1988).
7. Rosso, M., *Commun. Math. Phys.* **117**, 581–593 (1988).
8. Jimbo, M., *Lect. Notes in Phys.* **246**, 335–361 (1986).
9. Arnaudon, D. and Chakrabarti, A., *Commun. Math. Phys.* **139**, 461–478 (1991).
10. Roche, P. and Arnaudon, D., *Lett. Math. Phys.* **17**, 295–300 (1989).
11. Dobrev, V.K., *Prog. of Theor. Phys. Supp.* **102**, 137–158 (1990).
12. Dobrev, V.K. and Truini, P. in *Quantum Group Symposium at Group 21*, eds. Doebner, H.-D. and Dobrev, V.K., Heron Press, Sofia, 1997, 308–316.
13. Dobrev, V.K. and Truini, P., *J. Math. Phys.* **38**, 2631–2651 (1997).
14. Lascoux, A., Leclerc, B., and Thibon, J.-Y., *Commun. Math. Phys.* **181**, 205–263 (1996).
15. Ariki, S., *J. Math. Kyoto Univ.* **36**, 789–808 (1996).
16. Kashiwara, M., *Duke Math. J.* **63**, 465–516 (1991); *Duke Math. J.* **69**, 455–485 (1993).
17. Leclerc, B. and Thibon, J.-Y., *Internat. Math. Res. Notices* **9**, 447–456 (1996).
18. Varagnolo, M. and Vasserot, E., *Duke Math. J.* **100**, 267–297 (1999).

[1] Most of these results were first presented by R.C. King at the 3rd Wigner Symposium, Oxford, September 1993.

Multiloop Noncommutative Open String Theory

C.-S. Chu* and R. Russo†

*Center for Particle Theory, Department of Mathematical Sciences, University of Durham, DH1 3LE, UK, email: chong-sun.chu@durham.ac.uk
†Laboratoire de Physique Théorique de l'Ecole Normale Supérieure, 24 rue Lhomond, F-75231 Paris Cedex 05, France, email: rodolfo.russo@lpt.ens.fr

Abstract. The multiloop amplitudes for open bosonic string in presence of a constant B-field are derived from first principles. The basic ingredients of the construction are the commutation relations for the string modes and the Reggeon vertex describing the interaction among three generic string states. The modifications due to the presence of the B–field affect non–trivially only the zero modes. This makes it possible to write in a simple and elegant way the general expression for multiloop string amplitudes in presence of a constant B-field. The field theory limit of these string amplitudes is also considered. We show that it reproduces exactly the Feynman diagrams of noncommutative field theories. Issues of UV/IR are briefly discussed.

MULTILOOP NCOS AND REGGEON FORMALISM

The first basic ingredient for the construction of string amplitudes in the operator formalism are the commutation relations for the string modes. As usual, these commutation relations can be derived from the the tree level world–sheet action. We consider an open string ending on a D-brane in presence of a constant B-field. The open string mode expansion is

$$X^\mu(\tau,\sigma) = x_0^\mu + 2\alpha'(p_0^\mu\tau - p_0^\nu F_\nu{}^\mu\sigma) + \sqrt{2\alpha'}\sum_{n\neq 0}\frac{e^{-in\tau}}{n}\left(ia_n^k\cos n\sigma - a_n^\nu F_\nu{}^\mu \sin n\sigma\right), \quad (1)$$

where $F = B - dA$ is the modified Born-Infeld field strength. Canonical quantization yields the following commutation relations [1]

$$[a_n^\mu, x_0^\nu] = [a_n^\mu, p_0^\nu] = [p_0^\mu, p_0^\nu] = 0, \quad (2)$$

$$[a_m^\mu, a_n^\nu] = mM^{-1\mu\nu}\delta_{m+n}, \quad [x_0^\mu, p_0^\nu] = iM^{-1\mu\nu}, \quad [x_0^\mu, x_0^\nu] = i\Theta^{\mu\nu}, \quad (3)$$

where $M_{\mu\nu} = g_{\mu\nu} - (Fg^{-1}F)_{\mu\nu}$ is the open string metric and

$$(M^{-1})^{\mu\nu} = \left(\frac{1}{g+F}g\frac{1}{g-F}\right)^{\mu\nu}, \quad \Theta^{\mu\nu} = 2\pi\alpha'\left(\frac{1}{g+F}F\frac{1}{g-F}\right)^{\mu\nu}, \quad (4)$$

are the symmetric and antisymmetric part of the matrix $(\frac{1}{g+F})^{\mu\nu} = (M^{-1})^{\mu\nu} - \frac{\Theta^{\mu\nu}}{2\pi\alpha'}$. Although these commutation relations were derived at the tree level, they are valid at all loops [2, 3] and therefore one can use them directly to construct the higher loop string

CP589, *Fundamental Interaction Theories*, edited by J. Lukierski and J. Rembieliński

amplitudes. Due to the limitation of vertex operator formalism to go beyond 1-loop, it is necessary to introduce the Reggeon vertex formalism for string amplitudes.

The basic object in the Reggeon formalism is the 3-Reggeon vertex which describes, at the tree level, the interaction among three generic string states [4]

$$V_{3;0}^{\theta}(\zeta) = \int \frac{dp}{\sqrt{\det M}} \langle p,0;q=3| : \mathrm{e}^{\{\oint_0 dz(-X^v(\zeta+z)\partial_z X(z) - c^v(\zeta+z)b(z) + b^v(\zeta+z)c(z))\}} : . \quad (5)$$

Here the bra indicates the vacuum of the emitted string with momentum p and the label $q=3$ specifies the ghost number. X^v is the virtual propagating string. X and X^v both have an expansion of the form (1) with commutations relations given by (2), (3), while the virtual and the external strings simply commute among themselves. Notice that that (5) is almost identical to the standard 3-Reggeon vertex in the trivial background $B=0$. The only modification with the respect to the usual case is the appearance of a factor of $\det M$ in the measure of momentum integrals which is due to the fact that the open string metric is flat, but non trivial More generally one has to modify the usual ($B=0$) normalizations every time the volume of the space seen by the open strings appears: for instance, the open string vacuum has to be normalized as $\langle 0|0\rangle = \sqrt{\det M}\, V$ and, thus, the generic momentum state satisfies $\langle p|p'\rangle = \sqrt{\det M}\, \delta^d(p-p')$. However, even if (5) is formally unchanged, it contains a non-trivial dependence on B through the mode expansion (1) and the new commutation relations (2), (3). It should also be stressed that the zero-mode x_0 (or y_0 if the interaction is at $\sigma=\pi$) appears only in the expansion of the virtual string and this is the only source of the non-trivial dependence on Θ. We will employ the usual physical states having ghost number 1; thus, the (b,c) system is not affected by the background field F and one recovers the well-known results for the ghost contributions. Because of this, in what follows, ghosts will no longer be mentioned, and we will focus only on the F-dependent modifications coming from the orbital part.

The tree level N-Reggeon vertex is obtained by simply multiplying N 3-Reggeon vertices in different positions ζ, but with the common propagating string X^v. Finally one takes the vacuum expectation value in the Hilbert space of the propagating string in order to obtain a symmetric object in the N external states. The new Θ dependent part comes when one collects together the zero mode factors. In particular, if the external legs of all the original 3-Reggeon vertices are emitted from the border $\sigma=0$, one obtains the new phase factor

$$e^{ip^1 x_0} \dots e^{ip^N x_0} = e^{-\frac{i}{2}\sum_{r<s}^{N} p^r \Theta p^s} , \quad (6)$$

where momentum conservation has been used. As a result, we obtain the N-Reggeon vertex with all legs emitted from the $\sigma=0$ border

$$V_{N;0}^{\Theta} = \sqrt{\det M}\, V_{N;0}^{0} \exp\left(-\frac{i}{2}\sum_{i<j}^{N} p^i \Theta p^j\right), \quad (7)$$

where p_μ^i is the momentum *operator* of the i-th leg, in the direction flowing towards the boundary. Here $V_{N;0}^{0}$ indicates the N-Reggeon vertex derived for the usual "commutative" case $F=0$ in [5]. This vertex is a bra in the direct product of the N distinct Fock spaces for the external strings; like the 3-Reggeon vertex, it gives the scattering amplitude when the external legs are saturated with physical states. Note that the momentum dependent phase factor is exactly the same modification as the one introduced by the Moyal $*$-product in the tree level vertex of a noncommutative field theory [6].

The h-loop N-Reggeon vertex in the presence of a constant F-field can be constructed by sewing together pairs of legs in a tree level $(N+2h)$-Reggeon vertex. The sewing is achieved by using the BRST invariant operator $P(x)$, which is a function of L_0 and $L_{\pm 1}$ and of the ghosts. Since L_n does not involve the zero modes x_0, we conclude that the string propagator $P(x)$ and the sewing procedure are not modified by the presence of F-field. Therefore the only new feature for $\Theta \neq 0$ is in the zero modes part of the Reggeon vertex. And we obtain [2]

$$V^{\Theta}_{N;h} = \sqrt{\det M}\, \widetilde{V}^{0}_{N;h} \prod_{I=1}^{h} \int \frac{dp^I}{\sqrt{\det M}} \exp\left(\frac{1}{2} \sum_{I,J=1}^{h} p^I_\mu A^{\mu\nu}_{IJ} p^J_\nu + \sum_{I=1}^{h} B^\mu_I p^I_\mu + C \right), \tag{8}$$

where $\widetilde{V}^{0}_{N;h}$ [8, 5] contains the ghost contribution and only the nonzero mode piece of the orbital part of N-Reggeon vertex in absence of background. Here

$$A^{\mu\nu}_{IJ} = A^{0\mu\nu}_{\ IJ} - i\Theta^{\mu\nu} \mathcal{J}_{IJ}, \quad B^\mu_I = B^{0\mu}_{\ I} - i\Theta^{\mu\nu} P_{I\nu}, \quad C = C^0 - \frac{i}{2} \sum_{i<j}^{N} p^i \Theta p^j, \tag{9}$$

where p_I are the loop momenta with loop indices $I = 1, \cdots h$; p_i are the external momenta with legs labelled by $i = 1, \cdots, N$; $\mathcal{J}_{IJ}$ is the intersection matrix for the internal loops; P_I is the sum of the external momenta leaving the Ith loop, and μ, ν stands for the spacetime indices. A^0, B^0, C^0 are independent of Θ and are given by [8],

$$A^{0\mu\nu}_{\ IJ} = 2\alpha'(2\pi i \tau_{IJ})(M^{-1})^{\mu\nu}, \qquad B^{0\mu}_{\ I} = \frac{1}{2\pi} \sum_{i=1}^{N} \oint_0 dz \partial X^{(i)\mu}(z) \int_{z_0}^{V_i(z)} \omega_I, \tag{10}$$

$$\begin{aligned} C^0 &= -\frac{1}{2} \sum_{i=1}^{N} \oint_0 dz \partial X^{(i)}(z) p_0^{(i)} \ln V_i'(z) \\ &+ \frac{1}{2\alpha'} \sum_{i<j}^{N} \oint_0 dz \oint_0 dy \partial X^{(i)}(z) \ln[V_i(z) - V_j(y)] \partial X^{(j)}(y) \\ &+ \frac{1}{4\alpha'} \sum_{i,j=1}^{N} \oint_0 dz \oint_0 dy \partial X^{(i)}(z) \ln\left(\frac{E(V_i(z), V_j(y))}{V_i(z) - V_j(y)} \right) \partial X^{(j)}(y), \end{aligned}$$

where, due to (2) and (3), all indices are contracted by the open string metric M^{-1}. Here $V_i(z)$ is chosen to satisfy $V_i^{-1}(z) = 0$ for $z = z_i$. ω_I is the normalized Abelian differential and τ_{IJ} is the period matrix and $E(z,w)$ is the prime form. Their explicit expressions in term of the Schottky parameters can be found in [8].

Note that all the dependence in Θ is localized in the zero modes loop momentum integration. Carrying out the loop momentum integration, one obtains finally

$$V^{\Theta}_{N;h} = \left[\sqrt{\det M}\right]^{1-h} \widetilde{V}^{0}_{N;h} \frac{1}{\sqrt{\det \frac{-A}{2}}} \exp(-\frac{1}{2} B^T A^{-1} B + C), \tag{11}$$

where the determinant is taken over the space of Lorentz and loop indices (μI). The effects of Θ are summarized elegantly in (9).

Having the general form (11) of the multiloop Reggeon vertex, it is straightforward to write down explicitly the 1-loop amplitude and extract the 1-loop Green function. For example, the string amplitude for M-gauge bosons can be obtained by saturating the $V_{M;h}^{\Theta=0}$ with M external gluon states. Using the explicit formula (10), one obtains

$$A_M^{(h)}(p_1,\ldots,p_M) = C_h \mathcal{N}_0^M \int [dm]_h \frac{\prod_{i=1}^M d\rho_i/\rho_i}{dV_{abc}} \prod_{i<j} \exp\left(\sum_{i<j} 2\alpha' p_i^\mu p_j^\nu (G_{\mu\nu}(\rho_i,\rho_j)\right)$$

$$\times \left[\exp \sum_{i\neq j} \left(\sqrt{2\alpha'}\,\varepsilon_i^\mu \rho_i \partial_{\rho_i} G_{\mu\nu}(\rho_i,\rho_j) p_j^\nu + \frac{1}{2}\varepsilon_i^\mu \rho_i \rho_j \partial_{\rho_i}\partial_{\rho_j} G_{\mu\nu}(\rho_i,\rho_j)\varepsilon_j^\nu\right)\right], \quad (12)$$

where only terms linear in each polarization should be kept. In the general case of having other external states, one need to change correspondingly the form of the second line in (12), however the first line is universal. Here dV_{abc} is the projective invariant volume element which can be used to fix any three of the M punctures ρ_i or of the $2h$ fixed points ξ_μ, η_μ of the Schottky generators. For the measure $[dm]_1$ one can use the well–known expression of the $F=0$ case; the Θ dependence is localized, in the open string Green functions: written in terms of the Schottky representation of the annulus, the planar and nonplanar open string Green functions are [2]

$$G_P^{\mu\nu}(\rho,\rho') = I_0^P (M^{-1})^{\mu\nu} - \frac{i\Theta^{\mu\nu}}{4\alpha'}\varepsilon(\rho-\rho'), \quad (13)$$

$$G_{NP}^{\mu\nu}(\rho,\rho') = I_0^{NP} (M^{-1})^{\mu\nu} + \frac{(\Theta^2)^{\mu\nu}}{8\alpha'^2}\frac{1}{\ln k} \pm \frac{i\Theta^{\mu\nu}}{2\alpha'}\frac{\ln|\rho\rho'|}{\ln k}, \quad (14)$$

where the Θ-independent piece is given by

$$I_0^P(\rho,\rho') = \frac{\ln^2 \rho/\rho'}{2\ln k} + \ln\left|\sqrt{\frac{\rho}{\rho'}} - \sqrt{\frac{\rho'}{\rho}}\right| + \ln \prod_{n=1}^{\infty}\left|\frac{(1-k^n\rho/\rho')(1-k^n\rho'/\rho)}{(1-k^n)^2}\right|, \quad (15)$$

for the planar case, while for non–planar contractions one has

$$I_0^{NP}(\rho,\rho') = \frac{\ln^2|\rho/\rho'|}{2\ln k} + \ln\left(\sqrt{\frac{|\rho|}{|\rho'|}} + \sqrt{\frac{|\rho'|}{|\rho|}}\right) + \ln\prod_{n=1}^{\infty}\left|\frac{(1+k^n|\rho/\rho'|)(1+k^n|\rho'/\rho|)}{(1-k^n)^2}\right|. \quad (16)$$

In the last term of the nonplanar Green function in (14), positive sign is taken when $\rho > 0, \rho' < 0$ and negative sign is taken for the opposite case $\rho' > 0, \rho < 0$.

Remarks on the Green function

In the above, the open string amplitudes and the open string Green function in the presence of B-field were obtained directly using the basic commutation relations (2), (3) and the Reggeon formalism. A different approach of using the *closed* string Green function as input was adopted at the one loop level [12, 13, 14, 15, 16] to calculate the open string amplitude. The idea is to obtain the *open* string Green function by letting the arguments of the closed string Green function to approach the boundary and then use it as input (in, *e.g.* (12)) to calculate the open string amplitude. However as discussed in [14, 2], there are ambiguities associated with this approach.

1. Approaching the boundary

We first recall the procedure of [17] in obtaining the boundary correlator, in particular its antisymmetric part, from the bulk. The boundary correlator obtained there is

$$\mathcal{G}^{\mu\nu}(\theta,\theta') := \langle X^\mu(\theta) X^\nu(\theta')\rangle = -\alpha'(M^{-1})^{\mu\nu}\ln(\theta-\theta')^2 + \frac{i}{2}\Theta^{\mu\nu}\varepsilon(\theta-\theta'), \qquad (17)$$

where M^{-1} and Θ are given in (4). The antisymmetric piece was obtained from the term

$$\Theta^{\mu\nu}\ln\frac{z-\bar{z}'}{\bar{z}-z'} \qquad (18)$$

in the bulk correlator by letting z, z' to *approach* the boundary in an appropriate manner. Notice that restricting the arguments z, z' directly to the real axis yields $\ln 1 = 2n\pi i$ instead of the moduli dependent piece in (17). Note also that there is no uniform shift one can perform to the bulk correlator so that (17) is obtained when restricted to the boundary. A limiting procedure must be adopted. We also remark that radial ordering (on $|z|$) is employed for operators in the bulk, while a time ordering on θ is employed on the boundary. The two orderings agree on the positive θ axis, but is opposite to each other on the negative axis.

Thus there is a certain ambiguity in the antisymmetric part of (17) relating to the choice of the branch cut of the *ln*. This ambiguity can be fixed easily at the tree level and the form (17) is the correct one.

2. Choices of closed string Green functions

As noted in [14], there is an additional source of ambiguity at one loop: there is a freedom in the definition of the closed string Green function.

At the one loop level, it was first noted by [10] that it is not possible to strictly impose on the Green function the same boundary condition as imposed on the string coordinates $(\sigma = 0, \pi)$

$$\partial_\sigma X^\mu + \partial_\tau X^\nu F_\nu{}^\mu = 0, \quad \mu, \nu = 0, 1, \cdots, p, \qquad (19)$$

and hence there is a certain degree of freedom in the choice of what constraint is satisfied by G. If one insists on

$$\Delta G_{\mu\nu}(z,z') = -2\pi\alpha'\eta_{\mu\nu}\delta(z,z'), \qquad (20)$$

then since $\oint \partial_\perp G ds = -2\pi\alpha'$ that is fixed by the Gauss theorem, one obtains the following set of compatible boundary conditions

$$\frac{\partial}{\partial r}G_{\mu\nu}(z,z') - \frac{i}{r}\frac{\partial}{\partial\theta}F_\mu{}^\lambda G_{\lambda\nu}(z,z') = \begin{cases} -\frac{\alpha'}{a}\beta & \text{at } r = a, \\ -\frac{\alpha'}{a}(1-\beta) & \text{at } r = b, \end{cases} \qquad (21)$$

where β is arbitrary. The choice $\beta = 0$ was first derived by [10] using the method of image and was used in [16] in the computation of the tachyon amplitudes, while the choice $\beta = 1/2$ was adopted in [12]. One can also introduce a background charge so that a Green function

$$\Delta G_{\mu\nu}(z,z') = -2\pi\alpha'\eta_{\mu\nu}\delta(z,z') + \frac{2\pi\alpha'}{A}, \tag{22}$$

where $A = \pi(b^2 - a^2)$ is the area of the annulus, satisfying the boundary condition

$$\frac{\partial}{\partial r}G_{\mu\nu}(z,z') - \frac{i}{r}\frac{\partial}{\partial\theta}F_\mu{}^\lambda G_{\lambda\nu}(z,z') = 0, \quad r = a,b, \tag{23}$$

can be constructed. This closed string Green function was used in [13, 14, 15] to derive the open string Green function and in computing the 2 point function for noncommutative photons. The origin of this freedom in defining the Green function was analyzed from the point of view of the boundary state approach in [14], where the closed string Green functions was derived from the closed string worldsheet with boundary states inserted so as to create an open string worldsheet. There it was noted that the freedom in the definition of the closed string Green function is related to a freedom in interpreting the tachyon amplitudes in terms of contributions from the closed string Green function and contributions from the self-contraction C. A shifted

$$G'_{\mu\nu}(z,z') = G_{\mu\nu}(z,z') + \mathcal{M}_{\mu\nu}(z,z'), \tag{24}$$

with a $\mathcal{M}$ satisfying certain conditions [14] gives the same closed string tachyon amplitude. The reason that G and G' can gives the same tachyon amplitude is because of momentum conservation. For higher closed string states, the Green function may also be contracted with other available quantum numbers like polarization which doesn't have a conservation law. So the shift is indeed possible only for tachyon amplitudes. For the amplitude of higher massive closed string states, it is (13) and (14) which are to be used. As was shown in [14], they can be obtained from an appropriate limiting procedure of the bulk Green function that solves (22) and (23). Other form of Green function gives incorrect amplitudes.

This kind of ambiguity persists at higher loops. While it can be fixed easily at the tree level, it is more subtle at the loop level. When one go to higher loop, there are many more boundaries and the above mentioned ambiguities with the Green function will be much harder to resolve [18].

At tree and the 1-loop level, all the effects of Θ can be summarized in terms of a modified Green function. One may therefore also use other approaches to obtain the Green function and use it as input in , for example (12), to calculate the string amplitude. However this is no longer the case for two and higher loops. As we mentioned above, there are new modifications to the measure that is not presented at the tree and 1-loop level. We stress that this Θ-dependent modification to the measure cannot be obtained from the Green function approach. The advantage of the Reggeon operator formalism [2] is that one obtains the string amplitude (11) in one step and there is no need to isolate a Green function from it, which is where the ambiguities lie.

FIELD THEORY LIMIT

The relations between string and field theory amplitudes have been thoroughly studied since the early days of dual models [19]. It turned out that string amplitudes contain very precise information on various perturbative quantities of different field theories. In fact, even using the simple bosonic string as a starting point, it is possible to recover, with a suitable definition of the low energy limit, the results of the usual Feynman diagrams for scalar [20, 21], Yang–Mills [22] or gravity [23] field theories (see also the references in these papers).

In this section we briefly review the basic steps allowing to derive field theories amplitudes from string expressions. like the one in (12). In fact, the same procedure normally used to recover the Feynman diagrams of commutative theories can be also applied to the results derived in the previous sections. Of course the background F-field of the string calculations is now related to the appearance of the non commutative parameters Θ in the field results. However, the algorithm one uses to perform the low energy limit is essentially not affected by the presence of the F-field [12, 13, 14, 15, 16]: all important modifications related to the new feature of noncommutativity are already encoded in the string result and in particular in Eq. (9).

It is usually said that field theory results are recovered from string amplitudes simply by taking the limit $\alpha' \to 0$. Of course, this prescription has to be suitably interpreted to yield sensible answers. First, α' is a dimensionful parameter and what the above limit really means is that the typical energy scale of the external states is very small in comparison with the string scale ($\alpha' p_i \cdot p_j \to 0$). Moreover, it is not possible to blindly perform this limit on the amplitudes. In fact, string results are written as integrals over the moduli space of the Riemann surface representing the world–sheet. Also, the integrand contains various divergences which make the limiting procedure delicate. However, the general idea underlying the derivation of field amplitudes from string ones is to obtain also the answer for field diagrams in an integral form. In fact, the world–sheet moduli are strictly related to the usual Schwinger parameters introduced in diagrammatic perturbative computations. Moreover, only the corners of moduli space where the integrand diverges contribute to the field theory limit. Thus, a single string amplitude decomposes in a sum of different contributions coming from different corners of the full integration region. It turns out that each of these terms encodes the result of all Feynman diagrams of a given topology (for instance, in the Yang–Mills case the string approach automatically sums the contribution of ghost and gluon propagation).

Summarizing the low energy limit on string amplitudes is basically done in three steps:

- First it is necessary to express in terms of string quantities all the parameters appearing in the Lagrangian of the field theory one wants to reproduce. The dictionary between the two sets of parameters can be easily derived by looking at the simplest diagram in both theories; for instance, by matching the 3-point amplitudes, one can usually find the relation between the string and the field theory coupling constants.
- Then one has to focus on the different relevant corners of the region of integration. In each case, the string moduli are transformed into Schwinger parameters with a relation of the type $t = -\alpha' \ln f(z)$, for some function f which may depend on the

details of the worldsheet parameterization.

- Finally one can send $\alpha' \to 0$, but has to keep all field theory parameters finite (even if they are dimensionful, like the Schwinger proper parameters)

Taking the field theory limits

As it turns out, there are two ways one can derive the field theory limit. In [17], the Seiberg-Witten limit

$$\alpha' \sim \varepsilon, \quad g \sim \varepsilon^2, \quad F \sim \varepsilon, \tag{25}$$

with $\varepsilon \to 0$ is considered so that

$$M_{\mu\nu} = -(Bg^{-1}B)_{\mu\nu}, \quad \Theta^{\mu\nu} = -2\pi\alpha' \left(\frac{1}{F}\right)^{\mu\nu}, \tag{26}$$

are fixed in this limit. They show that in this limit the tree level, amplitudes which can be computed from (7), yields the result of a noncommutative field theory in a metric $M_{\mu\nu}$ and with the noncommutative parameter $\Theta^{\mu\nu}$ given by (26). For example,

$$S = \int \sqrt{\det M} \left[(\partial\Phi)^2 + V_\Theta(\Phi)\right]. \tag{27}$$

Since M and Θ are finite in this limit, all our multiloop formula also have a well defined limit. On the other hand, if one prefers (see for example [14, 2]), one can also rescale the string coordinates by a factor $\hat{X}^\mu = X^\nu (g-F)_\nu{}^\mu$ and use $\hat{X}$ to construct the corresponding string amplitude. For example, at the tree level, one gets

$$\hat{G}^{\mu\nu}(\theta,\theta') := \langle \hat{X}^\mu(\theta)\hat{X}^\nu(\theta')\rangle = -\alpha' g^{\mu\nu} \ln(\theta-\theta')^2 + \frac{i}{2}\theta^{\mu\nu}\varepsilon(\theta-\theta'), \tag{28}$$

where

$$\theta := (g+F)\Theta(g-F) = 2\pi\alpha' F \tag{29}$$

and the metric g is the closed string metric. For higher loops, one simply has to replace everywhere (e.g. in (9), (10), (13), (14)) $(M^{-1})^{\mu\nu}$, $\Theta^{\mu\nu}$ by $g^{\mu\nu}, \theta^{\mu\nu}$ in the string theoretic expressions. Thus one can also take the following *noncommutative field theory limit* [14, 2]

$$\alpha' \sim \varepsilon, \quad g \text{ fixed}, \quad F \sim 1/\varepsilon, \tag{30}$$

so that

$$g_{\mu\nu}, \quad \theta^{\mu\nu} = 2\pi\alpha' F^{\mu\nu} \tag{31}$$

are fixed in the limit. In this limit, the noncommutative field theory

$$S = \int \sqrt{\det g} \left[(\partial\Phi)^2 + V_\theta(\Phi)\right], \tag{32}$$

in a background metric g and with noncommutative parameter θ is resulted. We stressed that the two limits are complitely equivalent. No matter which limit one takes, there is always a pair of parameters (metric and the noncommutative parameter) in the string amplitude, and the field theory limit is always taken in such a way that they remain finite. In the end, we obtain a field theory with a background metric and a noncommutative parameter.

Examples

In what follows, we aim at producing a field theory limit with a flat Minkowskian metric $\eta_{\mu\nu}$, which is more often considered in the literatures of noncommutative quantum field theory. This can be most easily achieved from a string theory having a metric $g_{\mu\nu} = \eta_{\mu\nu}$ and using the rescaled $\hat{X}$ to construct the string amplitudes. The field theory noncommutativity parameter θ is then related to the string moduli F by equation (29).

In order to give some concrete examples, we will now consider two Feynman diagrams of the noncommutative scalar theory with cubic interaction in six dimensions. We will focus first on the non–planar contribution to the 2–point function at one loop and then on the non–planar 2-loop vacuum bubble. Many other examples are thoroughly described in [12, 14, 2] also for scalar theories with quartic interactions and for Yang–Mills theory. Eq. (12) can be used also as a starting point for studying scalar (i.e. tachyon) amplitudes simply by setting all the polarizations ε_i to zero. If we fix $\rho_1 = 1$ on the first border and the second puncture on the other border $\rho_2 \in [-1, -k]$, we have to use in the master formula the non–planar Green function (13), obtaining

$$A^{\rm NP}_{2,1}(p_1, -p_1) = \frac{\alpha'^{2-d/2}}{(4\pi)^{d/2}} \frac{g_3^2}{4} \int_0^1 \frac{dk}{k} e^{\alpha' m^2 \ln k} (-\ln k)^{-d/2} \int_{-1}^{-k} \frac{d\rho_2}{\rho_2} e^{[-2\alpha' p_{1\mu} p_{1\nu} G^{\mu\nu}_{\rm NP}(1,\rho_2)]}, \tag{33}$$

where we have already translated the string coupling constant into field theory one by using the relation $g_3 = 2^{5/2} g_{\rm op} (2\alpha')^{\frac{d-6}{4}}$, found from the matching of the 3-point functions. As usual, the logarithmic divergences in the integrand are related to the dimensionful Schwinger parameters via a factor of α'. In particular, at 1-loop one always associates $\ln k$ to the total length of the loop by taking, $\ln k = -T/\alpha'$ where T has to be kept finite as $\alpha' \to 0$. In this limit k goes to zero exponentially which means shrinking the annulus to a one-loop Feynman graph. After having replaced string quantities with field theory ones, the α' dependence of (33) simplifies and the whole amplitude is just proportional to a single power of α'. This means that, in order to have a finite answer for our field theory limit, it is necessary to introduce one more Schwinger parameter. Since we want to reproduce the irreducible diagram, we perform the $\alpha' \to 0$ limit by keeping fixed also $t_2 \ln \rho_2 = -t_2/\alpha'$ and get

$$\begin{aligned} A^{\rm NP}_{2,1}(p_1, -p_1) &= \frac{g_3^2}{4} \frac{1}{(4\pi)^{d/2}} \int_0^\infty \frac{dT}{T^{d/2}} e^{-m^2 T} \\ & \int_0^T dt_2 \exp\left[-p_1^2 t_2 \left(1 - \frac{t_2}{T}\right) + p_1^\mu p_1^\nu \frac{\theta^2_{\mu\nu}}{4T}\right]. \end{aligned} \tag{34}$$

Thus we recover the standard Schwinger proper time integral, but with the additional factor typical of the noncommutative diagrams. As noted in [24], for $\tilde{p}^2 \neq 0$ this serves as an effective UV cutoff and is at the origin of the UV/IR mixing.

In the above example the non-planarity of the diagram is related to the insertion of the external legs. From the string point of view this follows from the fact that the two insertions are done on the two different borders of the annulus. In field theory this means that the two 3-point vertices in the graph have different cyclical orientation (and so the Filk phase has opposite sign). At one loop, this kind of non–planarity is the only possible one. As we saw in the previous string calculation, at multiloop level a new feature appears, namely the internal legs can intersect and bring a non–trivial modification of the string measure (11). A first simple check of this string result is to show that, at low energies, it can reproduce the result of noncommutative Feynman diagrams.

The simplest field theory displaying this feature is the irreducible vacuum bubble that is obtained by sewing together two 3–point vertices with different relative orientations. As in the above 1–loop diagram, the two Filk factors combine and give an additional phase factor $e^{ir\theta q}$ with the respect to the usual computation. However, now we have to integrate over both momenta contracted with θ. Carrying out the r integral, we obtain

$$A_{0,2}^{\rm NP} = \int dq \int \frac{dt_i}{l^{d/2}} e^{-m^2 \sum_i t_i} \exp\left(-\frac{1}{l}[q^{\mu} \Delta_{\theta\mu\nu} q^{\nu}]\right), \tag{35}$$

where t_i are the Schwinger parameters related to the three propagators, $l := t_2 + t_3$. Finally Δ_θ is the following diagonal matrix

$$(\Delta_\theta)_{\mu\nu} := \delta_{\mu\nu}(t_1 t_2 + t_1 t_3 + t_2 t_3) - \frac{\theta^2_{\mu\nu}}{4}, \tag{36}$$

which matches exactly the leading contribution in the multiplier of the expression for $A^{\mu\nu}_{IJ}$ in (9), once the string parameters have been translated in Schwinger proper times [2].

REFERENCES

1. Chu, C.-S., and Ho, P.-M., *Nucl. Phys.*, **B550**, 151 (1999); hep-th/0001144.
2. Chu, C.-S., Russo, R., and Sciuto, S., *Nucl. Phys.*, **B585**, 193–218 (2000).
3. Dolan, L., and Nappi, C. R., hep-th/0009225.
4. Sciuto, S., *Nuovo Cimento Lett.*, **2**, 411 (1969).
   Di Vecchia, P., Nakayama, R., Petersen, J.L., Sciuto, S., *Nucl. Phys.*, **282**, 103 (1987).
5. Di Vecchia, P., Frau, M., Lerda, A., and Sciuto, S., *Phys. Lett.*, **B199**, 49 (1987).
6. Filk, T., *Phys. Lett.*, **B376**, 53 (1996).
7. Di Vecchia, P., Lerda, A., Magnea, L., Marotta, R., and Russo, R., *Nucl. Phys.*, **B469**, 235 (1996).
8. Di Vecchia, P., Pezzella, F., Frau, M., Hornfeck, K., Lerda, A., and Sciuto, S., *Nucl. Phys.*, **B322**, 317 (1989).
9. Frau, M., Pesando, I., Sciuto, S., Lerda, A., and Russo, R., *Phys. Lett.*, **B400**, 52 (1997)
10. Abouelsaood, A., Callan, C.G., Nappi, C.R., and Yost, S.A., *Nucl. Phys.*, **B280**, 599 (1987).
11. Andreev, O., *Phys. Lett.*, **B481**, 125 (2000).
12. Andreev, O., and Dorn, H., *Nucl. Phys.*, **B583**, 145 (2000).
13. Kiem, Y., and Lee, S., *Nucl. Phys.*, **B586**, 303 (2000).
14. Bilal, a., Chu, C., and Russo, R., *Nucl. Phys.*, **B582**, 65 (2000).
15. Gomis, J., Kleban, M., Mehen, T., Rangamani, M., and Shenker, S., *JHEP*, **08**, 011 (2000).
16. Liu, H., and Michelson, J., *Phys. Rev.*, **D62**, 066003 (2000).
17. Seiberg, N., and Witten, E., *JHEP*, **09**, 032 (1999).
18. Kiem, Y., Lee, S., and Park, J., *Nucl. Phys.*, **B594**, 169 (2001).
19. Scherk, J., *Nucl. Phys.*, **B31**, 222 (1971).
20. Frizzo, A., Magnea, L., and Russo, R., *Nucl. Phys.*, **B579**, 379 (2000).
21. Marotta, R., and Pezzella, F., *Phys. Rev.*, **D61**, 106006 (2000).

22. Frizzo, A., Magnea, L., and Russo, R., hep-ph/0012129.
23. Bern, Z., Dixon, L., Dunbar, D. C., Perelstein, M., and Rozowsky, J. S., *Nucl. Phys.*, **B530**, 401 (1998).
24. Minwalla, S., Raamsdonk, M.V., and Seiberg, N., hep-th/0002186, hep-th/9912072.

Classical and Quantum Polyhedra: A Fusion Graph Algebra Point of View

R. Coquereaux

Centre de Physique Théorique - CNRS - Luminy, Case 907, F-13288 Marseille Cedex 9 - France, email: coque@cpt.univ-mrs.fr

Abstract. Representation theory, for the classical binary polyhedral groups $\tilde{T}$, $\tilde{C}$ and $\tilde{I}$, is encoded by the affine Dynkin diagrams $E_6^{(1)}$, $E_7^{(1)}$ and $E_8^{(1)}$ (McKay correspondance). The quantum versions of these classical geometries are associated with representation theories described by the usual Dynkin diagrams E_6, E_7 and E_8. The purpose of these notes is to compare several chosen aspects of the classical and quantum geometries by using the study of spaces of paths and spaces of essential paths (Ocneanu theory) on these diagrams. To keep the size of this contribution small enough, most of our discussion will be limited to the cases of diagrams E_6 and $E_6^{(1)}$, i.e., to the classical and quantum tetrahedra. We shall in particular interpret the A_{11} labelling of E_6 vertices as a quantum analogue of the usual decomposition of spaces of sections for vector bundles above homogeneous spaces. We also show how to recover Klein invariants of polyhedra by paths algebra techniques and discuss their quantum generalizations.

INTRODUCTION

One way to understand the classification of modular invariant partition functions, for instance the *ADE* classification of $SU(2)$ conformal theories ([1], [17]) or its generalizations ([8], [7]), should be understood, following A. Ocneanu (many talks since 1995, for instance [15], and [16]) through the study of "quantum symmetries" on graphs, and in particular, on Dynkin *ADE* diagrams. In turn, this study leads, in each case, to new kinds of partition functions wich are not modular invariant but are nevertheless quite remarkable: the concept of "torus structure" of ADE graphs is due to A.Ocneanu (unpublished), the corresponding "twisted partition functions" have been recently discussed by ([19]), the expressions of the"toric matrices" for the quantum tetrahedron (E_6 model) have been presented in ([3]) and explicit calculations for the other models, using the techniques of this last paper should appear in [5]. The needed mathematical material is unfortunately not really standard and often not even available in published form. It happens, however, that many of these algebraic ("quantum") manipulations can be seen as a quantum analogue of finite group constructions. The purpose of these notes is to give a fresh look to the old-fashion representation theory of groups of symmetries of regular polyhedra, and to do it in such a way that generalization to the quantum case becomes (almost) straightforward. However, we shall not cover all the way, from *ADE* diagrams to Ocneanu quantum symmetries, even when discussing the classical analogue of these constructions: in particular the study of Racah-Wigner bi-algebras for Platonic groups will be left aside (this should be done within [4]) and we shall not even introduce the

CP589, *Fundamental Interaction Theories*, edited by J. Lukierski and J. Rembieliński

classical analogue of Ocneanu graphs... In particular, the twisted partition functions that one can associate to the different vertices of the Ocneanu graph will not be described. We hope nevertheless that this set of notes will provide a useful introduction to this fascinating subject.

The major part of what is going to be explained below is certainly known, at least in some circles. We believe, however, that our geometrical interpretation, for the "essential" labelling of vertices of exceptional Dynkin diagrams (like E_6) by vertices belonging to an appropriate A_N graphs (which is A_{11} in the case of E_6), as the quantum analogue of the decomposition of a space of sections of a homogeneous vector bundle, is probably new, and may be of interest for the expert.

Also, explicit calculations of projectors decomposing the representations $[2]^p$ of binary polyhedral groups into irreps have been probably carried out by several group theorists, chemists, or solid state physics practitionners, but we do not think that an actual implementation of a method using paths on graphs and spectral decomposition of the (classical or quantum) $\hat{R}$ matrix was described before. One should be aware, however, that in several cases, like for instance the calculation of Klein invariants, the results themselves have been known for more than a century!

CLASSICAL AND QUANTUM PLATONIC BODIES

As it is well known, geometry of classical Platonic bodies is encoded by their symmetry groups that we shall call respectively T (for the the tetrahedron), C (for the cube and its dual, the octahedron) and I (for the icosahedron and its dual, the dodecahedron). These are subgroups of $SO(3)$. By adding reflections, one may also consider the "full" polyhedral groups, which are subgroups of $O(3)$, but the objects of interest, for us, are the so-called binary polyhedral groups $\tilde{T},\tilde{C},\tilde{I}$, that are non abelian subgroups of $SU(2)$, the two-fold cover of $SO(3)$. They are defined as pre-image of the corresponding $SO(3)$ subgroups by using the sequence

$$0 \mapsto \mathbb{Z}_2 \mapsto SU(2) \mapsto SO(3) \mapsto 0$$

We concentrate our attention mostly on the particular example given by the binary *tetrahedral* group (of order $2 \times 12 = 24$) since we want to compare it, or better its representation theory, with a kind of quantum analogue. One reason to limit our study to $\tilde{T}$ is lack of space, the other is that the quantum geometry corresponding to C is somehow much more difficult to study that the one corresponding to T and I.

Classical Geometry

Several Realisations of the Group $\tilde{T}$

Elementary geometric considerations show that T itself is isomorphic with the group of even permutations on four elements (label the vertices of the tetrahedron by

$(1,2,3,4))$. Therefore

$$\#\tilde{T} = 2 \times \#T = 2 \times \frac{4!}{2} = 2 \times 12 = 24$$

Since $\tilde{T}$ is, by definition, a finite subgroup of $Spin(3)(\simeq SU(2))$, we can express all its elements in terms of $Cliff(\mathbb{R}^3)$, i.e., in terms of the "gamma matrices" of $\mathbb{R}^3$, namely the Pauli matrices τ_i. Setting $\gamma_x \doteq -\tau_2$, $\gamma_y \doteq \tau_1$ and $\gamma_z \doteq \tau_3$, we get $\gamma_i\gamma_j + \gamma_j\gamma_i = 2\delta_{ij}$ with

$$\gamma_x = \begin{pmatrix} 0 & i \\ -i & 0 \end{pmatrix} \quad \gamma_y = \begin{pmatrix} 0 & 1 \\ 1 & 0 \end{pmatrix} \quad \gamma_z = \begin{pmatrix} 1 & 0 \\ 0 & -1 \end{pmatrix}$$

It is easy to see that $\tilde{T}$ is generated, as a group, by

$$\begin{aligned} \mathbf{t} &= \gamma_x\gamma_y \\ \mathbf{s} &= \frac{1}{2}(\gamma_x\gamma_y + \gamma_y\gamma_z + \gamma_z\gamma_x - 1) \end{aligned}$$

Notice that $\mathbf{t}^2 = -1, \mathbf{t}^3 = -\mathbf{t}, \mathbf{t}^4 = 1$ and $\mathbf{s}^2 = -(\mathbf{s}+1)$, so that $\mathbf{s}^3 = 1$. Moreover $(\mathbf{st})^3 = 1$. Explicitly:

$$\mathbf{s} \doteq \begin{pmatrix} (-1+i)/2 & (-1+i)/2 \\ (1+i)/2 & -(1+i)/2 \end{pmatrix} \quad \mathbf{t} \doteq \begin{pmatrix} i & 0 \\ 0 & -i \end{pmatrix}$$

The reader may be interested in knowing that $\tilde{T}$ (resp. $\tilde{C}$ and $\tilde{I}$) is isomorphic with the group $\langle 2,3,n \rangle$ (Threlfall notation), when $n = 3$ (resp. $n = 4$ and $n = 5$). This notation refers, by definition, to the group generated by two elements A and B, with relations $A^3 = B^n = (AB)^2$. These groups, when $n = 3$ or $n = 5$ are also isomorphic with the groups $SL(2,F_3)$ or $SL(2,F_5)$, here F_n is the field with n elements; there is no such isomorphism when $n = 4$. It may be also nice to remember that the "extended" triplets $\langle 3,3,3 \rangle$, $\langle 2,4,4 \rangle$ and $\langle 2,3,6 \rangle$ are the only positive integer solutions of the equation :

$$\frac{1}{a} + \frac{1}{b} + \frac{1}{c} = 1$$

These extended triplets encode the affine Dynkin diagrams of type $E_6^{(1)}$, $E_7^{(1)}$, $E_8^{(1)}$ (lengths of the three legs counted from the triple point). Substracting 1 to a maximal element of these extended triplets give the triplets $\langle 2,3,3 \rangle$, $\langle 2,3,4 \rangle$ and $\langle 2,3,5 \rangle$ which solve the inequality[1]

$$\frac{1}{a} + \frac{1}{b} + \frac{1}{c} > 1$$

and encode the usual Dynkin diagrams for exceptionnal Lie groups. The same triplets also characterize the binary polyhedral groups since they give the relations obeyed by their generators in Threlfall notation.

---

[1] The other solutions of this inequality, if we exclude $a = 1, b \neq c$, are $\langle 1,n,n \rangle$ and $\langle 2,2,n \rangle$, corresponding to the A_{2n-1} and D_{n+2} Dynkin diagrams.

Representations of the Group T

One way to get the representations of T is to remember that $T \simeq A_4 \subset S_4$ and use the fact that representation theory of permutation groups (like S_4) is well known (use Young tableaux, for instance).

$T \simeq A_4$ has four irreducible inequivalent representations of respective dimensions $1, 1, 1, 3$. We call them σ_1, σ_1', σ_1'', σ_3. We check that the dimensions divide 12 (as they should) and that $1^2 + 1^2 + 1^2 + 3^2 = 12$, of course.

Representations of the Group $\tilde{T}$

The previous irreps are also irreducible representations for the corresponding binary group $\tilde{T} \subset SU(2)$, but the later also posesses other irreducible representations. Altogether, $\tilde{T}$ has seven irreducible inequivalent representations; their dimensions are

$$1, 1, 1, 3, 2, 2, 2.$$

The odd dimensional ones will be labelled as before and the last three will be called σ_2, σ_2' and σ_2''. The one called σ_2 is, by definition, the fundamental of $SU(2)$ restricted to $\tilde{T}$. With no surprise we check that all these dimensions divide 24 and that $1^2 + 1^2 + 1^2 + 3^2 + 2^2 + 2^2 + 2^2 = 24$. The representations σ_1, σ_2, σ_3 are self-conjugate (actually real) σ_1',σ_1'', σ_2',σ_2'' are complex and respectively conjugated to one another.

Tensor products of representations and McKay correspondance

In the case of $SU(2)$, it is well known that the tensor product of the representation of spin $1/2$ (i.e., dimension 2) by a representation of spin j (dimension $n = 2j + 1$) is equivalent to the sum of two representations of respective spin $j - 1/2$ and $j + 1/2$. In terms of dimensions, we have $2(2j+1) = (2j) + (2j+2)$; in terms of the representations themselves (working up to equivalence) we have:

$$\sigma_2 \otimes \sigma_n = \sigma_{n-1} \oplus \sigma_{n+1}$$

The irreps into which a given representation σ_n decomposes, upon tensorial multiplication by the fundamental representation σ_2 are given by the neighboors of σ_n on the following semi-infinite diagram called the A_∞ diagram.

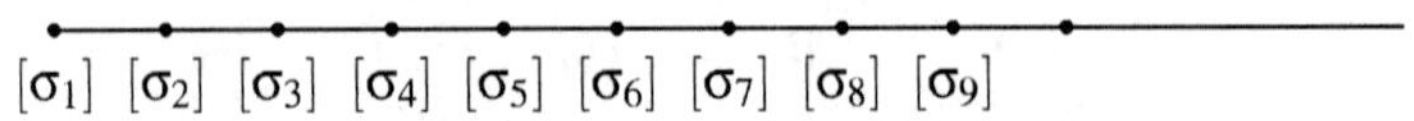

FIGURE 1. The diagram A_∞

Returning to the binary tetrahedral group $\tilde{T}$, we decide to encode in the same way the tensor product of the various irreps by the fundamental (the 2-dimensional). The calculation itself is a simple exercise in finite group theory and we shall not dwell on the matter... The point is that, if we decide to encode the results in terms of a graph with seven vertices (the seven irreps), this graph is nothing else than the Dynkin diagram of the exceptional affine Lie algebra $E_6^{(1)}$.

What comes as a surprise is that, if we perform the same construction with the binary groups of the cube and of the icosahedron, we obtain respectively the Dynkin diagrams of the affine Lie algebras $E_7^{(1)}$ and $E_8^{(1)}$. This observation is known as "McKay correspondance" ([14]).

The diagrams $E_6^{(1)}$ and $E_8^{(1)}$, labelled by the seven (nine) irreducible representations of the binary tetrahedral (icosahedral) group are displayed below.

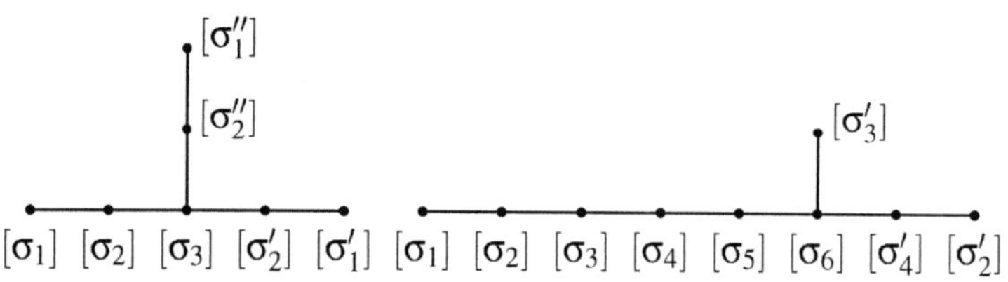

FIGURE 2. The diagrams $E_6{}^{(1)}$ and $E_8{}^{(1)}$

As explained before, this reads, for instance in the case of $E_6^{(1)}$, $\sigma_1' \otimes \sigma_2 = \sigma_2'$, or $\sigma_2 \otimes \sigma_3 = \sigma_2 \oplus \sigma_2' \oplus \sigma_2''$. The summands appearing on the right hand side are the neighbours, on the graph, of the chosen vertex. Of course, the dimensions should match (so, for instance $2 \times 3 = 2 + 2 + 2$).

For the above reason we decide to denote by $\mathcal{H}_{E_6^{(1)}}$, $\mathcal{H}_{E_7^{(1)}}$ and $\mathcal{H}_{E_8^{(1)}}$ the group algebras of these three binary groups. These are co-commutative finite dimensional semi-simple Hopf algebras. The corresponding dual objects (algebras of complex valued functions of these groups) are denoted by $\mathcal{F}_{E_6^{(1)}}$, $\mathcal{F}_{E_7^{(1)}}$ anf $\mathcal{F}_{E_8^{(1)}}$.

Structure of the Grothendieck Ring

We already know how to multiply representations of $\tilde{T}$ by σ_1 (the trivial representation) and by σ_2 (this is given by the graph $E_6^{(1)}$). In other words, we know the first two lines and the first two columns of the table of multiplication of characters. Another useful identity that one should get first is $1'1'' = 1$; this is a consequence of the fact that $1'1''$ is one dimensional (so it is either 1, $1'$ or $1''$) but it is real, since $\overline{1'} = 1''$. This data is sufficient to reconstruct the whole table of multiplication. We only have to use associativity of $\otimes$ and perform the calculations in the ring of virtual representations (hence allowing minus signs at intermediate steps in our calculations.)

Let us for instance compute $\sigma_3 \otimes \sigma_3$ (we shall just write 33 in the sequel):

$$\begin{aligned} 33 &= 3(22-1) = 322 - 3 = (32)2 - 3 = (2+2'+2'')2 - 3 = \\ &= 1+3+1'+3+1''+3-3 = 3+3+1+1'+1'' = (3)_2 + 1 + 1' + 1'' \end{aligned}$$

Here, the subindex 2 in $(3)_2$ means that 3 appears with multiplicity 2.

After some work, all other entries of the multiplication table can be worked out by simple manipulations analoguous to the above calculation of 33 and one obtains the following table (that we shall sometimes call the fusion table); it gives all the coupling constants C_{ijk} of the Grothendieck ring $Ch(G) = \mathbb{Z}(Irr(\tilde{T}))$; these constants are defined by $\sigma_i \otimes \sigma_j = C_{ijk}\sigma_k$.

	1	2	$1'$	$2'$	$1''$	$2''$	3
1	1	2	$1'$	$2'$	$1''$	$2''$	3
2	2	$1,3$	$2'$	$1'3$	$2''$	$1''3$	$22'2''$
$1'$	$1'$	$2'$	$1''$	$2''$	1	2	3
$2'$	$2'$	$1'3$	$2''$	$1''3$	2	13	$22'2''$
$1''$	$1''$	$2''$	1	2	$1'$	$2'$	3
$2''$	$2''$	$1''3$	2	13	$2'$	$1'3$	$22'2''$
3	3	$22'2''$	3	$22'2''$	3	$22'2''$	$11'1''3_2$

Notice that the final table only involves sums (no minus signs, of course!) of irreducible representations.

We should maybe write this table in the order $(1,1',1'',3;2,2',2'')$ to reflect the fact that the subset $1,1',1'',3$ (irreps of T) is stable under tensorial multiplication; this corresponds to the fact that the group T is a quotient of $\tilde{T}$.

The Fusion Matrices N_i

The (non necessarily symmetric) matrices N_i are defined by

$$(N_i)_{jk} = C_{ijk}$$

Still using the order $\{1,1',1'',2,2',2'',3\}$, we have :

$$N_1 = \begin{pmatrix} 1&0&0&0&0&0&0\\ 0&1&0&0&0&0&0\\ 0&0&1&0&0&0&0\\ 0&0&0&1&0&0&0\\ 0&0&0&0&1&0&0\\ 0&0&0&0&0&1&0\\ 0&0&0&0&0&0&1 \end{pmatrix} \quad N'_1 = \begin{pmatrix} 0&0&1&0&0&0&0\\ 1&0&0&0&0&0&0\\ 0&1&0&0&0&0&0\\ 0&0&0&0&0&1&0\\ 0&0&0&1&0&0&0\\ 0&0&0&0&1&0&0\\ 0&0&0&0&0&0&1 \end{pmatrix} \quad N''_1 = \begin{pmatrix} 0&1&0&0&0&0&0\\ 0&0&1&0&0&0&0\\ 1&0&0&0&0&0&0\\ 0&0&0&0&1&0&0\\ 0&0&0&0&0&1&0\\ 0&0&0&1&0&0&0\\ 0&0&0&0&0&0&1 \end{pmatrix}$$

$$N_2 = \begin{pmatrix} 0&0&0&1&0&0&0\\ 0&0&0&0&1&0&0\\ 0&0&0&0&0&1&0\\ 1&0&0&0&0&0&1\\ 0&1&0&0&0&0&1\\ 0&0&1&0&0&0&1\\ 0&0&0&1&1&1&0 \end{pmatrix} \quad N'_2 = \begin{pmatrix} 0&0&0&0&0&1&0\\ 0&0&0&1&0&0&0\\ 0&0&0&0&1&0&0\\ 0&0&1&0&0&0&1\\ 1&0&0&0&0&0&1\\ 0&1&0&0&0&0&1\\ 0&0&0&1&1&1&0 \end{pmatrix} \quad N''_2 = \begin{pmatrix} 0&0&0&0&1&0&0\\ 0&0&0&0&0&1&0\\ 0&0&0&1&0&0&0\\ 0&1&0&0&0&0&1\\ 0&0&1&0&0&0&1\\ 1&0&0&0&0&0&1\\ 0&0&0&1&1&1&0 \end{pmatrix}$$

$$N_3 = \begin{pmatrix} 0&0&0&0&0&0&1\\ 0&0&0&0&0&0&1\\ 0&0&0&0&0&0&1\\ 0&0&0&1&1&1&0\\ 0&0&0&1&1&1&0\\ 0&0&0&1&1&1&0\\ 1&1&1&0&0&0&2 \end{pmatrix}$$

It is easy to show that we obtain an isomorphism between the – commutative – ring of characters $Ch(\tilde{T})$ (or, equivalently, the ring generated by tensor powers of the irreducible representations) and the ring generated (over the integers $\mathbb{Z}$) by the matrices N_i. For instance, we have

$$\sigma_3 \otimes \sigma_3 = \sigma_3 + \sigma_3 + \sigma_1 + \sigma_{1'} + \sigma_{1''}$$

and, at the same time,

$$N_3.N_3 = 2N_3 + N_1 + N_{1'} + N_{1''}$$

In particular, the Dynkin diagram of $E_6^{(1)}$, considered as the fusion graph by the fundamental of the binary tetrahedral group (Mc Kay correspondance) can also be also read in terms of the fusion matrices N:

$$\begin{array}{ccc} N_2.N_1 = N_2 & N_2.N_2 = N_1 + N_3 & N_2.N_3 = N_2 + N_{2'} + N_{2''} \\ N_2.N_{2'} = N_{1'} + N_3 & N_2.N_{2''} = N_{1''} + N_3 & N_2.N_{1'} = N_{2'} \\ N_2.N_{1''} = N_{2''} & & \end{array}$$

From the Grothendieck Ring to the Character Table

The seven commuting 7×7 matrices N_i can be simultaneously diagonalized with a common similarity matrix X (which also gives the list of eigenvalues).

$$X = \begin{pmatrix} 1 & 1 & 1 & 1 & 1 & 1 & 1 \\ 1 & 1 & 1 & \omega^2 & \omega^2 & \omega & \omega \\ 1 & 1 & 1 & \omega & \omega & \omega^2 & \omega^2 \\ 2 & 0 & -2 & 1 & -1 & 1 & -1 \\ 2 & 0 & -2 & \omega^2 & -\omega^2 & \omega & -\omega \\ 2 & 0 & -2 & \omega & -\omega & \omega^2 & -\omega^2 \\ 3 & -1 & 3 & 0 & 0 & 0 & 0 \end{pmatrix}$$

A nice observation is that the character table is given by the same matrix X, i.e., , by the list (properly ordered!) of eigenvalues of the matrices N_i: each line corresponds to an irreducible representation and each column to a conjugacy class. The point is that we *did not* have to work out the conjugacy classes themselves; the structure constants of the Grothendieck ring (which are themselves, in the present case, encoded by the graph $E_6^{(1)}$) provide enough data to reconstruct the whole character table. This not so well known result is true for to any finite group.

Perron-Frobenius Data for the Graph $E_6^{(1)}$

We want now "reverse" the machine, i.e., forget everything we know about groups $\tilde{T}$ (or T) and try to reconstruct as much as we can form the combinatorial data provided by the $E_6^{(1)}$ diagram. The adjacency matrix $\mathcal{G}$ of an oriented graph is a matrix labelled by the vertices of $\mathcal{G}$ whose (i, j) element is equal to n whenever there are n edges from i to

j. When the graph is not oriented, each edge is considered as carrying both orientations, so that the matrix of the graph is symmetric.

In our case, we label the vertices of the graph $E_6^{(1)}$ by $1, 1', 1'', 2, 2', 2'', 3$, in this order; the corresponding adjacency matrix is clearly

$$\mathcal{G} = \begin{pmatrix} 0 & 0 & 0 & 1 & 0 & 0 & 0 \\ 0 & 0 & 0 & 0 & 1 & 0 & 0 \\ 0 & 0 & 0 & 0 & 0 & 1 & 0 \\ 1 & 0 & 0 & 0 & 0 & 0 & 1 \\ 0 & 1 & 0 & 0 & 0 & 0 & 1 \\ 0 & 0 & 1 & 0 & 0 & 0 & 1 \\ 0 & 0 & 0 & 1 & 1 & 1 & 0 \end{pmatrix}$$

This matrix is nothing else than the matrix N_2, i.e., the fundamental generator of the ring of N-matrices and it is associated with the fundamental representaion σ_2. The eigenvalues are $-2, -1, -1, 0, 1, 1, 2$. The biggest eigenvalue (also called "norm of the graph" or "Perron-Frobenius eigenvalue") is equal to 2. This is also true for the graphs $E_7^{(1)}$ and $E_8^{(1)}$. The corresponding eigenvector (we normalize the first entry to 1), also called "Perron-Frobenius vector of the graph" has components D that are positive integers. $D = 1, 1, 1, 2, 2, 2, 3$. We recognize the dimensions of the irreps of $\tilde{T}$.

The conclusion is that, from the graph alone, we can recover the dimension of the irreps. This is also true for the binary cubic and icosahedral groups. From the same graph, one then recovers, as already explained, the multiplication of representations by the fundamental (which is 2-dimensional) as well as the whole fusion algebra, by imposing associativity of the Grothendieck ring; the character table itsef is obtained by a simultaneous diagonalisation of the N_i matrices encoding the structure constants C_{ijk} of this ring.

Structure of Centralizer Algebras, Tower of Commutants for $\tilde{T}$

Using the structure of the Grothendieck ring for $\tilde{T}$, encoded by the graph $E_6^{(1)}$, we see immediately, by taking tensor powers of the fundamental, that

$$\begin{aligned}
[2]^1 &= 1[2] \\
[2]^2 &= 1[1] + 1[3] \\
[2]^3 &= 2[2] + 1[2'] + 1[2''] \\
[2]^4 &= 2[1] + 4[3] + 1[1'] + 1[1''] \\
[2]^5 &= 6[2] + 5[2'] + 5[2''] \\
[2]^6 &= 6[1] + 16[3] + 5[1'] + 5[1''] \\
\ldots &= \text{etc.}
\end{aligned}$$

We call C_p the centralizer algebras of the group $\tilde{T}$ in the representation $[2]^p$. It is clear that these algebras are not isomorphic with the Temperley-Lieb algebra T_p (which are isomorphic with the Schur centralizer algebras for $SU(2)$) as soon as $p \geq 3$. For instance $T_4 = M(2,\mathbb{C}) \oplus M(3,\mathbb{C}) \oplus \mathbb{C}$ since, in $SU(2)$, $[2]^4 = 2[1] + 3[3] + 1[5]$, but $C_4 = M(2,\mathbb{C}) \oplus M(4,\mathbb{C}) \oplus \mathbb{C} \oplus \mathbb{C}$ since, in $\tilde{T}$, $[2]^4 = 2[1] + 4[3] + 1[1'] + 1[1'']$. The above results leads immediately to the following structure (see Figure below) for the tower of commutants C_p's. As usual, inclusions are defined by the edges of this graph and an appropriate Pascal rule gives the dimensions. Notice that after a few steps, we get the folded $E_6^{(1)}$ diagram of reflected and repeated down to infinity. This picture – paths emanating from the endpoint vertex – can also be generated very simply by considering successive powers of the adjacency matrix N_2 of the Dynkin diagram $E_6^{(1)}$ acting on the (transpose) of the vector $(1,0,0,0,0,0,0)$ characterizing the leftmost vertex (identity representation).

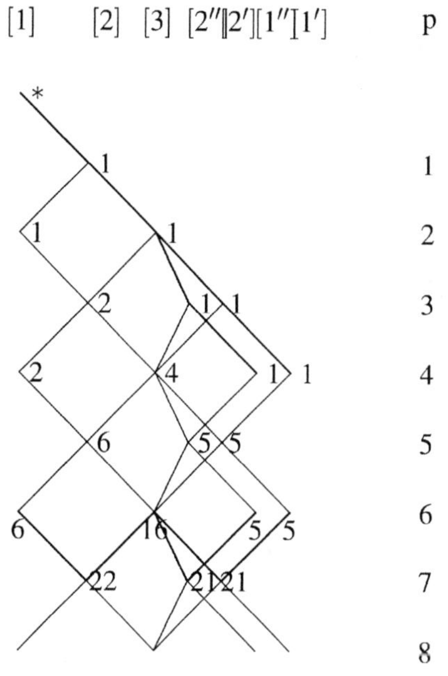

Essential Paths for the Graph $E_6^{(1)}$

We first define elementary paths on a graph as a sequence of successive vertices (here, for simplicity, we suppose that edges carry both orientations, i.e., no orientation at all). Elementary paths can therefore backtrack. Then we consider the Hilbert space *Paths* obtained by taking arbitrary linear combinations of elementary paths. The scalar product is defined by declaring that the basis of elementary paths is orthonormal. Since every elementary path has a lenght p, the vector space *Paths* is graded. In the case of $SU(2)$ (with graph A_∞) or $\tilde{T}$ (with graph $E_6^{(1)}$), irreducible representations appearing in the decomposition of $[2]^p$ can be characterized by paths on those graphs, emanating from the origin; they are also associated with particular projectors, that are, $2^p \times 2^p$ matrices. The notion of *essential* paths on a graph is due to Ocneanu ([16]).

Let us start with the case of $SU(2)$. A given irreducible representation of dimension d appears for the first time in the decomposition of $[2]^{d-1}$ and corresponds to a particular projector in the vector space $(\mathbb{C}^2)^{\otimes d-1}$ which is totally symmetric and therefore projects on the space of symmetric tensors. These symmetric tensors provide a basis of this particular representation space and are, of course, in one to one correspondance with symmetric polynomials in two complex variables u, v (representations of given degree).

However, irreducible representations of dimension d appear not only in the reduction of $[2]^{d-1}$ but also in the reduction of $[2]^f$, when $f = d+1, d+3, \ldots$. Such representations are equivalent with the symmetric representations previously described but they are nevertheless distinct, as explicit given representations and their associated projectors are not symmetric. Paths corresponding to irreducible symmetric representations are essential. For instance the representation $[3]$ that appears in $[2]^2$ corresponds to an essential path starting from the origin, but the three representations $[3]$ that appear in the reduction of $[2]^4$ do not correspond to essential paths. The notion of "essential path" formalizes and generalizes the above remarks.

When we move from the case of $SU(2)$ to the case of finite subgroups of $SU(2)$, in particular the binary polyhedral groups whose representation theories are described by the affine Dynkin diagrams $E_6^{(1)}$, $E_7^{(1)}$ and $E_8^{(1)}$, the notion of essential paths can be obtained very simply by declaring that a path on the corresponding diagram is essential if it describes an irreducible representation that appears in the reduction with respect to the chosen finite subgroup of an irreducible *symmetric* representation of $SU(2)$. For instance, the $[4]$ dimensional representation of $SU(2)$, obtained in the decomposition of $[2][3] = [2] + [4]$ is symmetric, and is associated with a Wenzl projector p_4 of the algebra T_4. In the case of the finite subgroup $\tilde{T}$, the corresponding projector of the centralizer algebra C_4 splits, and this corresponds to the reduction $[4] \to [2'] + [2'']$ into a sum of two inequivalent irreps. In $SU(2)$ we have therefore one essential path of length 3, emanating from the origin (it ends on $[4]$), but in $\tilde{T}$, this gives two essential paths, one ending on $[2']$ and the other on $[2'']$. In general, essential paths are linear combinations of elementary paths. Notice that essential paths for the finite subgroups of $SU(2)$ can be of arbitrary length since symmetric representations of $SU(2)$ can be of arbitrary degree (horizontal Young diagrams with an arbitrary number of boxes). This will not be so for their quantum analogues.

The number of essential paths starting from the origin and ending at a given vertex are readily obtained from the tower of centralizers by using a kind of "moderated" Pascal rule: the number of essential paths (with fixed origin) of lenght p reaching a particular vertex is obtained from the sum of the number of essential paths of length $p-1$ reaching the neighbouring points (as in Pascal rule) by substracting the number of essential paths of length $p-2$ reaching the chosen vertex. This observation was made by [16] in a general setting, and by J.B. Zuber ([20]) in the context of boundary conformal field theories.

The following picture – essential paths starting from the endpoint vertex – can be generated very simply as follows : Define a rectangular matrix E_1, with seven columns and infinitely many rows, whose $j-th$ row is $E_1(j) = N_2 E_1(j-1) - E_1(j-2)$, with $E_1(1) = (1,0,0,0,0,0,0), E_1(2) = N_2 E_1(1)$, and where N_2 is the adjacency matrix of the graph $E_6^{(1)}$. The entries of $E_1(j)$ give the number of paths of length $j-1$ ending on the different vertices. Essential paths starting from an arbitrary vertex of the Dynkin diagram can be constructed in the same way, by replacing E_1 by $E_2(1) = (0,1,0,0,0,0,0)$,

$E_3(1) = (0,0,1,0,0,0,0)$etc. . Information about essential paths (in particular their number) is therefore encoded by the a set of seven rectangular matrices (in the case of $E_6^{(1)}$) which have seven rows and infinitely many lines [2]. Their quantum analogues, however have only a finite number of lines. These matrices, introduced in [3], will be called "essential matrices".

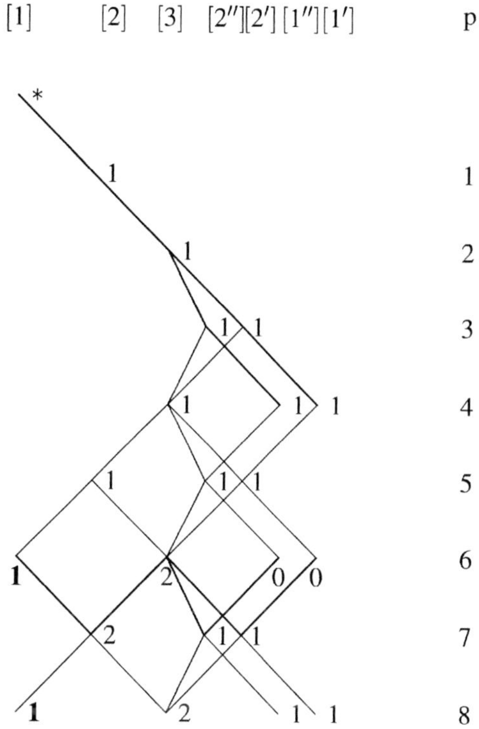

Projectors on Irreducible Representations

Our purpose, here, is to explain, in a nutshell, how to obtain explicitly a matrix expression for the projectors $\varpi_p[s]$ mapping the reducible representation space $[2]^p$ to one of its irreducible subrepresentations $[s]$, for $SU(2)$ or one of its finite subgroups. These are explicit $2^p \times 2^p$ matrices. We do it for $SU(2)$ first. Here are the steps:

- Find an explicit matrix realization for the Jones' projectors e_i's in the appropriate Jones-Temperley-Lieb algebra T_p.
- Express the minimal central projectors associated with the various blocks appearing in the algebra T_p in terms of the Temperley-Lieb' generators e_i.

[2] In the following picture, we decided arbitrarily to cut the graph at level $p = 8$

- Call A the classical antisymmetrizer of $SU(2)$ in the representation $[2]^2$ (it is a 4×4 matrix). One obtains the projectors $\varpi[s]$ from the minimal central projectors by replacing $e_1, e_2, \ldots$'s by $\varepsilon_1 = A \otimes 1\!\!1_2 \otimes 1\!\!1_2 \otimes 1\!\!1_2 \otimes \ldots$, $\varepsilon_2 = 1\!\!1_2 \otimes A \otimes 1\!\!1_2 \otimes 1\!\!1_2 \otimes \ldots$, etc. .

The sub-representation $[p+1]$ of $[2]^p$ is special since it is totally symmetric; the expression of the correspondig central projector of T_p is easy to obtain in this case since a recurrence formula exists for Wenzl projectors (see for instance [10]).

Now, we do it for the binary tetrahedral group.

- We first compute the projector decomposition of $[2]^p$ for $SU(2)$ as above.
- We use the fact that the binary tetrahedral group is generated by two explicit generators **s** and **t**. Considered as (group-like) elements of the group algebra, i.e., $\Delta\mathbf{s} = \mathbf{s} \otimes \mathbf{s}$ and $\Delta\mathbf{t} = \mathbf{t} \otimes \mathbf{t}$, one can compute their iterated coproduct in representation $[2]^p$, they are explicit $2^p \times 2^p$ matrices. The already obtained projectors of $SU(2)$ commute with them. To simplify the discussion, we take $p = 3$ and write $[2]^3 = [2_a] + [2_b] + [4]$ for $SU(2)$, and $[4] \mapsto [2'] + [2'']$ for the reduction to $\tilde{T}$. We want to obtain $\varpi_3[\sigma_{2'}]$ and $\varpi_3[\sigma_{2''}]$. Taking an arbitrary 8×8 matrix ϖ, we first impose that it belongs to the centralizer algebra, so it should commute with the iterated coproduct of **s** and **t** already calculated (linear equations for the matrix coefficients). This already restricts the number of unknown coefficients.
- We impose that the matrix ϖ should be orthogonal to the known $\varpi_3[\sigma_{2a}]$ and $\varpi_3[\sigma_{2b}]$ (again linear equations). This further restrict the number of coefficients.
- Finally we impose that ϖ should be a projector ($\varpi.\varpi = \varpi$). One finds three solutions : one is of rank 4 (it is $\varpi_3[\sigma_4]$ itself), the other two are of rank 2 and add up to $\varpi_3[\sigma_4]$. These are the projectors $\varpi_3[\sigma_{2'}]$ and $\varpi_3[\sigma_{2''}]$ that we were looking for.

Writing even a simple example in full details requires a lot of room, but the procedure should be clear.

Klein Invariants

Take a classical polyhedron, put its vertices V on the sphere; from the centroid of the polyhedron draw radial half-lines in direction of the points located at the center of the faces and at the middle of the edges. These half-lines intersect the sphere at points F, and E. Notice (Euler) that $\#F - \#E + \#V = 2$. Now make a stereographical projection and build a complex polynomial that vanishes precisely at the location of the projected vertices (or center of faces, or mid-edges): this polynomial is, by construction, invariant under the symmetry group of the polyhedron (at least projectively) since group elements only permute the roots. This is the historical method – see in particular the famous little book [13]. In the case of the tetrahedron, for instance, you get the three polynomials (in homogeneous coordinates): $V = u^4 + 2i\sqrt{3}u^2v^2 + v^4$, $E = uv(u^4 - v^4)$ and $F = u^4 - 2i\sqrt{3}u^2v^2 + v^4$. Actually V and F are only projectively invariant, but $X = 108^{1/4}E$, $Y = -VF = -(u^8 + v^8 + 14u^4v^4)$ and $Z = V^3 - iX^2 = (u^{12} + v^{12}) - 33(u^8v^4 + u^4v^8)$ are (absolute) invariants, of degrees $6, 8, 12$. Together with the relation $X^4 + Y^3 + Z^2 = 0$, they generate the whole set of invariants. Alternatively you can build the p-th power of the fundamental representation of the symmetry group of the chosen binary polyhedral group, and choose p such that there exists one essential path of length p starting at the origin of the graph of tensorisation by the fundamental representation (one of the affine ADE diagrams) that returns to the origin. Therefore you get a symmetric tensor (since the path is essential), hence a homogeneous polynomial of degree p; moreover this polynomial is invariant since the path goes back to the origin (the identity representation). By calculating explicitly the projectors corresponding to the (unique) essential path of $[2]^6$, $[2]^8$ and $[2]^{12}$ on the affine $E_6^{(1)}$ graph, one can recover the polynomials X, Y, Z.

Quantum Geometry

The main interest of the previous section was to show that a good deal of the geometry associated with symmetry groups of platonic solids could be carried out without using the groups themselves but only the exceptionnal affine Dynkin diagrams. Going to the quantum will now be relatively easy: we just replace the affine Dynkin diagrams by the usual Dynkin diagrams. This present section will be rather short. One reason is the limited amount of space available for these proceedings, another reason is that the techniques have already been presented in the previous section, and they can be translated directly without further ado. We shall only mention what are the differences with the previous (classical) situation.

Realisations of the Quantum Algebras

The quantum algebra analogues of the group algebras $\mathcal{H}_{E_6^{(1)}}$ and $\mathcal{H}_{E_8^{(1)}}$ should be objects called $\mathcal{H}_{E_6}$ and $\mathcal{H}_{E_8}$. The quantum analogue of $\mathcal{H}_{E_7^{(1)}}$ does not exist, for a reason explained later (but the E_7 diagram leads nevertheless to an interesting quantum geometry). Actually we shall not introduce such quantum algebras at all, but we proceed as if they had been constructed. Besides the basic reference [16], let us mention also the following papers: [19], that uses the formalism of boundary conformal field theories, [2], for a discussion in terms of nets of subfactors and [12], for a discussion in terms of braided modular categories. Let us finally mention the paper [11], that introduces planar algebras, a concept that probably allows one to accomodate many of these construcions.

Representations of These Quantum Algebras

Since we did not give any definition of these quantum tetrahedral or icosahedral algebras, we shall define their irreducible representations σ_p as mere symbols associated with vertices of the diagrams E_6 or E_8.

The norm of the adjacency matrix $\mathcal{G}$ of the three exceptionnal Dynkin diagram is no longer the integer 2 but the quantum integers $[2]_q$ i.e., $2\cos(\pi/\kappa)$ for $\kappa = 12, 18, 30$. κ is also the Coxeter number of the diagram[3].

The components of the corresponding (normalized) eigenvector D (Perron-Frobenius) with the first entry normalized to 1 (extremity of the longest leg), are not positive integers but they provide a definition for the (quantum) dimensions of the irreducible representations. For E_6, $D = qdim(\sigma_0, \sigma_1, \sigma_2, \sigma_5, \sigma_4, \sigma_3) = ([1], [2], [3], [2], [1], [3]/[2] = \sqrt(2))$.

To keep the size of this paper small enough and specify our conventions for the labelling of vertices, we only give the graphs E_6 and E_8. Much more information concerning the "quantum tetrahedron" can be found in the paper [3]. The diagrams E_6 and E_8, with our labelling for vertices (increasing labels starting from endpoints) are given as follows

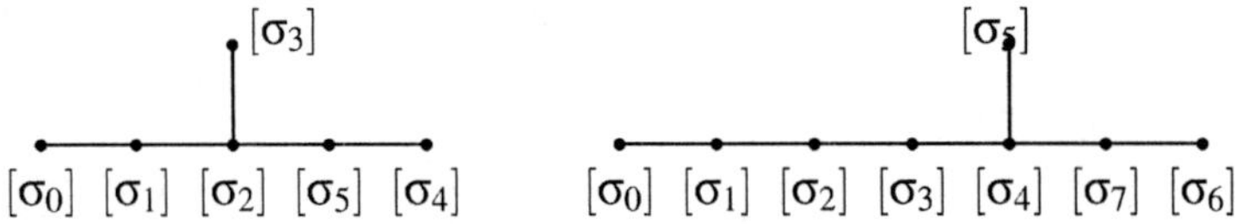

FIGURE 3. The diagrams E_6 and E_8

Tensor Products of Representations and Structure of the Grothendieck Ring

The representation σ_0 associated with the end-point of the longest leg is the unit. The Dynkin diagram defines multiplication by the algebraic generator σ_1. The only task is to complete the table (which is 6×6 in the case of the E_6 diagram) as we did in the classical case. This is rather straightforward and works perfectly, but for the fact that the obtained structure constants are positive integers for E_6 and E_8, but not for E_7. This was observed long ago by [17]. Here is the "fusion table" one gets for E_6:

	0	3	4	1	2	5
0	0	3	4	1	2	5
3	3	04	3	2	15	2
4	4	3	0	5	2	1
1	1	2	5	02	135	24
2	2	15	2	135	0224	135
5	5	2	1	24	135	02

[3] The quantum numbers are $[n] = \frac{q^n - q^{-n}}{q - q^{-1}}$. For E_6, $q = exp(i\pi/12)$.

The Fusion Matrices N_i

These six 6×6 matrices can be worked out easily, as we did in the case of the binary tetrahedral group. They can be found in [3].

From the Grothendieck Ring to the Character Table

Again, there is no drastic difference with the classical situation and one gets a quantum character table... but for the fact that conjugacy classes are not even defined! This defines a kind of Fourier transform which is both finite and quantum; its relation with the action of the modular group should be further investigated.

Structure of Centralizer Algebras and Tower of Commutants for $\mathcal{H}_{E_6}$

We can take the powers of σ_1 and decompose them into irreducibles, as we did in the classical case, just by using the adjacency matrix and the above fusion table. The coefficients – multiplicities – define the would-be centralizer algebras; they could be defined as multi-towers algebras associated with the chosen Dynkin diagram thought of as a Bratelli diagram, see [10]. Here are the first eight rows (one can go down to infinity):

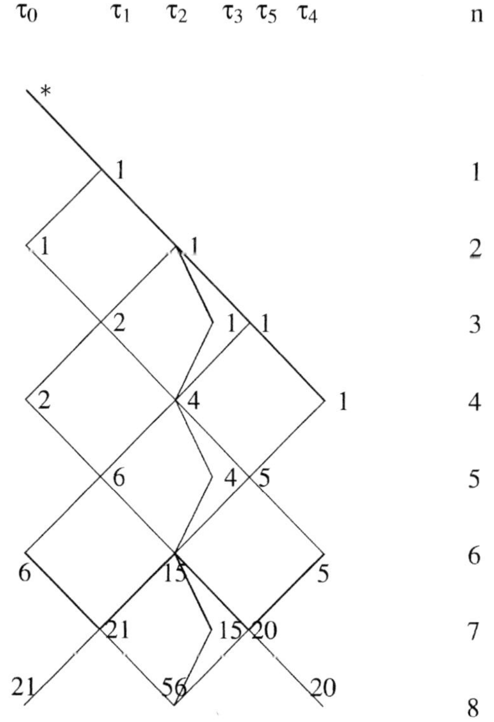

Essential Paths

Here we need to know the precise general definition (Ocneanu) of essential paths if we want to find them explicitly (see [16], [3]), but if we want only to count them, the method explained in the classical case works. The only difference with the classical case is that essential matrices, which are rectangular, with infinitely many lines in the classical case, are defined as matrices with only $\kappa - 1$ lines in the quantum situation (if the matrices were not truncated at that level, the line κ would be filled with 0 and the next ones would contain negative integers). This reflects the fact that essential paths having a bigger length do not exist. For the E_6 diagram, $\kappa = 12$.

The six rectangular matrices E_a of size 11×6 describing essential paths on E_6 are explicitly given in [3]. Columns of the essential matrices are labelled by the length p of the paths, so p runs from 0 to 10. This can be seen as a kind of labelling by the vertices of the Dynkin diagram A_{11}. Essential matrices have therefore columns labelled by E_6 and lines labelled by A_{11}. This is actually more than a simple remark since the algebra generated by the eleven 6×6 square matrices $F_i(a,b) \doteq E_a(i,b)$ provide a representation of the fusion algebra of the graph A_{11}. A pictural description of essential paths for all *ADE* Dynkin diagrams can be found in the appendix of [16]. The essential matrix describing (essential) paths leaving the origin leads to the following picture:

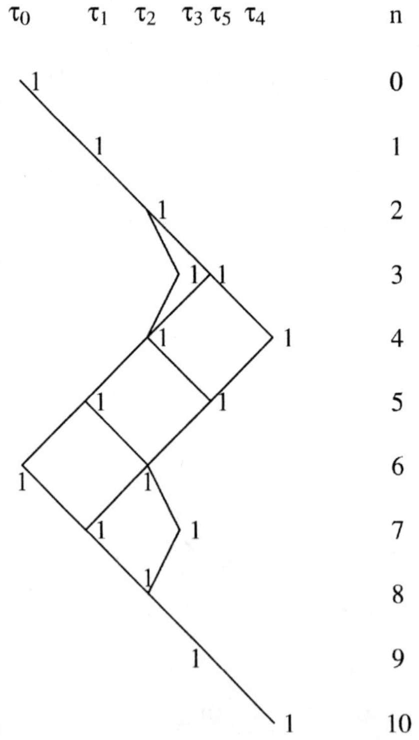

Projectors on Irreducible Representations

The method is essentially the same as in the classical situation, but for the fact that symmetrizer and antisymmetizer of $SU(2)$ have to be replaced by their quantum analogues, which can be obtained from the spectral decomposition of the flipped R-matrix of the quantum $SU(2)$ Hopf algebra (the one that obeys the braid group relation). However, When p increases (when we reach the triple point) one needs to know how to split the Wenzl projector. This was done, in the classical situation, by using explicitly the generators of the binary polyhedral groups and the expression of their iterated coproducts. Here, the problem is still open, since we do not have an explicit realization for the quantum algebras $\mathcal{H}$.

Klein Invariants

In the quantum situation, Klein invariants are *defined* ([16]) as essential paths starting from the origin and coming back to the origin. From the first essential matrix of E_6 or from the above equivalent corresponding graph of essential paths, we see that such an invariant exist (its length n is equal to 6). One can actually compute it explicitly, as a path, i.e., as a linear combination of elementary paths on the graph E_6. It would be nice to exhibit also a kind of homogeneous (but non-commutative) polynomial that implements it. This has not be obtained so far.

CLASSICAL AND QUANTUM INDUCTION-RESTRICTION

The purpose of the first coming two subsections is to investigate induction-restriction theory of representations in two classical situations (one finite, and the other infinite dimensional); namely we take the group $\tilde{T}$ as a subgroup of $SU(2)$ or as a subgroup of $\tilde{I}$. As we know, representation theory of these three classical objects is encoded by the Dynkin diagrams $E_6^{(1)}$, $E_8^{(1)}$ and A_∞.

The purpose of the last subsection is to investigate a similar situation in a quantum (but finite) case, encoded by the two graphs E_6 and A_{11}.

Classical Induction-Restriction: $E_8^{(1)}$ Versus $E_6^{(1)}$, i.e., $\tilde{I}$ Versus $\tilde{T}$

Classical Branching Rules $\tilde{I} \to \tilde{T}$

Both groups are finite subgroups of $SU(2)$ and we have inclusions: $\tilde{I} \subset \tilde{T} \subset SU(2)$. In both cases (diagrams $E_8^{(1)}$ and $E_6^{(1)}$) we have given the dimensions of the corresponding representations. Representations of $SU(2)$ can be restricted to its subgroups, and, in the same way, irreducible representations of $\tilde{I}$ can be restricted to $\tilde{T}$, and decomposed into irreducible representations of the later. The following table gives the branching rules

$\tilde{I} \to \tilde{T}$:

$$\begin{array}{ccccccccc} 1 & \to & 1; & 2 & \to & 2; & 3 & \to & 3 \\ 4 & \to & 2'+2''; & 5 & \to & 3+1+1''; & 6 & \to & 2+2'+2'' \\ 4' & \to & 3+1; & 3' & \to & 3; & 2' & \to & 2 \end{array}$$

It is easy to obtain the above table: one should just compaire multiplications of irreps of $SU(2), \tilde{I}$, or $\tilde{T}$ by the 2-dimensional fundamental representation. For instance, $2 \times 3 = 2+4$ in $SU(2)$, but $2+2'+2''$ in $\tilde{T}$ and $2+4$ in $\tilde{I}$; therefore, with respect to the branching $\tilde{I} \to \tilde{T}$ we have $4 \to 2'+2''$. Then we have $2 \times 4 = 3+5$ in $SU(2)$, but $2 \times (2'+2'') = 1+3+1'+3+1''+3$ in $\tilde{T}$, whereas $2 \times 4 = 3+5$ in $\tilde{I}$; therefore, we get $5 \to 3+1'+1''$ for the branching $\tilde{I} \to \tilde{T}$. In the same way we get $2 \times 5 = 4+6$ both in $SU(2)$, and in $\tilde{I}$, the restriction to $\tilde{T}$ reads $2 \times (3+1'+1'') = 2+2'+2''+2'+2''$, but we already know that $4 \to 2'+2''$ therefore, we find $6 \to 2+2'+2''$. Next we have $2 \times 6 = 5+4'+3'$ in $\tilde{I}$, so that restriction of both sides to $\tilde{T}$ gives $1+3+1'+3+1''+3 = 3+1'+1''+(4'+3')_{\tilde{T}}$ and one finds $4'+3' \to 3+3+1$, so that the only possibility is $4' \to 3+1$ and $3' \to 3$. The last branching rules can be obtained by restricting $2 \times 4' = 6+2'$ to $\tilde{T}$: one gets $2 \times (3+1) = 2+2'+2''+(2')_{\tilde{T}}$ i.e., $2+2'+2''+2 = 2+2'+2''+(2')_{\tilde{T}}$ and therefore the restriction $2' \to 2$.

Sections of Classical Vector Bundles Over $\tilde{I}/\tilde{T}$

Since $\tilde{T}$ is a subgroup of $\tilde{I}$, we can write the binary icosahedral group as a principal bundle over the quotient $\tilde{I}/\tilde{T}$, with structure group (typical fiber) $\tilde{T}$, the binary tetrahedral group. For each representation of the structure group, in particular for each irreducible representation ρ (and carrier space V_ρ) of $\tilde{T}$ (we know that there are 7 of them), we can build an associated vector bundle $\tilde{I} \times_{\tilde{T}} V_\rho$, with basis $\tilde{I}/\tilde{T}$ and typical fiber the vector space V_ρ. Now we may consider the spaces of sections Γ_ρ of those bundles, which are functions on the finite homogeneous space $\tilde{I}/\tilde{T}$ and valued in the corresponding vector spaces. In the particular case of the trivial representation of $\tilde{T}$ (called 1), the carrier vector space is $\mathbb{C}$ and the sections of Γ_1 just coïncide with the space of complex valued functions on the finite set $\tilde{I}/\tilde{T}$, whose cardinality is $5 = 120/24$.

Here we are in a finite dimensional situation, but Peter Weyl theory of induced representations still applies. Let us take the example of Γ_3, the space of sections of $\tilde{I} \times_{\tilde{T}} V_3$; this vector bundle is a collection of five vector spaces of dimension 3 (one above each of the five points of the coset); the dimension of Γ_3 is therefore $3 \times 5 = 15$. This fifteen dimensional space is the carrier space of a natural representation of $\tilde{I}$ (the one induced by this particular vector bundle), but this representation is not irreducible: its decomposition, in irreps of $\tilde{I}$, can be obtained from the previously given table of branching rules: 3 (of $\tilde{T}$) appears on the right hand side of the branching rules corresponding to the irreps $3, 5, 4'$ and $3'$ of $\tilde{I}$, from this information, one deduces that $\Gamma_3 = [3] \oplus [5] \oplus [4'] \oplus [3']$ (whose sum is indeed $5 \times 3 = 15$, as it should). This induction process leads to the following table, for the decomposition of the various spaces of sections (the vector spaces Γ_ρ) in irreducible representations of $\tilde{I}$:

ρ	$dim(\Gamma_\rho)$	Γ_ρ
1	$5\times 1=5$	$1+4'$
2	$5\times 2=10$	$2+6+2'$
3	$5\times 3=15$	$3+5+4'+3'$
$2'$	$5\times 2=10$	$4+6$
$1'$	$5\times 1=5$	5
$2''$	$5\times 2=10$	$4+6$
$1''$	$5\times 1=5$	5

In particular, the space of functions on the finite set (five points) $\tilde{I}/\tilde{T}$, that we may call $Fun(\tilde{I}/\tilde{T}) \equiv \Gamma_1$ decomposes into irreps of $\tilde{I}$ as $4'+1$.

In our case (space of sections of vector bundles above the finite *left* homogeneous space $\tilde{I}/\tilde{T}$) we see that $\sum dim(\Gamma_p) = 5+10+15+10+5+10+5 = 60$, which is one half of the order of the group $\tilde{I}$ (by considering both left and right bundles, we would get $120 = \#\tilde{I}$). The main interest, for us, of the previous remarks, is that, if we knew only the dimensions of the spaces of sections, without knowing neither the order of $\tilde{T}$ nor the order of $\tilde{I}$, we could recover the order of $\tilde{I}$ (namely 120) by taking the double of the sum of those dimensions, and the cardinality of the quotient (namely $dim\Gamma_1 = 5$) by summing the dimensions appearing in the decomposition of the space of sections associated with the trivial representation; the order of $\tilde{T}$ itself is then obtained by taking from the quotient $120/5 = 24$. Once the cardinality of the quotient is obtained (5), one can recover the dimensions of the irreducible representations ρ themselves by taking the ratio $dim(\Gamma_\rho)/dim(\Gamma_1)$.

Classical Induction-Restriction: A_∞ Versus $E_6^{(1)}$ (i.e., $SU(2)$ Versus $\tilde{T}$)

We restrict the representations of $SU(2)$ to irreps of $\tilde{T}$. We build the principal bundle $SU(2)$ as a $\tilde{T}$ bundle over the quotient $SU(2)/\tilde{T}$ (which is a three dimensional manifold) and consider the (seven) associated vector bundles relative to the seven irreps of $\tilde{T}$. We have therefore one such vector bundle for every point of the extended Dynkin diagram $E_6^{(1)}$. The only difficulty is to compute the branching rules. One method is to proceed step by step, i.e., to use the information provided by the two Dynkin diagrams encoding tensor multiplication by the fundamental representation, computing tensor products of irreps both for $SU(2)$ and its finite subgroup and comparing the results. The easiest method is to use essential matrices (they have infinitely many raws, in the present case); another technique – which amounts to the same but is aesthetically more appealing – is to draw the essential paths on $E_6^{(1)}$. The only relevant essential matrix (for our present purpose) is the one labelled by the trivial representation of $\tilde{T}$ i.e., by the space of essential paths emanating from the leftmost point of the Dynkin diagram $E_6^{(1)}$. For instance the line $p=8$ of that graph (referring to $[2]^8$ i.e., to the representation of dimension 9 of $SU(2)$, and associated with an essential path of length 8 on the graph A_∞) tells us that $[9] \rightarrow [1]+2[3]+[1']+[1'']$ in the branching $SU(2)$ versus $\tilde{T}$. In other words, in order to perform reduction, we read the first essential matrix "horizontally", i.e., we look at representations of $SU(2)$ given by *symmetric* polynomials of degree n and branch them to this particular subgroup. In order to perform induction, we look at the same essential matrix, but "vertically", i.e., we choose a particular irrep of $\tilde{T}$ (a particular column) and see for which values it appears in branching rules of $[p+1]$.

Sections of Vector Bundles Over $SU(2)/\tilde{T}$

Call V_ρ the vector space carrying the irreducible representation ρ of $\tilde{T}$ and Γ_ρ the space of sections of the homogeneous vector bundle $SU(2)\times_{\tilde{T}} V_\rho$. The spaces of sections of these vector bundles can be decomposed as follows into irreducible representations of $SU(2)$ (subscript give the multiplicity):

$$\begin{aligned}
\Gamma_1 &= 1+7+9+13_2+15+17+\ldots = Fun(SU(2)/\tilde{T}) \\
\Gamma_2 &= 2+6+8_2+10+12_2+14_3+16_2+\ldots \\
\Gamma_3 &= 3+5+7_2+9_2+11_3+13_3+15_4+17+\ldots \\
\Gamma_2' &= 4+6+8+10_2+12_2+14_2+16_3+\ldots \\
\Gamma_1' &= 5+9+11+13+15+17_2+\ldots \\
\Gamma_2'' &= 4+6+8+10_2+12_2+14_2+16_3+\ldots \\
\Gamma_1'' &= 5+9+11+13+15+17_2+\ldots
\end{aligned}$$

The degree of the homogenous polynomials providing a basis for a representation space of dimension d is $d-1$, so that representations of degree $0,6,8,12,14,16,\ldots$ appear in Γ_1, as it should: one recovers the fact that these representations of $SU(2)$ indeed contain $\tilde{T}$ - invariant subspaces (Klein polynomials for the tetrahedron).

One can make the same kind of comments as in the previous section for instance $dim(\Gamma_3)/dim(\Gamma_1) = 3$, but we now have to maneuver infinite sums and one should use generating functions (we shall not do it here). In the case of Γ_3, for instance, we can also write

$$[3]\otimes([1]\oplus[7]\oplus[9]\oplus 2[13]\oplus\ldots) = [3]\oplus[5]\oplus 2[7]\oplus\ldots$$

The reader may wander why we did not introduce also "essential matrices" in the previous subsection (with columns labelled by irreps of the binary tetrahedral group and lines labelled by irreps of the binary icosahedral group). There is no reason: we could have done it as well.

Quantum Induction-Restriction: A_{11} Versus E_6

We now replace the diagram A_∞ that describe irreps of $SU(2)$ by the diagram A_{11} and the classical binary tetrahedral group by its would-be quantum counterpart described by the diagram E_6.

Both examples studied in the corresponding classical two sections actually provide interesting – and complementary – classical analogies: the first ($E_8^{(1)} \to E_6^{(1)}$) because it is finite dimensional, and the other $A_\infty \to E_6^{(1)}$ because A_{11} looks indeed as a "truncated" A_∞.

From the embedding $\tilde{T} \subset \tilde{I}$, one can deduce an embedding of the corresponding group algebras (finite dimensional Hopf algebras) $\mathcal{H}_{\tilde{T}} = \mathbb{C}\tilde{T} \subset \mathcal{H}_{\tilde{I}} = \mathbb{C}\tilde{I}$ but although we do not plan to give here a construction of the "would-be groups" (or would-be group algebras) that one could associate with the two genuine Dynkin diagrams E_6 and A_{11}, we want nevertheless to consider the first as a kind of sub-object of the next by following a discussion very similar to the one presented in the corresponding section describing the classical situation and proceed as if we had an embedding $\mathcal{H}_{E_6} \subset \mathcal{H}_{A_{11}}$.

One takes $\hat{q} = exp(i\pi/12)$ (so that if $q = \hat{q}^2$, then $q^{12} = 1$). Irreps of A_{11} are representations called $\tau_0, \tau_1, \ldots \tau_{10}$. They have q-dimension respectively equal to $[1], [2], [3], [4], [5], [6], [7] = [5], [8] = [4], [9] = [3], [10] = [2], [11] = [1]$. There exists actually a non semi-simple Hopf algebra defined as a finite dimensional quotient of the enveloping quantum algebra of $SU(2)$, when q is a twelve root of unity, and which is such that the above list of τ_i indeed labels its irreducible representations of non-zero quantum dimension; however this knowledge will not be used here. The "representations" τ_i are therefore just abstract symbols that one can associate with the various points of the diagram A_{11}, whose own fusion table could have been worked out as discussed previously, and the "q-dimensions" is just a name for the entries of the normalized eigenvector associated with the norm of the adjacency matrix of this diagram (the norm being, by definition, its biggest eigenvalue). Remember that both graph A_{11} and E_6 have same norm. A priori, the ring of representations that we are considering here have a "dimension function" valued in a $\mathbb{Z}$-ring linearly generated by the q-integers $[1], [2], [3], [4], [5], [6]$. This is a clearly the case both for A_{11} and E_6.

It may be useful to note that

$$[1] = 1, [2] = \frac{\sqrt{2}}{\sqrt{3}-1}, [3] = \frac{2}{\sqrt{3}-1}, [4] = \frac{\sqrt{6}}{\sqrt{3}-1}, [5] = \frac{1+\sqrt{3}}{\sqrt{3}-1}, [6] = \frac{2\sqrt{2}}{\sqrt{3}-1}$$

Quantum Branching Rules $A_{11} \to E_6$

The branching rules from A_{11} to E_6 (that gave restriction in one direction and induction in the other) are gotten from the "q-symmetric" representations, or, equivalently, from the essential matrices of the graph E_6.

The only relevant essential matrix, for our present purpose, is the one labelled by the "trivial representation" (leftmost point of the graph E_6). The following table summarizes the results for the reduction $A_{11} \to E_6$. One should remember that the q-dimension corresponding to irreps τ_p of A_{11}, (i.e., vertices of A_{11}) is $[p+1]$.

$\tau_0 \to$	σ_0	$\tau_1 \to$	σ_1	$\tau_2 \to$	σ_2
$\tau_3 \to$	$\sigma_3 + \sigma_5$	$\tau_4 \to$	$\sigma_2 + \sigma_4$		
$\tau_5 \to$	$\sigma_1 + \sigma_5$				
$\tau_6 \to$	$\sigma_0 + \sigma_2$	$\tau_7 \to$	$\sigma_1 + \sigma_3$		
$\tau_8 \to$	σ_2	$\tau_9 \to$	σ_5	$\tau_{10} \to$	σ_4

Sections of Quantum Vector Bundles Over A_{11}/E_6

Using the previous table, and using a formal analogy, we associate a quantum vector bundle [4] to each point of the E_6 graph and decompose its space of sections Γ_{σ_p}, using induction, exactly as we did in the classical case (for instance we see that σ_0 can be obtained *from* the reduction of τ_0 and τ_6). We obtain:

$$\begin{aligned}
\Gamma_{\sigma_0} &= \tau_0+\tau_6 \\
\Gamma_{\sigma_1} &= \tau_1+\tau_5+\tau_7 \\
\Gamma_{\sigma_2} &= \tau_2+\tau_4+\tau_6+\tau_8 \\
\Gamma_{\sigma_5} &= \tau_3+\tau_5+\tau_9 \\
\Gamma_{\sigma_4} &= \tau_4+\tau_{10} \\
\Gamma_{\sigma_3} &= \tau_3+\tau_7
\end{aligned}$$

This information can be also displayed as

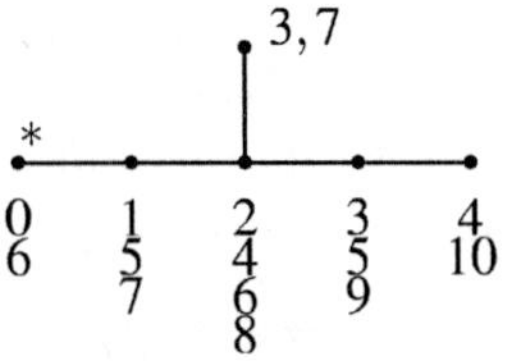

The quantum dimension of Γ_{σ_0} is

$$[1]+[7]=[1]+[5]=1+\frac{\sin(5\pi/12)}{\sin(\pi/12)}=\frac{2\sqrt{3}}{\sqrt{3}-1}$$

Morally this is the quantum dimension of the space of "functions" on the quantum space A_{11}/E_6. One may then check that, by dividing the q-dimension of each space of sections Γ_{σ_p} by the abobe q-dimension of Γ_{σ_0}, one recovers exactly the q-dimensions of the "typical fibers", i.e., the q-dimensions of the irreducible representations σ_p, already obtained from the normalized Perron-Frobenius vector associated with the graph E_6. We have therefore a perfect analogy with the classical situation.

REFERENCES

1. Cappelli, A., Itzykson, C., and Zuber, J.B., *Commu. Math. Phys.*, **13**, 1 (1987).
2. Böckenhauer, J., and Evans, D., *Commun. Math. Phys.*, **200**, 57-103 (1999).
3. Coquereaux, R., *Notes on the quantum tetrahedron*, `math-ph/0011006`.

---

[4] There are several ways to define quantum principal bundles and associated quantum vector bundles, (see for instance [6]), but we do not use these technical definitions here.

4. Coquereaux, R., Garcia, A., and Trinchero, A., *Racah-Wigner bi-algebras and Ocneanu quantum symmetries*, in preparation.
5. Coquereaux, R., and Schieber, G., *Ocneanu quantum symmetries and twisted partition functions for ADE graphs*, in preparation.
6. Coquereaux, R., Garcia, A., and Trinchero, R., *Czech J. Phys.*, **50**, 29–36 (2000), hep-th/9908007.
7. Di Francesco, P., and Zuber, J.-B., in *Recent developments in Conformal Field Theory*, Trieste Conference, 1989, Eds.: S. Randjbar-Daemi, E. Sezgin and J.-B. Zuber, World Scientific, 1990; Di Francesco, P., *Int. J. Mod. Phys.*, **A7**, 407–500, (1992).
8. Di Francesco, P., and Zuber, J.-B., *Nucl. Phys.*, **B338**, 602 (1990).
9. Di Francesco, P., Matthieu, P., and Senechal, D., *Conformal Field Theory*, Springer 1997.
10. Goodman, F.M., de la Harpe, P., and Jones, V.F.R., *Coxeter graphs and towers of algebras*, MSRI publications 14, Springer (1989).
11. Jones, V.F.R., *Planar algebras*, V.F.R Jones home page.
12. Kirillov, A., and Ostrik, V., *On q-analog of McKay correspondence and ADE classification of SL2 conformal field theories*, math.QA/0101219.
13. Klein, F., *Lectures on the Icosahedron and the solution of the equation of the fifth degree*, Reprog. Nachdr. d. Ausg. Leipzig (1884), Teubner. Dover Pub, New York, Inc IX, 289p (1956).
14. McKay, J., *Graphs, singularities and finite groups*, Proc. Symp. Pure Math. **37**, 183 (1980).
15. Ocneanu, A., *Paths on Coxeter diagrams: from Platonic solids and singularities to minimal models and subfactors*, Talks given at the Centre de Physique Théorique, Luminy, Marseille, 1995.
16. Ocneanu, A., *Paths on Coxeter diagrams: from Platonic solids and singularities to minimal models and subfactors*, Notes taken by S. Goto, Fields Institute Monographs, Eds. Rajarama Bhat et al., AMS 1999.
17. Pasquier, V., *Nucl.Phys.*, **B285**, 162 (1987).
18. Petkova, V.B., and Zuber, J.B., *BCFT: from the boundary to the bulk*, Proceedings Nonperturbative Quantum Effects 2000, hep-th/0009219.
19. Petkova, V.B., and Zuber, J.B., *The many faces of Ocneanu cells*, hep-th/0101151.
20. Zuber, J.B., *CFT, BCFT, ADE and all that*, Lectures at Bariloche Summer School, Argentina, Jan 2000, hep-th/0006151. To appear in AMS Contemporary Mathematics, Eds.: R. Coquereaux, A. Garcia and R. Trinchero.

Spacetime and Fields, a Quantum Texture

S. Doplicher

Dipartimento di Matematica University of Rome "La Sapienza" 00185 Roma, Italy, email: dopliche@mat.uniroma1.it

Abstract. We report on joint works, past and in progress, with K.Fredenhagen and with J.E.Roberts, on the quantum structure of spacetime in the small which is dictated by the principles of Quantum Mechanics and of General Relativity; we comment on how these principles point to a deep link between coordinates and fields. This is an expanded version of a lecture delivered at the 37th Karpacz School in Theoretical Physics, February 2001.

SPACETIME UNCERTAINTY RELATIONS

At large scales spacetime is a pseudo Riemaniann manifold locally modelled on Minkowski space. But the concurrence of the principles of Quantum Mechanics and of Classical General Relativity points at difficulties at the small scales, which make that picture untenable. For if we try to locate an event in say a spherically symmetric way around the origin in space with accuracy a, according to Heisenberg principle an uncontrollable energy E of order $1/a$ has to be transferred, which will generate a gravitational field with Schwarzschild radius $R \simeq E$ ($\hbar = c = G = 1$). Hence $a \gtrsim R \simeq 1/a$ and $a \gtrsim 1$, i.e. in CGS units

$$a \gtrsim \lambda_P \simeq 1.6 \cdot 10^{-33}\text{cm}. \tag{1}$$

If however we measure one of the space coordinates of our event with great precision a, but allow large uncertainties L in the knowledge of the other coodinates, the energy $1/a$ may spread over a thin disk of radius L and thus generate a gravitational potential that would vanish everywhere as $L \to \infty$.

One has therefore to expect Space Time Uncertainty Relations emerging from first principles, already at a semiclassical level. Carrying through such an analysis [1, 2] one finds indeed that, if the smallest and largest space uncertainties of an event are denoted by a, b respectively, and the time uncertainty by τ , the gravitational potential generated by the energy $1/\min(a,\tau)$ localized at some instant with accuracies a, b, τ, is at most of the order

$$|V| \simeq \frac{1}{b \cdot \min(a,\tau)}. \tag{2}$$

Now our basic requirement is that the localization experiment should not deform spacetime in such a way that no signal from the region we wish to observe can reach infinity in space, otherwise this would put the observed event out of reach for any distant observer; namely

CP589, *Fundamental Interaction Theories*, edited by J. Lukierski and J. Rembieliński

$$g_{00} = 1 + 2V > 0, \tag{3}$$

where V is the potential generated by the energy transferred with the localization measurement itself; hence by (2)

$$b \cdot \min(a, \tau) \gtrsim 1. \tag{4}$$

The Space Time Uncertainty Relations strongly suggest that spacetime has a Quantum Structure at small scales, expressed, in generic units, by

$$[q_\mu, q_\nu] = i\lambda_P^2 Q_{\mu\nu}, \tag{5}$$

where Q has to be chosen not as a random toy mathematical model, but in such a way that (4) follows from (5). Further we want to impose (full) Lorentz invariant conditions on $Q_{\mu\nu}$, so that our models are compatible with Special Relativity; since in (5) Q is dimensionless, the commutator will effectively vanish for large distances compared to the Planck scale.

But we do not insist on covariance under general coordinate transformations, which, at a quantum level and at small scales, cannot be supported by conceptual experiments, as the freely falling laboratory, in presence of fields which vary significantly over Planckian distances. Moreover, for the sake of Elementary Particle Physics, an asymptotically flat background is an appropriate idealization, for the distribution of masses in the Universe should not affect significantly the outcome of collision experiments in our laboratories.

The noncommutativity of the operators $q_0, ..., q_3$ can be measured by the fundamental invariants

$$Q_{\mu\nu}Q^{\mu\nu};$$

$$[q_0, ..., q_3] := \det \begin{pmatrix} q_0 & \cdots & q_3 \\ \vdots & \ddots & \vdots \\ q_0 & \cdots & q_3 \end{pmatrix} := \tag{6}$$

$$\varepsilon^{\mu\nu\lambda\rho} q_\mu q_\nu q_\lambda q_\rho = -(1/2) Q_{\mu\nu} (*Q)^{\mu\nu}$$

but the second is invariant only under the proper Lorentz transformations and only its square is invariant under space and time reflections as well.

If we (temporarily) assume that the components of Q commute with one another, and let $\mathbf{e}, \mathbf{m}$ denote the triples of electric respectively magnetic components, we have

$$(-1/2) Q_{\mu\nu} Q^{\mu\nu} = \mathbf{e}^2 - \mathbf{m}^2; \tag{7}$$

since $\mathbf{e}$ and $\mathbf{m}$ respectively govern the space - time and space - space uncertainty relations, symmetry and (4) suggest the condition

$$Q_{\mu\nu} Q^{\mu\nu} = 0. \tag{8}$$

Therefore the basic Quantum Condition must read

$$[q_0, ..., q_3]^2 = S, \tag{9}$$

where S is a Lorentz invariant.

We will see later how more general choices for S are important, but (4) suggest a multiple of I. If we also require that the Q commute with the q, we get the Basic Model introduced and discussed in detail in [1] that we will breefly report on in the next Section.

Other approaches to uncertainty relations affected by gravity and related phenomena can be found e.g. in [7],..., [15]. We do not attempt to give a complete list of references related to this subject, which became quite numerous in the last three years; approaches based on the quantum deformations of the Poincaré Algebra received a lot of attention, cf [17] and references therein.

THE BASIC MODEL

In the notation introduced above the quantum conditions of the Basic Model may be rewritten as

$$[q_\mu, Q_{\lambda\rho}] = 0, \tag{10}$$

$$\mathbf{e}^2 = \mathbf{m}^2, \mathbf{e}\cdot\mathbf{m} = \pm I. \tag{11}$$

In this model the following weaker form of (4) is implemented, cf [1, 2] :

$$\Delta q_0 \cdot \sum_{j=1}^{3} \Delta q_j \gtrsim 1; \quad \sum_{1\leq j<k\leq 3} \Delta q_j \Delta q_k \gtrsim 1. \tag{12}$$

Relations (11) define an algebraic manifold with two connected components each isomorhic to the coset space of the proper Lorentz group modulo boosts along a fixed direction and rotations around it, i.e. to $SL(2,\mathbb{C})/\mathbb{C}_*$, where $\mathbb{C}_*$ is embedded in $SL(2,\mathbb{C})$ as the 1,1 component of diagonal matrices. Each pair (**e'**,**m'**) as in (11) can be obtained from a pair such that $\mathbf{e} = \pm\mathbf{m}$ by a boost with velocity, say **v**, orthogonal to **e**, hence by (11) **e** and **m** are vectors in the unit sphere S^2 in three dimensional space, and **v** is a tangent vector to S^2; summarizing

$$\Sigma_+ \simeq \Sigma_- \simeq SL(2,\mathbb{C})/\mathbb{C}_* \simeq TS^2. \tag{13}$$

While classical Spacetime is described by the commutative C* algebra of continuos functions vanishing at infinity, Quantum Spacetime will be described by a noncommutative C* algebra $\mathcal{E}$, to which the q are affiliated in the sense of [18], cf [1], i.e each representation of $\mathcal{E}$ determines operators q_μ fulfilling our Quantum condition, and all "regular" representations appear this way.

We adopted in [1] the following paradigm, which may well apply to more general cases (cf [4]): if we interpret (5) as defining a bundle of Lie algebras, in that case over Σ, the regular representations will be those which are integrable to a representation of the corresponding bundle of simply connected Lie groups; the C* algebra $\mathcal{E}$ will then arise as a continuos field of group C* algebras.

In the basic model the fibers are just Heisenberg groups with a non degenerate C - number commutator matrix (the generic point in Σ), hence we get a continuos field of the algebra of all compact operators (on a separable infinite Hilbert space) which can be proved to be trivial (see [1]), i.e.

$$\mathcal{E} \simeq \mathcal{C}_0(\Sigma, \mathcal{K})\,. \tag{14}$$

These findings fit very well in the theory of strict deformation quantization [19].

This C* algebra carries a natural action of the (full) Poincaré group $\mathcal{P}$ as automorphisms, which is actually determined by its extension to the affiliated q's, fulfilling the natural relations

$$\alpha_L(q) = L^{-1}q, \qquad L \in \mathcal{P}. \tag{15}$$

Thus $\mathcal{E}$ is a Quantum space but its global symmetries are the classical ones, as expected since at large scales the model turns classical again, and the Poincaré tranformations are global motions, acting the same way in the small and in the large. This situation parallels the familiar one in nonrelativistic Quantum Mechanics, where the Schroedinger Operators q and p do not commute, but the Galilei invariance is expressed by an action of the classical Galilei Group as automorphisms (representations up to a phase appear only in the unitary implementations). Actually this structure is indeed a special case of our present model, cf below.

The classical concept of points in a space has to be replaced by pure states with minimal uncertainties, i.e. pure states which are optimally localized in the sense that the quantity

$$(\Delta q_0)^2 + \ldots + (\Delta q_3)^2 \tag{16}$$

is minimal; this is a frame dependent condition, which picks a point $\mathbf{e} = \pm\mathbf{m}$ in the spectrum of the Q's, i.e. a point in the base $S \times \{\pm 1\}$ if we think of Σ as a tangent manifold, so that the q's fulfilling (5) now become the Schroedinger operators q, p for a particle in two dimensions (the four dimensional translations acting as Galilei transformations), and (16) minimal implies that its value is 2 and that our state is the ground state of the harmonic oscillator for those Schroedinger operators.

Such states ought to have a preferred role in discussing the large scale limit of the Quantum space; since in (14) the Planck length appears only in the exponential in the Weyl relations, which force the fiber to be $\mathcal{K}$, and in the large scale limit $\mathcal{K}$ deforms to $\mathcal{C}_0(\mathbb{R}^4)$, we see that the quantum spacetime becomes $\mathbb{R}^4 \times \Sigma$ in the large scale limit, while, if only optimally localized states are considered, the limit is rather

$$\mathbb{R}^4 \times S^2 \times \{\pm 1\}\,.$$

Thus the discrete space $\{\pm 1\}$ appears because the spectrum of the centre of the algebra generated by the Q's is not connected (a consequence of imposing symmetry under reflections too), and while the continuos factor in the ghost manifold is not compact, only a compact manifold, actually a sphere with radius the square of the Planck length, plays a role if we are testing with optimally localized states.

The paradigm we adopted in attaching a C* algebra to relations (5) in our model tells us how to calculate functions $f(q)$: as in the von Neumann -Wigner-Moyal calculus, write f as the Fourier transform of its ordinary anti-Fourier transform, and replace the exponentials by the Weyl operators $\exp i(\alpha q)$; the multiplication of these exponentials is precisely governed by the bundle of Lie groups associated to the models; thus this paradigm can, and will be, applied in some more general context. Moreover space integration at time t and spacetime integration can be easily defined and related to the trace in each fiber, so that we can introduce Free Fields on QST, the free Hamiltonian, which turns out to be unchanged by the quantum deformation, and interaction Hamiltonians, i.e. we can lay down the stup to apply the usual perturbation expansion (cf [1]).

While integration over space or spacetime poses no problem in this model, integration over Σ does, since we tacitly assumed that our fields do not depend on the points of Σ ; but we have no bounded invariant measure on Σ so we cannot integrate to get an invariant result.

The way out chosen in [1] was to integrate over the base $S^2 \times \{\pm 1\}$ of Σ, thus keeping only rotation invariance; but in the end we face a more serious difficulty. Namely the perturbation expansion is found to be exactly that of a non local theory on the classical Minkowski space.

Of course (4) suggests that causality breaks down at short distances: but it should be recovered at large scales with respect to Planck length (say at QCD scales, 10^{-17} cm.), while the acausal effects of ordinary nonlocal theories might cumulate after summing the perturbation expansion in a disruptive way.

Strangely enough, these lessons of [1] have been largely neglected; well after the appearence of [1] we assisted to a flow of papers on QFT models on a QST which is characterized by (4) with a fixed C-number tensor on the right, disregarding the physical meaning of noncommutativity and the need of Lorentz invariance, but extensively applying the calculation aspect of what we exposed, summarized by the use of the "star product".

The negative conclusions referred to above might lead us to reanalyse the concept of interaction over QST; in particular the ordinary Wick product should not be allowed: if e.g. we are to define the Wick square of the field A, we can evaluate $A(q)A(q')$ on distinct variables q, q', but then we cannot set $q - q' = 0$, since these operators obey similar relations to (4); we can however evaluate a conditional expectation defined by an optimally localized state on $q - q'$; as this is the ground state of an harmonic oscillator, it introduces a Gassian nonlocality factor which violates again causality, but might fully regularize the theory, and in fact might give rise to a gaussian fall off of cross sections at large energies [5].

The problems with causality lead us [5] to enquire about light propagation, i.e. classical ED on QST; while the local gauge group of Classical ED on Minkowski space is the unitary group of $\mathcal{C}_0(\mathbb{R}^4) + \mathbb{C} \cdot I$, it is that of $\mathcal{E} + \mathbb{C} \cdot I$ in the case of QST. Treating it the usual way we found that ED is characterized by nonlinearly selfinteracting equations, for which a plane wave is a stationary solution, but superpositions of two different plane waves are not, with a propagation into massive modes; in principle, this effect ought to be detectable splitting a monochromatic laser beam into a superposition of states with different momenta with the help of a partially reflecting mirror (or detecting the light of a distant galaxy split by a gravitational lense); but the fraction of energy density which would go into these massive modes, calculated to the lowest order in the Planck length, turns out to be a fraction of the order lower than 10^{-130}.

More seriously, such a theory has a huge gauge group, so it is difficult to propose testable effects, no matter how tenuous, in terms of gauge invariant quantities.

For recent discussions of possible testable effects of Quantum Gravity cf e.g. [21, 22].

Another drastic consequence is the nonvanishing of the current divergence, due to quantum gravitational anomalies.

But the Gauge Principle expresses the point nature of interactions, and is the basic principle lying behind locality in ordinary QFT, so it might well by itself provide a rigid substitute to causality in QFT on QST. This hope motivates a long standing attempt to a general formulation of gauge theories on a noncommutative manifolds using the absolute differential calculus ([5, 6]; this calculus emerged also in other papers appeared meanwhile, cf e.g. [16]). One might expect that a proper noncommutative approach might lead to a new picture of interactions at Planck scale, which avoids the unpleasant features met when we just replace products with * products.

An approach to gauge theories on noncommutative spaces based on the notion of covariant coordinates has been proposed in [24] and references therein.

DEFORMED MODELS

The remaining sections are based on work in progress with K.Fredenhagen.

The basic model discussed in the previous section has the virtue of simplicity, but has the vice of not implementing the Space Time Uncertainty Relations (4) in full, but only their weaker consequence (12). But if we relax the drastic simplification (10) that the Q's are central, and instead we only assume that they commute with one another, keeping the other Quantum Conditions, there is room for deformed models where (4) are fully implemented.

One such model ([5]; partly announced in [4]) can be formulated introducing self adjoint central operators, which form two antisymmetric 2-tensors H, T and a four vector C, and adding to the q's two scalar commuting generators R, S, and imposing

$$\begin{aligned} [q_\mu, q_\nu] &= i(H_{\mu\nu} + T_{\mu\nu}R)\,, \\ [q_\mu, R] &= iC_\mu S\,, \\ [q_\mu, S] &= -iC_\mu R\,, \\ [R, S] &= 0\,, \end{aligned} \tag{17}$$

where $S^2 + R^2$ is central and can be set equal to I, and, by Jakobi identity,

$$C_\mu T_{\nu\lambda} + C_\nu T_{\lambda\mu} + C_\lambda T_{\mu\nu} = 0, \quad for \quad all \quad \mu, \nu, \lambda. \tag{18}$$

Furthermore, the contraction of T and of its Hodge dual $*T$ with itself and with H should vanish, while H fulfills the same conditions as Q in the basic model.

One can verify that in simple irreducible representations the full STUR (4) are fully implemented. The H, T, C will be translation invariant while the R, S will be Lorentz scalars but not translation invariant; the full Lorentz group $\mathcal{L}$ will act transitively on the spectrum of the centre, so that here, at large scales, the classical limit of our QST will be

$$\mathbb{R}^4 \times \mathcal{L} \tag{19}$$

a manifold with 10 dimensions, which would effectively reduce here too if we restricted attention to optimally localized states.

A DYNAMICAL PICTURE OF QUANTUM SPACETIME

The models of QST outlined above try to implement in the noncommutative nature of the underlying geometry some of the minimal limitations on the localization of an event which are imposed by our present knowledge of the principles of Physics. Developing QFT in the appropriate way on this underlying geometry rather than on Minkowski space should avoid some of the contradictions we would be otherwise bound to meet. But we might expect that the very structure of spacetime in the small, hence the algebraic structure of the underlying model, should depend on the dynamics, and thus on the quantum state. We propose a general scenario where this would be the case.

Let us first note that if we develop QFT over a fixed geometric background described by a (noncommutative C*) algebra $\mathcal{E}$, carrying an action of the Poincaré group by automorphisms, the (bounded functions of the) field operators (or more generally [23] local observables) should take values in a quasilocal C* algebra $\mathfrak{A}$, and fields would be (functions from QST to $\mathfrak{A}$) described by elements of (or affiliated to) the tensor product

$$\mathcal{E} \otimes \mathfrak{A}. \tag{20}$$

But a more realistic picture of QST might well involve operators in (20) which cannot be easily split in the two factors; the commutators of the q's would then appear as functions of the fields, more specifically of the metric $g_{\mu\nu}$, coupled to all fields by Einstein Equation, the fields themselves being at the same time functions of the q's. Thus the commutation relations between the q's should appear as part of the equations of motion:

$$\begin{aligned} \left[q_\mu, q_\nu\right] &= iQ_{\mu\nu}(g), \\ R_{\mu\nu} - (1/2)Rg_{\mu\nu} &= 8\pi T_{\mu\nu}(\psi), \\ F(\psi) &= 0, \end{aligned} \tag{21}$$

where $T_{\mu\nu}(\psi)$ is the energy momentum tensor of the fields involved, except gravitation, and the last line is symbolic for their equation of motion. Of course the action of translations on the q's will no longer be just the addition of multiples of the identity, since the q's depend on the metric g on which translations act as well.

As an attempt to investigate the form of (21,a), suppose we adopt a semiclassical approach, replacing the right hand side of (21,b) by its expectation value in a given state and let g be a classical solution; if we perform a measurement to localize an event in this state, we should repeat the considerations of [1], cf Section 1 above, in the background g; in the approximation of linearized gravity, with V as in equation (3), we should now impose

$$g_{00} + 2V > 0; \tag{22}$$

hence

$$g_{00} \cdot b \cdot \min(a, \tau) \gtrsim 1\,. \tag{23}$$

If we forget for a moment not only general covariance, according to which g should have no intrinsic meaning, but even Lorentz covariance, we could fulfill (23) requiring

$$[q_\mu, q_\nu] = iQ_{\mu\nu} g_{00}^{-1}\,, \tag{24}$$

where $Q_{\mu\nu}$ does not depend of g, for instance is defined by (16) and conditions following there.

According to General Relativity the Ricci tensor R is physically significant but the metric tensor g is not; yet it has been proposed [15] that Quantum Mechanics might alter this view, a possibility to be kept in mind while trying to rewrite a more convincing covariant extrapolation of (24).

The first natural guess would be to replace g_{00}^{-1} in (24) by a scalar depending only on the local variations of g, as the scalar curvature R; hence, using Einstein Equation (21,b) we would write

$$[q_\mu, q_\nu] = -8\pi\alpha i Q_{\mu\nu} g^{\lambda\rho} T_{\lambda\rho}(\psi)\,, \tag{25}$$

where a further constant factor has been allowed; or, even more generally, we could replace the Quantum Conditions by

$$\begin{aligned} Q_{\mu\nu} Q^{\mu\nu} &= 0\,, \\ [q_0, \ldots, q_3]^2 &= (\alpha R)^4\,. \end{aligned} \tag{26}$$

Equations (25) and (26) do not reduce to our background model where $R = 0$, so we are tempted to replace αR by $I + \alpha R$; we limit ourselves here to support the scenario expressed in (21) without committing ourselves to a choice, but point out (maybe only as a curiosity) that if our state is stricly localized in a tiny region, expectations of observables which are spacelike to that tiny region will be the same as in the vacuum and there we will find the following semiclassical approximation to (25)

$$[q_\mu, q_\nu] = -8\pi\alpha i Q_{\mu\nu} g^{\lambda\rho} < T_{\lambda\rho} >_0; \tag{27}$$

we might here insert the empirical evidence that $< T_{00} >_0$ is not zero but equal to the cosmological constant Λ ; in a relativistic vacuum

$$< T_{\lambda\rho} >_0 = \Lambda \cdot diag(1,-1,-1,-1);$$

now if g is a spherically symmetric stationary solution with $g_{j0} = g_{0j} = 0, j = 1,2,3$, and $-\mathbf{g}$ is its space part, (27) takes the form

$$[q_\mu, q_\nu] = -8\pi\alpha\Lambda i Q_{\mu\nu}(g_{00}^{-1} + tr(\mathbf{g}^{-1}));$$

for the Schwarzschild solution, for instance, the last term in brackets would be equal to $g_{00}^{-1} + g_{00} + 2$; but if there is a preferred frame (that of the Cosmic Background Radiation) where $< T_{00} >_0 = \Lambda$, $< T_{jj} >_0 = 0$, $j = 1,2,3$, we would get exactly (24).

These comments do not pretend to be neither satisfactory nor in a final shape (we used in our heuristic argument strict locality, which is bound to fail at Planck distances); yet it might well turn out that the quantum nature of spacetime does say something on the problem of the cosmological constant; for the presence of T in the right hand side of our spacetime commutation relations should imply an effective repulsion at short distances, and since quantum spacetime links aspects in the small (ultraviolet) to aspects in the large (infrared), this short range repulsion might well give rise to long range effects.

HINTS OF RELATIONS TO STRING THEORY

With the notation of Section 1, our Space Time Uncertainty Relations read

$$a \cdot b \gtrsim, \qquad \tau \cdot b \gtrsim 1; \tag{28}$$

the second one had been actually proposed earlier on the basis of a qualitative argument in String Theory [7], and derived later in the context of D-branes [11, 8]. Other recent findings in that domain lead to relations similar to our first relation too [16].

Other superficial coincidences can be noted: $U(1)$ gauge theory on QST described by the basic model is actually a $U(\infty)$ gauge theory (more precisely, the gauge group will be the unitary group of $\mathcal{E} + \mathbb{C} \cdot I$, namely a bundle over Σ with costant fiber the torus $\mathbb{T}$ times the group of unitaries which are perturbations of I by a compact operator), while $U(N)$ gauge theory in the limit $N \to \infty$ is believed to merge with String Theory; the QST version of Wick product seems to lead to a gaussian cutoff; eventually, if we take seriously the deformed model described in Section 3, the classical limit is a 10 dimensional manifold.

These facts might be no more than fortuitous coincidences, but suggest that the physical principles underlying the proposal of Quantum Spacetime might even turn out to provide the fundamental physical motivations which are still lacking in String Theory.

REFERENCES

1. Doplicher, S., Fredenhagen, K., and Roberts, J.E., *Commun. Math. Phys.*, **172**, 187 - 220 (1995).

2. Doplicher, S., Fredenhagen, K., and Roberts, J.E., *Phys. Lett.*, **B331**, 39 - 44 (1994).
3. Doplicher, S., *Quantum Physics, Classical Gravity, and Noncommutative Spacetime*, Proceedings of the XIth International Conference of Mathematical Physics, Ed. D.Iagolnitzer, 324 - 329, World Sci. 1995.
4. Doplicher, S., *Annales Inst. Henri Poincaré* vol. **64**, 543 - 553, (1996).
5. Work in progress with K. Fredenhagen.
6. Fredenhagen, K., address to the Goslar Meeting, 1998; *Quantum fields and noncommutative space-time*, Proceedings of the Hesselberg Meeting, 1999, to appear.
7. Yoneya, T., *Duality and Indeterminacy Principle in String Theory* , in *Wandering in the Fields*, Eds.: K. Kawarabayashi, A. Uwaka, World Sc. (1987).
8. Yoneya, T., *String Theory and Space-Time Uncertainty Principle*, hep-th/0004074.
9. Mead, C.A., *Phys. Rev.*, **135B**, 849-862 (1964).
10. Amati, D., Ciafaloni, M., and Veneziano, G., *Phys. Lett.*, **B216**, 41 (1989).
11. Amelino-Camelia, G., Ellis, J., Mavromatos, N.E., and Nanopoulos, D.V., *On the Spacetime Uncertainty Relations of Liouville Strings and D-Branes*, hep-th/9701144.
12. Lukierski, J., Nowicki, A., and Ruegg, H., *Phys. Lett.*, **B293**, 344-352 (1992).
13. Kempf, A., *J. Math. Phys.*, **35**, 4483-4496 (1994).
14. Maggiore, M., *Phys. Rev.*, **D49**, 5182-5187 (1994).
15. Chong-Sun-Chu, Pei-Ming Ho, and Yeong-Chuan Kao, *Worldvolume uncertainty relations for D-branes*, hep-th/9904133.
16. Cho, S., Hinterding, R., Madore, J., and Steinacker, H., *Int. J. Mod. Phys.*, **D9**, 161-199 (2000).
17. Kosiński, P., Lukierski, J., and Maślanka, P., *Noncommutative parameters of quantum symmetries and Star Products*, hep-th/0012056.
18. Woronowicz, S.L., *Commun. Math. Phys.*, **136**, 399-432 (1991).
19. Rieffel, M.A., *On the Operator Algebra for the Spacetime Uncertainty Relations*, in *Operator Algebras and Quantum Field Theory*, Eds.: S. Doplicher, R. Longo, J.E. Roberts and L. Zsido, I.P. 1997.
20. Ahluwalia, D.V., *Principle of equivalence and wave-particle duality in quantum gravity*, gr-qc/0009033.
21. Amelino-Camelia, G., *Gravity mediated interferometers as probes of a low-energy effective quantum gravity* gr-qc/9903080.
22. Ellis, J., Mavromatos, N.E., and Nanopoulos, C.V., *Probing Models of Quantum Space-Time Foam*, gr-qc/9909085.
23. Haag, R., *Local Quantum Physics*, Texts and Monographs in Physics, Springer 1994.
24. Madore, J., Schraml, S., Schupp, P., and Wess, J., *Gauge theory on noncommutative spaces*, hep-th/0001203;
    Jurco, B., Mueller, L., Schraml, S., Schupp, P., and Wess, J., *Construction of non abelian gauge theories on noncommutative spaces*, hep-th/0104153.

Adelic Strings and Noncommutativity

B. Dragovich

Institute of Physics, P.O.Box 57, 11001 Belgrade, Yugoslavia, email: dragovic@phy.bg.ac.yu

Steklov Mathematical Institute, Gubkin St. 8, 117966 Moscow, Russia

Abstract. We consider adelic approach to strings and spatial noncommutativity. Path integral method to string amplitudes is emphasized. Uncertainties in spatial measurements in quantum gravity are related to noncommutativity between coordinates. p-Adic and adelic Moyal products are introduced. In particular, p-adic and adelic counterparts of some real noncommutative scalar solitons are constructed.

INTRODUCTION

There is a common belief that an appropriate description of the Planck scale phenomena needs some new physical principles and nonstandard mathematical methods. To this end, we consider adelic approach to string theory and spatial noncommutativity. The space of adeles $\mathbf{A}$ is a mathematical instrument, which unifies real numbers (with archimedean geometry) and all p-adic numbers (with their nonarchimedean geometries). Note that all numerical experimental data belong to the field of rational numbers $\mathbf{Q}$, and that $\mathbf{Q}$ is a dense subfield of the field of real numbers $\mathbf{R}$ and the fields of p-adic numbers $\mathbf{Q}_p$ (p denotes a prime number, i.e. $p = 2,3,5,\cdots$). There is a sense to expect that basic mathematical methods and fundamental physical laws are invariant under interchange of $\mathbf{R}$ and $\mathbf{Q}_p$ [1], and such invariance has a place in adelic formalism. Possible spatial p-adic effects become sensitive in the vicinity of the characteristic length of a quantum system. Since the Planck length

$$\ell_0 = \sqrt{\frac{\hbar G}{c^3}} \sim 10^{-33} cm \tag{1}$$

is the natural one for quantum gravity, the Planck scale physics should exhibit some p-adic effects. Since 1987, there have been interesting investigations of p-adic and adelic string models (for an early review, see [2, 3]). As a basis for systematic investigations, p-adic [4] and adelic [5] quantum mechanics are formulated. It is well-known that an application of quantum-mechanical principles to general relativity leads to the uncertainties in measurements of very short distances in the form

$$\Delta x^i \Delta x^j \geq \ell_0^2, \quad i,j = 1,2,\cdots,n, \tag{2}$$

where n is spatial dimensionality. This fact requires a reconsideration of many our usual notions about the spacetime structure approaching to the Planck scale. The uncertainty

CP589, *Fundamental Interaction Theories*, edited by J. Lukierski and J. Rembieliński

(2) has to be a consequence of the corresponding noncommutativity between operators of space coordinates in the Hilbert space. This conclusion is an analogue of the similar situation in ordinary quantum mechanics: the uncertainty $\Delta x \Delta k \geq \frac{\hbar}{2}$ is a direct consequence of the noncommutativity in the form of the Heisenberg algebra $[\hat{x}, \hat{k}] = i\hbar$, where x and k are coordinates of the phase space. Thus, we see that the uncertainty (2) implies noncommutative geometry given by the commutation relation

$$[\hat{x}^i, \hat{x}^j] = i\hbar\theta^{ij}, \;\; \theta^{ij} = -\theta^{ji}, \tag{3}$$

where $\theta^{ij} = \theta\varepsilon^{ij}$ $(\varepsilon^{ij} = 1$ if $i < j)$ and $\theta = 2\ell_0^2/\hbar$. If the uncertainty (2) holds also for the case $i = j$, i.e. $\Delta x^i \geq \ell_0$, then it leads to the direct restriction on application of real numbers below ℓ_0 and gives rise to employment of p-adic numbers. However, it is not clear how this kind of uncertainty can be related to some spatial noncommutativity, and consequently we will omit here further discussion of this subject. One of the very interesting and fruitful recent developments in string theory (for a reviev, see [6, 7]) has been noncommutative geometry and the corresponding noncommutative field theory. This subject started to be very actual after Connes, Douglas and Schwarz shown [8] that gauge theory on noncommutative torus describes compactifications of M-theory to tori with constant background three-form field. Noncommutative field theory may be regarded as a deformation of the ordinary one in which field multiplication is replaced by the Moyal (star) product

$$(f \star g)(x) = \exp\left[\frac{i\hbar}{2}\theta^{ij}\frac{\partial}{\partial y^i}\frac{\partial}{\partial z^j}\right] f(y)g(z)|_{y=z=x}, \tag{4}$$

where $x^1, x^2, \cdots, x^d$ denote coordinates of noncommutative space, and $\theta^{ij} = -\theta^{ji}$ are noncommutativity parameters. There are many properties of D-brane dynamics which may be studied by noncommutative field theory. In particular, it enables to investigate a mixing of the UV and IR effects, and the tachyon condensation. Replacing the ordinary product between coordinates by the Moyal product (4), we have

$$x^i \star x^j - x^j \star x^i = i\hbar\theta^{ij}, \tag{5}$$

which resembles the usual Heisenberg algebra. Comparing the above equations it follows that the algebra (3) can be realized by ordinary coordinates using the Moyal product between them. In the Section *p-Adic Numbers and Adeles* we provide reader with some very basic facts on p-adic numbers and adeles. Section *Adelic Strings* is devoted to adelic strings. In the last section we consider some p-adic and adelic aspects of spatial noncommutativity.

P-ADIC NUMBERS AND ADELES

In order to introduce p-adic numbers it is suitable to start from $\mathbf{Q}$, since $\mathbf{Q}$ is the simplest field of numbers of characteristic 0 and it contains results of all physical measurements. Any non-zero rational number can be presented as infinite expansions into the two quite different forms. The usual one is to the base 10, i.e.

$$\sum_{k=n}^{-\infty} a_k 10^k, \quad a_k = 0, \cdots, 9, \tag{6}$$

and the other one is to the base p (p is a prime number) and reads

$$\sum_{k=m}^{+\infty} b_k p^k, \quad b_k = 0, \cdots, p-1, \tag{7}$$

where n and m are some integers. These representations have the usual repetition of digits, but expansions are in the mutually opposite directions. The series (6) and (7) are convergent with respect to the usual absolute value $|\cdot|_\infty$ and p-adic absolute value $|\cdot|_p$, respectively. Allowing arbitrary distributions of digits, we obtain standard representation of real numbers (6) and p-adic numbers (7). $\mathbf{R}$ and $\mathbf{Q}_p$ exhaust all number fields which contain $\mathbf{Q}$ as a dense subfield. They have many distinct geometric and algebraic properties. Geometry of p-adic numbers is the nonarchimedean one. There are mainly two kinds of analysis on $\mathbf{Q}_p$ based on two different mappings: $\mathbf{Q}_p \to \mathbf{Q}_p$ and $\mathbf{Q}_p \to \mathbf{C}$. We use here both of them, in classical and quantum p-adic models, respectively. Elementary p-adic functions are given by the same series as in the real case, but their regions of convergence are usually different. For instance, $\exp x = \sum_{n=0}^{\infty} \frac{x^n}{n!}$ and $\ln x = \sum_{n=1}^{\infty} (-1)^{n+1} \frac{(x-1)^n}{n}$ converge if $|x|_p < |2|_p$ and $|x-1|_p < 1$, respectively. Derivatives of p-adic valued functions are also defined as in the real case, but using p-adic norm instead of the absolute value. As a definite p-adic valued integral we take difference of the corresponding antiderivative in end points. Usual complex-valued p-adic functions are: (i) an additive character $\chi_p(x) = \exp 2\pi i \{x\}_p$, where $\{x\}_p$ is the fractional part of $x \in \mathbf{Q}_p$, (ii) a multiplicative character $\pi_s(x) = |x|_p^s$, where $s \in \mathbf{C}$, and (iii) locally constant functions with compact support, like, e.g. $\Omega(|x|_p) = 1$ if $|x|_p \leq 1$ and $\Omega(|x|_p) = 0$ otherwise. There is well defined Haar measure and integration. For instance,

$$\int_{\mathbf{Q}_p} \chi_p(\alpha x^2 + \beta x) dx = \lambda_p(\alpha) |2\alpha|_p^{-\frac{1}{2}} \chi_p\left(-\frac{\beta^2}{4\alpha}\right), \quad \alpha \neq 0, \tag{8}$$

where $\lambda_p(\alpha)$ is an arithmetic function [3]. For much more on p-adic numbers and p-adic analysis one can see, e.g. [3, 9, 10]. An adele $x_{\mathbf{A}}$ [9] is an infinite sequence

$$x_{\mathbf{A}} = (x_\infty, x_2, \cdots, x_p, \cdots), \tag{9}$$

where $x_\infty \in \mathbf{R}$ and $x_p \in \mathbf{Q}_p$ with the restriction that for all but a finite set $\mathbf{S}$ of primes p we have $x_p \in \mathbf{Z}_p$. Here $\mathbf{Z}_p = \{x \in \mathbf{Q}_p : |x|_p \leq 1\}$ is the ring of p-adic integers. All adeles make a ring with respect to componentwise addition and multiplication. It is convenient to present the ring of adeles $\mathbf{A}$ in the form

$$\mathbf{A} = \cup_{\mathbf{S}} \mathcal{A}(\mathbf{S}), \quad \mathcal{A}(\mathbf{S}) = \mathbf{R} \times \prod_{p \in \mathbf{S}} \mathbf{Q}_p \times \prod_{p \notin \mathbf{S}} \mathbf{Z}_p. \tag{10}$$

$\mathbf{A}$ is also locally compact topological space with the Haar measure. There are two kinds of analysis over $\mathbf{A}$, which generalize the corresponding analyses over $\mathbf{R}$ and $\mathbf{Q}_p$.

ADELIC STRINGS

A notion of p-adic string was introduced by Volovich in [11], where the hypothesis on the existence of nonarchimedean geometry at the Planck scale was made, and string theory with p-adic numbers was initiated. In particular, generalization of the usual Veneziano and Virasoro-Shapiro amplitudes with complex-valued multiplicative characters over various number fields was proposed and p-adic valued Veneziano amplitude was constructed by means of p-adic interpolation. Very successful p-adic analogues of the Veneziano and Virasoro-Shapiro amplitudes were proposed in [12] as the corresponding Gel'fand-Graev [9] beta functions. Using this approach, Freund and Witten obtained [13] an attractive adelic formula, which states that the product of the crossing symmetric Veneziano (or Virasoro-Shapiro) amplitude and its all p-adic counterparts equals unit (or a definite constant). This gives possibility to consider an ordinary four-point function, which is rather complicate, as an infinite product of its inverse p-adic analogues, which have simple forms. These first papers induced an interest in various aspects of p-adic string theory (for a review, see [2, 3]). A recent interest in p-adic string theory has been mainly related to the generalized adelic formulas for four-point string amplitudes [14], the tachyon condensation [15], and the new promising adelic approach [16]. Like in the ordinary string theory, the starting point of p-adic strings is a construction of the corresponding scattering amplitudes. Recall that the ordinary crossing symmetric Veneziano amplitude can be presented in the following forms:

$$A_\infty(a,b) = g^2 \int_{\mathbf{R}} |x|_\infty^{a-1} |1-x|_\infty^{b-1} dx \tag{11}$$

$$= g^2 \left[\frac{\Gamma(a)\Gamma(b)}{\Gamma(a+b)} + \frac{\Gamma(b)\Gamma(c)}{\Gamma(b+c)} + \frac{\Gamma(c)\Gamma(a)}{\Gamma(c+a)} \right] \tag{12}$$

$$= g^2 \frac{\zeta(1-a)}{\zeta(a)} \frac{\zeta(1-b)}{\zeta(b)} \frac{\zeta(1-c)}{\zeta(c)} \tag{13}$$

$$= g^2 \int \mathcal{D}X \exp\left(-\frac{i}{2\pi} \int d^2\sigma \partial^\alpha X_\mu \partial_\alpha X^\mu \right) \prod_{j=1}^{4} \int d^2\sigma_j \exp\left(i k_\mu^{(j)} X^\mu \right), \tag{14}$$

where $\hbar = 1$, $T = 1/\pi$, and $a = -\alpha(s) = -1 - \frac{s}{2}$, $b = -\alpha(t)$, $c = -\alpha(u)$ with the condition $s+t+u = -8$, i.e. $a+b+c = 1$. To introduce the corresponding p-adic Veneziano amplitude there is a sense to consider p-adic analogues of all the above four expressions. p-Adic generalization of the first expression was proposed in [12] and it reads

$$A_p(a,b) = g_p^2 \int_{\mathbf{Q}_p} |x|_p^{a-1} |1-x|_p^{b-1} dx, \tag{15}$$

where $|\cdot|_p$ denotes p-adic absolute value. In this case only string world-sheet parameter x is treated as p-adic variable, and all other quantities have their usual (real) valuation. An attractive adelic formula of the form

$$A_\infty(a,b) \prod_p A_p(a,b) = 1 \tag{16}$$

was found [13], where $A_\infty(a,b)$ denotes the usual Veneziano amplitude (11). A similar product formula holds also for the Virasoro-Shapiro amplitude. These infinite products are divergent, but they can be successfully regularized. Unfortunately, there is a problem to extend this formula to the higher-point functions. p-Adic analogues of (11) and (12) were also proposed in [11] and [17], respectively. In these cases, world-sheet, string momenta and amplitudes are manifestly p-adic. Since these string amplitudes are p-adic valued functions, it is not enough clear their physical meaning. Expression (13) is based on Feynman's functional integral method, which is a useful tool in all quantum theory and has successful p-adic generalization [18]. Its p-adic counterpart, proposed in [16], has been elaborated [19] and deserves further study. Note that in this approach, p-adic string amplitude is complex-valued, while not only the world-sheet parameters but also target space coordinates and string momenta are p-adic variables. Such p-adic generalization is a natural extension of the formalism of p-adic [4] and adelic [5] quantum mechanics to string theory. Our starting point to unified path integral approach to ordinary and p-adic N-point bosonic string amplitudes at the tree level is

$$A_v(k_1,\cdots,k_N) = g_v^{N-2}\prod_{j=1}^{N}\int d^2\sigma_j \int \chi_v\left(-\frac{1}{h}\int \mathcal{L}(X^\mu,\partial_\alpha X^\mu)d^2\sigma\right)\mathcal{D}_v X, \qquad (17)$$

where $v=\infty,2,\cdots,p,\cdots$, $\mu=0,1,\cdots,25$, $\alpha=0,1$, and $\chi_\infty(a)=\exp(-2\pi i a)$, $\chi_p(a)=\exp(2\pi i\{a\}_p)$. The above Lagrangian is

$$\mathcal{L} = -\frac{T}{2}\partial_\alpha X^\mu(\sigma,\tau)\partial^\alpha X_\mu(\sigma,\tau)+\sqrt{-1}\sum_{j=1}^{N}k_\mu^{(j)}X^\mu(\sigma,\tau)\delta(\sigma-\sigma_j)\delta(\tau-\tau_j). \qquad (18)$$

In fact, our approach is adelic and based on the following assumptions: (*i*) spacetime and matter are adelic at the Planck (M-theory) scale, (*ii*) Feynman's path integral method is an inherent ingredient of quantum theory, and (*iii*) adelic quantum theory is a more complete theory than the ordinary one. Consequently, a string is an adelic object which has simultaneously real and all p-adic characteristics. Here the term p-adic (real) string is related to string with dominant p-adic (real) properties. The target space and world-sheet are adelic spaces. Adelic Feynman's path integral is an infinite product of the ordinary one and all p-adic counterparts [21]. The corresponding adelic string amplitude is

$$A(k_{\mathbf{A}}^{(1)},\cdots,k_{\mathbf{A}}^{(N)})$$
$$= A_\infty(k_\infty^{(1)},\cdots,k_\infty^{(N)})\prod_{p\in\mathbf{S}}A_p(k_p^{(1)},\cdots,k_p^{(N)})\prod_{p\notin\mathbf{S}}A_p(k_p^{(1)},\cdots,k_p^{(N)}), \qquad (19)$$

where $k_{\mathbf{A}}^{(i)}$ is an adele, i.e.

$$k_{\mathbf{A}}^{(i)} = (k_\infty^{(i)},k_2^{(i)},\cdots,k_p^{(i)},\cdots) \qquad (20)$$

with the restriction that $k_p^{(i)}\in\mathbf{Z}_p$ for all but a finite set $\mathbf{S}$ of primes p. The topological ring of adeles $k_{\mathbf{A}}^{(i)}$ provides a framework for simultaneous and unified consideration of real and p-adic string momenta. Adelic string amplitude contains nontrivial p-adic modification of the ordinary one. An evaluation of the above p-adic and adelic amplitudes will be presented in detail elsewhere [19].

P-ADIC AND ADELIC NONCOMMUTATIVITY

There is a noncommutative scalar soliton [22]

$$\phi(x^1,x^2) = 2\exp\left(-\frac{(x^1)^2+(x^2)^2}{\theta}\right) \tag{21}$$

which is the simplest nontrivial solution (trivial solutions are $\phi = 0$ and $\phi = 1$) of the equation

$$(\phi\star\phi)(x) = \phi(x), \tag{22}$$

where $\star$ denotes the Moyal product (4) with $\theta^{ij} = \theta\varepsilon^{ij}/\hbar$. The solution (21) of the equation (22) extremises energy in noncommutative scalar field theory [22] with the potential

$$V(\phi) = \frac{1}{2}m^2\phi\star\phi - \frac{1}{3}\phi\star\phi\star\phi, \tag{23}$$

where $m = 1$ and the kinetic term is neglected in the limit $\theta \to \infty$. An intriguing similarity between the above noncommutative scalar soliton and a solitonic brane solution [15] in an effective p-adic string theory [23] is discussed in [24]. This two-dimensional noncommutative scalar field model can be extended to the more general case with

$$V(\phi) = \frac{1}{2}m^2\phi_\star^2 - \frac{c_{k+1}}{k+1}\phi_\star^{k+1}, \tag{24}$$

where $\phi_\star$ denotes that fields are multiplied by the star product, and $\phi \equiv \phi(x^1,\cdots,x^n)$ with even n spatial directions. The corresponding equation

$$c_{k+1}\phi_\star^k(x) = m^2\phi(x) \tag{25}$$

has the solution

$$\phi(x) = 2^{\frac{n}{2}}\left(\frac{m^2}{c_{k+1}}\right)^{\frac{1}{k}}\exp\left(-\frac{1}{\theta}\sum_{i=1}^{n}(x^i)^2\right). \tag{26}$$

The above formulas (21) - (26) are related to the case with real numbers. Now we introduce their p-adic and adelic generalization. Let the Moyal product for p-adic valued functions f and g be

$$(f\star g)(x) = \exp\left[\frac{\sqrt{-1}}{2}\theta\varepsilon^{ij}\frac{\partial}{\partial y^i}\frac{\partial}{\partial z^j}\right]f(y)g(z)|_{y=z=x}, \tag{27}$$

where $x,y,z \in \mathbf{Q}_p$. Then the equations (21) - (26) also hold in the p-adic case. The region of convergence of exponential functions is given by inequality $|x^i|_p < |2\theta|_p^{1/2}$.

Thus, the equations (22) and (25) have real $\phi(x_\infty)$ and p-adic $\phi(x_p)$ solutions of the same functional form ϕ, i.e. they are invariant under interchange of $\mathbf{R}$ and $\mathbf{Q}_p$. Moreover, they have natural adelic solutions

$$\phi_{\mathbf{A}}(x_{\mathbf{A}}) = (\phi(x_\infty), \phi(x_2), \cdots, \phi(x_p), \cdots), \tag{28}$$

where parameters $m, \theta, c_{k+1} \in \mathbf{Q}$. Hence, one can consider not only real, but also p-adic and adelic noncommutative scalar solitons. This subject will be presented in more details in [25]. It is worth noting that one can introduce [26] the Moyal product in complex-valued p-adic quantum mechanics and it reads

$$(f * g)(x) = \int_{\mathbf{Q}_p^n} \int_{\mathbf{Q}_p^n} dk dk' \, \chi_p(-\frac{1}{h}(x^i k_i + x^j k'_j) + \frac{1}{2h} k_i k'_j \theta^{ij}) \tilde{f}(k) \tilde{g}(k'), \tag{29}$$

where n denotes spatial dimensionality. This is a direct p-adic analogue of the integral form of the Moyal product in the real case. The corresponding adelic version of (29) is

$$\begin{aligned} (f * g)_{\mathbf{A}}(x_{\mathbf{A}}) &= (f * g)_\infty(x_\infty) \prod_p (f * g)_p(x_p) \qquad (30) \\ &= \prod_v \int_{\mathbf{Q}_v^n} \int_{\mathbf{Q}_v^n} dk_v dk'_v \, \chi_v(-\frac{1}{h}(x_v^i k_{vi} + x_v^j k'_{vj}) + \frac{1}{2h} k_{vi} k'_{vj} \theta_v^{ij}) \tilde{f}_v(k_v) \tilde{g}_v(k'_v), \end{aligned}$$

where $x_p^i \in \mathbf{Z}_p$ and $\theta_p^{ij} \in \mathbf{Z}_p$ for almost all primes p.

ACKNOWLEDGMENTS

Author wishes to thank organizers of the XXXVII Karpacz Winter School of Theoretical Physics: New Developments in Fundamental Interaction Theories, (February 6-15, 2001), for their invitation to participate and give a talk. The work on this paper was supported in part by RFFI grant 990100866.

REFERENCES

1. Volovich, I. V., *Number Theory as the Ultimate Physical Theory*, Preprint CERN-TH 4781, (1987).
2. Brekke, L., and Freund, P. G. O., *Phys. Rep.*, **233**, 1-66 (1993).
3. Vladimirov, V. S., Volovich, I. V., and Zelenov, E. I., *p-Adic Analysis and Mathematical Physics*, World Scientific, Singapore, 1994.
4. Vladimirov, V. S, and Volovich, I. V., *Commun. Math. Phys.*, **123**, 659-676 (1989).
5. Dragovich, B., *Theor. Math. Phys.*, **101**, 1404-1414 (1994); *Int. J. Mod. Phys.*, **A 10**, 2349-2365 (1995).
6. Schwarz, J. H., *Recent Progress in Superstring Theory*, hep-th/0007130.
7. Sen, A., *Recent Developments in Superstring Theory* , hep-lat/0011073.
8. Connes, A., Douglas, M. R., and Schwarz, A., *JHEP*, **9802**, 003 (1998), hep-th/9711162.
9. Gel'fand, I. M, Graev, M. I., and Pyatetski-Shapiro, I. I., *Representation Theory and Automorphic Functions*, Saunders, London, 1966.
10. Schikhof, W. H., *Ultrametric Calculus*, Cambridge U.P., Cambridge, 1984.
11. Volovich, I. V., *Class. Quantum Grav.*, **4**, L83-L87 (1987).

12. Freund, P. G. O., and Olson, M., *Phys. Lett.*, **B 199**, 186-190 (1987).
13. Freund, P. G. O., and Witten, E., *Phys. Lett.*, **B 199**, 191-194 (1987).
14. Vladimirov, V. S., *Adelic Formulas for Gamma and Beta Functions of One-Class Quadratic Fields: Applications to 4-Particle Scattering string amplitudes*, math-ph/0004017.
15. Ghoshal, D., and Sen, A., *Tachyon Condensation and Brane Descent Relations in p-Adic String Theory*, hep-th/0003278.
16. Dragovich, B., *On Adelic Strings* , hep-th/0005200.
17. Aref'eva, I. Ya., Dragovich, B., and Volovich, I. V., *Phys. Lett.*, **B 209**, 445-450 (1988).
18. Djordjević, G. S., and Dragovich, B., *Mod. Phys. Lett.*, **A 12**, 1445-1463 (1997).
19. Dragovich, B., Rodić, P. and Volovich, I. V., *New Amplitudes for p-Adic and Adelic Bosonic Strings*, in preparation.
20. Green, M. B., Schwarz, J. H., and Witten, E., *Superstring Theory, I*, Cambridge U.P., Cambridge, 1987.
21. Djordjević, G. S., Dragovich, B., and Nešić, Lj., *Adelic Path Integrals for Quadratic Lagrangians*, hep-th/0105030.
22. Gopakumar, R., Minwalla, S., and Strominger, A., *Noncommutative Solitons*, hep-th/0003160.
23. Frampton, P. H., and Okada, Y., *Phys. Rev. Lett.*, **60**, 484-488 (1988).
24. Dragovich, B., and Volovich, I. V., "*p*-Adic Strings and Noncommutativity", in *Noncommutative Structures in Mathematics and Physics*, edited by J. Wess and S. Duplij, Kluwer Ac. Publishers, Dordrecht, 2001, pp. 391-400.
25. Dragovich, B., and Sazdović, B., *Real, p-Adic and Adelic Noncommutative Scalar Solitons*, in preparation.
26. Djordjević, G., Dragovich, B. and Nešć, Lj., "Adelic Quantum Mechanics: Nonarchimedean and Noncommutative Aspects", in *Noncommutative Structures in Mathematics and Physics*, edited by J. Wess and S. Duplij, Kluwer Ac. Publishers, Dordrecht, 2001, pp. 401-415.

The Real Quantum Plane as Part of $2d$-Minkowski Space

G. Fiore*, J. Madore[1†] and M. Maceda[†]

*Università di Napoli, I-80125 Napoli
†Université de Paris-Sud, F-91405 Orsay

Abstract. Using the frame formalism we consider some possible metrics on the real quantum plane. We require that the metric be real and symmetric. In practice this means that we use the freedom of noncommutative geometry to impose a different 'σ-symmetry', which is chosen so that a complex metric is 'σ-real' and an un-symmetric metric is 'σ-symmetric'. The notion of reality and symmetry are changed so that the definition of hermitian does not change. An analysis is then made of a set of possible metrics.

INTRODUCTION AND NOTATION

Let $\mathcal{A}$ be a $*$-algebra with differential calculus $\Omega^1(\mathcal{A})$ [1] and suppose that it has a frame [2, 3], a set of 1-forms θ^i dual to a set of inner derivations $e_i = \mathrm{ad}\lambda_i$ and which therefore commutes with the elements of the algebra:

$$\theta^i f = f\theta^i. \tag{1}$$

The differential calculus will be real [4, 5] if the λ_i are anti-hermitian. Using the frame we can set

$$df = e_i f\theta^i, \tag{2}$$

from which it follows that the module structure of $\Omega^1(\mathcal{A})$ is given by

$$f dg = (fe_i g)\theta^i, \qquad dg f = (e_i g) f\theta^i.$$

If a frame exists the module $\Omega^1(\mathcal{A})$ is free of rank n as a left or right module. It can therefore be identified with the direct sum

$$\Omega^1(\mathcal{A}) = \bigoplus_1^n \mathcal{A}, \tag{3}$$

of n copies of $\mathcal{A}$. In this representation θ^i is given by the element of the direct sum with the unit in the i-th position and zero elsewhere. We shall refer to the integer n as the dimension of the geometry.

---

[1] email:John.Madore@th.u-psud.fr

CP589, *Fundamental Interaction Theories*, edited by J. Lukierski and J. Rembieliński

Let π be the product in $\Omega^*(\mathcal{A})$ and set

$$\pi(\theta^i \otimes \theta^j) = P^{ij}{}_{kl}\theta^k \otimes \theta^l, \qquad P^{ij}{}_{kl} \in \mathcal{Z}(\mathcal{A}).$$

Since π is a projection we have

$$P^{ij}{}_{mn}P^{mn}{}_{kl} = P^{ij}{}_{kl} \tag{4}$$

and the product $\theta^i\theta^j$ satisfies

$$\theta^i\theta^j = P^{ij}{}_{kl}\theta^k\theta^l\,. \tag{5}$$

If the θ^i anti-commute then

$$P^{ij}{}_{kl} = \frac{1}{2}(\delta^i_k\delta^j_l - \delta^j_k\delta^i_l)\,. \tag{6}$$

Since the exterior derivative of θ^i is a 2-form it can necessarily be written as

$$d\theta^i = -\frac{1}{2}C^i{}_{jk}\theta^j\theta^k\,,$$

where, because of (5), the structure elements can be chosen to satisfy the constraints

$$C^i{}_{jk}P^{jk}{}_{lm} = C^i{}_{lm}\,.$$

From the generators θ^i we can construct a 1-form

$$\theta = \lambda_i\theta^i \tag{7}$$

in $\Omega^1(\mathcal{A})$ which plays the role [1] of a Dirac operator:

$$df = -[\theta, f]\,.$$

We introduce the coefficients K_{ij} by the equation

$$d\theta + \theta^2 = -\frac{1}{2}K_{ij}\theta^i\theta^j\,. \tag{8}$$

If we write then the identity $d^2 = 0$ as

$$d(\theta f - f\theta) = [d\theta, f] + [\theta, [\theta, f]] = [d\theta + \theta^2, f] = 0\,,$$

we see that the K_{ij} must lie in $\mathcal{Z}(\mathcal{A})$. Again from (5) they can be chosen to satisfy the constraints

$$K_{jk}P^{jk}{}_{lm} = K_{lm}.$$

It will also be convenient to introduce the quantities

$$C^{ij}{}_{kl} = \delta^i_k\delta^j_l - 2P^{ij}{}_{kl}\,. \tag{9}$$

Then from (4) we find that

$$C^{ij}{}_{kl}C^{kl}{}_{mn} = \delta^i_m \delta^j_n . \tag{10}$$

From the general consistency of the differential calculus it follows that

$$2P^{ij}{}_{kl}\lambda_i\lambda_j - F^i{}_{kl}\lambda_i - K_{kl} = 0,$$

for some array of numbers $F^i{}_{jk}$. We introduce [6] a flip σ:

$$\Omega^1(\mathcal{A}) \otimes_{\mathcal{A}} \Omega^1(\mathcal{A}) \xrightarrow{\sigma} \Omega^1(\mathcal{A}) \otimes_{\mathcal{A}} \Omega^1(\mathcal{A}) . \tag{11}$$

In terms of the frame it is given by $S^{ij}{}_{kl} \in \mathcal{Z}(\mathcal{A})$ defined by

$$\sigma(\theta^i \otimes \theta^j) = S^{ij}{}_{kl}\theta^k \otimes \theta^l .$$

It must satisfy the reality constraint [5], which takes the simple form

$$(S^{ji}{}_{kl})^* S^{lk}{}_{mn} = \delta^i_m \delta^j_n \tag{12}$$

if $(\theta^i)^* = \theta^i$.

A covariant derivative on the module $\Omega^1(\mathcal{A})$ must satisfy both a left and a right Leibniz rule. We use the ordinary left Leibniz rule and define the right Leibniz rule as

$$D(\xi f) = \sigma(\xi \otimes df) + (D\xi)f , \tag{13}$$

for arbitrary $f \in \mathcal{A}$ and $\xi \in \Omega^1(\mathcal{A})$.

For every differential calculus and flip one can construct the linear connection

$$\omega^i{}_{jk} = \lambda_l(S^{il}{}_{jk} - \delta^l_j \delta^i_k) . \tag{14}$$

The connection 1-form is given by

$$\omega^i{}_k = \lambda_l S^{il}{}_{jk}\theta^j + \delta^i_k \theta . \tag{15}$$

When $F^i{}_{jk} = 0$ the curvature of the covariant derivative D defined in (14) can be readily calculated.

We define frame components of the metric by

$$g^{ij} = g(\theta^i \otimes \theta^j) .$$

They lie necessarily in the center $\mathcal{Z}(\mathcal{A})$ of the algebra. The condition that (14) be metric-compatible can be written as

$$S^{im}{}_{ln} g^{np} S^{jk}{}_{mp} = g^{ij}\delta^k_l . \tag{16}$$

One can understand this seemingly odd condition by introducing a 'covariant derivative' D_iX^j of a 'vector' X^j. The covariant derivative $D_i(X^jY)$ of the product of X^j by a 'field' Y must be then defined as

$$D_i(X^jY) = D_iX^jY + S^{jl}{}_{im}X^m D_l Y ,$$

since there is a 'flip' as the index on the derivation crosses the index on the first 'vector'. If we apply again this rule to $Y = Y^k Z$, with Y^k also a 'vector' and Z another 'field' we find

$$D_i(X^j Y^k Z) = D_i(X^j Y^k)Z + S^{jl}{}_{im} X^m Y^p S^{kn}{}_{lp} D_n Z.$$

Since g^{jk} is a bivector, the 'crossing rule' is the same as for $X^j Y^k$:

$$D_i(g^{jk} Z) = D_i g^{jk} Z + S^{jl}{}_{im} g^{mp} S^{kn}{}_{lp} D_n Z.$$

The condition that the metric be constant under parallel transport then implies (16) and

$$D_i g^{jk} = 0.$$

We shall require that the metric be symmetric in the sense

$$g \circ \pi = 0, \tag{17}$$

that it annihilates the 2-forms. We shall impose also the condition

$$\pi \circ (\sigma + 1) = 0, \tag{18}$$

that the antisymmetric part of a symmetric tensor vanish. This can be considered as a condition on the product or on the flip. In ordinary geometry it is the definition of π; a 2-form can be considered as an antisymmetric tensor. Because of this condition the torsion is a bilinear map [7]. The most general solution can be written in the form

$$1 + \sigma = (1 - \pi) \circ \tau, \tag{19}$$

where τ is arbitrary. Suppose that τ is invertible. Then because of the identity

$$1 = \pi + (1 + \sigma) \circ \tau^{-1},$$

one can identify the second term on the right-hand side as the projection onto the symmetric part of the tensor product. The choice $\tau = 2$ yields the value $\sigma = 1 - 2\pi$. If τ is not invertible then there arises the possibility that part of the tensor product is neither symmetric nor antisymmetric.

We use σ to impose the reality condition [5]

$$S^{ij}{}_{kl} g^{kl} = (g^{ji})^* \tag{20}$$

on the metric, valid in this form for a real frame. This is a combination of a 'twisted' symmetry condition and the ordinary condition of reality on a complex matrix. It can also be written as an ordinary condition of symmetry and a 'twisted' definition of reality. Using σ one can also impose [5] a reality condition on the curvature. We refer elsewhere [21] for more details.

The extended quantum plane is the $*$-algebra $\mathcal{A}$ generated by the hermitian elements $x^i = (x, y)$ with their inverses and the relation

$$xy = \tilde{q} yx, \tag{21}$$

as well as the usual relations between inverses. We define, for $\tilde{q}^4 \neq 1$,

$$\lambda_1 = -\varepsilon_1 \frac{\tilde{q}^4}{\tilde{q}^4 - 1} x^{-2} y^2, \qquad \lambda_2 = \varepsilon_2 \frac{\tilde{q}^2}{\tilde{q}^4 - 1} x^{-2}.$$

There is an ambiguity in this definition due to the fact that the defining relations (21) are homogeneous and which we have reduce to a sign: $\varepsilon_a = \pm 1$. The extra minus is a 'historical convenience'. The important fact is that the λ_a are singular in the limit $\tilde{q} \to 1$ and that they are anti-hermitian if $\tilde{q}$ is of unit modulus. We find for $\tilde{q}^2 \neq -1$

$$\begin{aligned} e_1 x &= \varepsilon_1 \frac{\tilde{q}^2}{(\tilde{q}^2 + 1)} x^{-1} y^2, & e_1 y &= \varepsilon_1 \frac{\tilde{q}^4}{\tilde{q}^2 + 1} x^{-2} y^3, \\ e_2 x &= 0, & e_2 y &= -\varepsilon_2 \frac{\tilde{q}^2}{\tilde{q}^2 + 1} x^{-2} y. \end{aligned} \tag{22}$$

These derivations are again extended to arbitrary polynomials in the generators by the Leibniz rule. Using them we find

$$dx = \frac{\tilde{q}^2}{(\tilde{q}^2 + 1)} x^{-1} y^2 \varepsilon_1 \theta^1, \qquad dy = \frac{\tilde{q}^2}{\tilde{q}^2 + 1} x^{-2} y (\tilde{q}^2 y^2 \varepsilon_1 \theta^1 - \varepsilon_2 \theta^2) \tag{23}$$

and solving for the θ^i we obtain

$$\varepsilon_1 \theta^1 = (\tilde{q}^2 + 1) x y^{-2} dx, \qquad \varepsilon_2 \theta^2 = -(\tilde{q}^2 + 1) x (x y^{-1} dy - dx).$$

The module structure which follows from the condition (1) that the θ^i commute with the elements of the algebra is equivalent to the Wess-Zumino relations [22]

$$\begin{aligned} x dx &= \tilde{q}^2 dx x, & x dy &= \tilde{q} dy x + (\tilde{q}^2 - 1) dx y, \\ y dx &= \tilde{q} dx y, & y dy &= \tilde{q}^2 dy y. \end{aligned} \tag{24}$$

One can show that they are invariant under the coaction of the quantum group $SL_q(2, \mathbb{C})$. This invariance was encoded in the choice of λ_a.

Consider the change of generators defined by

$$u = \varepsilon_2 x^2, \qquad v = \varepsilon_1 (\sqrt{\tilde{q}} x y^{-1})^2.$$

We shall see that each of the four possible choices of sign combinations corresponds to an identification of x and y as the coordinates of one of the four regions on $\mathbb{R}^2$ defined by the light cone of a metric with Minkowski signature. If one sets also $q = \tilde{q}^{-4}$ then one finds that (24) becomes

$$
\begin{aligned}
udu &= q^{-1}duu, \quad udv = q^{-1}dvu, \\
vdu &= q^{-1}duv, \quad vdv = q^{-1}dvv.
\end{aligned} \tag{25}
$$

The algebra is still defined by a quadratic relation $uv = qvu$. In terms of the new generators the θ^i become

$$\theta^1 = q^{-1/2}(uv^{-1})^{-1}du, \qquad \theta^2 = q^{1/2}(uv^{-1})dv.$$

What we have done in fact is use the λ_a^{-1} as generators of the algebra and the differential calculus; otherwise nothing has been changed. The form θ is most conveniently expressed in terms of the λ_a. Since

$$\lambda_1 = \frac{q^{1/2}}{q-1}v^{-1}, \qquad \lambda_2 = -\frac{q^{1/2}}{q-1}u^{-1}, \tag{26}$$

we find that

$$\theta = \frac{1}{q-1}(q\lambda_2^{-1}d\lambda_2 - \lambda_1^{-1}d\lambda_1).$$

It is an anti-hermitian closed form. The volume element is a product of two exact forms:

$$\theta^1\theta^2 = dudv.$$

The structure of the differential algebra is given by the relations

$$(\theta^1)^2 = 0, \qquad (\theta^2)^2 = 0, \qquad \theta^1\theta^2 + q\theta^2\theta^1 = 0. \tag{27}$$

This can be written in the form (5). The reality of the differential implies that the structure elements must satisfy the conditions

$$((C^i{}_{jk})^* + C^i{}_{kj})P^{jk}{}_{lm} = 0,$$

from which follows that

$$(C^i{}_{21})^* = -C^i{}_{12} = q^{-1}C^i{}_{21}, \qquad (C^i{}_{12})^* = -C^i{}_{21} = qC^i{}_{12},$$

are given by

$$C^1{}_{12} = (q^{-1}-1)\lambda_2, \qquad C^2{}_{12} = (q^{-1}-1)\lambda_1, \tag{28}$$

The $C^i{}_{jk}$ do not depend on the sign ambiguities. With the generators

$$t = \frac{1}{\sqrt{2}}(u+v), \qquad r = \frac{1}{\sqrt{2}}(u-v), \tag{29}$$

the four possible sign combinations can be written as

$$\varepsilon_1 = \varepsilon_2: \quad \mathrm{sgn}(t) = \varepsilon_1, \qquad \varepsilon_1 = -\varepsilon_2: \quad \mathrm{sgn}(r) = \varepsilon_2.$$

We shall later in Section 5.1 introduce a light-cone and interpret these relations in terms of space-like and time-like.

Introduce the notation

$$X = \begin{pmatrix} t \\ r \end{pmatrix}, \qquad \Xi = \begin{pmatrix} dt \\ dr \end{pmatrix}, \qquad Q = \begin{pmatrix} \cos(\pi\gamma) & i\sin(\pi\gamma) \\ i\sin(\pi\gamma) & \cos(\pi\gamma) \end{pmatrix} \qquad q = e^{2\pi i\gamma}.$$

Then Q is unitary. The commutation relations in $\Omega^*(\mathcal{A})$ can be written in the form

$$X^t(Q\sigma_2)X = 0, \qquad X\Xi^t = \Xi(Q^2X)^t, \qquad \Xi^t Q\Xi = 0. \tag{30}$$

The σ_2 is the Pauli matrix.

THE METRICS AND THEIR CONNECTIONS

We now consider some possible metrics on the real quantum plane. We require that the metric be 'σ-real' and 'σ-symmetric'. This means that we use the extra freedom of noncommutative geometry to impose a different symmetry, which is chosen so that a complex metric becomes real and a non-symmetric metric is symmetric. The notion of reality and symmetry are changed so that the definition of hermitian does not change. We are especially interested in real solutions, which satisfy therefore also (20). We have found that there are several types of solutions, four of which we shall describe in the following subsections. One can show that there are no solutions with $\tau = 2$. A complete classification has been given [24] of the solutions to the braid equation as well [25, 26] as of those which satisfy a weaker modified equation.

If one considers locality as of importance only in the commutative limit then there is no restriction on the coefficients of the metric, except that they be local functions in this limit. If one considers locality as of importance even before the limit but is willing to accept a metric which is real and symmetric only in the commutative limit then the most general line element one can obtain is of the form

$$ds^2 = g_{ij}\theta^i \otimes_S \theta^j .$$

The subscript S indicates a symmetrized tensor product; the g_{ij} is a real symmetric matrix and the moving frame θ^i is defined by

$$\theta^1 = vu^{-1}du, \qquad \theta^2 = uv^{-1}dv.$$

The line element becomes then

$$ds^2 = g_1 v^2 u^{-2} du^2 + 2g_2 du dv + g_4 u^2 v^{-2} dv^2 . \tag{31}$$

The product here is the symmetrized tensor product; not the exterior product.

The associated metric connection is given by the structure functions

$$C^1{}_{12} = u^{-1}, \qquad C^2{}_{12} = -v^{-1}.$$

If we interpret the matrix g_{ij} as the components of the Killing metric on $SO(2)$ or $SO(1,1)$ then we can use it to calculate the connection form. The result will be of the form

$$\omega^i{}_j = A^i{}_{jk} u^{-1} \theta^k + B^i{}_{jk} v^{-1} \theta^k,$$

with $g_{ik}\omega^k{}_j$ antisymmetric in the two indices. The Gaussian curvature K is a second-order homogeneous polynomial in the variables u^{-1} and v^{-1}:

$$K = \kappa_{11} u^{-2} + 2\kappa_{12} u^{-1} v^{-1} + \kappa_{22} v^{-2}.$$

Solutions

A family of solutions can be found with a Minkowski-signature metric. These are the most interesting solutions. With the convenient normalization of the metric so that $g^3 = q^{-1/2}$ the flip is given by the matrix

$$S = \begin{pmatrix} q & -q^{-1/2}\zeta & -q^{1/2}\zeta & q^{-1}(q^2-1)^{-1}\zeta^2(q^2+1) \\ 0 & 0 & q & -q^{-1/2}\zeta \\ 0 & q^{-1} & 0 & q^{-3/2}\zeta \\ 0 & 0 & 0 & q^{-1} \end{pmatrix}.$$

It tends to the ordinary flip as $q \to 1$ and for $\zeta = 0$ is a solution to the braid equation. The corresponding metric given by

$$g^{ij} = \begin{pmatrix} (q-1)^{-1}\zeta & q^{1/2} \\ q^{-1/2} & 0 \end{pmatrix}. \tag{32}$$

It is σ-symmetric for all g^1 and real if $g^1 = 0$. In this case σ is given by

$$S = \begin{pmatrix} q & 0 & 0 & 0 \\ 0 & 0 & q & 0 \\ 0 & q^{-1} & 0 & 0 \\ 0 & 0 & 0 & q^{-1} \end{pmatrix}. \tag{33}$$

The σ and π are related as in (19) with (using the same conventions)

$$T = \begin{pmatrix} 1+q & 0 & 0 & 0 \\ 0 & 2 & 0 & 0 \\ 0 & 0 & 2 & 0 \\ 0 & 0 & 0 & 1+q^{-1} \end{pmatrix}. \tag{34}$$

The fact that T is not proportional to the identity is due to the fact that the map $(1+\sigma)/2$ is not a projector and that we would like it to act as such and be the complementary to π. The metric is of indefinite signature and in 'light-cone' coordinates. If we use the expression $q = e^{2\pi i\gamma}$ we find that

$$g_S^{ij} = \cos(\pi\gamma)\begin{pmatrix} 0 & 1 \\ 1 & 0 \end{pmatrix}, \qquad g_A^{ij} = i\sin(\pi\gamma)\begin{pmatrix} 0 & 1 \\ -1 & 0 \end{pmatrix}. \tag{35}$$

The inverse metric components are defined by the equation

$$g_{ij}g^{jk} = \delta_i^k .$$

This matrix also can be split. If we rescale so that the symmetric part is of the standard form we find

$$\eta_{ij} = \begin{pmatrix} 0 & 1 \\ 1 & 0 \end{pmatrix}, \qquad B_{ij} = i\tan(\pi\gamma)\begin{pmatrix} 0 & 1 \\ -1 & 0 \end{pmatrix}.$$

The metric connection has vanishing curvature. The linear connection (14) is given by

$$\omega^i{}_j = (1-q)\begin{pmatrix} 1 & 0 \\ 0 & -q^{-1} \end{pmatrix}\theta.$$

Because of the identities

$$d\theta = 0, \qquad \theta^2 = 0,$$

the curvature vanishes; with the choice (33) of flip the quantum plane is flat. In the commutative limit the line element is given by

$$ds^2 = g_{ij}\theta^i \otimes_S \theta^j = 2\theta^1 \otimes_S \theta^2 = 2du \otimes_S dv = dt^2 - dr^2 .$$

The frame is singular along the light cone through the origin. Suppose $\varepsilon_1 = \varepsilon_2 = 1$. If in a representation one forces x and y to be hermitian then the u and v must be positive operators. One concludes then that $t > |r|$; the geometry describes only the forward light-cone through the origin. The other three regions are given by the other three possible combinations of signs.

PATCHING

To each of the four regions defined by the light cone through the origin in two dimensions we have associated an algebra and differential calculus. With the metric we have found, none is complete as 'manifold'. However we could expect to obtain a complete 'manifold' if we could smoothly patch the four regions together to form one. From the form of the metric we see that this can be done using the generators (t,r) or (u,v) but that the generators (x,y) are singular on the cone. The patching is done [23] by extending the domain of definition of u for example to negative eigenvalues. The frame θ^i is also singular on the cone but the equivalent frame du^i is quite regular. We can write $\theta^i = \Lambda^i{}_j du^j$ where

$$\Lambda^i{}_j = \sqrt{q}\begin{pmatrix} u^{-1}v & 0 \\ 0 & uv^{-1} \end{pmatrix}$$

is a local Lorentz transformation in the commutative limit.

ACKNOWLEDGMENTS

The authors would like to thank A. Chakrabarti for enlightening conversations. One of them (JM) was supported by the Deutsche Forschungsgemeinschaft and he would like to thank Dieter Lüst for his hospitality at the Institut für Physik, Berlin, were part of this research was carried out. Also MM would like to thank the National Council of Science and Technology (CONACYT, México) for financial support.

REFERENCES

1. Connes, A., *Noncommutative Geometry*, Academic Press, 1994.
2. Madore, J., Kaluza-Klein aspects of noncommutative geometry, in *Differential Geometric Methods in Theoretical Physics*, edited by A. I. Solomon, pages 243–252, World Scientific Publishing, 1989, Chester, August 1988.
3. Dimakis, A., and Madore, J., *J. Math. Phys.*, **37**, 4647 (1996).
4. Connes, A., *J. Math. Phys.*, **36**, 6194 (1995).
5. Fiore, G., and Madore, J., *Euro. Phys. Jour.*, C, (1998).
6. Mourad, J., *Class. and Quant. Grav.*, **12**, 965 (1995).
7. Dubois-Violette, M., Madore, J., Masson, T., and Mourad, J., *J. Math. Phys.*, **37**, 4089 (1996).
8. Cerchiai, B.L., Hinterding, R., Madore, J., and Wess, J., *Euro. Phys. Jour.*, **C8**, 533 (1999).
9. Connes, A., and Lott, J., The metric aspect of non-commutative geometry, in *New Symmetry Principles In Quantum Field Theory*, edited by J. Frölich et al., volume 295 of *NATO Advanced Study Institute Series. B, Physics*, pages 53–93, Plenum Press, New York, 1997, Cargese, July, 1991.
10. Landi, G., *An Introduction to Noncommutative Spaces and their Geometries*, volume 51 of *Lecture Notes in Physics. New Series M, Monographs*, Springer-Verlag, 1997.
11. Figueroa, H., Gracia-Bondía, J. M., and Váilly, J. C., *Elements of Noncommutative Geometry*, Birkhauser Advanced Texts, Birkhäuser Verlag, Basel, 2000.
12. Fichtmüller, M., Lorek, A., and Wess, J., *Z. Physik C - Particles and Fields*, **71**, 533 (1996).
13. Lorek, A., Weich, W., and Wess, J., *Z. Physik C - Particles and Fields*, **76**, 375 (1997).
14. Podleś, P., and Woronowicz, S. L., *Commun. Math. Phys.*, **130**, 381 (1990).
15. Majid, S., *J. Math. Phys.*, **34**, 2045 (1993).
16. Aschieri, P., and Castellani, L., *Int. J. Mod. Phys.*, **A11**, 4513 (1996).
17. Podleś, P., *Commun. Math. Phys.*, **181**, 569 (1996).
18. Aschieri, P., Castellani, L., and Scarfone, A. M., *Euro. Phys. Jour.*, **C7**, 159 (1999).
19. Majid, S., *J. Geom. Phys.*, **30**, 113 (1999).
20. Kosiński, P. , Lukierski, J., and Maślanka, P., *Phys. Rev.*, **D62**, 025004 (2000).
21. Fiore, G., Maceda, M., and Madore, J., (to be published) (2000).
22. Wess, J., and Zumino, B., *Nucl. Phys.*, (Proc. Suppl.) **18B**, 302 (1990).
23. Schmüdgen, K., (to appear) (2000).
24. Hietarinta, J., *J. Math. Phys.*, **34**, 1725 (1993).
25. Gerstenhaber, M., and Giaquinto, A., (to appear) (1997).
26. Chakrabarti, A., (to be pubished) (2000).
27. Faddeev, L. D., Reshetikhin, N. Y., and Takhtajan, L. A., *Lenin. Math. Jour.*, **1**, 193 (1990).
28. Fiore, G., and Madore, J., *J. Geom. Phys.*, **33**, 257 (2000).
29. Cerchiai, B. L., Fiore, G., and Madore, J., *Commun. Math. Phys.*, (to appear) (2000).
30. Cotta-Ramusino, P., and Rinaldi, M., Link-diagrams, Yang-Baxter equation and quantum holonomy, in *Quantum Groups with applications to Physic*, edited by M. Gerstenhaber and J. D. Stasheff, volume 134, pages 19–44, Amer. Math. Soc., Providence, Rhode Island, 1992.
31. Madore, J., *An Introduction to Noncommutative Differential Geometry and its Physical Applications*, Number 257 in London Mathematical Society Lecture Note Series, Cambridge University Press, second edition, 1999.

Regularization and Renormalization of QFT from Noncommutative Geometry

H. Grosse

Boltzmanngasse 5, A-1090 Vienna, Austria, email: grosse@doppler.thp.univie.ac.at

Abstract. We review first regularization methods based on matrix geometry and show how an ultraviolet cut-off for scalar fields respecting symmetries results. Sections of bundles over the sphere can be quantized too. This procedure even allows to regularize supersymmetry without violating it. This work was extended recently to include quantum group covariant regularizations.
In a second part recent attempts to renormalize fourdimensional deformed quantum field theory models is reviewed. For scalar models the well-known IR-UV mixing does not allow to use standard techniques. The same applies to the Yang-Mills model in four dimensions. Only enough symmetry, as it occurs in the Wess-Zumino model, allows to avoid this problem.
Nevertheless there is some hope that the Yang-Mills model can be handled too. We used the Seiberg-Witten map to transform the noncommutative gauge field to a commutative one and used the degree of freedom of this map to obtain counter terms for the renormalization procedure.

INTRODUCTION

Almost ten years ago, I learnt from John Madore what he called the Fuzzy sphere. I was immediately enthusiastic: If you work within a geometry which does not have points, singularities of quantum field theory models may be cured. We applied these ideas first to simple two-dimensional models [1].
Lateron, together with Klimcik and Presnajder we treated all kind of models from scalar fields to spinor and gauge fields, even supersymmetric models were successfully treated. Lateron we dealt with models defined on $CP(2)$, on a cylinder respectively on an hyperboloid. All this was within what is now called the Lie-algebra-deformation.
Very recently, together with John Madore and Harold Steinacker we went to q-deformed models. We obtained a cut-off procedure based on a sequence of embedded Podles spheres and were able to formulate a differential calculus which led to the formulation of gauge models too. This way we connected the Yang-Mills models and the Chern-Simons model to those obtained from string theory.
Despite many attempts four dimensional quantum field theory on commutative space-time is still in bad shape. For two dimensional models constructive methods as well as new algebraic methods led to enormous insights. Even threedimensional models are quite well understood. In four dimensions we have to rely on renormalized perturbation theory, and attempts to cure the diseases go from unification ideas to adding additional dimensions, strings etc. It is natural therefore to ask for alternatives.
One idea is to change the geometry. The final goal would be to include full quantized gravity, but on the way we may well study quantum field theoretic models on quantized

CP589, *Fundamental Interaction Theories,* edited by J. Lukierski and J. Rembieliński

space-time. This led recently to some surprises and new problems have shown up. Nevertheless there is some hope that the Yang-Mills model can be handled and the first steps have been done in joined work with Wulkenhaar and the group at the Technical University of Vienna. Following the Muniche group we used the Seiberg-Witten map to expand the models on noncommutative space-time into a commutative gauge field. The idea is to use the degree of freedom of the SW-map to include the necessary counter terms for loop renormalization. This program is under development.

There is an old and simple argument that a smooth space-time manifold contradict quantum physics. If one localizes an event within a region of extension l an energy of the order hc/l is transfered. This energy generates a gravitational field. A strong gravity field prevents on the other hand signals to reach an observer. If one inserts the energy density to the rhs of Einsteins equation and puts a length r characterizing the curvature of space-time we get

$$1/r^2 = (G/c^4)(hc/l)(1/l^3)\,. \tag{1}$$

Next we put the two length scales to be equal, since it is certainly operational impossible to localize an event beyond this resulting Planck length. To the best of our knowledge the first time this argument was put to precise mathematics in the work by Doplicher, Fredenhagen and Roberts. [2] They obtained what is now called the canonical deformation but averaged over two-spheres. I believe that this averaging will not improve the divergences problems, although it seems that nobody did do explicite calculations.

We used in our work part of the ideas of noncommutative geometry. There are a number of books giving an overview [3], [4], [5]. We replace first the algebra of functions over a manifold by a noncommutative deformed algebra. Three kinds of deformations are treated. If the commutator of coordinates is put equal to a constant antisymmetric tensor, we call it the canonical deformation. If it is put to be a linear function of coordinates we call it a Liealgebra deformation, and if it is put to be a quadratic expression we call it a quantum group deformation. In the last case the Yang-Baxter equation has to be fulfilled.

Next step concerns the differential calculus. We replace vector fields by derivations on the algebra, we define differential forms by duality. Although it is somewhat tricky to do this for the q-deformed sphere, in all three cases one succeeds. Hodge duality, Lie-derivatives, the Laplace operator etc., in summary most of the steps of ordinary differential geometry can be simulated.

An essential step concerns fields. Classically they are sections of bundles, but also modules over the algebra. This last notion can be taken over to the noncommutative situation. We study finitely generated projective modules over the deformed algebra and are able to quantize scalar, spinor and gauge fields.

Finally we write down actions and integrate over the algebra, so we use certain trace functionals. Next steps concerns cohomology problems, the formulation of the spectral triples with the help of a Dirac operator and the use of cyclic cocyles. The final version of Alain Connes ideas concerns the spectral action which allows to unify all the interactions within one principle.

REGULARIZATION OF QUANTUM FIELD THEORY

A) Scalar Fields

The ideas of deforming or quantizing the commutative algebra of functions over a manifold can best be explained for the simple example of the two-sphere and leads to the Fuzzy Sphere. The euclidean action of a scalar field is given by

$$S[\Phi] = \frac{1}{4\pi} \int_{S^2} d\Omega[(J_i\Phi)^2 + \mu^2(\Phi)^2 + POL(\Phi)] \tag{2}$$

and is invariant under isometies or rotations of the sphere. The generators are angular momentum operators, they close under $su(2)$. $\phi(x)$ can be expanded according to the infinite set of irreducible representations of $su(2)$.
Next we truncate this tower: Consider vector spaces transforming according to the first N representations. They can be identified with mappings from the representation space $N/2$ to itself. These mappings are $N+1$ times $N+1$ dimensional matrices, the noncommutative product of these is taken as the product within the algebra. In addition we have to give a precise description of the embedding of these algebras for different N, which gives a precise meaning also to the limit. For details see [6]. The Lie-algebra of the generators of this algebra is easy to describe. They form the $su(2)$ Lie- algebra with suitable rescaling, such that the Casimir operator still fulfills the defining relation for the two-sphere as an operator.

$$[\hat{X}_i, \hat{X}_j] = i\lambda\varepsilon_{ijk}\hat{X}_k \ , \ \sum_{i=1}^{3} \hat{X}_i^2 = R^2 \ , \tag{3}$$

$$R\lambda^{-1} = \sqrt{\frac{N}{2}\left(\frac{N}{2}+1\right)} . \tag{4}$$

Since we work on a matrix algebra it is easy to introduce a differential calculus. All derivations are inner, they are given by the adjoint action with the generators themself. For the two-sphere we take the generators introduced above and can develop a derivation based differetial calculus. Our aim next is to make sense out of a functional integral of the type

$$\langle F[\Phi]\rangle = \frac{\int D\Phi e^{-S[\Phi]} F[\Phi]}{\int D\Phi e^{-S[\Phi]}} . \tag{5}$$

The action (2) will be replaced by

$$S[\phi] = 1/NTr([\hat{X}_i,\phi][\hat{X}_i,\phi] + POL(\phi)) , \tag{6}$$

where everything depends on N. As for the measure we take just the product measure for the finite number of degrees of freedom. This makes the functional integral well-defined. In the limit where N tends to infinity, the old divergences show up. Feynman rules can be developed, a tadpole graph for the ϕ^4 model diverges logarithmically in the limit.
In the same spirit it is possible to quantize Kähler manifolds like $CP(n)$, models on the

cylinder and on hyperboloids have been treated too.
In addition it is possible to quantize sections of line bundles over S^2, which are characterized by the Chern number. This way we construct projective modules which lead in a certain limit to the sections of the line bundles over S^2. We start from the Hopf-fibration of S^3 over S^2 with fiber $U(1)$. Let χ_1 and χ_2 be components of a spinor. Ristrict the sum of the squares of these two complex numbers to be R. This defines the sphere S^3. We study next expansions in terms of these two complex coordinates and their complex conjugates. Define $X_i = \chi^\dagger \sigma_i \chi$. The squares of X_i sum up to R^2 and define therefore S^2. If in an expansion an equal number of χ's and $\chi^\dagger$'s occur, it becomes a well-defined function on S^2. Otherwise it is a section of a bundle.
We next quantize this scheme: Replace the complex quantities by creation and annihilation operators of two bosons:

$$[A_i, A_j^\dagger] = \delta_{ij}, \qquad i,j = 1,2. \tag{7}$$

This means we use the Jordan-Schwinger representation of $su(2)$. Next we define N-dimensional subspaces F_N of the Fockspace given by a fixed number of $N-1$ creation operators. They form an irreducible representation of $su(2)$. We study now maps from one such subspace to another one. If they are of equal dimension, square matrices will map one to the other and the sequence of them forms again the Fuzzy Sphere. If they are of different dimensions and their difference equals twice the topological charge, we quantize in a certain sense sections of the bundles. This way we obtain a sequence of embedded modules formed from nonsquare matrices of special size.
The Dirac operator and spinors were obtained from a supersymmetric treatment. We extended $su(2)$ to the supergroup $osp(2/1)$ and obtained a quantization of superfields through an embedded sequence of graded matrix algebras.

B) The Fuzzy Q-Sphere

The above described Fuzzy Sphere is invariant under the action of $SO(3)$, or equivalently under the action of $U(so(3))$. Together with John Madore and Harold Steinacker we defined a sequence of finite algebras, which have analogous properties, but which are covariant under the quantized universal enveloping algebra $U_q(su(2))$ [7]. This has been done for real q as well as for q being a root of unity. In the later case certain restrictions have to be obeyed. Covariance of an algebra A under $U_q(su(2))$ means that there exists an action

$$U_q(su(2)) \times A \to A, (u,a) \mapsto u \triangleright a, \tag{8}$$

such that $u \triangleright (ab) = (u_{(1)} \triangleright a)(u_{(2)} \triangleright b)$ for $a, b \in A$. Here $\Delta(u) = u_{(1)} \otimes u_{(2)}$ is the Sweedler notation for the coproduct.
We may follow now the undeformed scheme and define q-deformed creation and annihilation operators.

$$A^{\dagger i} A_j = \delta^i_j + q \hat{R}^{ik}_{jl} A_k A^{\dagger l}. \tag{9}$$

Next one considers again N dimensional subspaces of Fock space which form irreducible representations of $U_q(su(2))$. It is possible to define the q-deformed spheres in terms of $Z_i = A^{\dagger\alpha}\varepsilon_{\alpha\beta}\sigma_i^{\beta\gamma}A_\gamma$ by

$$\varepsilon_k^{ij} Z_i Z_j = constant Z_k \tag{10}$$

and to prove that the full matrix algebra is generated and to map the generators from the universal enveloping algebra to the others. After a study of the reality structure we introduced an invariant integral and studied a differential calculus. There exists a unique threedimensional module of 1-forms. As opposed to the classical case, an additional radial one-form shows up. This leads to the addition of a scalar field.
Next it is possible to write down actions for scalar fields as well as for gauge fields. Three possible actions can be formed for the later.

$$S_1 = \int A * A + 2A\Theta, \tag{11}$$

$$S_2 = \int 2A^3 + 3(AdA + A * A) + 6A\Theta + 2\Theta^3, \tag{12}$$

$$S_3 = \int (dA + A^2) * (dA + A^2), \tag{13}$$

were * denotes the Hodge star operation and Θ a special one-form which plays the role of the Maurer-Cartan form. The commutator of algebra elements with this special form gives one-forms.

$$df = [\Theta, f]. \tag{14}$$

As for the step from one-forms to two-forms one has to take the commutator but subtract the Hodge star of the one form in order to get two-forms.

$$d\alpha = [\Theta, \alpha]_+ - *\alpha, \tag{15}$$

where the anticommutator enters. The step to three-forms is given again by the commutator with this special form.
It is interesting to note that special linear combinations of S_i 's correspond to the Yang-Mills action and to the Chern-Simons action. A special combination of both resulted from string theory after taking a particular limit [8].
Very recently we studied second quantization of a field theory on the q-deformed fuzzy sphere for q real. For this case it was necessary to perform a path integral over modes, which generate a quasiassociative algebra. This way we kept the symmetry and obtained a smooth limit for q going to one.

RENORMALIZATION OF DEFORMED QUANTUM FIELD THEORY

A) IR-UV Mixing

Recently after the work of Seiberg and Witten [9] many attempts have been done to study field theory on noncommutative space-time. In analogy to quantum mechanics one assumes that the four dimensional coordinates obey a commutator relation like

$$[x^\mu, x^\nu] = -2i\theta^{\mu\nu}. \tag{16}$$

This leads to the simplest noncommutative space-time and allows explicite calculations. Plane waves $u_p = e^{ip_\mu x^\mu}$ obey the algebra

$$u_p u_q = e^{i\theta^{\mu\nu} p_\mu q_\nu} u_{p+q}, \tag{17}$$

were p and q denote commutative d-dimensional momenta. Derivations on this algebra are defined through the multiplication with these momenta, an integral on this algebra is defined by mapping to the $p = 0$ part. For the noncommutative torus p and q runs over an integer lattice. The later occured through compactification of M-theory on noncommutative 2-tori [10] and corresponds to turning on a background three-form. Superpositions of plane waves with smooth functions $f(x)$ gives the Weyl operators $W[f]$. A standard question is which product of functions allows to encode this algebra. The answer is given by the Moyal-Weyl product

$$W[f]W[g] = W[f * g], (f * g)(x) - e^{i\theta^{\mu\nu}\partial_\mu\partial_\nu} f(x)g(y) \tag{18}$$

with x equal to y.
Since divergences come from singularities due to point like interactions, there was hope that fields on these deformed algebras may be better behaved. Unfortunately divergences still persist. Loop integrals in nc theories differ only by phase factors $e^{i\theta^{\mu\nu} p_\mu q_\nu}$ [11]. For $p = q$, which corresponds to planar diagrams, the same integral with the same divergences occurs as in undeformed models.
Next question concerns renormalizability: It turns out that the Yang-Mills model is one-loop renormalizable, divergences can be absorbed such that Ward-Slavnov identities are fulfilled.
For higher loop contributions a phenomenon similar to the IR-UV mixing of the ϕ^4 theory occurs. Although the one-loop integrals for nonplanar diagrams is finite for generic external momenta, it diverges for zero external momenta. This causes infrared problems even in massive theories. Inserted in higher loop contributions it gives rise to divergences which cannot be absorbed by standard procedures [12]. They would give rise to counter terms which are of different structure as the initial action was. We therefore conclude that the regulating phase becomes inefficient for vanishing momenta, which manifests itself as IR divergences. Higer loop contributions have been analysed by Chepelev and Roiban [13]. They represented Feynman graphs by ribbons, have drawn them on a genus-g Riemann surface with boundary and established

a convergence theorem except for two kind of dangerous cases. So called rings stacked on the same cycle led to singularities (which may be summed-up). Commutants occur if exceptional momenta lead to a vanishing of the oscillating phase factor. This leads to non-local counterterms.
One possible way out concerns additional symmetries. We used a superfield formulation and established a proof, that the Wess-Zumino model does not have the above mentioned disease [14]. Since only logarithmically divergent diagrams occur, and we showed earlier that even iterated integrals of this type are harmless, the standard renormalizability proof can be adapted.

B) Yang-Mills Fields

Recently nc Yang-Mills theory attracted many people. Especially Seiberg and Witten argued that there should exist an equivalence of regularization schemes (point splitting versus Pauli-Villars), and that there should exist a map which relates the nc gauge field with nc gauge parameter to counterparts on ordinary space-time. This map is meant to be given as a formal power expansion series in $\theta^{\mu\nu}$. The Muniche group [15] used these ideas and argued that this is the way to obtain a finite number of degrees of freedom in non-Abelian nc Yang-Mills theories.
If one inserts the Seiberg-Witten map to the Yang-Mills action it leads to a gauge field theory with an infinite number of vertices and Feynman graphs with unbounded degree of divergences, which seems to rule out a perturbative renormalization. An explicite quantum field theoretical investigation using this map for nc Maxwell theory led to the surprising result that the one-loop photon self-energy is gauge-invariant and gauge independent. It was not renormalizable. However, it was possible to absorb the divergences into gauge invariant extension terms added to the classical action involving θ, which was interpreted as coming from a modified scalar product.
It turns out, that the extended action is actually a part of the freedom which exists in the Seiberg-Witten map. It means that a renormalization of constants which are free in that map can be used to absorb the one-loop divergences. This extends to a complete proof of all-order renormalizability of the photon self-energy. This freedom in the Seiberg-Witten map can be regarded as a field redefinition. This short summary is the result of a collaboration with Raimar Wulkenhaar and the group at the Technical University in Vienna consisting of Bichl, Grimstrup, Popp and Schweda [16].
We may discuss the steps in more detail. Denote by $\hat{\lambda}$, $\hat{A}$ and $\hat{F}$ gauge parameter, gauge connection and field strength for the deformed model, and by λ, A and F the appropriate quantities for the undeformed case. For the Seiberg-Witten map one requires that

$$\delta_{\hat{\lambda}[\lambda,A]}\hat{A}[A] = \hat{A}[\delta_\lambda A]\,, \tag{19}$$

which means that the gauge transformed nc field should be equal to the nc field of the gauge transformed commutative one. This requirement leaves a great degree of freedom open. Various methods are avalable to get the mapping from A to $\hat{A}$, or better to get the expansion of the later in terms of an expansion of A. This may be inserted into the

Maxwell action and leads to an action describing the interaction of a gauge field with an external θ field.
Quantization may be done in two ways: One may fix the gauge of the commutative gauge field or of the nc one. Both resulting actions are invariant under abelian BRST transformations and the Slavnov-Taylor idetities are fulfilled. One develops Feynman rules for both cases and evaluates loop corrections. The one-loop corrections to the photon propagator turns out to be independent of this gauge fixing procedure, but clearly the initial action turns out to be incomplete. One has to add four counterterms which were not present originally.
This degree of freedom is compatible with the degree of freedom which occurs in the Seiberg-Witten map. From a general Ansatz we deduce, that the difference beween the actions $S_n[A]$ and $S_n[A']$ for two gauge fields, which differ in n-th order in θ is given by

$$S'_n[A] = S_n[A] + 1/g^2 \int d^4x F^{\mu\nu} D_\nu A^n_\mu, \tag{20}$$

where D_ν denotes the covariant derivative and A^n_μ the difference of the two gauge field configurations. Next one has to show that all counter terms necessary to renormalize the model are of this type. This can be done using the Ward-identities.
In summary we conclude that ideas from noncommutative geometry allow to develop new regularization schemes, which have interesting features; they preserve, for example symmetries. Renormalization problems are still present. But the new models and their study may still lead to surprises and gives hope that finally a consistent quantization scheme of relativistic quantum field theory (with slight modifications) may be found. The final goal, to include quantum gravity too, is still far beyond our ability.

ACKNOWLEDGMENTS

I would like to thank J. Madore, H. Steinacker, R. Wulkenhaar, M. Schweda and his group for enjoyable collaborations and the organizers of this school for invitation.

REFERENCES

1. Grosse, H., and Madore, J., *Phys. Lett.*, **B283**, 218 (1992).
2. Doplicher, S., Fredenhagen, K., and Roberts, J.E., *Commun. Math. Phys.*, **172**, 187 (1995).
3. Connes, A., *Noncommutative Geometry*, Academic Press, San Diego 1994.
4. Madore, J., *An introduction to noncommutative differential geometry and its physical applications*, Cambridge University Press, 1999.
5. Gracia-Bondia, J.M., Varilly, J.C., and Figueria, H., *Elements of noncommutative geometry*, Birkhäuser, Boston 2000.
6. Grosse, H., Klimcik, C., and Presnajder, P., *Int. J. Theor. Phys.* **35**, 231 (1996), *Commun. Math. Phys.* **178**, 507 (1996), *Commun. Math. Phys.* **185**, 155 (1997).
7. Grosse, H., Madore, J., and Steinacker, H., *Field Theory on the q-deformed Fuzzy Sphere I,II* hep-th/0005273, hep-th/0103164.
8. Alekseev, A.Yu., Recknagel, A., and Schomerus, V., *JHEP*, **9909**, 023 (1999); hep-th/0003187.
9. Seiberg, N., and Witten, E., *JHEP*, **9909**, 032 (1999); hep-th/9908142.

10. Connes, A., Douglas, M.R., and Schwarz, A., *JHEP*, **9802**, 003 (1998); hep-th/9711162.
11. Filk, T., *Phys. Lett* **B376**, 53 (1996).
12. Minwalla, S., Van Raamsdonk, M., and Seiberg, N., *JHEP*, **0002**, 002 (2000); hep-th 9912072.
13. Chepelev, I., and Roiban, R., *JHEP*, **0103**, 001 (2001); hep-th/0008090.
14. Bichl, A., Grimstrup, J., Popp, L., Schweda, M., Grosse, H., and Wulkenhaar, R., *JHEP*, **10**, 046 (2000); hep-th/0007050.
15. Jurco, B., et al., hep-th/006246; hep-th/0102129; hep-th/0104153.
16. Bichl, A., Grimstrup, J., Grosse, H., Popp, L., Schweda, M., and Wulkenhaar, R., *Renormalization of the noncommutative photon self-energy to all orders via Seiberg-Witten map*; hep-th/0104097.

Modified Basis and Quantum $\mathcal{R}$-Matrices Corresponding to Belavin-Drinfeld Triples

A.P. Isaev* and O.V. Ogievetsky†

*Bogoliubov Laboratory of Theoretical Physics, JINR,
141980 Dubna, Moscow region, Russia, email: isaevap@thsun1.jinr.ru
†Center of Theoretical Physics, Luminy, 13288 Marseille, France
and
P. N. Lebedev Physical Institute, Theoretical Department, Leninsky Pr. 53, 117924 Moscow, Russia

Abstract. We intrepret a set of data defining a Belavin - Drinfeld triple in terms of some suitable basis of the universal enveloping algebra. A formula for a universal $\mathcal{R}$-matrix, corresponding to any Belavin - Drinfeld triple is given with the use of this modified basis. Several simple examples are considered explicitly.

INTRODUCTION

Let $\mathcal{R}$ be the standard universal Drinfeld-Jimbo R-matrix for the Lie algebra $\mathbf{g}$. This R-matrix satisfies the Yang-Baxter equation

$$\mathcal{R}_{12}\,\mathcal{R}_{13}\,\mathcal{R}_{23} = \mathcal{R}_{23}\,\mathcal{R}_{13}\,\mathcal{R}_{12}\ . \tag{1}$$

One can show [1] that an operator

$$\tilde{\mathcal{R}}_{12} = F_{21}\,\mathcal{R}_{12}\,F_{12}^{-1}\ , \tag{2}$$

satisfies the Yang-Baxter equation (1) as well, if the twisting operator F_{12} satisfies the cocycle equation

$$F_{12}\,(\Delta\otimes id)\,F = F_{23}\,(id\otimes\Delta)\,F\ . \tag{3}$$

Our presentation is based on [6]. We give a formula for the twisting operator F_{12} for an arbitrary Belavin-Drinfeld triple. These triples have been proposed in [2] to classify classical quasitriangular r-matrices for semisimple Lie algebras. The Belavin-Drinfeld triple (Γ_1,Γ_2,τ) for a simple Lie algebra $\mathbf{g}=\mathbf{g}^+\oplus\mathbf{h}\oplus\mathbf{g}^-$ consists of the following data: Γ_1,Γ_2 are subsets of the set Γ of simple roots of the algebra $\mathbf{g}$ and τ is a one-to-one mapping: $\Gamma_1\to\Gamma_2$ such that $<\tau(\alpha),\tau(\beta)>=<\alpha,\beta>$ for any $\alpha,\beta\in\Gamma_1$ and, for any natural k, $\tau^k(\alpha)\neq\alpha$ whenever $\tau^k(\alpha)$ is defined. The largest k for which there exists an α for which $\tau^k(\alpha)$ is defined, is called the degree of the triple.

Using these data we represent the twisting operator F_{12} in a factorized form

$$F_{12} = F_{12}^{(N)}\cdot F_{12}^{(N-1)}\cdots F_{12}^{(2)}\cdot F_{12}^{(1)}\cdot K\ , \tag{4}$$

CP589, *Fundamental Interaction Theories,* edited by J. Lukierski and J. Rembieliński

where the factors $F^{(k)}$ are special canonical elements defined by the powers of the one-to-one map τ; the operator K belongs to $q^{\mathbf{h}\otimes\mathbf{h}}$.

The main role in our approach is played by a certain modified Cartan-Weyl basis for $U_q(\mathbf{g})$.

The modified generators, corresponding to simple roots are introduced in Section 2. In Section 3 we give an interpretation of Belavin-Drinfeld triples in terms of the modified basis and introduce a modified Cartan-Weyl basis. Section 4 contains a construction of the twisting operator F_{12}. The simplest examples are presented in Section 5.

We assume that the deformation parameter q is not a root of unity.

MODIFIED BASIS FOR QUANTUM UNIVERSAL ENVELOPING ALGEBRAS.

Consider a quantum universal enveloping algebra $U_q(\mathbf{g})$ with relations (see e.g. [4])

$$[h_i,h_j]=0\,,\ \ [h_i,e_j]=a_{ij}e_j\,,\ \ [h_i,f_j]=-a_{ij}f_j\,,$$

$$[e_i,f_j]=\delta_{ij}\frac{K_i-K_i^{-1}}{q^{d_i}-q^{-d_i}}\,, \qquad (5)$$

and Serre relations (Serre relations for generators f_i are the same as for generators e_i)

$$\sum_{k=0}^{1-a_{ij}}(-1)^k\left[\begin{array}{c}1-a_{ij}\\ k\end{array}\right]_{q^{d_i}}(e_i)^k e_j (e_i)^{1-a_{ij}-k}=0\,, \qquad (6)$$

where

$$\left[\begin{array}{c}n\\ k\end{array}\right]_q=\frac{[n]_q!}{[k]_q!\,[n-k]_q!}\,,\quad [k]_q=\frac{q^k-q^{-k}}{q-q^{-1}}\,,$$

a_{ij} is the Cartan matrix for $\mathbf{g}$, $K_i=q^{d_ih_i}$ and d_i are smallest positive integers such that $d_ia_{ij}=a^{(s)}_{ij}$ is a symmetric matrix. The algebra $U_q(\mathbf{g})$ is a Hopf algebra with the comultiplication

$$\Delta(h_i)=h_i\otimes 1+1\otimes h_i\,,\ \ \Delta(e_i)=e_i\otimes K_i+1\otimes e_i\,,\ \ \Delta(f_i)=f_i\otimes 1+K_i^{-1}\otimes f_i\,. \qquad (7)$$

Any operator $K\in q^{\mathbf{h}\otimes\mathbf{h}}$,

$$K=q^{(\sum_{ij}b_{ij}h_i\otimes h_j)}\,, \qquad (8)$$

with an arbitrary numerical matrix b_{ij}, obviously satisfies the cocycle equation (3),

$$K_{12}\,(\Delta\otimes id)\,K=K_{23}\,(id\otimes\Delta)\,K\,, \qquad (9)$$

and, therefore, one can twist the comultiplication by K:

$$\widetilde{\Delta}(a):=K\,\Delta(a)\,K^{-1}\,. \qquad (10)$$

We change the basis in the algebra $U_q(\mathbf{g})$ by introducing new generators

$$E_i = X_i e_i\,, \quad F_i = f_i Y_i\,, \tag{11}$$

where $X_i \in q^{\mathbf{h}}$, $Y_i \in q^{\mathbf{h}}$. The new generators (11) are related with the generators e_i and f_i by

$$K\,(e_i \otimes f_i)\,K^{-1} = (E_i \otimes F_i)\,. \tag{12}$$

Eqn. (12) implies that, up to multiplicative constants $\{\mu_i\}$ ($X_i \mapsto \mu_i X_i$, $Y_i \mapsto \mu_i^{-1} Y_i$), one has

$$X_i = q^{-(hba)_i}\,, \quad Y_i = q^{(h\bar{b}a)_i}\,, \tag{13}$$

where $(hba)_i = \sum_{mn}(h_m b_{mn} a_{ni})$ and $\bar{b}_{mn} = b_{nm}$ is the transposed matrix.

The comultiplication (10) for the new generators (11), (13) is:

$$\begin{aligned} \tilde{\Delta}(E_i) &= K\,\Delta(E_i)\,K^{-1} = E_i \otimes R_i^{+} + 1 \otimes E_i\,, \\ \tilde{\Delta}(F_i) &= K\,\Delta(F_i)\,K^{-1} = F_i \otimes 1 + R_i^{-} \otimes F_i\,. \end{aligned} \tag{14}$$

where

$$R_i^{\pm} = K_i^{\pm 1}\, q^{-(h(b-\bar{b})a)_i}\,, \quad R_i^{+} = K_i^2 R_i^{-}\,. \tag{15}$$

The relations (5) and the Serre relations (6) (the Serre relations in the new basis are different for E_i and F_i) for the quantum algebra $U_q(g)$ in terms of the new generators (11) take the form

$$[E_i, F_j] = \delta_{ij}\frac{R_i^{+} - R_i^{-}}{q^{d_i} - q^{-d_i}}\,, \tag{16}$$

$$R_i^{\pm} E_j = q^{\pm a_{ij}^{(s)} + A_{ij}} E_j R_i^{\pm}\,, \quad R_i^{\pm} F_j = q^{\mp a_{ij}^{(s)} - A_{ij}} F_j R_i^{\pm}\,, \tag{17}$$

$$\sum_{k=0}^{1-a_{ij}} (-1)^k \begin{bmatrix} 1-a_{ij} \\ k \end{bmatrix}_{q^{d_i}} q^{-kA_{ij}} (E_i)^k E_j (E_i)^{1-a_{ij}-k} = 0\,, \tag{18}$$

$$\sum_{k=0}^{1-a_{ij}} (-1)^k \begin{bmatrix} 1-a_{ij} \\ k \end{bmatrix}_{q^{d_i}} q^{kA_{ij}} (F_i)^k F_j (F_i)^{1-a_{ij}-k} = 0\,, \tag{19}$$

with a skewsymmetric matrix $A_{ij} = (\bar{a}(b-\bar{b})a)_{ij}$.

MODIFIED BASIS AND BELAVIN - DRINFELD TRIPLES

The data from the Belavin-Drinfeld triple can be conveniently interpreted in terms of the modified basis for a suitable matrix b_{ij} [6]:

Proposition. *Let Γ be the set of simple roots of* $\mathbf{g}$, *Γ_1 and Γ_2 subsets of Γ and τ a one-to-one mapping: $\Gamma_1 \to \Gamma_2$. Then the following equations for the matrix b_{ij}*

$$R_{\alpha_i}^{+} = R_{\tau(\alpha_i)}^{-} \;\; \forall\, \alpha_i \in \Gamma_1 \tag{20}$$

where $R^{\pm}_{\alpha_i} \equiv R^{\pm}_i$, *admit a solution if and only if the triple* $(\Gamma_1, \Gamma_2, \tau)$ *is the Belavin-Drinfeld triple.*

Proof. See [6].

Remark 1. Commuting both sides of eq. (20) with e_m (or f_m) one finds that eq. (20) is equivalent (here it is important that q is not a root of unity) to the following condition on the skewsymmetric matrix $A_{mn} = (\overline{a}(b-\overline{b})a)_{mn}$

$$A_{im} + A_{m\tau(i)} + a^{(s)}_{im} + a^{(s)}_{\tau(i)m} = 0\,, \tag{21}$$

where the subscript m runs over all simple roots while i numerates only roots from Γ_1. Eq. (21) implies

$$A_{im} = A_{\tau(i)\tau(m)}\,, \tag{22}$$

and we conclude that the map τ does not change the defining relations (16) – (19) for the modified basis.

Denote by Δ_+ the system of all positive roots of $\mathbf{g}$ with respect to Γ. A construction of a Cartan-Weyl basis in terms of the modified generators E_i and F_i repeats the usual procedure for $U_q(\mathbf{g})$ (see [7]).

Recall the notion of a normal (convex) order in Δ_+: the set Δ_+ is ordered normally if any root γ which is a sum of roots α and β is placed between α and β.

We write $\alpha < \beta$ if the root α is located to the left of the root β. For $\alpha < \beta$, the interval between roots α and β is denoted by $\{\alpha, \beta\}$.

Given a normal order in Δ_+, the modified Cartan-Weyl basis is constructed by the following inductive procedure. The generators for the simple roots are already defined. For a composite root γ, take a minimal interval $\{\alpha, \beta\}$, $\alpha < \beta$, with $\gamma = \alpha + \beta$ ("minimal" means that there is no subinterval $\{\widetilde{\alpha}, \widetilde{\beta}\} \subset \{\alpha, \beta\}$ for which $\gamma = \widetilde{\alpha} + \widetilde{\beta}$). Assume that generators E_α, E_β, F_α and F_β were defined at previous steps. Then generators E_γ and F_γ are defined by

$$\begin{aligned} E_\gamma &= [E_\alpha, E_\beta]_\mu = E_\alpha E_\beta - \mu E_\beta E_\alpha\,, \\ F_\gamma &= [F_\alpha, F_\beta]_\nu = F_\alpha F_\beta - \nu F_\beta F_\alpha\,. \end{aligned} \tag{23}$$

where

$$\mu = q^{-<\alpha,\beta>+<\alpha,A\beta>}\,, \quad \nu = q^{-<\alpha,\beta>-<\alpha,A\beta>}$$

and A is the operator with the matrix A_{ij}:

$$<\alpha_i, A\alpha_j> = A_{ij}\,.$$

If there are several possible minimal intervals $\{\alpha, \beta\}$ for which $\gamma = \alpha + \beta$, the definitions (23) give proportional results.

Remark 2. For the case $A_{ij} = 0$ the definition (23) of composite roots does not coincide with the definition in [7] since we use the comultiplication (7) which is different from the comultiplication in [7].

TWISTING OPERATORS F_{12} FOR BELAVIN-DRINFELD TRIPLES

For a given simple Lie algebra **g** fix a normal order in Δ_+.

We need the expression for the inverse of the universal R- matrix for the algebra $U_q(\mathbf{g})$:

$$\mathcal{R}^{-1} = \vec{\prod}_{\beta\in\Delta_+} \exp_{q_\beta}\left(-\lambda a_\beta\left(e_\beta \otimes f_\beta\right)\right) \cdot K^{(0)}\,, \tag{24}$$

where $q_\alpha = q^{<\alpha,\alpha>}$, $\lambda = q - q^{-1}$ and $K^{(0)} \in q^{\mathbf{h}\otimes\mathbf{h}}$. The product in eq. (24) is the ordered product corresponding to the chosen normal order of roots. For precise values of the constants a_β see [8], [7]. The function $\exp_q$ is the standard q-exponent,

$$\exp_q(u) = \sum_{k=0}^{\infty} \frac{u^k}{k_q!}\,, \quad k_q = \frac{q^k - 1}{q-1}\,; \tag{25}$$

it satisfies an equation: $\exp_q(uq) = (1 + (q-1)u)\exp_q(u)$.

Let $(\Gamma_1, \Gamma_2, \tau)$ be a Belavin-Drinfeld triple of degree N. Define elements $F^{(k)}$ by

$$F_{12}^{(k)} = \vec{\prod}_{\beta\in\Delta_+^{(k)}} \exp_{q_\beta}\left(-\lambda a_\beta\left(E_\beta \otimes F_{\tau^k(\beta)}\right)\right)\,, \tag{26}$$

where in the ordered product we keep terms corresponding to only those roots β for which $\tau^k(\beta)$ is defined. This is reflected in the notation $\beta \in \Delta_+^{(k)}$.

The expression (26) can be written in the form

$$F_{12}^{(k)} = (1 \otimes T^k)\left(K\mathcal{R}^{-1}(K^{(0)})^{-1}K^{-1}\right)\,, \tag{27}$$

where eqs. (12) have been used and the action of the operator T on the elements F_β is defined by $T(F_\beta) = F_{\tau(\beta)}$ wherever $\tau(\beta)$ is defined; $T(F_\beta) = 0$ otherwise. The operator K corresponds to the solution of eqs. (20) for a given Belavin Drinfeld triple.

Theorem. *For the quantum algebra $U_q(\mathbf{g})$ and the Belavin-Drinfeld triple $(\Gamma_1, \Gamma_2, \tau)$ of degree N the universal twisting element F_{12} is*

$$F_{12} = F_{12}^{(N)} \cdot F_{12}^{(N-1)} \cdots F_{12}^{(2)} \cdot F_{12}^{(1)} \cdot K \equiv \widetilde{F}_{12} \cdot K \tag{28}$$

with the factors $F_{12}^{(k)}$ defined in (26).

See [6] for the sketch of the proof of the Theorem.

EXAMPLES

i) $U_q(sl(3))$ case.

Here we have only one nontrivial Belavin-Drinfeld triple:

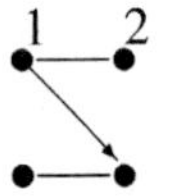

Fig.1

This Cremmer-Gervais type triple has degree 1 and the basic relations (20) which define this triple are reduced to one equation $R_1^+ = R_2^-$. The antisymmetric matrix A_{ij} is

$$A_{ij} = \delta_{i,j+1} - \delta_{j,i+1} \implies (b-\bar{b})_{ij} = \frac{1}{3} A_{ij} \,, \tag{29}$$

with $1 \leq i,j \leq 2$. The corresponding universal twisting element (28) has the form

$$F_{12} = F_{12}^{(1)} \cdot K = \exp_{q^2}(-\lambda E_1 \otimes F_2) \cdot K \,, \tag{30}$$

$$K = q^{[\frac{1}{6}(h_2 \otimes h_1 - h_1 \otimes h_2) + \frac{1}{2}(b+\bar{b})_{ij} h_i \otimes h_j]} \,.$$

Note that the parameter λ in (30) can be made arbitrary by a rescaling of the generators E_i and F_i. The formulas which are analogous to (30) have been considered also in [5], [10].

The remarkable fact is that one can find the different solution for the twisting operator F_{12} (for the Belavin-Drinfeld triple presented in Fig.1):

$$F_{12} = q^{\sum_{i,j} c_{ij} h_i \otimes h_j} \exp_{q^{-2}}(\lambda f_2 \otimes e_1) \,, \tag{31}$$

where e_i and f_i form the usual basis for $U_q(sl(3))$ (5), (6). One can check that F_{12} given by (31) satisfies the cocycle condition (3) for the following matrix c_{ij}:

$$c_{ij} = \begin{pmatrix} a & a \\ a+1/3 & a \end{pmatrix} \implies c_{ij} - c_{ji} = \frac{1}{3} A_{ij} \,, \tag{32}$$

where a is an arbitrary parameter. This twist (which is an analogue of the twist proposed in [3]) is represented in the form $\tilde{K} \cdot \Phi(f_i \otimes e_j)$ while our twist operator (30) (which is written in terms of the modified basis) has the form $\tilde{\Phi}(X_i e_i \otimes f_j Y_j) \cdot K$ where Φ and $\tilde{\Phi}$ are some functions. It is clear that R-matrix (2) in the limit $q \to 1$ gives the same classical r-matrix for both twist operators (30) and (31), but a coincidence of the quantum R-matrices (2) for these twists is not evident.

ii) Cremmer-Gervais $U_q(sl(4))$ case.
For this case the triple is given by the following diagram

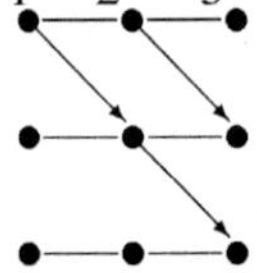

Fig.2

It has degree 2. The basic relations (20) which define this triple are $R_1^+ = R_2^-$, $R_2^+ = R_3^-$. The matrix A_{ij} is given by (29), now with $1 \leq i, j \leq 3$. The corresponding universal twisting element (28) has the form

$$F_{12} = F_{12}^{(2)} \cdot F_{12}^{(1)} \cdot K \,, \tag{33}$$

where

$$F_{12}^{(2)} = \exp_{q^2}(-\lambda E_1 \otimes F_3) \,, \tag{34}$$

$$F_{12}^{(1)} = \exp_{q^2}(-\lambda E_1 \otimes F_2) \exp_{q^2}(q^{-1}\lambda [E_{12}] \otimes [F_{23}]_{q^2}) \exp_{q^2}(-\lambda E_2 \otimes F_3) \,. \tag{35}$$

Here $[E_{12}] = E_1 E_2 - E_2 E_1$ and $[F_{23}]_{q^2} = F_2 F_3 - q^2 F_3 F_2$. The generalization of the formulas (33) – (35) for the Cremer-Gervais $U_q(sl(n))$- case (for $n \geq 5$) is obvious.

Remark. One can directly check that (30), (31) and (33) obeys the cocycle conditions (3). For (30) and (31) this check requires only the basic equation for the q-exponent, $\exp_q(y)\exp_q(x) = \exp_q(x+y)$ if $xy = qyx$. For (33) one needs two more quantum identities. The first one is the famous pentagon identity (see e.g. [11] and references therein)

$$\exp_q(u)\exp_q(v) = \exp_q(v)\exp_q([u,v])\exp_q(u) \,, \tag{36}$$

where the operators u and v satisfy the commutation (Serre) relations

$$u[u,v] = q[u,v]u \,, \quad v[u,v] = q^{-1}[u,v]v \,.$$

The second identity is

$$\exp_{q^2}(E)\exp_{q^2}(-R^+)\exp_{q^2}(F) = \exp_{q^2}(F)\exp_{q^2}(-R^-)\exp_{q^2}(E) \,, \tag{37}$$

where E, F and $R^{\pm}$ generate the algebra: $[R^+, R^-] = 0$ and

$$[E,F] = (R^+ - R^-) \,, \quad R^{\pm} E = q^{\pm 2} E R^{\pm} \,, \quad R^{\pm} F - q^{\mp 2} F R^{\pm} \,.$$

iv) $U_q(sl(2) \oplus sl(2))$ case.
As the last example we consider the case of semisimple but not simple Lie algebra $sl(2) \oplus sl(2)$. The defining relations for this algebra can be written in the form (5), with the "Cartan" matrix $a_{ij} = 2\delta_{ij}$. The Serre relations are trivial: $[e_1, e_2] = 0 = [f_1, f_2]$. Consider a triple corresponding to a diagram

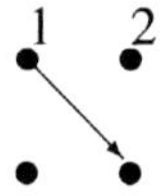

Fig.3

The condition $R_1^+ = R_2^-$ gives

$$A_{ij} = 2(\delta_{i,j+1} - \delta_{j,i+1}) \implies (b - \bar{b})_{ij} = \frac{1}{2} A_{ij} \,,$$

The universal twisting element (28) which is related to the Fig.3 is represented in the form similar (30):

$$F_{12} = F_{12}^{(1)} \cdot K = \exp_{q^2}(-\lambda E_1 \otimes F_2) \cdot K \,, \tag{38}$$

$$K = q^{\left[\frac{1}{4}(h_2 \otimes h_1 - h_1 \otimes h_2) + \frac{1}{2}(b+\bar{b})_{ij} h_i \otimes h_j\right]} \,.$$

ACKNOWLEDGMENTS

We are indebted to V. Fock, S. Khoroshkin, P. Pyatov and V. Tolstoy for valuable discussions and comments. The work was supported in part by the RFBR grant No. 00-01-00299 and the CNRS grant PICS-608.

REFERENCES

1. Drinfeld, V. G. *Leningrad Math. J.*, **1**, 1419–1457(1990).
2. Belavin, A. A., and Drinfeld, V. G., *Sov. Sci. Rev.*, **C4**, 93–166 (1984).
3. Etingof, P., Schedler, T., and Schiffmann, O., *Explicit quantization for dynamical r-matrices for finite dimensional semisimple Lie algebras*, (1999), math.QA/9912009.
4. Chari, V., and Pressley, A., *A Guide to Quantum Groups*, Cambridge University Press (1994).
5. Hodges, T., *Nonstandard quantum groups associated to certain Belavin-Drinfeld triples*, (1996), q-alg/9609029.
6. Isaev, A.P., and Ogievetsky, O., *On quantization of r-matrices for Belavin-Drinfeld triples*, (2000), math.QA/0010190.
7. Khoroshkin, S.M., and Tolstoy, V.N., *Commun. Math. Phys.*, **141**, 599 (1991); *Lett. Math. Phys.*, **24**, 231 (1992); hep-th/9404036.
8. Rosso, M., *Commun. Math. Phys.*, **124**, 307–318 (1989).
9. Arnaudon, D., Buffenoir, E., Ragoucy, E., and Roche, Ph., *Lett. Math. Phys.*, **44**, 201–214 (1998); q-alg/9712037.
10. Kulish, P.P., and Mudrov, A. I., *Lett. Math. Phys.*, **47**, 139–148 (1999), q-alg/9804006.
11. Faddeev, L.D., and Kashaev, R.M., *Mod. Phys. Lett*, **A9**, 427 (1994), hep-th/9310070; Kashaev, R.M., *Heisenberg Double and Pentagon Relation*, q-alg/9503005.

Noncommutative Gauge Theories and Kontsevich's Formality Theorem

B. Jurčo[1*], P. Schupp* and J. Wess*

*Universität München, Theoretische Physik, Theresienstr. 37, 80333 München, Germany

Abstract. The equivalence of star products that arise from the background field with and without fluctuations and Kontsevich's formality theorem allow an explicitly construction of a map that relates ordinary gauge theory and noncommutative gauge theory (Seiberg-Witten map.) Using noncommutative extra dimensions the construction is extended to noncommutative nonabelian gauge theory for arbitrary gauge groups; as a byproduct we obtain a "Mini Seiberg-Witten map" that explicitly relates ordinary abelian and nonabelian gauge fields. All constructions are also valid for non-constant B-field, and even more generally for any Poisson tensor.

INTRODUCTION

The topic of this contribution is the type of noncommutative gauge theory that arises in string theory [1]. Let us briefly recall how noncommutative gauge theory arises in string theory: Consider an open string σ-model with background term

$$S_{B+da} = \frac{1}{2i}\int_D B_{ij}\varepsilon^{ab}\partial_a x^i \partial_b x^j - i\int_{\partial D} d\tau a_i(x(\tau))\partial_\tau x^i(\tau)\,, \tag{1}$$

where the constant nondegenerate background field B has been perturbed by adding a gauge potential $a_i(x)$. Classically we have the naive gauge invariance $\delta a_i = \partial_i\lambda$, but in quantum theory this depends on the choice of regularization: Pauli-Villars preserves the classical symmetry but point-splitting leads to noncommutative gauge transformations $\hat{\delta}\hat{A}_i = \partial_i\hat{\lambda} + i\hat{\lambda}\star\hat{A}_i - i\hat{A}_i\star\hat{\lambda}$. Since a sensible quantum theory should be independent of the choice of regularization there should be field redefinitions $\hat{A}(a)$, $\hat{\lambda}(a,\lambda)$ (Seiberg-Witten maps) that map the classical to noncommutative transformations:

$$\hat{A}(a) + \hat{\delta}_{\hat{\lambda}}\hat{A}(a) = \hat{A}(a+\delta_\lambda a)\,. \tag{2}$$

We shall discuss: (i) a direct construction of the maps $\hat{A}(a)$, $\hat{\lambda}(a,\lambda)$ to all orders in the deformation parameter, that works for any $\star$-product and is covariant under general coordinate transformations, (ii) a suitably general definition of noncommutative gauge theory in the present context, (iii) a generalization to arbitrary nonabelian gauge groups, and (iv) a mini Seiberg-Witten map from abelian to nonabelian gauge theory. See [8]

---

[1] email: jurco@theorie.physik.uni-muenchen.de

CP589, *Fundamental Interaction Theories,* edited by J. Lukierski and J. Rembieliński

for more details. In related work a path integral approach [9] and operator methods [10] have been discussed.

KONTSEVICH FORMULA AFTER CATTANEO AND FELDER

Cattaneo and Felder [2] have given an explicit path-integral formula for the star product of Kontsevich for an arbitrary Poisson structure on an open subset of $\mathbb{R}^n$. In the symplectic case (non-degenerate Poisson tensor) their formula reduces to the ordinary quantum mechanical Feynman path-integral.

The path-integral representation of an $\star$-product $f \star g$ of two functions f and g on $\mathbb{R}^n$ is described as follows. Let $X^i(u)$, $i = 1, ..., n$ be bosonic fields on the 2-dim disc D. We also need auxiliary bosonic fields η_{ia}, $i = 1, ..., n$ and $a = 1, 2$. Components of the Poisson tensor on $\mathbb{R}^n$ will be denoted by θ^{ij}. Then

$$(f \star g)(x) = \int \mathcal{D}X\, \mathcal{D}\eta\, f(X(0))g(X(1))e^{\frac{i}{\hbar}S[X,\eta]}, \tag{3}$$

with action

$$\begin{aligned} S[X,\eta] &= \int_D \eta_i(u) \wedge dX^i(u) + \frac{1}{2}\theta^{ij}(X(u))\eta_i(u) \wedge \eta_j(u) \\ &= \int_D d^2u\, \varepsilon^{ab} \left(\eta_{ia}\partial_b X^i + \frac{1}{2}\theta^{ij}(X)\eta_{ia}\eta_{jb} \right), \end{aligned} \tag{4}$$

where $\eta_i \equiv \eta_{ia}du^a$. The boundary conditions are as follows: $X(\infty) = x$ and for each i the 2-vector η_{ia} must be perpendicular to the boundary of the disc; 0, 1, and ∞ are three consecutive points on the boundary of the disc. The model described by this action is called Poisson sigma model [3]

To evaluate the path integral one has to take gauge fixing and renormalization into account. The star product $f \star g$ is given by the semiclassical expansion around the particular classical solution $X(u) = x$ and $\eta(u) = 0$. Associativity of the star product is a direct consequence of the topological nature of the Poisson sigma model.

The Poisson sigma model action

$$S[X,\eta] = \int_D \eta_i(u) \wedge dX^i(u) + \frac{1}{2}\theta^{ij}(X(u))\eta_i(u) \wedge \eta_j(u), \tag{5}$$

is invariant under the following infinitesimal gauge transformations

$$\delta_\beta X^i = \theta^{ij}\beta_j, \tag{6}$$

$$\delta_\beta \eta_{ia} = -\partial_a\beta_i - \partial_i\theta^{jk}\eta_{ja}\beta_k, \tag{7}$$

with parameter β_i vanishing on the boundary of the disc. Direct computation of the commutators of gauge transformations reveals that the structure constants of the gauge group depend explicitly on the fields X^i and furthermore the algebra of gauge transformations closes only on shell. This precludes the usual BRST approach to quantization and one has to resort to the more general Batalin-Vilkovisky (BV) quantization prescription.

Within this approach it makes sense to associate with any collection of multivector fields α_i on $\mathbb{R}^n$ and any collection of fuctions f_k on $\mathbb{R}^n$ some correlation function $U_n(\alpha_1,...,\alpha_n)(f_1,...,f_m)$. The Ward identities for such correlation function express what is undertsood under the name formality. The star product is an example

$$g\star h=\sum\frac{(i\hbar)^n}{n!}U_n(\theta,\ldots,\theta)(g,h)\,. \tag{8}$$

COVARIANT COORDINATES AND NONCOMMUTATIVE GAUGE FIELDS

In the point-splitting prescription the perturbation da effectively leads to a shift of coordinates: $x^i \to x^i + \theta^{ij}\hat{A}_j$. The shifted coordinates are covariant under noncommutative gauge transformations. Covariant coordinates and functions can be introduced more abstractly: Let $\mathcal{A}_x$ be an associative algebra "noncommutative space" with product $\star$. The gauge transformation of a field $\psi \in$ (left module of) $\mathcal{A}_x$ is $\hat{\delta}\psi = i\hat{\lambda}\star\psi$, where the gauge parameter $\hat{\lambda} \in \mathcal{A}_x$ is an arbitrary function on the noncommutative space. (In reference to their commutative analog we shall call the elements of $\mathcal{A}_x$ functions). Since the product of a function and a field is not covariant on a noncommutative space, $\hat{\delta}(f\star\psi) = f\star\hat{\delta}\psi = f\star(i\hat{\lambda}\star\psi) \neq i\hat{\lambda}\star(f\star\psi)$, one needs to introduce covariant functions $\mathcal{D}f = f + \mathcal{A}(f)$, where $\mathcal{A} \in \mathrm{Hom}(\mathcal{A}_x,\mathcal{A}_x)$ transforms such that $\hat{\delta}(\mathcal{D}f\star\psi) = i\hat{\lambda}\star(\mathcal{D}f\star\psi)$, i.e., $\hat{\delta}\mathcal{A}(f) = [i\hat{\lambda} \stackrel{\star}{,} f] + [i\hat{\lambda} \stackrel{\star}{,} \mathcal{A}(f)]$. Other covariant objects include the generalized field strength $\mathcal{F}(f,g) = [\mathcal{D}f \stackrel{\star}{,} \mathcal{D}g] - \mathcal{D}([f \stackrel{\star}{,} g])$: it maps two functions to a new function that depends on the gauge potential and transforms covariantly; $\mathcal{F} \in \mathrm{Hom}(\mathcal{A}_x{}^{\wedge 2},\mathcal{A}_x)$.

For constant θ one recovers the standard expressions [1]. The abstract objects $\mathcal{A}$ and $\mathcal{D}$ are important to understand covariance under general coordinate transformations and are e.g. needed to construct covariant versions of scalar fields $\phi(x)$.

NONCOMMUTATIVE GAUGE FIELDS ABSTRACTLY

Finite projective modules take the place of fiber bundles in the noncommutative realm. This is also the case here, but may not have been apparent since we have been working with component fields. As we have argued, $\mathcal{A} \in C^1$, $\mathcal{F} \in C^2$ with $C^p = \mathrm{Hom}(\mathcal{A}_x{}^{\wedge p},\mathcal{A}_x)$, $C^0 \equiv \mathcal{A}_x$. These p-cochains take the place of forms on a noncommutative space, the wedge product is replaced by a cup product $\wedge : C^p \otimes_{\mathcal{A}_x} C^q \to C^{p+q}$ and the exterior differential is replaced by the standard coboundary operator $\mathbf{d}_\star$ in the Lie-algebra cohomology; $\mathbf{d}_\star^2 = 0$, $\mathbf{d}_\star 1 = 0$. We can introduce a connection on a (finite) projective right $\mathcal{A}_x$-module $\mathcal{E}$ as a linear map $\nabla : \mathcal{E}\otimes_{\mathcal{A}_x} C^p \to \mathcal{E}\otimes_{\mathcal{A}_x} C^{p+1}$ which satisfies the Leibniz rule

$$\nabla(\eta\psi) = (\tilde{\nabla}\eta)\psi + (-)^p\eta\,\tilde{\mathbf{d}}_\star\psi, \tag{9}$$

for all $\eta \in \mathcal{E}\otimes_{\mathcal{A}_x} C^p$, $\psi \in C^r$, and where $\tilde{\nabla}\eta = \nabla\eta - (-)^p\eta\,\tilde{\mathbf{d}}_\star 1$, $\tilde{\mathbf{d}}_\star(a\wedge\psi) = (\mathbf{d}_\star a)\wedge\psi + (-)^q a\wedge(\tilde{\mathbf{d}}_\star\psi)$ for all $a \in C^q$, and $\tilde{\mathbf{d}}_\star 1$ is the identity operator on $\mathcal{A}_x$. Let (η_a) be a generating family for $\mathcal{E}$; any $\xi \in \mathcal{E}$ can then be written as $\xi = \sum \eta_a \psi^a$ with $\psi^a \in \mathcal{A}_x$ (with only a finite number of terms different from zero). For a free module the ψ^a are unique, but we shall not assume that. Let the generalized gauge potential be defined by the action of $\tilde{\nabla}$ on the elements of the generating family: $\tilde{\nabla}\eta_a = \eta_b \mathcal{A}^b_a$. In the following we shall suppress indices and simply write $\xi = \eta.\psi$, $\tilde{\nabla}\eta = \eta.\mathcal{A}$ etc. We compute

$$\nabla\xi = \nabla(\eta.\psi) = \eta.(\mathcal{A} \wedge \psi + \tilde{\mathbf{d}}_\star \psi) = \eta.(\mathcal{D} \wedge \psi). \tag{10}$$

Evaluated on a function $f \in \mathcal{A}_x$ this yields a covariant function times the matter field: $(\mathcal{D} \wedge \psi)(f) = (f + \mathcal{A}(f)) \star \psi = (\mathcal{D}f) \star \psi$. Similarily

$$\nabla^2\xi = \eta.(\mathcal{A} \wedge \mathcal{A} + \mathbf{d}_\star \mathcal{A}).\psi = \eta.\mathcal{F}.\psi, \tag{11}$$

with the field strength

$$\mathcal{F} = \mathbf{d}_\star \mathcal{A} + \mathcal{A} \wedge \mathcal{A}, \tag{12}$$

which agrees with our previous definition.

SEIBERG-WITTEN MAP FROM EQUIVALENCE OF STAR PRODUCTS

We can construct an explicit map that relates the ordinary gauge potential a to the generalized noncommutative gauge potential $\mathcal{A}$ by realizing that the latter defines an equivalence map $\mathcal{D} = \mathrm{id} + \mathcal{A}$ between star products $\star$ and $\star'$: $\mathcal{D}f \star \mathcal{D}g = \mathcal{D}(f \star' g)$. In the nondegenerate case we can argue for this from Moser's lemma [11]: The symplectic structures B and $B' = B + da$ and the corresponding Poisson structures θ and θ' are related by a coordinate transformation ρ^* generated by the vector field $\theta^{ij} a_j \partial_i$: $\rho^*\{f,g\}' = \{\rho^* f, \rho^* g\}$. It then follows from a theorem due to Kontsevich [12] that the corresponding star products $\star$ and $\star'$ are equivalent. In first order in θ the equivalence map and $\mathcal{A}$ are given by Moser's vector field, i.e., in terms of the ordinary gauge potential a_i. In the next sections we shall construct this equivalence map $\mathcal{D}$ and the generalized noncommutative gauge potential $\mathcal{A}$ explicitly as a function of the ordinary gauge potentials a.

Abelian Gauge Fields. Take an abelian gauge potential $a = a_i dx^i$, the corresponding field strength $f = da$ and abelian gauge transformations $\delta_\lambda a = d\lambda$ as given data. Let us introduce the coboundary operator $\mathbf{d}_\theta = -[\cdot, \theta]_S$, where $[\,,\,]_S$ is the Schouten-Nijenhuis bracket of polyvector fields and $\theta = \frac{1}{2}\theta^{ij}\partial_i \wedge \partial_j$ is the Poisson bivector and let us define $\mathbf{a}_\theta = a_i \mathbf{d}_\theta x^i = \theta^{ji} a_i \partial_j$ (Moser's vector field) and $\mathbf{f}_\theta = \mathbf{d}_\theta \mathbf{a}_\theta = -\frac{1}{2}(\theta \cdot f \cdot \theta)^{ij} \partial_i \wedge \partial_j$. We can introduce a one parameter deformation θ_t of the Poisson structure θ: $t \in [0,1]$, $\theta_0 = \theta$, $\theta_1 =: \theta'$ by the differential equation $\partial_t \theta_t = \mathbf{f}_{\theta_t}$. The t-evolution is generated by the vector field $\mathbf{a}_\theta$, because $\mathbf{f}_{\theta_t} = -[\mathbf{a}_{\theta_t}, \theta_t]_S$. The Poisson structures θ and θ' are related by the flow $\rho^*_a = e^{\mathbf{a}_{\theta_t} + \partial_t} e^{-\partial_t}\Big|_{t=0}$. Let

$$A_a = \rho^*_a - \mathrm{id}, \qquad \tilde{\lambda} = \sum_{n=0}^{\infty} \frac{(\mathbf{a}_{\theta_t} + \partial_t)^n(\lambda)}{(n+1)!}\Bigg|_{t=0}, \tag{13}$$

then: $A_{a+d\lambda} = A_a + \mathbf{d}_\theta\tilde{\lambda} + \{A_a, \tilde{\lambda}\}$. This is the semi-classical version of the Seiberg-Witten condition (2).

We can quantize the semi-classical solution by lifting the Poisson structures to star products and by lifting polyvector fields to polydifferential operators. This is done by Kontsevich's formality maps U_n [12]. We lift Moser's vector field $\mathbf{a}_\theta$ to the differential operator

$$\mathbf{a}_\star = \sum \frac{(i\hbar)^n}{n!} U_{n+1}(\mathbf{a}_\theta, \theta, \ldots, \theta), \tag{14}$$

whose transformation properties follow from the formality condition (FC) [12, 6]:

$$\delta\mathbf{a}_\theta = \mathbf{d}_\theta\lambda \overset{\text{(FC)}}{\Rightarrow} \delta\mathbf{a}_\star = \frac{1}{i\hbar}\mathbf{d}_\star\bar{\lambda}; \quad \bar{\lambda} \equiv \sum \frac{(i\hbar)^n}{n!} U_{n+1}(\lambda, \theta, \ldots, \theta). \tag{15}$$

The analog of $\mathbf{d}_\theta$ is the coboundary operator $\mathbf{d}_\star = -[\,\cdot\,, \star]_G$, where $[\ ,\ ]_G$ is the Gerstenhaber bracket. $\mathbf{f}_\theta$ is lifted via formality to the bidifferential operator

$$\mathbf{f}_\star = \mathbf{d}_\star\mathbf{a}_\star \overset{\text{(FC)}}{=} \sum \frac{(i\hbar)^{(n+1)}}{n!} U_{n+1}(\mathbf{f}_\theta, \theta, \ldots, \theta) \tag{16}$$

and θ_t to a t-dependent star product

$$g \star_t h = \sum \frac{(i\hbar)^n}{n!} U_n(\theta_t, \ldots, \theta_t)(g, h) \overset{\text{(FC)}}{\Rightarrow} \partial_t(g \star_t h) = \mathbf{f}_{\star_t}(g, h). \tag{17}$$

As an operator equation: $\partial_t(\star_t) = \mathbf{f}_{\star_t}$. Since $\mathbf{f}_{\star_t} = -[\mathbf{a}_{\star_t}, \star_t]_G$, the t-evolution is generated by the differential operator $\mathbf{a}_\star$. The star products $\star$ and $\star'$ are related by the equivalence map or "quantum flow" $\mathcal{D}_a = e^{\mathbf{a}_{\star_t} + \partial_t} e^{-\partial_t}\Big|_{t=0}$. Let

$$\mathcal{A}_a = \mathcal{D}_a - \mathrm{id}, \qquad \Lambda(\lambda, a) = \sum \frac{(\mathbf{a}_{\star_t} + \partial_t)^n(\bar{\lambda})}{(n+1)!}\Bigg|_{t=0}, \tag{18}$$

then

$$\mathcal{A}_{a+d\lambda} = \mathcal{A}_a + \frac{1}{i\hbar}(\mathbf{d}_\star\Lambda - \Lambda \star \mathcal{A}_a + \mathcal{A}_a \star \Lambda). \tag{19}$$

This is the full Seiberg-Witten map for the generalized noncommutative gauge potential $\mathcal{A}$ for an arbitrary Poisson structure $\theta(x)$. In components:

$$\mathcal{A}_a(x^i) = \theta^{ij}a_j + \frac{1}{2}\theta^{kl}a_l(\partial_k(\theta^{ij}a_j) - \theta^{ij}f_{jk}) + \ldots, \tag{20}$$

$$\Lambda = \lambda + \frac{1}{2}\theta^{ij}a_j\partial_i\lambda + \frac{1}{6}\theta^{kl}a_l(\partial_k(\theta^{ij}a_j\partial_i\lambda) - \theta^{ij}f_{jk}\partial_i\lambda) + \ldots. \tag{21}$$

Nonabelian Gauge Fields. The given nonabelian data is $A_\mu(x) = A_{\mu b}(x)T^b$, with $[T^a, T^b] = iC^{ab}_c T^c$, $\delta_\Lambda A_\mu = \partial_\mu\Lambda + i[\Lambda, A_\mu]$. Our goal is to find maps $\hat{A}(A_\mu)$, $\hat{\Lambda}(A_\mu, \Lambda)$, such that $\hat{A}(A_\mu + \delta_\Lambda A_\mu) = \hat{A}(A_\mu) + \hat{\delta}_{\hat{\Lambda}}\hat{A}(A_\mu)$, with a nonabelian noncommutative gauge transformation $\hat{\delta}$. A naive generalization of the abelian construction immediately runs into several problems: We could try to let $B \to B + F =: B'$ and correspondingly $\Theta \to \Theta'$, but even if $dB = 0$ (Θ Poisson) $dB' \neq 0$ and Θ' will not be Poisson, since $dF = -A \wedge A \neq 0$. Furthermore, the nonabelian analog of Moser's vector field, $\Theta^{\mu\nu} A_\nu D_\mu$, is not a derivation, essentially because A_ν is matrix-valued. A better strategy is the following: Introduce a larger space that is the product of noncommutative space-time with coordinates $x^\mu, x^\nu, \ldots$ and an internal space with coordinates $t^a, t^b, \ldots$, which are the symbols of the generators T^a, T^b, ... The enlarged space has a Poisson structure which is the direct sum of external $\Theta^{\mu\nu}$ and internal $\vartheta^{ab} = C^{ab}_c t^c$ Poisson structures. Then we pick appropriate *abelian* gauge potentials and parameters that are linear in the internal coordinates and quantize everything with the previous method. After quantization of the internal space (and taking an appropriate representation) the symbols t^a become the generators of the Lie algebra with structure constants C^{ab}_c, giving an nonabelian theory on commutative space-time. After quantizing everything we obtain the desired nonabelian noncommutative gauge theory. Such theories have also been discussed in [13].

As in the abelian case the space-time components of the noncommutative gauge potential are determined via the Seiberg-Witten map by the term of lowest order in Θ: $\mathcal{A}_a(x^\mu) = \Theta^{\mu\nu} A_\nu(a) + O(\Theta^2)$. From (18) we can compute the nonabelian gauge potential $A_\nu(a)$ and the nonabelian gauge parameter Λ to all orders in C^{ab}_c and in the internal components of the abelian gauge potential a_b:

$$A_\nu(a) = \sum \frac{1}{n!} t^c (M^{n-1})^b_c (a_{\nu b} - (n-1) f_{\nu b}), \tag{22}$$

with $a_{\nu b} t^b = a_\nu$ and the matrix $M^a_b = C^{ac}_b a_c$,

$$\Lambda(\lambda, a) = \sum \frac{1}{(n+1)!} t^a (M^n)^b_a \lambda_b. \tag{23}$$

(In the special gauge of vanishing internal gauge potential $a_b = 0$ this becomes simply $A_\mu(a) = a_\mu$, $\Lambda = \lambda$.)

REFERENCES

1. Seiberg, N. and Witten, E., *JHEP* **9909**, 032 (1999) [hep-th/9908142].
2. Cattaneo, A. S. and Felder, G., *Commun.Math.Phys* **212**, 591 (2000) [math.QA/9902090]
3. Schaller, P. and Strobl, T., *Mod. Phys. Lett. A* **9**, 3129 (1994) [hep-th/9405110].
4. Madore, J., Schraml, S., Schupp, P. and Wess, J., *Eur. Phys. J.* **C16**, 161 (2000) [hep-th/0001203].
5. Jurco, B. and Schupp, P., *Eur. Phys. J.* **C14**, 367 (2000) [hep-th/0001032].
6. Jurco, B., Schupp, P. and Wess, J., *Nucl. Phys.* **B584**, 784 (2000) [hep-th/0005005].
7. Jurco, B., Schraml, S., Schupp, P. and Wess, J., *Eur. Phys. J.* **C17**, 521 (2000) [hep-th/0006246].
8. Jurco, B., Schupp, P. and Wess, J., *Nonabelian noncommutative gauge theory via noncommutative extra dimensions*, [hep-th/0102129].
9. Okuyama, K., *JHEP* **0003**, 016 (2000) [hep-th/9910138].
10. Cornalba, L., *D-brane physics and noncommutative Yang-Mills theory*, [hep-th/9909081].
11. Moser, J., *Trans. Amer. Math. Soc.* **120**, 286 (1965).
12. Kontsevich, M., *Deformation quantization of Poisson manifolds, I*, [q-alg/9709040].
13. Bonora, L., Schnabl, M., Sheikh-Jabbari, M. M. and Tomasiello, A., *Nucl. Phys.* **B589**, 461 (2000) [hep-th/0006091].

On Conservation Laws for Models in Discrete, Noncommutative and Fractional Differential Calculus

M. Klimek

Institute of Mathematics and Computer Science, Technical University of Częstochowa, Dąbrowskiego 73, 42-200 Częstochowa, Poland, email: klimek@matinf.pcz.czest.pl

Abstract. We present the general method of derivation the explicit form of conserved currents for equations built within the framework of discrete, noncommutative or fractional differential calculus. The procedure applies to linear models with variable coefficients including also nonlinear potential part. As an example an equation on quantum plane, nonlinear Toda lattice model and homogeneous equation of fractional diffusion in 1+1 dimensions are studied.

INTRODUCTION

The aim of the paper is to report on the results concerning conservation laws for equations on discrete and noncommutative spaces as well as for models built using standard differential derivatives and fractional nonlocal ones.
In principle the construction generalizes the known Takahashi-Umezawa procedure for linear equations in classical field theory [1]. We have checked that in a wide class of models built within the framework of various differential calculi the common form of Leibniz's rule for exterior partial derivatives appears. The first step of construction is then the modification of Leibniz's rule, the explicit solution of certain operator equation in respective algebra of exterior partial derivatives and finally application of this solution in derivation of conserved currents.

The method is then applied to second order equation on quantum plane with variable coefficients and nonlinear potential, to a nonlinear Toda lattice model and to homogeneous fractional 1+1 diffusion process. Various other examples of construction of the conservation law and conserved charges together with proofs of propositions mentioned below can be found in earlier papers [2, 3, 4, 5, 6, 7, 8, 9, 10, 11].

CP589, *Fundamental Interaction Theories*, edited by J. Lukierski and J. Rembieliński

CONSTRUCTION OF CONSERVATION LAWS FOR A CLASS OF EQUATIONS WITH VARIABLE COEFFICIENTS

Linear Equations with Variable Coefficients

We consider equations of the form:

$$\Lambda(\partial)\Phi = 0\,, \tag{1}$$
$$\Lambda(\partial) = \Lambda_0 + \sum_{l=1}^{N} \Lambda_{\mu_1 \ldots \mu_l} \partial^{\mu_1} \ldots \partial^{\mu_l}\,, \tag{2}$$

where coefficients (may be matrices) fulfill the condition:

$$\partial^{\dagger\;\mu_1} \Lambda_{\mu_1 \ldots \mu_l} = 0\,, \tag{3}$$

for $l = 1, \ldots, N$ and with the conjugated derivative:

$$\partial^{\dagger\;\mu} = -\partial^{\nu} \zeta_{\nu}^{-\;\mu}\,. \tag{4}$$

The ζ^- is the inverse transformation operator connected with the Leibniz's rule of respective differential calculus used in construction of the model. We include into considerations the nonlinear equations provided the only nonlinear term does not depend on derivatives that means:

$$\Lambda_0 = \Lambda_0(\Phi)\,. \tag{5}$$

Area of Application

The admissible differential calculus has the following Leibniz's rule for exterior partial derivatives:

$$\partial^i (fg) = (\partial^i f) g + (\zeta^i_j f) \partial^j g\,, \tag{6}$$

where the transformation operator ζ is linear, multiplicative and invertible:

$$\zeta^i_j (fg) = (\zeta^i_k f)(\zeta^k_j g)\,, \qquad \zeta_j^{-\;i} (fg) = (\zeta_j^{-\;k} f)(\zeta_k^{-\;i} g)\,, \tag{7}$$
$$\zeta^k_j \zeta_k^{-\;i} = \zeta_j^{-\;k} \zeta^i_k = \delta^i_j\,. \tag{8}$$

The following differential calculi obey the Leibniz's rule of the above form:

- discrete models and mixed discrete-continuous models with derivatives given by one of the following formulas [2, 3, 4]:

$$\partial^i f_k = (\zeta^i - 1) f_k = f_{k+1} - f_k\,, \tag{9}$$
$$\partial^i f_k = (x^i_{k+1} - x^i_k)^{-1} (f_{k+1} - f_k)\,, \tag{10}$$

  where the transformation operator ζ is the shift operator and the inverse ζ^- the backshift operator in a given direction on the lattice

- supersymmetric models built using the supersymmetric differential calculus including standard derivatives on Minkowski space and anticommuting Grassman derivatives with respect to spinor coordinates [10, 12, 13]:

$$\zeta^{(spin\ 1)}_{(spin\ 2)}(spinor) = -\delta^{(spin\ 1)}_{(spin\ 2)}(spinor)\,, \tag{11}$$

$$\zeta^{-} = \zeta\,, \tag{12}$$

- quantum Minkowski spaces by Podleś and Woronowicz with the following transformation operator and its inverse [5, 7, 10, 14, 15, 16]:

$$\zeta^{i}_{j}x^{k} = R^{ik}_{aj}x^{a} - (RZ)^{ik}_{j}\,, \qquad \zeta^{-\ i}_{j}x^{k} = R^{ki}_{ja}x^{a} + Z^{ki}_{j}\,, \tag{13}$$

- braided linear spaces by Majid including q-Minkowski space [6, 8, 9, 17, 18]:

$$\zeta^{i}_{j}x_{k} = R^{li}_{kj}x_{l}\,, \qquad \zeta^{-\ i}_{j}x_{k} = \tilde{R}^{li}_{kj}x_{l}\,, \tag{14}$$

- some fractional differintegrable equations which use the nonlocal fractional derivatives with respect to part of coordinates of space-time and standard derivatives for the rest of coordinates. The Riemann-Liouville partial fractional derivative is given by the formula [19, 20, 21]:

$$D^{\alpha_k}_{k}f(\vec{x}) := \frac{1}{\Gamma(m_k+1-\alpha_k)}(\partial_{x_k})^{m_k+1}\int_0^{x_k}(x_k-s)^{-\alpha_k+m_k}f(\vec{x}+(s-x_k)\vec{e}_k)ds\,, \tag{15}$$

  where $m_k \leq Re\alpha_k < m_k+1$ and $(\vec{e}_k)_l = \delta_{kl}$. The upper index in the formula denotes the fractional order of the partial derivative while the lower one says that it was taken with respect to coordinate x_k.

Let $x_1,...,x_m$ be a subset of coordinates in our n-dimensional model for which the fractional partial derivatives appear in the equation. The algebra of functions is defined by the convolution formula [11]:

$$f * g(\vec{x}) := \int_0^{x_1}...\int_0^{x_m} f\left(\vec{x}-\sum_{l=1}^{m}s_l\vec{e}_l\right) g\left(\vec{x}+\sum_{l=1}^{m}(s_l-x_l)\vec{e}_l\right) ds_1...ds_m \tag{16}$$

and corresponding Leibniz's rule for fractional and standard derivatives looks in this algebra as follows (with $\beta_k \in [0,1]$ for $k = 1,...,m$):

$$D^{\alpha_k}_{k}f * g = \beta_k(D^{\alpha_k}_{k}f) * g + (1-\beta_k)f * D^{\alpha_k}_{k}g\,, \tag{17}$$

$$\partial_j(f*g) = (\partial_j f)*g + f*\partial_j g\,, \tag{18}$$

provided functions f and g obey the respective assumptions concerning their behaviour in the neighbourhood of 0 [11]. Clearly the transformation operator in this case is the identity operator $\zeta^{i}_{j} = \zeta^{-\ i}_{j} = \delta^{i}_{j}$.
Let us point out that in the case of fractional models we assume in addition to condition (3) that the coefficients of the equation depend on the subset of coordinates $x_{m+1},...,x_n$ for which only classical derivatives appear in the equation (1).

Modification of the Leibniz's Rule

We modify the Leibniz's rule as follows:

$$\partial^k[(\zeta_k^{-\ i} f)g] = f\left(-\overleftarrow{\partial}^{\dagger\ i} + \partial^i\right) g, \tag{19}$$

where the conjugated derivative $\partial^\dagger$ is defined by (4). We see that after modification we deal with the Leibniz's rule where the right-hand side is analogous to the classical differential calculus (where the conjugated derivative is given by: $\partial^{\dagger\ i} = -\partial^i$):

$$\partial^i(fg) = f\left(-\overleftarrow{\partial}^{\dagger\ i} + \partial^i\right) g. \tag{20}$$

The conjugated derivatives form the conjugated equation:

$$\Phi'\Lambda(\overleftarrow{\partial}^{\dagger}) = \Phi'\left[\Lambda_0(\Phi) + \sum_{l=1}^{N} \overleftarrow{\partial}^{\dagger\ \mu_1} \ldots \overleftarrow{\partial}^{\dagger\ \mu_l} (\zeta_{\mu_l}^{-\ \alpha_l} \ldots \zeta_{\mu_1}^{-\ \alpha_1} \Lambda_{\alpha_1 \ldots \alpha_l})\right] = 0, \tag{21}$$

which is linear with respect to solution Φ' and the potential Λ_0 depends on the solution of initial nonlinear equation.
Let us notice that for the fractional models the conjugated equation looks as follows:

$$\Phi'\Lambda(-\overleftarrow{D}, -\overleftarrow{\partial}) = 0, \tag{22}$$

as the conjugation in this differential calculus means: $(D_k^{\alpha_k})^\dagger = -D_k^{\alpha_k}$, $\partial_j^\dagger = -\partial_j$ (for $\beta_k = \frac{1}{2}\ k = 1, \ldots, m$).

The Γ and $\hat{\Gamma}$ Operators

In the construction of conserved currents the solution of certain operator equation in respective exterior differential algebra is necessary. It is determined by the following proposition [10, 11]:

Proposition 1. *The unique solution of the operator equation:*

$$\sum_\mu (-\overleftarrow{\partial}^{\dagger\ \mu} + \partial^\mu) \circ \Gamma_\mu(\partial, \overleftarrow{\partial}^{\dagger}) = \Lambda(\partial) - \Lambda(\overleftarrow{\partial}^{\dagger}) \tag{23}$$

in the class of polynomials of derivatives $\overleftarrow{\partial}^{\dagger}$ *and* ∂ *for the equation operator* Λ *fulfilling the restriction*

$$\partial^{\dagger\ \mu_1} \Lambda_{\mu_1 \ldots \mu_l} = 0 \tag{24}$$

is of the form:

$$\Gamma_\mu(\partial, \overleftarrow{\partial}^{\dagger}) = \alpha_f(\zeta_\mu^{-\ \alpha} \Lambda_\alpha) + \tag{25}$$

$$+\alpha_f \sum_{l=1}^{N-1} \sum_{k=0}^{l} \overleftarrow{\partial}^{\dagger\,\mu_1} \ldots \overleftarrow{\partial}^{\dagger\,\mu_k} \left(\zeta_\mu^{-\,\alpha} \zeta_{\mu_k}^{-\,\alpha_k} \ldots \zeta_{\mu_1}^{-\,\alpha_1} \Lambda_{\alpha_1 \ldots \alpha_k \alpha \mu_{k+1} \ldots \mu_l}\right) \partial^{\mu_{k+1}} \ldots \partial^{\mu_l}.$$

The factor $\alpha_f = 2$ for fractional part of operator of the equation (1) and $\alpha_f = 1$ for other types of derivatives.

Proof: the explicit proofs of the proposition respectively for discrete and mixed discrete, noncommutative, fractional and mixed fractional models are enclosed in papers [4, 7, 8, 10, 11].

We modify the Γ operator due to the deformation of the Leibniz's rule and obtain the operator $\hat{\Gamma}$:

$$\hat{\Gamma}_\mu(\partial, \overleftarrow{\partial}^{\dagger}) = \alpha_f \, \overleftarrow{\zeta}_\mu^{-\,j} \, (\zeta_j^{-\,\alpha} \Lambda_\alpha) + \tag{26}$$

$$+\alpha_f \sum_{l=1}^{N-1} \sum_{k=0}^{l} \overleftarrow{\partial}^{\dagger\,\mu_1} \ldots \overleftarrow{\partial}^{\dagger\,\mu_k} \overleftarrow{\zeta}_\mu^{-\,j} \left(\zeta_j^{-\,\alpha} \zeta_{\mu_k}^{-\,\alpha_k} \ldots \zeta_{\mu_1}^{-\,\alpha_1} \Lambda_{\alpha_1 \ldots \alpha_k \alpha \mu_{k+1} \ldots \mu_l}\right) \partial^{\mu_{k+1}} \ldots \partial^{\mu_l},$$

which for a pair of arbitrary functions f and g is connected with the Γ operator by the equation:

$$\sum_\mu \partial^\mu f \hat{\Gamma}_\mu(\partial, \overleftarrow{\partial}^{\dagger}) g = \sum_\mu f \left(-\overleftarrow{\partial}^{\dagger\,\mu} + \partial^\mu\right) \circ \Gamma_\mu(\partial, \overleftarrow{\partial}^{\dagger}) g. \tag{27}$$

Let us notice that in the case of fractional differential equations the modified $\hat{\Gamma}$ operator coincides with the Γ operator as the transformation operator for this class of models is simply the identity operator.

Conservation Law

Proposition 2. *Let us assume that functions Φ and Φ' solve respectively the initial and conjugated equations with coefficients fulfilling condition (3):*

$$\Lambda(\partial)\Phi = 0, \qquad \Phi' \Lambda(\overleftarrow{\partial}^{\dagger}) = 0. \tag{28}$$

Then the current:

$$J_\mu = \Phi' \hat{\Gamma}_\mu(\partial, \overleftarrow{\partial}^{\dagger}) \Phi, \tag{29}$$

where the operator $\hat{\Gamma}_\mu$ is the modified Γ operator from Proposition 1 obeys the conservation law:

$$\sum_\mu \partial^\mu J_\mu = 0. \tag{30}$$

Proof: see [4, 7, 8, 10, 11] for corresponding type of equation.

APPLICATIONS

The Second Order Equation on Quantum Plane with Variable Coefficients and Nonlinear Potential

The commutation rules for coordinates and derivatives on quntum plane look as follows:

$$yx = qxy\,, \qquad \partial^y\partial^x = q^{-1}\partial^x\partial^y\,. \tag{31}$$

The Leibniz's rule is defined by the corresponding R-matrix [17, 18]. It yields the following formulas for the inverse transformation ζ^- acting on the monomials of the first order:

$$\zeta_x^{-\;x}x = q^{-2}x, \quad \zeta_x^{-\;x}y = q^{-1}y, \quad \zeta_y^{-\;x}x = (q^{-2}-1)y, \quad \zeta_y^{-\;x}y = 0, \tag{32}$$

$$\zeta_y^{-\;y}y = q^{-2}y, \qquad \zeta_y^{-\;y}x = q^{-1}x, \quad \zeta_x^{-\;y}y = 0, \qquad \zeta_x^{-\;y}x = 0. \tag{33}$$

The conjugated derivatives are constructed according to (4):

$$\partial^{\dagger\;x} = -\partial^x\zeta_x^{-\;x} - \partial^y\zeta_y^{-\;x}\,, \qquad \partial^{\dagger\;y} = -\partial^x\zeta_x^{-\;y} - \partial^y\zeta_y^{-\;y}\,. \tag{34}$$

We shall consider the nonlinear equation of the second order with the coefficients depending on x and y:

$$[\Lambda(\partial^x,\partial^y) + \Lambda_0(\Phi)]\Phi = [\phi(y)\partial^x\partial^x + \psi(x)\partial^y\partial^y + \Lambda_0(\Phi)]\Phi = 0\,. \tag{35}$$

The condition (3) is fulfilled for the above equation:

$$\partial^{\dagger\;x}\Lambda_{xx} + \partial^{\dagger\;y}\Lambda_{yx} = 0\,, \qquad \partial^{\dagger\;y}\Lambda_{yy} + \partial^{\dagger\;x}\Lambda_{xy} = 0\,. \tag{36}$$

The conjugated equation takes the following form:

$$\Phi'[\Lambda(\overleftarrow{\partial}) + \Lambda_0(\Phi)] = \Phi'\left[\overleftarrow{\partial}^{\dagger\;x}\overleftarrow{\partial}^{\dagger\;x}\phi(q^{-2}y) + \overleftarrow{\partial}^{\dagger\;y}\overleftarrow{\partial}^{\dagger\;y}\psi(q^{-2}x) + \Lambda_0(\Phi)\right] = 0\,. \tag{37}$$

The Γ operator is yielded by the construction in Proposition 1:

$$\Gamma_x = \overleftarrow{\partial}^{\dagger\;x}\phi(q^{-2}y) + \phi(q^{-1}y)\partial^x\,, \qquad \Gamma_y = \overleftarrow{\partial}^{\dagger\;y}\psi(q^{-2}x) + \psi(q^{-1}x)\partial^y\,. \tag{38}$$

Then the components of the modified $\hat{\Gamma}$ operator are given by the formulas:

$$\hat{\Gamma}_x = \overleftarrow{\partial}^{\dagger\;x}\overleftarrow{\zeta}_x^{-\;x}\phi(q^{-2}y) + \overleftarrow{\zeta}_x^{-\;x}\phi(q^{-1}y)\partial^x\,, \tag{39}$$

$$\hat{\Gamma}_y = \overleftarrow{\partial}^{\dagger\;x}\overleftarrow{\zeta}_y^{-\;x}\phi(q^{-2}y) + \overleftarrow{\zeta}_y^{-\;x}\phi(q^{-1}y)\partial^x + \overleftarrow{\partial}^{\dagger\;y}\overleftarrow{\zeta}_y^{-\;y}\psi(q^{-2}x) + \overleftarrow{\zeta}_y^{-\;y}\psi(q^{-1}x)\partial^y \tag{40}$$

and the current:

$$J_x = \Phi'\hat{\Gamma}_x\Phi\,, \qquad J_y = \Phi'\hat{\Gamma}_y\Phi\,, \tag{41}$$

obeys the conservation law by Proposition 2:

$$\partial^x J_x + \partial^y J_y = 0\,, \tag{42}$$

provided functions Φ' and Φ are the solutions of the corresponding initial and conjugated equation.

Nonlinear Toda Lattice Model

The auxiliary linear equations for discrete nonlinear Toda lattice model were derived by Dimakis and Müller-Hoissen within their framework of bicovariant differential calculus [22, 23]:

$$\dot{\chi}_k = \lambda(e^{q_{k-1}-q_k}\chi_{k-1} - \chi_k)\zeta, \qquad \chi_{k+1} - \chi_k = -\lambda(\dot{\chi}_k + \dot{q}_k\chi_k)\zeta. \tag{43}$$

From the linearized set of equation we obtained the following linear consistency condition for the auxiliary field χ [10]:

$$\ddot{\chi}_k + \dot{q}_k\dot{\chi}_k - e^{q_{k-1}-q_k}(\chi_{k-1} - \chi_k) - (\chi_{k+1} - \chi_k) = 0. \tag{44}$$

In our notation the consistency condition looks as follows:

$$\chi_{k-1} - \chi_k = \partial\chi_k, \qquad \chi_{k+1} - \chi_k = \partial^{\dagger}\chi_k, \tag{45}$$

$$q_{k-1} - q_k = \partial q_k, \qquad \dot{f} = \partial^t f, \tag{46}$$

$$\left[\partial^t\partial^t + (\partial^t q)\partial^t - e^{(\partial q)}\partial - \partial^{\dagger}\right]\chi = 0. \tag{47}$$

It is a second order linear equation with coefficients depending on gauge field q, fulfilling the condition (3) which becomes the nonlinear Toda lattice equation:

$$\partial^{\dagger\,t}\Lambda_t + \partial^{\dagger}\Lambda_x = -\ddot{q}_k - e^{q_k - q_{k+1}} + e^{q_{k-1}-q_k} = 0. \tag{48}$$

We build the $\hat{\Gamma}$ operator following the general construction:

$$\hat{\Gamma}_t = \Gamma_t = -\overset{\leftarrow t}{\partial} + \partial^t + (\partial^t q), \quad \hat{\Gamma}_x = -\overset{\leftarrow -}{\zeta}\, e^{-(\partial^{\dagger} q)} \quad \hat{\tilde{\Gamma}}_x = \overset{\leftarrow}{\zeta}. \tag{49}$$

The components of the $\hat{\Gamma}$ operator yield the respective components of the current:

$$J_t^{\delta} = \delta\chi'\hat{\Gamma}_t\chi, \quad J_x^{\delta} = \delta\chi'\hat{\Gamma}_x\chi, \quad \tilde{J}_x^{\delta} = \delta\chi'\hat{\tilde{\Gamma}}_x\chi, \tag{50}$$

which obeys the conservation law including the conjugated derivative:

$$\partial^t J_t^{\delta} + \partial J_x^{\delta} + \partial^{\dagger}\tilde{J}_x^{\delta} = 0, \tag{51}$$

provided the field χ fulfills the consistency condition (47) and the field $\delta\chi'$ its conjugation:

$$\delta\chi'\left[\overset{\leftarrow t}{\partial}\overset{\leftarrow t}{\partial} - \overset{\leftarrow t}{\partial}(\partial^t q) - \overset{\leftarrow \dagger}{\partial} e^{-(\partial^{\dagger})q} - \overset{\leftarrow}{\partial}\right] = 0 \tag{52}$$

After integrating the time-component of the conserved current we arrive at the charges:

$$Q^{\delta} = \int dx\ J_t = \int dx\ \delta\chi'\hat{\Gamma}_t\chi, \tag{53}$$

which are also conserved:

$$\partial^t Q = 0. \tag{54}$$

The contents of the above charges is studied in detail in [11].

Fractional Diffusion Equation in 1+1 Dimensional Space

Let us recall the homogenous version of the fractional diffusion equation discussed in [24, 25, 26, 27, 28, 29]:

$$D_t^{\alpha}\phi(x,t) = \lambda^2\partial_x^2\phi(x,t)\,, \tag{55}$$

where $t > 0$, $x \in R$ and $0 < \alpha < 1$ describes the process of ultraslow diffusion while the value $1 < \alpha < 2$ is used for intermediate processes. The modified diffusion equation and its stationarity-conservation law was studied in [11] for the case of 1+1 and d+1 dimensional space.

Let us focus now on the case of ultraslow diffusion in homogeneous form which is a good illustration of our method. The operator of the equation contains both types of derivatives: fractional with respect to time and standard for the spatial dimension:

$$\Lambda(D_t^{\alpha},\partial_x) = D_t^{\alpha} - \lambda^2\partial_x^2\,. \tag{56}$$

The product of functions for this model is defined according to (16) and looks as follows:

$$f * g(x,t) = \int_0^t f(x,t-s)g(x,s)ds\,. \tag{57}$$

Using the properties of the multiplication by convolution and of the fractional derivative (15) we construct the operator Γ with components:

$$\hat{\Gamma}_x = \Gamma_x = \lambda^2\overleftarrow{\partial}_x - \lambda^2\partial_x\,, \qquad \hat{\Gamma}_t = \Gamma_t = 2\,. \tag{58}$$

Then the current:

$$J_x = \phi'\Gamma_x * \phi = \phi'\lambda^2\overleftarrow{\partial}_x * \phi - \phi' * \lambda^2\partial_x\phi\,, \tag{59}$$

$$J_t = \phi'\Gamma_t * \phi = 2\phi' * \phi\,, \tag{60}$$

obeys the stationarity-conservation equation for $t \geq 0$ in the area $\psi(x) = \psi'(x) = 0$ (where these functions define the initial conditions given below (63)):

$$\partial_x J_x + D_t^{\alpha} J_t = 0\,, \tag{61}$$

provided the function ϕ is the solution of initial equation (55) while ϕ' solves its conjugation:

$$\phi'(x,t)\Lambda(-\overleftarrow{D}_t^{\alpha}, -\overleftarrow{\partial}_x) = 0\,. \tag{62}$$

The restriction of the area where the stationarity-conservation law can be constructed is determined by the asymptotic properties of solutions around $t = 0$. Let us notice that using the Laplace-Fourier transform method we arrive at the initial condition of general form:

$$D_t^{-(1-\alpha)}\phi(x,t)\,|_{t=0} = \psi(x)\,, \qquad D_t^{-(1-\alpha)}\phi'(x,t)\,|_{t=0} = \psi'(x)\,. \tag{63}$$

The above formula allows to apply the Leibniz's rule for fractional derivative (17) in the area where $\psi(x) = \psi'(x) = 0$ [11]. Let us assume the initial condition: $\frac{\psi(x)}{a} = \frac{\psi'(x)}{a'} = \delta(x)$

with a and a' being arbitrary numbers. It gives the maximal area of possible application of the fractional Leibniz's rule, namely it is fulfilled then for $x \neq 0$ and yields the currents obeying the stationarity-conservation law for the same part of space-time.

Having obtained the general form of stationary current (59,60) we can discuss the possible symmetries of equation (55) which can be used in construction of different solutions of the initial diffusion problem. The set includes the spatial momentum $P_x = \partial_x$ and the time momentum $P_t = D_t^\alpha$ as these operators commute with the operator of diffusion equation (55).

The stationarity-conservation laws for currents including new solutions are fulfilled for transformed solutions in the area $x \neq 0$. In this way we obtain the stationary currents connected with symmetries of the fractional diffusion equation:

$$J_x^x = \phi' \Gamma_x * P_x \phi, \qquad J_t^x = \phi' \Gamma_t * P_x \phi, \tag{64}$$

$$J_x^t = \phi' \Gamma_x * P_t \phi, \qquad J_t^t = \phi' \Gamma_t * P_t \phi. \tag{65}$$

The stationarity-conservation equation (61) can be reformulated using the definition of Riemann-Liouville derivative so as to obtain the standard conservation equation namely:

$$\partial_x J_x' + \partial_t J_t' = 0, \tag{66}$$

which is fulfilled for $x \neq 0$ and the components of the new current look as follows:

$$J_x' = J_x, \qquad J_t' = J_t * \Phi_\alpha = \frac{1}{\Gamma(1-\alpha)} J_t * t^{-\alpha}. \tag{67}$$

REFERENCES

1. Takahashi, Y., *An Introduction to Field Quantization*, Pergamon Press, Oxford 1969.
2. Klimek, M., *J. Phys. A: Math. & Gen.*, **26,** 955-967 (1993).
3. Klimek, M., *Czech. J. Phys.*, **44,** 1049-1057 (1994).
4. Klimek, M., *J. Phys. A: Math. & Gen.*, **29,** 1747-1758 (1996).
5. Klimek, M., " The symmetry algebra and conserved currents for Klein-Gordon equation on quantum Minkowski space", in *Quantum Groups and Quantum Spaces* edited by R. Budzyński, W. Pusz and S. Zakrzewski, Banach Center Publications **40** (1997), Inst. of Math., Polish Acad. Sci., Warszawa 1997.
6. Klimek, M., *Czech. J. Phys.*, **47,** 1199-1206 (1997).
7. Klimek, M., *Commun. Math. Phys.*, **192,** 29-45 (1998).
8. Klimek, M., *J. Math.Phys.*, **40,** 4165-4176 (1999).
9. Klimek, M., *Czech. J. Phys.*, **50,** 105-114 (2000).
10. Klimek, M., *Conservation laws for a class of nonlinear equations with variable coefficients*, math-ph/0011030.
11. Klimek, M., *Stationarity-conservation laws for certain linear fractional differential equations*, math-ph/0104025, submitted to J. Phys. A: Math. & Gen.
12. Srivastava, P.P., *Supersymmetry, Superfields and Supergravity: an Introduction*, Adam Hilger for IOP Publishing Ltd, Bristol 1986.
13. Buchbinder, I.L., and Kuzenko, S.M.,*Ideas and Methods of Supersymmetry and Supergravity*, IOP Publishing Ltd, Bristol 1998.
14. Podleś, P., and Woronowicz, S. L., *Commun. Math. Phys.*, **178,** 61-82 (1996).
15. Podleś, P., and Woronowicz, S. L., *Commun. Math. Phys.*, **130,** 381-431 (1990).
16. Podleś, P., *Commun. Math. Phys.*, **181,** 569-585 (1996).

17. Majid, S., "Introduction to braided geometry and q-Minkowski space", in *Quantum groups and their applications* edited by Castellani L. and Wess J., Proceedings of the International School of Physics "Enrico Fermi" Varenna 1994.
18. Majid, S., *Foundations of quantum group theory,* Cambridge University Press, Cambridge 2000
19. Oldham, K.B., and Spanier, J., *The Fractional Calculus,* Academic Press, New York 1974.
20. Miller, K.S., and Ross, B., *An Introduction to the Fractional Calculus and Fractional Differential Equations,* John Wiley & Sons, New York 1993.
21. Samko, S.G., Kilbas, A.A., and Marichev, O.I. *Fractional Derivatives and Integrals. Theory and Applications,* Gordon and Breach, Amsterdam 1993.
22. Dimakis, A., and Müller- Hoissen, F., *J. Phys. A: Math. & Gen.,* **29,** 5007-5018 (1996).
23. Dimakis, A., and Müller-Hoissen, F., *Bi-differential calculi and integrable models,* `math-ph/9908015`.
24. Nigmatullin, R.R., *Phys. Stat. Solidi,* **B133,** 425 (1986).
25. Wyss, W., *J. Math. Phys.,* **27,** 2782 (1986).
26. Schneider, W.R., and Wyss, W., *J. Math. Phys.,* **30,** 134 (1989).
27. Compte, A., *Phys. Rev. E,* **53,** 4191 (1996).
28. Mainardi, F., *Chaos, Solitons & Fractals,* **7,** 1461 (1996).
29. Metzler, R., and Klafter, J., *Phys. Rep.,* **339,** 1 (2000).

D-Branes on Spheres and a q-Matrix Model

J. Pawełczyk[1]

Institute of Theoretical Physics, Warsaw University, Hoża 69, PL-00-681 Warsaw, Poland, email: Jacek.Pawelczyk@fuw.edu.pl

Abstract. Using properties of the DBI action we find stable D-branes on $AdS_3 \times S^3$ spaces wrapping spheres. The branes are stable due to nonzero anti-symmetric tensor field background. Next we discuss a matrix model description of the branes based on quantum group symmetries.

INTRODUCTION

Recently it has been discovered that in a special limit the dynamics D-branes can be described by a finite matrix model [1]. Together with works on NCG in string theory [2] this has sparked new interest in strings propagating in antisymmetric tensor fields backgrounds. D-branes in such background have also shed light on the connection between quantum groups and WZW models [3].

In this work we discuss branes on $AdS_3 \times S^3$ spaces with NSNS and RR antisymmetric tensor field background. The branes partially wrap S^3. In the first part of this work we use DBI action to describe the dynamics of branes. Contrary to the naive expectation they appear to be stable and moreover their position is quantized. Next we construct a matrix model of D0-branes based on a quantum group symmetry which reproduces some of the properties of the wrapped D2-branes.

BRANES IN $ADS_3 \times S^3$

In this section we shall analyze equation of motion resulting from the DBI action. We require the string coupling constant to be small and radius of spheres to be large in which case supergravity is a good approximation to string theory and D-branes can be described completely classically by the DBI action.

NSNS Background

The example of string theory background which involves the level k SU(2) WZW model is the near horizon limit of the F1, NS5 system (see e.q. [4]). Below we write

[1] Work supported by Polish State Committee for Scientific Research (KBN) under contract 5 P03B 150 20 (2001-2002)

CP589, *Fundamental Interaction Theories*, edited by J. Lukierski and J. Rembieliński

only the relevant terms

$$ds^2/\alpha' = k\, d\Omega_3^2, \quad H^{NSNS}/\alpha' = 2k\, \varepsilon_3, \quad e^{2\phi} = const., \tag{1}$$

where k denotes the number of NS5-branes and equals to the level k of the appropriate SU(2) WZW model, ε_3 is the volume element of the unit 3-sphere.

The effective action of the D-branes is given by DBI expression [5]

$$S_{DBI} = -T_p \int_{\text{Vol}} e^{-\phi} \sqrt{-\det[(X^*G + 2\pi\alpha' F + X^*B)_{ab}]}\,. \tag{2}$$

In the following we shall discuss classical configurations of D2-brane embedded in S^3 of (1)[2]

On any 2-d submanifold of S^3 the B filed is well (but not uniquely) defined. We have the freedom of changing B by an exact 2-form - in our case this will be compensated by the choice of the solution for F so that $B+F$ will be given uniquely. In the coordinates in which the metric on S^3 is $ds^2 = k[d\vartheta^2 + \sin^2(\vartheta) d\Omega_2^2]$ we have (for the chart covering $\vartheta = 0$) $B = k\alpha'(\vartheta - \nu - \frac{1}{2}\sin(2\vartheta))\, \varepsilon_2$ where ε_2 is the volume form of the unit S^2.

We shall find extrema of the DBI action corresponding to a brane wrapped on S^2 sitting at a constant angle ϑ. The Euler-Lagrange equations are respected by

$$2\pi\alpha' F = -k\alpha'(\vartheta - \nu)\, \varepsilon_2, \quad \vartheta(x) = \vartheta = const.. \tag{3}$$

with all the other components of F equals zero. We shall set $\nu = 0$ requiring that the charge and the tension of the $\vartheta = 0$ brane be zero. It is worth to note although that the classical solution exists for any ϑ but it is subject to an extra quantization condition discussed below. It is also worth to note that contrary to the naive expectation the static brane it stable [5].

Recall that D-branes are defined to be the ends of the open strings. The string couple to the external sources (gauge A and B field) as follows

$$\exp[\frac{i}{2\pi\alpha'}(\int_{\partial\Sigma} 2\pi\alpha' X^*A + \int_{\Sigma} X^*B)]\,. \tag{4}$$

The above formula can not be well defined globally for topologically non-trivial A and B fields. Consider a configuration of D-brane embedded into submanifold M_D of the target space manifold M [6, 7]. For the world sheet with one boundary the appropriate definition of (4) is

$$\exp\left[\frac{i}{2\pi\alpha'}\left(-\int_{D^2} \widehat{X}^*(2\pi\alpha' F + B) + \int_{\widehat{\Sigma}} \widehat{X}^* H\right)\right], \tag{5}$$

where $\partial\widehat{\Sigma} = \Sigma + D^2$, $\widehat{X}$ is an extension of $X(\partial\Sigma)$ to a 2-disc D^2 such that $\widehat{X}(D^2) \subset M_D$ ($F \neq 0$ only on the D-brane manifold M_D). Notice that one must be able to define B on

[2] It can be also a partially wrapped D3 brane.

any $\widehat{X}(D^2)$ thus we must have $[H]_{M_D} = 0$. The value of the integral (5) should not depend on the way one make the extension hence

$$\frac{i}{2\pi\alpha'}\left[\int_{C^2}(2\pi\alpha' F + B) - \frac{i}{2\pi\alpha'}\int_{C^3} H\right] = 2i\pi m\,, \quad m \in \mathbf{Z}\,, \tag{6}$$

where $C^2 \in H_2(M_D)$ and $C^3 \in H_3(M, M_D)$. Here a note is necessary concerning topology of the problem. For the argument we need an exact sequence of homologies

$$\cdots \to H_3(M_D) \to H_3(M) \to H_3(M, M_D) \to H_2(M_D) \to H_2(M) \to \ldots. \tag{7}$$

If we assume that $H_3(M_D) = H_2(M) = 0$ then all cycles of $H_3(M)$ are in $H_3(M, M_D)$. Then one can write $\int_{C^3} H = \int_{C^2} B \bmod \frac{i}{2\pi}\int_{C^3_M} H = 2\pi k m$ for $C^3_M \in H_3(M)$. Thus the quantization condition reads

$$\int_{C^2} F = 2\pi n\,, \quad n \in \mathbf{Z} \tag{8}$$

and n is defined modulo k. When we apply this condition to (3) we get

$$\vartheta = \frac{\pi n}{k}\,. \tag{9}$$

RR Background

The relevant supergravity background in the near horizon limit is [4]

$$\begin{aligned} ds^2 &= \ell^2\alpha'\left(u^2(-dt^2 + (dx)^2) + \frac{du^2}{u^2} + d\Omega_3^2\right), \\ H^{\mathrm{RR}} &- 2Q_5\alpha'(\varepsilon_3 + *_6\varepsilon_3), \quad e^{-2\phi} = \frac{1}{g_6^2}\frac{Q_5}{Q_1}, \end{aligned}$$

where $*_6$ denotes Hodge duality in $AdS_3 \times S^3$ space, ε_3 is the volume element of the unit 3-sphere, and $\ell^2 = g_6\sqrt{Q_1 Q_5}$. The background is characterized by the constant dilaton and the non-trivial R-R 3-form fields. We will find it convenient to represent the metric on S^3 as $d\Omega_3^2 = d\vartheta^2 + \sin^2\vartheta d\Omega_2^2$: a 2-sphere S^2 of the radius $\sin\vartheta$ is located at latitude angle ϑ.

At leading order in α', D3-brane world-volume effective action in the above background is given by

$$S_{\mathrm{DBI}} = T_3 Q_5 \ell^2 \alpha'^2\left[-\int_{\mathrm{Vol}} d^4\sigma\sqrt{-\det[(X^*G + 2\pi\alpha' F)_{ab}]} \pm 2\pi F \wedge X^* C_2^{\mathrm{RR}}\right], \tag{10}$$

where we rescaled $F \to \ell^2 F$ and factored out all powers of α', ℓ^2, Q_5 from the metric and C_2^{RR}. Locally, we may integrate $H^{\mathrm{RR}} = dC_2^{\mathrm{RR}}$ and obtain the 2-form potential C_2^{RR} on S^3: $C_2^{\mathrm{RR}} = (\vartheta - \nu - \frac{1}{2}\sin(2\vartheta))\varepsilon_2$, where ν is an integration constant and ε_2 denotes the volume-form of the unit S^2. It is not possible to define $C_2^{\rm RR}$ globally over the whole §3, as $H^{\rm RR}$ is a non-trivial element of $H^3(S^3,R) = \mathbf{Z}$.

Consider now a D3-brane embedded in $AdS_3 \times S^3$, whose world-volume is extended along $\sigma^\mu = (t,u,S^2)$ and is located at fixed ϑ. Among the world-volume gauge field strength, only the F_{0u} component couples linearly to the background R-R potential $C_2^{\rm RR}$ through the Chern-Simons term. The coupling in turn leads to $F_{0u} \neq 0$, which is needed for a non-trivial extremum to exist. As the partially wrapped D3-brane is extended along u-direction, the D3-brane behaves effectively as a fundamental string(F-string). The non-trivial extremum (10) is

$$2\pi F = \mp \cos\vartheta\,, \qquad \vartheta = \text{constant}. \tag{11}$$

In the S-dual situation, where nontrivial background involves NS-NS tensor field, the ϑ moduli space has been discrete as a result of the requirement that the effective D-brane carries an appropriate R-R charge. In the present case, the effective fundamental string in AdS_3 carries NS-NS charge, which can be derived from terms in (10) linear in the NS-NS 2-form B_2. Taking into account of the non-trivial background of F, we find:

$$T_3 \int_{S^2} \frac{\delta S_{\rm DBI}}{\delta(2\pi\alpha' F_{ab})} X^* B_{ab} = \mp T_3 Q_5 \alpha' {\rm Vol}(S^2)\,(\vartheta - \nu)\, X^* B_{0u}\,, \tag{12}$$

where $T_{Dp} = 1/((2\pi)^p \alpha'^{(p+1)/2} g_{st})$ and the last expression has been derived at the classical extremum. We can also fix the integration constant ν by demanding that the NS-NS charge should be zero for $\vartheta = 0$. This sets $\nu = 0$. The condition that the last expression of (12) equals the integer multiple of the fundamental string charge yields

$$\vartheta_n = \frac{n}{Q_5}\pi, \qquad 0 \leq n \leq Q_5\,. \tag{13}$$

The result the same are for the $AdS_3 \times S^3$ with nontrivial NS-NS background. This is just manifestation of S-duality.

MATRIX MODEL

In [1] a matrix model describing multiple of D0-branes was described. We can apply those results to the $AdS_3 \times S^3$ spaces discussed above. One must keep in mind that the results are reliable only in the leading $1/k$ expansion. Then one can show that the equation of motion for system of N D0-branes is solved by

$$[X^i, X^j] = i\varepsilon^{ijk} X^k\,, \tag{14}$$

where X's are $N \times N$ matrices of positions of N D0-branes. For fixed central element of the above algebra (the second Casimir) this defines the so-called fuzzy sphere [9]. It appeared that the sphere has the dynamics and quantum number just as the branes from the previous section [8, 5]. However the applied approximation restricts the sized of the fuzzy brane to be much smaller than k.

The above results was rederived in [8] using CFT description of D-branes. It was noticed that at large k the ground ring of boundary fields form a closed algebra isomorphic to the above matrix algebra. For finite k this algebra cease to be associative. Here quantum groups come to rescue - one can recover the associativity of the algebra in price of twisting it. This is the idea behind [10] and below we shall follow it. For the notation we refer to this work.

We wish to construct a matrix model which is manifestly covariant under this $U_q(so(4))$ (an isometry of a "quantum 3-sphere"), which after "spontaneous" breaking (see below) to a "vector" $U_q(su(2))^V$ algebra will reproduce Eqs. (14). Here $q = \exp(\frac{\pi i}{k+2})$.

$\mathrm{U}_q(so(4))$

The results presented below have been obtained in collaboration with Harold Steinacker.

The aim of this section is to construct a quantum analog of the isometry group of a quantum 3 sphere, formally defined by (23). As an algebra, the quantum group $U_q(so(4))$ is simply the tensor product of two commuting $U_q(su(2))$ algebras, $U_q(so(4)) = U_q(su(2))^L \otimes U_q(su(2))^R$. It carries naturally the structure of a (quasitriangular) Hopf algebra, with coproduct

$$\begin{aligned} \Delta: U_q^L \otimes U_q^R &\rightarrow (U_q^L \otimes U_q^R) \otimes (U_q^L \otimes U_q^R), \\ u^L \otimes u^R &\mapsto (u_1^L \otimes u_a^R) \otimes (u_2^L \otimes u_b^R) \end{aligned} \tag{15}$$

and the obvious remaining structures. The embedding of the "vector" algebra $U_q(su(2))^V$ is quite clear:

$$\begin{aligned} d: \quad U_q(su(2))^V &\rightarrow U_q^L \otimes U_q^R, \\ u &\mapsto (u_1 \otimes u_2) = \Delta(u), \end{aligned} \tag{16}$$

where $\Delta(u)$ is the coproduct of $U_q(su(2))$. The problem with this definition is that the embedding d is not compatible with the coproduct, since the two factors in the middle of (15) become flipped. This will mean that products will not transform appropriately under $U_q(su(2))^V$.

The solution to this problem is provided by twisting. For the general theory of twisting we refer to [11], and to [12], Section 2.3 for an introduction. Consider instead of (15) the modified coproduct

$$\begin{aligned} \Delta_{\mathcal{F}}: U_q^L \otimes U_q^R &\rightarrow (U_q^L \otimes U_q^R) \otimes (U_q^L \otimes U_q^R), \\ u^L \otimes u^R &\mapsto \mathcal{F}(u_1^L \otimes u_1^R) \otimes (u_2^L \otimes u_2^R) \mathcal{F}^{-1}, \end{aligned} \tag{17}$$

where $\mathcal{F} = \mathcal{R}_{32}$. This indeed defines a quantum group, i.e. a quasitriangular Hopf algebra. We will denote this twisted quantum group with $U_q(so(4))_{\mathcal{F}}$. By construction, it satisfies

$$\Delta_{\mathcal{F}} \circ d = (d \otimes d) \circ \Delta . \tag{18}$$

This means that $U_q(so(4))_{\mathcal{F}}$ is compatible with the embedding (16) of the quantum group $U_q(su(2))^V$.

Moreover, we can define a modified action of $U_q(so(4))_{\mathcal{F}}$ on matrices X,Y as follows

$$(X,Y) \mapsto u \triangleright_{\mathcal{F}} (X,Y) = (u^L X S(u^R), \mathcal{R}_2 u^R \mathcal{R}_b^{-1} Y S(\mathcal{R}_1 u^L \mathcal{R}_a^{-1})), \tag{19}$$

extended to products via the new coproduct (17).

Using these results, one sees immediately that functionals of the form

$$S_1^{(n)} = \mathrm{tr}_q((XY)^n) \; = \; \mathrm{tr}((XY)^n q^{-H}), \quad S_2^{(n)} = \mathrm{tr}_q((YX)^n) \; = \; \mathrm{tr}((YX)^n q^{-H}), \tag{20}$$

are invariant under $U_q(so(4))_{\mathcal{F}}$ for natural n. Explicitly, in case of interest (2 dimensional modules X, Y) $q^{-H} = \mathrm{diag}(q, q^{-1})$.

So far, we found actions which are invariant under $U_q(so(4))_{\mathcal{F}}$, and preserve the vector symmetry $U_q(su(2))^V$. However, we should be able to construct a 4 dimensional real module. To achieve this, one can first identify the X and Y degrees of freedom via

$$Y = -\mathcal{R}_2 q^H S(X) S(\mathcal{R}_1) . \tag{21}$$

It is easy to verify that this induces indeed the action (19) on Y.

Invariant Actions and Equations of Motion

Below we construct $U_q(so(4))_{\mathcal{F}}$ covariant tensors. Let us introduce two fields

$$M = \sigma_\mu M^\mu, \quad \widetilde{M} = \tilde{\sigma}_\mu M^\mu , \tag{22}$$

($\sigma_\mu = (i, \sigma_i)$, $\tilde{\sigma}_\mu = (-iq^{-3/2}, q^{1/2}\sigma_i)$.) transforming as X and Y, respectively, thus using (20) we can conveniently build tensors of $U_q(so(4))_{\mathcal{F}}$.

We impose the constraint

$$(M^4)^2 + g_{ij} M^i M^j = R^2 \cdot 1 , \tag{23}$$

for $R > 0$ and treat (23) as the formal quantum 3-sphere (here we shall not discuss its algebraic meaning). Define field-strength tensors

$$F_L^k \;=\; (-i(qM^4M^k - q^{-1}M^kM^4) + \varepsilon_{ij}^k M^i M^j) , \tag{24}$$

$$F_R^k \;=\; (i(q^{-1}M^4M^k - qM^kM^4) + \varepsilon_{ij}^k M^i M^j) , \tag{25}$$

which are vectors of $U_q(su(2))^L$ and $U_q(su(2))^R$, respectively. We call them L (R) field-strength as, in the limit $q = 1$, they are equal and correspond to ordinary field strength as defined in [8].

Now we can easily construct the invariant actions. We need to consider a system of N D0-branes on (23). Although it is not essential here we shall assume that M^μ are $N \times N$ matrices which decomposes as the direct sum of the representations of another $U_q(su(2))$ algebra. Thus the invariants are expressed as q-traces of invariant tensors of $U_q(so(4))_{\mathcal{F}}$.

$$\mathrm{tr}_q(g_{ij}F_L^iF_L^j), \quad \mathrm{tr}_q(g_{ij}F_R^iF_R^j), \quad \mathrm{tr}_q(\varepsilon_{ij}^nF_{Ln}F_L^iF_L^j), \quad \mathrm{tr}_q(g_{ij}F_L^iF_L^jg_{kl}F_R^kF_R^l), \tag{26}$$

where g and ε are invariant tensors of $U_q(su(2)) \subset U_q(so(4))_{\mathcal{F}}$.

We shall not derive the most general equations of motion as we do not propose any definite action. However notice that

$$F_L^k = F_R^k = 0, \tag{27}$$

nullify any of the term (26) and solves possible equations of motion. Eqs. (27) are equivalent to

$$\begin{aligned} [M^4, M^k] &= 0, \\ \varepsilon_{ij}^k M^iM^j &= i(q-q^{-1})M^4M^k. \end{aligned} \tag{28}$$

If one takes $M^4 = a \cdot 1$ then $x^k = -\frac{M^k}{i(q-q^{-1})a}$ respects the q-fuzzy sphere equation valid for arbitrary k.

$$\varepsilon_{ij}^k x^ix^j = x^k. \tag{29}$$

In the limit $k \to \infty$ this is the same as (14).

ACKNOWLEDGMENTS

The author thanks J.Lukierski and the organizers of the XXXVII Winter School of Theoretical Physics in Karpacz for hospitality and stimulating atmosphere during the school. I am grateful to Chong-Sun Chu, J.Kowalski-Glikman, H.Grosse, B.Jurco, J.Madore, H.Steinacker and W. Zakrzewski for inspiring discussions.

REFERENCES

1. Myers, R.C., *J. High-Energy Phys.*, **9912**, 022 (1999), hep-th/9910053;
Kabat, D., and Taylor, W., *Adv. Theo. Math. Phys.*, **2**, 181 (1998), hep-th/9711078;
Rey, S.-J., hep-th/9711081
2. Douglas, M.R., and Hull, C., *J. High Energy Phys.*, **9802**, 008 (1998), hep-th/9711165; Connes, A., Douglas, M.R., and Schwarz, A., *J. High Energy Phys.*, **9802**, 003 (1998), hep-th/9711162; Schomerus, V., *JHEP*, **9906**, 030 (1999), hep-th/9903205; Seiberg, N., and Witten, E., hep-th/9908142.
3. Alekseev, A. Yu., Recknagel, A., and Schomerus,V., hep-th/9908040.
4. Maldacena, J., and Strominger, A., *JHEP*, **9812**, 005 (1998), hep-th/9804085.
5. Bachas, C., Douglas, M., and Schweigert, C., *J. High-Energy Phys.*, **0005**, 048 (2000), hep-th/0003037;
Pawełczyk, J., hep-th/0003057.

6. Alekseev, A. Yu., and Schomerus, V., *Phys. Rev.*, **D60**, 061901 (1999), hep-th/9812193.
7. Gawedzki, K., hep-th/9904145.
8. Alekseev, A. Yu., Recknagel, A., and Schomerus,V., *J. High-Energy Phys.*, **0005**, 010 (2000), hep-th/0003187.
9. Madore, J., *Class.Quant.Grav.*, **9**, 69 (1992).
10. Grosse, H., Madore, J., and Steinacker,H., "Field Theory on the *q*–deformed Fuzzy Sphere I", hep-th/0005273.
11. Chari, V., and Pressley, A., "A guide to quantum groups", Cambridge University press, 1994.
12. Majid, S., "Foundations of Quantum Group Theory", Cambridge University Press, 1995.

Perturbative Issues in Noncommutative Field Theories

M.M. Sheikh-Jabbari

The Abdus Salam ICTP, Strada Costiera, 11, Trieste, Italy, email: jabbari@ictp.trieste.it

Abstract. In this talk we review the field theories on noncommutative spaces. First we study the classical issues and particularly the Noether theorem. Then, we work out the one loop effective action for noncommutative Φ^4 theory and also for NCQED. In the end we discuss some possible phenomenological consequences of noncommutativity in space-time.

INTRODUCTION; NONCOMMUTATIVE MOYAL SPACES

In the usual quantum mechanics we have the well-known commutation relations:

$$\begin{aligned} [\hat{X}_i, \hat{P}_j] &= i\hbar\delta_{ij}, \qquad \text{and} \\ [\hat{X}_i, \hat{X}_j] &= [\hat{P}_i, \hat{P}_j] = 0. \end{aligned} \tag{1}$$

However there is no evidence that at very short distances (or very high energies) these relations should still be true. Then a natural generalization of the above is to take the coordinates which do not commute any more,

$$[\hat{X}_i, \hat{X}_j] = i\theta_{ij}, \tag{2}$$

where θ_{ij} is a *constant* of dimension $[L]^2$. Such noncommutative coordinates have recently been motivated by string theory [1, 2]. An immediate remark is that introducing this kind of commutation relation between coordinates the Lorentz invariance is spoiled explicitly. We should remember however that we assumed this feature to appear only at very short distances, i.e. for $\theta \to 0$ we should recover the Lorentz symmetry. This is one of the main constraints on our noncommutative field theories: at least at classical level, in the limit $\theta \to 0$ we should find a previously known commutative field theory. In general (2) can be extended to space-time coordinates:

$$[\hat{X}_\mu, \hat{X}_\nu] = i\theta_{\mu\nu}. \tag{3}$$

Hereafter we call a space with the above commutation relations as a noncommutative space.

To construct the perturbative field theory formulation, it is more convenient to use fields which are some functions and not operator valued objects. To pass to such fields while keeping property (3) one should redefine the multiplication law of functional

CP589, *Fundamental Interaction Theories*, edited by J. Lukierski and J. Rembieliński

(field) space. This new multiplication is induced from (3) through the so called Weyl-Moyal correspondence [3, 4]:

$$\hat{\Phi}(\hat{X}) \longleftrightarrow \Phi(x)\ ;$$

$$\begin{aligned}\hat{\Phi}(\hat{X}) &= \int_{\alpha} e^{i\alpha\hat{X}}\, \phi(\alpha)\, d\alpha, \\ \phi(\alpha) &= \int e^{-i\alpha x}\, \Phi(x)\, dx,\end{aligned} \tag{4}$$

where α and x are real variables. Then,

$$\begin{aligned}\hat{\Phi}_1(\hat{X})\, \hat{\Phi}_2(\hat{X}) &= \int_{\alpha,\beta} e^{i\alpha\hat{X}}\, \phi(\alpha)\, e^{i\beta\hat{X}}\, \phi(\beta)\, d\alpha\, d\beta \\ &= \int_{\alpha,\beta} e^{i(\alpha+\beta)\hat{X} - \frac{1}{2}\alpha_\mu\beta_\nu[\hat{X}_\mu,\hat{X}_\nu]}\, \phi_1(\alpha)\, \phi_2(\beta)\, d\alpha\, d\beta\,,\end{aligned} \tag{5}$$

and hence,

$$\hat{\Phi}_1(\hat{X})\, \hat{\Phi}_2(\hat{X}) \longleftrightarrow \Big(\Phi_1 \star \Phi_2\Big)(x)\,, \tag{6}$$

$$\Big(\Phi_1 \star \Phi_2\Big)(x) \equiv \left[e^{\frac{i}{2}\theta_{\mu\nu}\partial_{\xi_\mu}\partial_{\eta_\nu}}\, \Phi_1(x+\xi)\, \Phi_2(x+\eta) \right]_{\xi=\eta=0}. \tag{7}$$

This suggests that we can work on a usual commutative space for which the multiplication operation is modified to the so called star product (7). Then, one can define a bracket out of the $\star$-product

$$\{f,\, g\}_{MB} = f \star g - g \star f\,. \tag{8}$$

It is easy to check that the Moyal bracket of two coordinates x_μ and x_ν gives exactly the desired commutation relations, (3):

$$[x^\mu, x^\nu]_{MB} \;=\; i\,\theta^{\mu\nu}\,. \tag{9}$$

NONCOMMUTATIVE FIELD THEORY AT CLASSICAL LEVEL

As we have seen in the previous section, the way to treat the noncommutative theories is to modify the usual product of fields with the star product. So, for example, the action for the noncommutative analog of the real Φ^4 theory will be:

$$S[\Phi] \;=\; \int d^4x \left[\frac{1}{2}\partial_\mu\Phi \star \partial^\mu\Phi - \frac{m^2}{2}\Phi \star \Phi - \frac{\lambda}{4!}\Phi \star \Phi \star \Phi \star \Phi \right]. \tag{10}$$

Thanks to the properties of the $\star$-product under the integral sign (see the Appendix), the quadratic part of the action is the same as in the commutative case. Therefore, the only thing which is modified is the interaction. This is a very important point to keep in mind that the free theory is *the same* as in the commutative case. However we should remind that this is not true for topologically non-trivial spaces [7].

Since in the $\star$-product derivatives are involved, it would be more convenient to use the Fourier modes, in terms of which the action can be written as

$$\begin{aligned} S &= \int d^4k \frac{-1}{2} \phi(k)(k^2+m^2)\phi(-k) + S_{int}, \\ S_{int} &= -\frac{\lambda}{3\cdot 4!} \int d^4k_1 \dots d^4k_4\, \phi(k_1)\, \phi(k_2)\, \phi(k_3)\, \phi(k_4)\, (2\pi)^4 \delta^{(4)}(k_1+k_2+k_3+k_4) \times \\ &\times \ [\cos k_1\theta k_2 \cos k_3\theta k_4 + \cos k_1\theta k_3 \cos k_2\theta k_4 + \cos k_1\theta k_4 \cos k_2\theta k_3], \end{aligned} \tag{11}$$

where $k_i\theta k_j = k_i^\mu \theta_{\mu\nu} k_j^\nu$. Therefore the only difference which appears in the noncommutative theory, compared to the commutative one, is that for every vertex in the noncommutative Φ^4 theory we should multiply by an additional factor:

$$\frac{1}{3}[\cos k_1\theta k_2 \cos k_3\theta k_4 + \cos k_1\theta k_3 \cos k_2\theta k_4 + \cos k_1\theta k_4 \cos k_2\theta k_3]. \tag{12}$$

Conjugate Momentum and the Equations of Motion

The classical equations of motion, similar to the commutative case, are obtained by minimizing the action, i.e.

$$\frac{\delta S}{\delta \Phi} = 0. \tag{13}$$

Then the equation of motion for the scalar field theory with a Φ^4 interaction can be written as:

$$(\partial_\mu \partial^\mu + m^2)\Phi = \frac{\lambda}{3!} (\Phi \star \Phi \star \Phi)(x). \tag{14}$$

In order to find the conjugate momentum we should first distinguish two major cases:

- $\theta_{0i} = 0$,
- $\theta_{0i} \neq 0$.

$\underline{\theta_{0i} = 0}$

In this case the only place where we encounter time derivatives is the kinetic term, so the conjugate momentum is the same as in the commutative case.

$\underline{\theta_{0i} \neq 0}$

This case is more delicate since we have an infinite number of time derivatives in the interaction term. It is obvious that there is something non-trivial in this case; the conjugate momentum depends on the interaction terms. The infinite number of time derivatives suggests us that the theory is nonlocal in time so causality may be violated [5]. It was also shown that at quantum level unitarity is not preserved any more [6]. For these reasons, from now on, we will restrict ourselves only to the case with $\theta_{0i} = 0$.

Noether Theorem

To complete our analysis on the noncommutative field theories at classical level, we study the Noether theorem. Suppose that our action has a *global continuous* symmetry. For an infinitesimal transformation we can write:

$$S[\Phi] \;=\; S[\Phi+\varepsilon\,\mathcal{F}(\Phi)]\,,\ \text{ with } \varepsilon=\text{constant}. \tag{15}$$

Taking now an x-dependent ε we define the current J through the relation:

$$S[\Phi+\varepsilon(x)\,\mathcal{F}]-S[\Phi] \;\equiv\; -\int J^{\mu}\left(\Phi(x)\right)\partial_{\mu}\varepsilon(x)\,. \tag{16}$$

By definition, the action is stationary for *any* field variation around the classical path, i.e. $\frac{\delta S}{\delta\Phi}=0$. In particular for $\delta\Phi=\varepsilon(x)\,\mathcal{F}$ Eq. (16) becomes:

$$\int J^{\mu}\left(\Phi(x)\right)\partial_{\mu}\varepsilon(x)\Big|_{C.P.} \;=\; 0, \tag{17}$$

where C.P. stands for the classical path. Integrating by parts we find:

$$\int \partial_{\mu}J^{\mu}\left(\Phi(x)\right)\varepsilon(x)\,d^4x \;=\; 0, \tag{18}$$

for any $\varepsilon(x)$. So the current J is conserved. This result is very general and it can be applied for any kind of noncommutative theory. The notion of conserved current is a little different from the commutative case. Due to the star product properties

$$\int [f,g]_{MB}\,d^4x \;=\; 0, \tag{19}$$

so the most we can say from Eq. (18) is:

$$\partial_{\mu}J^{\mu} \;=\; [f,g]_{MB}\,, \tag{20}$$

for some proper functions $f,\ g$. This result is somehow natural since in the limit $\theta\to 0$ the Moyal bracket vanishes and we recover the classical result $\partial_{\mu}J^{\mu}=0$.

Let us see now what happens to the charge which in the commutative case was conserved

$$Q \;=\; \int J^0\,d^3x\,. \tag{21}$$

Since we are considering only the case $\theta_{0i}=0$, we can repeat the argument; however, now we have

$$\int [f,g]_{\theta_{0i}=0}d^3x \;=\; 0\,. \tag{22}$$

This means that if we integrate (20) over the spatial coordinates we get:

$$\partial_0\int J^0\,d^3x+\int\vec{\nabla}\cdot\vec{J}d^3x \;=\; 0, \tag{23}$$

and from here we can say that, as in the commutative case, the charge Q is conserved. Note that this is true only for $\theta_{0i} = 0$, while for $\theta_{0i} \neq 0$ even the notion of conserved charge is ill-defined.

In general, the symmetries in noncommutative space can be classified into two major categories:

1. Internal Symmetries

These are the symmetries which are not acting on space-time, e.g. the gauge and global $U(1)$ symmetries. The conserved current for such symmetries is exactly the same as in the commutative case, but the product of the fields in the corresponding current is now replaced by star product.

2. External (Space-time) Symmetries

These are basically the symmetries like translations, rotations, boosts and scaling (Weyl) and conformal symmetries. This group itself should be decomposed into two categories:

2.1) those which do not change θ. These are mainly translations, rotations in the plane along $\theta_{\mu\nu}$ and boosts in the directions perpendicular to θ-plane. For example, if we consider only $\theta_{23} \neq 0$ (and all the other components of θ to be zero), then we have rotations in the $x^2 - x^3$ plane and boost in the x^1 direction. The other Lorentz generators are explicitly broken, for a given constant θ_{23}, which we have considered here. Let us work out the particular case of translations in more detail [4]:

$$\begin{aligned} x_\mu &\longrightarrow x_\mu + \varepsilon_\mu\,, \\ \Phi &\longrightarrow \Phi + \delta\Phi\,, \\ \delta\Phi &= \varepsilon^\mu \partial_\mu \Phi\,. \end{aligned} \tag{24}$$

For the action of the form:

$$S = \int d^4x\, \mathcal{L}(\Phi, \partial\Phi)\,, \tag{25}$$

where

$$\mathcal{L} = \frac{1}{2}\left(\partial_\mu \Phi \star \partial^\mu \Phi - m^2 \Phi \star \Phi\right) + V_\star(\Phi) \tag{26}$$

we find:

$$\delta S|_{\delta\Phi = \varepsilon^\mu \partial_\mu \Phi} = \int d^4x \left[\frac{1}{2}\partial_\mu \left(\partial^\mu \Phi \star \partial_\nu \Phi\, \varepsilon^\nu + \varepsilon^\nu \partial_\nu \Phi \star \partial^\mu \Phi\right) - \partial_\mu \left(\varepsilon^\mu \mathcal{L}\right)\right]. \tag{27}$$

If we take Φ to be on the classical path, i.e. $\delta S = 0$, we can write:

$$\int \partial_\mu \left(T_{\mu\nu}\right) \varepsilon^\nu\, d^4x = 0\,, \tag{28}$$

where

$$T_{\mu\nu} = \frac{1}{2}\left(\partial_\mu \Phi \star \partial_\nu \Phi + \partial_\nu \Phi \star \partial_\mu \Phi\right) - g_{\mu\nu} \mathcal{L}\,. \tag{29}$$

However we should remind that the divergence of $T_{\mu\nu}$ is not zero, e.g. for the particular case of $V_\star(\Phi) = \frac{\lambda}{4!}\Phi^{\star 4}$, using the equations of motion, we have:

$$\partial_\mu T^{\mu\nu} = \frac{\lambda}{4!}\left[[\Phi, \partial^\nu\Phi]_{MB}, \Phi^{\star 2}\right]_{MB}, \tag{30}$$

which, of course, along the earlier discussions on the conserved charges is not going to destroy the energy-momentum conservation, for $\theta_{0i} = 0$ cases.
2.2) The symmetries which change θ. This group consists of rotations in the direction transverse to θ, scaling and conformal symmetries. As for the scaling, $x \to \kappa x$, and hence $\theta \to \kappa^2\theta$. In general, since the action besides the fields has also θ-dependence, namely $S = S[\Phi;\theta]$, to work out the Noether theorem in this case one should also take care of that, i.e.

$$\delta S = \frac{\delta S}{\delta\Phi}\delta\Phi + \frac{\delta S}{\delta\theta^{\mu\nu}}\delta\theta^{\mu\nu}.$$

The extra terms appearing as a result of θ variations will change the Noether current and even in general one cannot associate a conserved current to these symmetries [11].

QUANTIZATION OF NONCOMMUTATIVE FIELD THEORIES

Now, let us proceed with the quantization issue. As usual there are two equivalent ways of quantization:

Canonical quantization:
Since the quadratic part of the action and hence the momentum conjugate of the fields (of course only for the $\theta_{0i} = 0$ cases) are the same as in the commutative case, we can consistently take the usual Hilbert space and the usual commutation relations, namely,

$$[\Phi(x,t),\, \Phi(x',t)] = 0\,,\quad [\Phi(x,t),\, \Pi(x',t)] = i\delta(x-x')\,,\quad [\Pi(x,t),\, \Pi(x',t)] = 0\,. \tag{31}$$

Path integral quantization:
In this method, instead of the Hilbert space, we should specify the path integral measure. Since the measure is basically composed of the variation of the fields in *different* points, it is not affected by star product and therefore we remain with the same measure as in the commutative case. This is of course compatible with our arguments for the canonical quantization method.

Loop Calculations in Noncommutative Φ^4 Theory in $d = 4$

Using the action (11), we can easily perform loop calculations, just we should take care of the momentum dependent interaction terms. Let us work out the two point function. Considering these phase factors, in general one can distinguish two types of integrals: one without noncommutative phase factor, *the planar part*, and the other with noncommutative phase factor running in the loop, *non-planar integrals* [8], i.e.

$$\begin{aligned}\Gamma^{(2)}_{1\ planar} &= \frac{-\lambda}{3(2\pi)^4}\int\frac{d^4k}{k^2+m^2},\\ \Gamma^{(2)}_{1\ nonplanar} &= \frac{-\lambda}{6(2\pi)^4}\int\frac{d^4k}{k^2+m^2}e^{ik\theta p},\end{aligned}\tag{32}$$

where $k\theta p = k_\mu\theta^{\mu\nu}p_\nu$. The planar integral is proportional to the one loop mass correction of the commutative theory, and is quadratically divergent at high energies. We would like to stress that the coefficient $\frac{1}{3}$ in front of the planar integral is not there in the commutative case. Since these numeric factors are very crucial in the renormalizability of the theory we should explicitly check that these factors are not going to spoil the renormalizability. In order to see the effect of the phase factor in the second integral, we rewrite the expressions for the two integrals in terms of Schwinger parameters

$$\frac{1}{k^2+m^2} = \int_0^\infty d\alpha\, e^{-\alpha(k^2+m^2)}.\tag{33}$$

Now we can easily perform the momentum Gaussian integrals. In order to regularize the α integrals we introduce the cut-off factor $exp(-\frac{1}{\Lambda^2\alpha})$ into the integrals, and then we have:

$$\begin{aligned}\Gamma^{(2)}_{1\ planar} = \frac{-\lambda}{48\pi^2}\int\frac{d\alpha}{\alpha^2}e^{-\alpha m^2-\frac{1}{\Lambda^2\alpha}} &= \frac{-\lambda}{48\pi^2}\left(\Lambda^2 - m^2\ln(\frac{\Lambda^2}{m^2})+O(1)\right),\\ \Gamma^{(2)}_{1\ nonplanar} = \frac{-\lambda}{96\pi^2}\int\frac{d\alpha}{\alpha^2}e^{-\alpha m^2-\frac{p\circ p+\frac{1}{\Lambda^2}}{\alpha}} &= \frac{-\lambda}{96\pi^2}\left(\Lambda^2_{eff} - m^2\ln(\frac{\Lambda^2_{eff}}{m^2})+O(1)\right),\end{aligned}\tag{34}$$

where

$$\Lambda^2_{eff} = \frac{1}{\frac{1}{\Lambda^2}+p\circ p}\ ,\qquad p\circ p = (p_\mu\theta^{\mu\nu})(p^\rho\theta_{\rho\nu})\ .\tag{35}$$

In the limit $\Lambda\to\infty$, the nonplanar one loop graph remains finite, effectively regulated by the noncommutativity of space-time. In this limit, the effective cutoff $\Lambda^2_{eff} = \frac{1}{p\circ p}$ goes to infinity when either $\theta\to 0$ or $p\to 0$.

The one loop 1PI quadratic effective action is

$$\begin{aligned}S^{(2)}_{1PI} = \int d^4p \quad &\frac{-1}{2}\left\{p^2+M^2+\frac{\lambda}{96\pi^2(p\circ p+\frac{1}{\Lambda^2})} - \frac{\lambda M^2}{96\pi^2}\ln\left(\frac{1}{M^2(p\circ p+\frac{1}{\Lambda^2})}\right)+\right.\\ &\left.\cdots+O(\lambda^2)\right\}\phi(p)\phi(-p),\end{aligned}\tag{36}$$

where $M^2 = m^2+\frac{\lambda\Lambda^2}{48\pi^2}-\frac{\lambda m^2}{48\pi^2}\ln\left(\frac{\Lambda^2}{m^2}\right)$... is the renormalized mass. Consider the two cases

a) $p\circ p \ll \frac{1}{\Lambda^2}$, and in particular the zero momentum limit. Here $\Lambda_{eff}\approx\Lambda$, and we recover the effective action of the commutative theory,

$$S^{(2)}_{1PI} = \int d^4p\frac{-1}{2}\left(p^2+M'^2\right)\phi(p)\phi(-p)\ ,\tag{37}$$

where $M'^2 = M^2 + 3\frac{\lambda\Lambda^2}{96\pi^2} - \frac{3\lambda m^2}{96\pi^2}\ln\left(\frac{\Lambda^2}{m^2}\right)\dots$. If M is fine tuned to be cutoff independent, then M' and also $S^{(2)}_{1PI}$ diverge as $\Lambda \to \infty$.

b) $p \circ p \gg \frac{1}{\Lambda^2}$ and in particular the limit $\Lambda \to \infty$. Here $\Lambda^2_{eff} = \frac{1}{p \circ p}$, and

$$S_{eff} = \int d^4p \frac{-1}{2} \left[p^2 + M^2 + \frac{\lambda}{96\pi^2 p \circ p} - \frac{\lambda M^2}{96\pi^2}\ln\left(\frac{1}{m^2 p \circ p}\right) + \ldots + O(\lambda^4)\right]\phi(p)\phi(-p)\,. \qquad (38)$$

The fact that the limit $\Lambda \to \infty$ does not commute with the low momentum limit $p \to 0$ demonstrates the interesting mixing of the UV ($\Lambda \to 0$) and IR ($p \to 0$) in this theory.

To summarize, at one loop level, the planar part of diagrams shows the same UV divergence structure as in the corresponding commutative theory, while the non-planar pieces are UV finite. So, altogether the counter-terms (responsible for the cancellation of UV infinities) have the same structure, but with different numeric factors, compared to the commutative counter-part. However, surprisingly it turns out that this difference in the numeric factors is not going to destroy the (UV) renormalizability of the theory. As for the non-planar diagrams, although being UV finite, they involve IR divergences. This is what is known as UV/IR mixing, which is a general feature of the noncommutative field theories [8, 9]. This UV/IR mixing can be understood intuitively: Let us suppose the x and y coordinates to be noncommuting, $[x,y] = i\theta$. From the usual operator algebra one can easily conclude that $\Delta x \Delta y \geq \theta$. So, an increasing precision in x direction ($\Delta x \to 0$ or UV limit) is naturally related to the $\Delta y \to \infty$ or the IR limit.

Although in [8] some arguments have been presented to remove these new IR divergences, they are not yet well-understood.

One can proceed with the loop calculations at two loops level. It has been shown that the ϕ^4 theory is (UV) renormalizable. Again we remain with this IR/UV mixing issue. The important result of the two loops calculations is that the θ parameter *does not* receive any quantum corrections and we expect this result to remain at all loops order [4].

THE NONCOMMUTATIVE QED

As the second example we study the noncommutative gauge theories and NCQED. In order to do that we first start by constructing the action.

i) Pure gauge theory

The action for the pure gauge theory is

$$S = \frac{1}{4\pi}\int F_{\mu\nu} \star F^{\mu\nu} d^4x = \frac{1}{4\pi}\int F_{\mu\nu} F^{\mu\nu} d^4x\,, \qquad (39)$$

with

$$F_{\mu\nu} = \partial_{[\mu}A_{\nu]} + ie\{A_\mu, A_\nu\}_{MB}\,. \qquad (40)$$

In the above e is the gauge coupling constant. One can show that the above action enjoys the noncommutative gauge transformations [2, 12, 15]

$$A_\mu \to A'_\mu = U(x) \star A_\mu \star U^{-1}(x) + \frac{i}{e} U(x) \star \partial_\mu U^{-1}(x) ,$$

$$U(x) = exp\star(i\lambda), \quad U^{-1}(x) = exp\star(-i\lambda) , \tag{41}$$

where

$$\begin{aligned} exp\star(i\lambda(x)) &\equiv 1 + i\lambda - \frac{1}{2}\lambda\star\lambda - \frac{i}{3!}\lambda\star\lambda\star\lambda + \dots , \\ U(x)\star U^{-1}(x) &= 1 . \end{aligned} \tag{42}$$

Since here we are only interested in NCQED, we choose λ and A_μ to be in $U(1)$ algebra. However, this can easily be extended to $U(n)$ valued functions giving rise to NCU(n) theory.

ii)Fermionic Part

Fermions can be added to the above gauge theory, developing the definition of "covariant derivative". In the NCQED, there are two different kinds of covariant derivatives related by charge conjugation. In other words, there are two different types of fermions which are mapped into each other by charge conjugation, hence they can be called positively or negatively charged fermions [12, 13]. The explicit form of the covariant derivative for the positive charge is

$$D^+_\mu \psi(x) \equiv \partial_\mu\psi(x) - ie(A_\mu \star \psi)(x) , \tag{43}$$

while for the particles with negative charge it is

$$D^-_\mu \psi(x) \equiv \partial^\mu\psi(x) + ie(\psi\star A_\mu)(x) .$$

Here we only consider the D^+ case; the other case, D^-, can be recovered by just sending θ to $-\theta$. The fermionic part of NCQED action is then

$$S_f = \int d^4x \bar{\psi}\star(-i\gamma^\mu D^+_\mu - m)\psi . \tag{44}$$

It is easy to verify that this action is also invariant under NCU(1) transformations defined by $\psi \to U\star\psi$ and (41). Using the definition of the covariant derivative one can verify that:

$$\{D^\pm_\mu, D^\pm_\nu\}_{MB} = \mp ieF_{\mu\nu} . \tag{45}$$

Before proceeding with the loop calculations, let us summarize the behaviour of NCQED under the discrete symmetries, P, C and T; for a complete discussion we refer the reader to [13].

Parity

Under parity, $\theta_{0i} \to -\theta_{0i}$ and $\theta_{ij} \to \theta_{ij}$. Therefore NCQED remains invariant under parity if we consider the noncommutative space (and not space-time).

Charge conjugation

Under charge conjugation, $\theta_{\mu\nu} \to -\theta_{\mu\nu}$ and $A_\mu \to -A_\mu$. Hence NCQED with θ is mapped onto a NCQED with $-\theta$.

Time reversal

Under time reversal, $\theta_{0i} \to \theta_{0i}$ and $\theta_{ij} \to -\theta_{ij}$. So, QED on noncommutative space violates T invariance.

Altogether, QED in noncommutative space violates CP and also T, but, it preserves the whole CPT. The CPT conservation is true also for general noncommutative space-time.

Feynman Rules for NCQED

To perform the loop calculations we should fix the gauge. This can be done, as in the usual gauge through ghosts and Fadeev-Popov method. Here we will fix the Feynman gauge, so that:

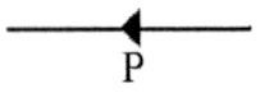

$$= \frac{i}{\not{p} - m + i\varepsilon}\,, \tag{46}$$

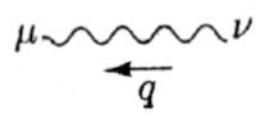

$$= \frac{g^{\mu\nu}}{i(p^2 + i\varepsilon)}\,, \tag{47}$$

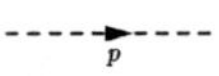

$$= \frac{-1}{i(p^2 + i\varepsilon)}\,. \tag{48}$$

For the interaction terms, we see that they are similar to those of non-Abelian gauge theories [12, 14], in the sense that we get cubic and quadric interaction vertices for the gauge fields besides the usual vertices found in the usual QED. Here we just present the Feynman rules, to show their similarities and differences with respect to the non-Abelian case.

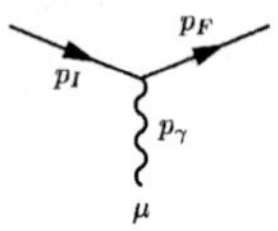

$$= ie\gamma^{\mu} e^{\frac{i}{2} p_I \theta p_F}$$

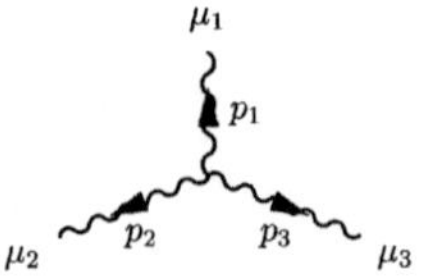

$$\begin{aligned} = & -2e\sin\left(\frac{1}{2} p_1 \theta p_2\right) \\ & \times\Big[(p_1 - p_2)^{\mu_3} g^{\mu_1\mu_2} \\ & + (p_2 - p_3)^{\mu_1} g^{\mu_2\mu_3} \\ & + (p_3 - p_1)^{\mu_2} g^{\mu_3\mu_1}\Big] \end{aligned}$$

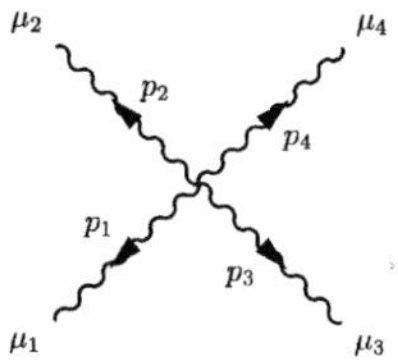

$$
\begin{aligned}
= \; & -4ie^2\Big[(g^{\mu_1\mu_3}g^{\mu_2\mu_4} - g^{\mu_1\mu_4}g^{\mu_2\mu_3}) \\
& \times \sin\left(\frac{1}{2}p_1\theta p_2\right)\sin\left(\frac{1}{2}p_3\theta p_4\right) \\
& \Big[(g^{\mu_1\mu_4}g^{\mu_2\mu_3} - g^{\mu_1\mu_2}g^{\mu_3\mu_4}) \\
& \times \sin\left(\frac{1}{2}p_3\theta p_1\right)\sin\left(\frac{1}{2}p_2\theta p_4\right) \\
& \Big[(g^{\mu_1\mu_2}g^{\mu_3\mu_4} - g^{\mu_1\mu_3}g^{\mu_2\mu_4}) \\
& \times \sin\left(\frac{1}{2}p_1\theta p_4\right)\sin\left(\frac{1}{2}p_2\theta p_3\right)
\end{aligned}
$$

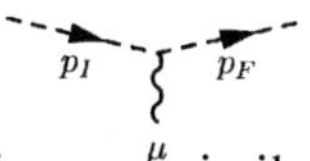

$$
= \; 2iep_F^{\mu}\sin(\frac{1}{2}p_I\theta p_F)\,. \tag{49}
$$

We observe that all vertices are similar to those in non-Abelian gauge theories in which the structure constant is replaced by $2sin(\frac{1}{2}p\theta p')$. This can be seen if we notice that the structure constants appear because of the commutation $[A_a, A_b] = if_{abc}A^c$ in non-Abelian theories. Hence we expect the appearance of the factor $2sin(\frac{1}{2}p\theta p')$ as a consequence of the Moyal bracket, i.e.

$$
\begin{aligned}
[A_\mu, A_\nu]_{MB} &= A_\mu \star A_\nu - A_\nu \star A_\mu \\
&= \int d^4p\, d^4p'\, A_\mu(p)A_\nu(p')\left(e^{\frac{i}{2}p\theta p'} - e^{\frac{-i}{2}p\theta p'}\right)e^{i(p+p').x} \\
&= \int d^4p\, d^4p'\, 2iA_\mu(p)A_\nu(p')\sin(\frac{1}{2}p\theta p')e^{i(p+p').x}\,.
\end{aligned} \tag{50}
$$

We note that the three photon vertex is an explicit sign of the *CP* violation we mentioned earlier.

A very important and peculiar property of noncommutative particles (e.g. electrons) is that they carry a momentum dependent *electric* dipole moment [14]. This can be seen directly from the electron-photon vertex. If we consider the non-relativistic limit we can expand the noncommutative phase factor, namely:

$$
\Gamma^\mu = \gamma^\mu(1 - \frac{i}{2}p_I\theta p_F)\,. \tag{51}
$$

The first term essentially shows the electric charge (at classical level). However, the second term corresponds to an electric dipole moment [14]:

$$
d_e^i = \frac{1}{2}e\theta^{ij}p_j\,, \tag{52}
$$

where p_j is the electron momentum. Also we note that the terms of higher order in momentum contribute to higher-order poles moment.

NCQED at One Loop

Now, let us study the loop corrections to NCQED and particularly check the extra terms in one loop effective action which depend on θ.

Two Point Functions

Photon propagator

As it is clear from the basic Feynman diagrams presented earlier, there are four graphs contributing to the photon propagator at one loop level:

i) one in which the electron is running in the loop. It is easy to check that the noncommutative phase factor is canceled out in this diagram. Therefore it is essentially the same as usual QED.

ii) A photon is running in the loop, and the loop is composed of two 3-photon vertices.

iii) The ghost loop, which is quite similar to the *ii)* case.

iv) The loop diagram which is obtained by joining two of the legs of a 4-photon vertex.

Performing the loop calculations, and regularizing the loop integrals in the the same way as we discussed for noncommutative Φ^4 theory, and sending the UV cut-off Λ to infinity, the photon propagator is found to be [12]

$$i\Pi^{\mu\nu}(q) \sim i\frac{e^2}{16\pi^2}\left\{\frac{10}{3}\left(g^{\mu\nu}q^2 - q^\mu q^\nu\right)\ln(q^2\tilde{q}^2) + 32\frac{\tilde{q}^\mu\tilde{q}^\nu}{\tilde{q}^4} - \frac{4}{3}\frac{q^2}{\tilde{q}^2}\tilde{q}^\mu\tilde{q}^\nu\right\}, \tag{53}$$

where

$$\tilde{q}^\mu = \theta^{\mu\nu}q_\nu\,. \tag{54}$$

We note that since $\tilde{q}\cdot q = 0$, this propagator is manifestly satisfying the Ward identity, $p_\mu\Pi^{\mu\nu} = 0$.

Electron-propagator

Since the noncommutative phase factors are canceled out in the loop, the electron two point function at one loop is exactly the same as in the usual QED.

Electron-Photon Vertex

In the usual QED the more interesting loop contribution is actually appearing as corrections to anomalous magnetic moments. So, here we try to present the same calculation for NCQED. There are two diagrams contributing to electron-photon-electron vertex at one loop. The first involves three basic electron-photon vertices (this has a counter-part in usual QED). The second has a 3-photon vertex and two electron-photon vertices. Using the tools developed for noncommutative Φ^4 loop integrals and the corresponding regularization method, it is a straightforward but tedious calculation to find the electron-photon vertex. For a more detailed analysis we refer to [14] and here we just quote the results:

$$\langle\vec{\mu}\rangle = \frac{e}{m}\left[(1+\frac{\alpha}{\pi})\vec{S} + \frac{\alpha\gamma_{Euller}}{6\pi}\, m^2\vec{\theta}\right] , \tag{55}$$

$$\langle\vec{P}\rangle = \frac{1}{4}e(\vec{\theta}\times\vec{p})\,(1+\frac{3}{\pi}\alpha\gamma_{Euller}) , \tag{56}$$

where

$$(\vec{\theta})_i = \varepsilon_{ijk}\theta_{jk} .$$

The $\vec{\theta}$ dependence of the magnetic dipole moment, Eq.(55), is remarkable because unlike the usual case (the first part of (55)) it is *CP* odd.

SOME PHENOMENOLOGICAL CONSEQUENCES

Now that we have studied the noncommutative field theories and in particular NCQED, we would like to see whether we can impose any bounds on the noncommutativity parameter θ, using the present data. Here without entering to the details, we just mention the results of some cases which so far have been explored:

Pair Production in Constant Electro-Magnetic Field Background

In the usual QED it is well known that the constant electric (and also electro-magnetic) background fields lead to the production of the electron-positron pairs and in this way the background field loses its energy to the particles (e.g. see [16]). Performing the same calculations in the noncommutative case we find that at classical level essentially noncommutativity has *no* effect. However, at one loop, since there is a θ dependence in the magnetic dipole moment (55), the pair production rate becomes [15]

$$I_{1\text{st loop}} = \frac{\alpha EB}{\pi}\; exp\big[-\frac{\pi m^2}{eE}(1-\frac{e\alpha\gamma_E}{3\pi}\vec{\theta}.\vec{B})\big]\times\frac{\cosh(\frac{\pi g B}{2E})}{\sinh(\frac{\pi B}{E})} , \tag{57}$$

where g is the usual gyro-magnetic factor at one loop. As we see the noncommutative effects reduce the rate, however, the effect is so small that we cannot impose any bound on θ from the pair production rate.

Hydrogen Atom Spectrum and Lamb-Shift

In order to study the hydrogen atom in noncommutative space, we need to develop the formulation of noncommutative quantum mechanics [17]. But, without entering to explicit calculations, because we have a preferred direction in our space, namely, $\vec{\theta}$, we expect to find some deviations from the usual spherically symmetric spectrum. In particular, we expect to see a Lamb-shift ($2P\to 2S$ transition) which is polarized [17].

Comparing this shift with the present data and fitting the noncommutative results into the corresponding precision bounds, we find

$$\theta \lesssim (10^4\, GeV)^{-2}\,, \tag{58}$$

which is better than the bound found from scattering results [18].

Primordial Magnetic Field and Noncommutativity in Space

As the last application or consequence of noncommutativity in space we would like to mention some cosmological effects. As the first such effects, we put forward the idea that the primordial magnetic field is a direct consequence of space noncommutativity [19]. To understand that we note that in the magnetic dipole expression (55), we have a spin independent part, or more precisely, there is a term in the magnetic dipole moment which is *parallel* for all the particles. As a result when we consider the effect of all the particles, although the usual spin dependent part upon the summation over all the particles goes away, this $\vec{\theta}$ part remains. This can act a source for the seed magnetic field which can be amplified through the usual astrophysical arguments, i.e. the dynamo effect [19].

Besides the effects on magnetic fields, there have been some speculations on how the noncommutativity can be cooked up in an inflationary scenario [20].

Also recently, it was shown that the noncommutativity will have interesting and remarkable effects in non-zero temperature field theory. Certainly this will have many cosmological consequences, as well as other applications. We will elaborate on this point in future works.

Appendix: Some useful identities in $\star$-calculus

Let f,g be two arbitrary functions on noncommutative R^d:

$$f(x) = \int f(k) e^{ik.x} d^d k, \quad g(x) = \int g(k) e^{ik.x} d^d k.$$

Then

$$(f*g)(x) = \int f(k) g(l) e^{-ik\theta l/2} e^{i(k+l).x} d^d k d^d l\,,$$

where $k\theta l = k^\mu \theta_{\mu\nu} l^\nu$. From the above relation it is straightforward to see:

1) $g*f = f*g|_{\theta\to-\theta}$, and hence $\{f,g\}_{M.B.} = f*g|_\theta - f*g|_{-\theta}$.

2) $\int (f*g)(x) d^d x = \int (g*f)(x) d^d x = \int fg(x) d^d x$.

3) If we denote complex conjugation by $c.c.$, then $(f*g)^{c.c.} = g^{c.c.} * f^{c.c.}$.

If h is another arbitrary function:

4) $(f*g)*h = f*(g*h) \equiv f*g*h$.

5) $\int (f*g*h)(x) d^d x = \int (h*f*g)(x) d^d x = \int (g*h*f)(x) d^d x$.

6) $(f*g*h)|_\theta = (h*g*f)|_{-\theta}$.

In other words the integration on the space coordinates, x, has the cyclic property, and it has all the properties of the Tr in the matrix calculus.

From 2) we learn that the kinetic part of the actions (which are quadratic in fields) is the same as their commutative version. So the free field propagators in commutative and noncommutative spaces are the same.

ACKNOWLEDGMENTS

I would like to thank the organizers of the 37th Karpacz winter school, and particularly to J. Lukierski, for warm hospitality and also for providing a friendly atmosphere for fruitful discussions. I am also grateful to A. Tureanu for reading the manuscript.

REFERENCES

1. Ardalan, F., Arfaei, H., and Sheikh-Jabbari, M.M., *Mixed Branes and M(atrix) Theory on Noncommutative Torus*, hep-th/9803067; *JHEP*, **9902**, 016 (1999), hep-th/9810072; *Nucl. Phys.*, **B576**, 578 (2000), hep-th/9906161.
   Chu, C-S., Ho, P-M., *Nucl. Phys.*, **B550**,151 (1999), hep-th/9812219; *Nucl. Phys.*, **B568**, 447 (2000), hep-th/9906192.
2. Seiberg, N., and Witten, E., *JHEP*, **9909**, 032 (1999), hep-th/9908142.
3. Alvarez-Gaume, L., and Wadia, S. R., *Gauge Theory on a Quantum Phase Space*, hep-th/0006219.
4. Micu, A., and Sheikh-Jabbari, M.M., *JHEP*, **01**, 025 (2001), hep-th/0008057.
5. Seiberg, N., Susskind, L., and Toumbas, N., *JHEP*, **0006**, 044 (2000), hep-th/0005015.
6. Gomis, J., and Mehen, T., *Space-Time Noncommutative Field Theories and Unitarity*, hep-th/0005129.
7. Chaichian, M., Demichev, A., and Presnajder, P., *Nucl. Phys.*, **B 567**, 360 (2000);
   Chaichian, M., Demichev, A., Presnajder, P., and Tureanu, A., *Space-Time Noncommutativity, Discreteness of Time and Unitarity*, hep-th/0007156.
8. Minwalla, S., Van Raamsdonk, M., and Seiberg, N., *Noncommutative Perturbative Dynamics*, hep-th/9912072.
9. Matusis, A., Susskind, L., and Toumbas, N., *The IR/UV Connection in the Noncommutative Gauge Theories*, hep-th/0002075.
10. Mark Van Raamsdonk, Nathan Seiberg, *JHEP*, **0003**, 035 (2000), hep-th/0002186.
11. Gerhold, A., Grimstrup, J., Grosse, H., Popp, L., Schweda, M., and Wulkenhaar, R., *The Energy-Momentum Tensor on Noncommutative Spaces - Some Pedagogical Comments*, hep-th/0012112.
12. Hayakawa, M., *Perturbative analysis on infrared and ultraviolet aspects of noncommutative QED on* R^4, hep-th/9912167.
13. Sheikh-Jabbari, M.M., *Phys. Rev. Lett.*, **84**, 5265 (2000), hep-th/0001167.
14. Riad, I. F., Sheikh-Jabbari, M.M., *JHEP*, **08**, 045 (2000), hep-th/0008132.
15. Chair, N., and Sheikh-Jabbari, M. M., *Phys. Lett.*, **B504**, 141 (2001), hep-th/0009037.
16. Nikishov, A. I., *Sov. Phys. JETP*, **30**, 660 (1970); *Nucl. Phys.*, **B21**, 346 (1970).
    Kruglov, S.I., *Pair Production and Solutions of the Wave Equation for Particles with Arbitrary Spin*, hep-ph/9908410.
17. Chaichian, M., Sheikh-Jabbari, M. M., and Tureanu, A., *Phys. Rev. Lett.*, **86**, 2716 (2001), hep-th/0010175.
18. Hewett, J. L., Petriello, F. J., and Rizzo, T. G., *Signals for Noncommutative Interactions at Linear Colliders*, hep-ph/0010354.
19. Mazumdar, Anupam, Sheikh-Jabbari, M. M., *Noncommutativity in Space and Primordial Magnetic Field*, To Appear in *Phys. Rev. Lett.*, hep-ph/0012363.
20. Chu, C.-S., Greene, B. R., and Shiu, G., *Remarks on Inflation and Noncommutative Geometry*, hep-th/0011241.

Quantum Field Theory on the q–Deformed Fuzzy Sphere

H. Steinacker[1]

Laboratoire de Physique Théorique et Hautes Energies, Université de Paris-Sud, Bâtiment 211, F-91405 Orsay, email: Harold.Steinacker@th.u-psud.fr
Sektion Physik der Ludwig–Maximilians–Universität, Theresienstr. 37, D-80333 München

Abstract. We discuss the second quantization of scalar field theory on the q–deformed fuzzy sphere $S^2_{q,N}$ for $q \in \mathbb{R}$, using a path–integral approach. We find quantum field theories which are manifestly covariant under $U_q(su(2))$, have a smooth limit $q \to 1$, and satisfy positivity and twisted bosonic symmetry properties. Using a Drinfeld twist, they are equivalent to ordinary but slightly "nonlocal" QFT's on the undeformed fuzzy sphere, which are covariant under $SU(2)$.

INTRODUCTION

In this paper, we first give a short introduction to the q–deformed fuzzy sphere, and then discuss some aspects of second quantization on this space. This is essentially a short introduction to the more extensive discussion in [1]. Much of the considerations concerning the second quantization generalize to other, higher–dimensional q–deformed spaces.

The q–deformed fuzzy sphere $S^2_{q,N}$ is a q–deformed version of the "ordinary" fuzzy sphere S^2_N [2]. The algebra of functions on $S^2_{q,N}$ is isomorphic to the matrix algebra $Mat(N+1,\mathbb{C})$, but viewed as a $U_q(su(2))$–module algebra. It admits additional structure compatible with covariance under the Drinfeld–Jimbo quantum group $U_q(su(2))$, such as an invariant integral and a differential calculus. It can be defined for both $q \in \mathbb{R}$ and $|q| = 1$, however we restrict ourselves to the case $q \in \mathbb{R}$ here. Then $S^2_{q,N}$ is precisely the "discrete series" of Podles spheres [3]. Moreover, we only consider scalar fields for simplicity. A much more detailed description of $S^2_{q,N}$ has been given in [4]. This space is of interest in the context of D–branes on the $SU_k(2)$ WZW model, as discussed by Alekseev, Recknagel and Schomerus [5]. These authors extract an "effective" algebra of functions on the D–branes from the OPE of the boundary vertex operators, which is twist–equivalent [4] to the space of functions on $S^2_{q,N}$ for q a root of unity.

[1] This work was supported in part by the DFG fellowship STE 995/1-1

CP589, *Fundamental Interaction Theories*, edited by J. Lukierski and J. Rembieliński

THE SPACE $S^2_{Q,N}$

Consider the spin $\frac{N}{2}$ representation of $U_q(su(2))$,

$$\rho : U_q(su(2)) \to Mat(N+1,\mathbb{C}),$$

which acts on $\mathbb{C}^{N+1}$. It can be used to define the quantum adjoint action of $U_q(su(2))$ on the set of matrices $Mat(N+1,\mathbb{C})$, by

$$u \triangleright_q M = \rho(u_1) M \rho(Su_2).$$

The usual matrix algebra $Mat(N+1,\mathbb{C})$ thereby becomes a $U_q(su(2))$–module algebra, which means that $u \triangleright_q (ab) = (u_{(1)} \triangleright_q a)(u_{(2)} \triangleright_q b)$ for $a, b \in Mat(N+1,\mathbb{C})$. Here $\Delta(u) = u_{(1)} \otimes u_{(2)}$ denotes the coproduct of $u \in U_q(su(2))$. $S^2_{q,N}$ is defined to be precisely this $U_q(su(2))$–module algebra $Mat(N+1,\mathbb{C})$, together with some additional structure. It is easy to see that under the (adjoint) action of $U_q(su(2))$, it decomposes into the irreducible representations

$$S^2_{q,N} = Mat(N+1,\mathbb{C}) = (1) \oplus (3) \oplus \ldots \oplus (2N+1), \tag{1}$$

where $(2K+1)$ is the spin K representation of $U_q(su(2))$. This is the analog of the decomposition of functions on the sphere into spherical harmonics, which it is truncated on the fuzzy spheres. Let $\{x_i\}_{i=+,-,0}$ be the weight basis of the spin 1 components in (1), so that $u \triangleright_q x_i = x_j \pi i^j_i(u)$ for $u \in U_q(su(2))$. One can show that they satisfy the relations

$$\begin{aligned} \varepsilon^{ij}_k x_i x_j &= \Lambda_N x_k, \\ g^{ij} x_i x_j &= R^2. \end{aligned}$$

Here

$$\Lambda_N = R\, \frac{[2]_{q^{N+1}}}{\sqrt{[N]_q [N+2]_q}},$$

$[n]_q = \frac{q^n - q^{-n}}{q - q^{-1}}$, and ε^{ij}_k and g^{ij} are the q–deformed invariant tensors. For example, $\varepsilon^{33}_3 = q^{-1} - q$, and $g^{1-1} = -q^{-1}$, $g^{00} = 1$, $g^{-11} = -q$. In [4], these relations were derived using a Jordan–Wigner construction. For $q = 1$, the relations of S^2_N are recovered.

Integration. The unique invariant integral of a function $f \in S^2_{q,N}$ is given by its quantum trace over $Mat(N+1,\mathbb{C})$,

$$\int f := \frac{4\pi R^2}{[N+1]_q} \mathrm{Tr}_q(f) = \frac{4\pi R^2}{[N+1]_q} \mathrm{Tr}(f\, q^{-H}),$$

normalized such that $\int 1 = 4\pi R^2$. Here H is the Cartan generator of $U_q(su(2))$. Invariance means that $\int u \triangleright_q f = \varepsilon(u) \int f$.

Real structure. In order to define a *real* noncommutative space, one must specify a star structure on the algebra of functions. The star of an element f is simply defined to be the hermitean adjoint of the matrix $f \in S^2_{q,N} = Mat(N+1,\mathbb{C})$. In terms of the generators x_i, this becomes

$$x_i^* = g^{ij} x_j,$$

since q is real.

$S^2_{q,N}$ admits additional structure, in particular a differential calculus. While the calculus is very interesting in the context of gauge theories, we shall not discuss it here. The interested reader is referred to [4]. However, we do need a Laplacian in order to write down Lagrangians and actions. While it can naturally be defined using the differential calculus as $\Delta = *_H d *_H d$, we give an ad–hoc definition here for simplicity. Assume that $\{\psi_{K,n}(x)\}_{K,n} \subset S^2_{q,N}$ is a weight basis of the spin K representation of $U_q(su(2))$, so that

$$u \triangleright_q \psi_{K,n}(x) = \psi_{K,m}(x) \pi^m_n(u). \tag{2}$$

It can be normalized such that

$$\int \psi_{K,n}(x) \psi_{K',m}(x) = \delta_{K,K'}\, g^K_{n,m}.$$

The Laplacian is then given by

$$\Delta \psi_{K,n}(x) \; = \frac{1}{R^2} [K]_q [K+1]_q \, \psi_{K,n}(x).$$

SCALAR FIELD THEORY ON $S_{Q,N}$

We can now write down Lagrangians and actions defining scalar field theory on $\S_{q,N}$. Consider for example

$$S[\Psi] = -\int \frac{1}{2} \Psi \Delta \Psi + \lambda \Psi^4 = S_{\mathrm{free}}[\psi] + S_{int}[\psi],$$

where

$$\Psi(x) = \sum_{K,n} \psi_{K,n}(x)\, a^{K,n}. \tag{3}$$

The free action can be rewritten as[2]

$$S_{\mathrm{free}}[\Psi] = -\sum_{K,n} \frac{1}{2} D_K \, \tilde{g}^K_{nm} a^{K,m} a^{K,n}.$$

In general, actions will be polynomials in the variables $a^{K,n}$ which are invariant under $\tilde{U}_q(su(2))$.

---

[2] the tilde labels objects associated with $\tilde{U}_q(su(2))$, which is another copy of $U_q(su(2))$ but with reversed coproduct, see [1].

We want to discuss the second quantization of such models, as in [1]. On the undeformed fuzzy sphere, this is fairly straightforward [7, 2]: the coefficients $a^{K,n}$ are considered as complex numbers or more precisely as coordinate functions[3] on the representation space $\mathbb{R}^{2K+1}$, so that the actions can be considered as polynomials in the algebra $\mathcal{A} = \otimes_{K=0}^{N} Fun(\mathbb{R}^{2K+1})$. The "path integral" is then simply the product of the ordinary integrals over the coefficients $a^{K,n}$, i.e. over $\prod_K \mathbb{R}^{2K+1}$. This defines a quantum field theory which has a $SO(3)$ rotation symmetry, because the path integral is invariant.

In the q–deformed case, this is not as easy, and needs some discussion. We certainly want the models to have a $U_q(su(2))$ symmetry at the quantum level. This means that the coefficients $a^{K,n}$ in (3) must be considered as representations of $U_q(su(2))$. In order to be able to do calculations, we also require that the $a^{K,n}$ generate some kind of algebra $\mathcal{A}$; this is almost a tautology. This strongly suggests that $\mathcal{A}$ should be a $U_q(su(2))$–module algebra. We do not have in mind here an algebra of field operators, which in fact would not be appropriate in the Euclidean case even for $q = 1$. Rather, $\mathcal{A}$ should be an analog of the algebra of coordinate functions on configuration space as above for $q = 1$, i.e. some deformed version of $\otimes_{K=0}^{N} Fun(\mathbb{R}^{2K+1})$. Our goal is to define correlation functions of the fields (3), which after "Fourier transform" amounts to defining

$$\langle a^{K_1,n_1} a^{K_2,n_2} ... a^{K_k,n_k} \rangle =: \langle P(a) \rangle \ \in \mathbb{C}\,, \tag{4}$$

perhaps by some kind of a path integral $\langle P(a) \rangle = \frac{1}{\mathcal{N}} \int \mathcal{D}a\, e^{-S[\Psi]} P(a)$. $P(a)$ will denote some polynomial in the variables $a^{K,n}$ from now on.

It follows immediately from these considerations that $\mathcal{A}$ cannot be commutative, because the coproduct of $U_q(su(2))$ is not cocommutative. In particular, the $a^{K,n}$ cannot be ordinary complex numbers. Therefore an ordinary integral over commutative modes $a^{K,n}$ would violate $U_q(su(2))$ invariance at the quantum level. In some sense, this means that on q–deformed spaces, a second quantization is required by consistency. There is one more essential requirement: $\mathcal{A}$ should have the same Poincaré series as classically, i.e. the dimension of the space of polynomials at a given degree should be the same as in the undeformed case. This is in fact precisely the content of a symmetrization postulate, and it is of course an essential physical requirement at least for low energies, in order to have the correct number of degrees of freedom. It means that the "amount of information" contained in the n–point functions should be the same as for $q = 1$, so that a smooth limit $q \to 1$ is conceivable. In other words, we want to consider ordinary bosons[4]. While some proposals have been given in the literature [8] how to define QFT on spaces with quantum group symmetry, none of them seems to satisfy these requirements.

On a more formal level, we impose the following requirements [1]:

(1) *Covariance:*

$$\langle u \,\tilde{\triangleright}_q P(a) \rangle = \varepsilon_q(u)\, \langle P(a) \rangle,$$

which means that the $\langle P(a) \rangle$ are invariant tensors of $\tilde{U}_q(su(2))$,

---

[3] with star structure $a_i^* = g_c^{ij} a_j$

[4] we do not consider fermions here

(2) *Hermiticity:*

$$\langle P(a)\rangle^* = \langle P^*(a)\rangle\,,$$

for a suitable involution $*$ on $\mathcal{A}$,

(3) *Positivity:*

$$\langle P(a)^* P(a)\rangle \geq 0\,,$$

(4) *Symmetry*
under permutations of the fields, by which we mean that the polynomials in the $a^{K,n}$ can be ordered as usual, i.e. the Poincaré series of $\mathcal{A}$ should be underformed.

A slight refinement will be needed later.

Unfortunately, there is no obvious candidate for an associative $U_q(su(2))$–module algebra $\mathcal{A}$ with the same Poincaré series as $\otimes_{K=0}^{N} Fun(\mathbb{R}^{2K+1})$ (except for small N). We will therefore construct a suitable quasiassociative algebra $\mathcal{A}$ which is a star–deformation of the commutative $\otimes_{K=0}^{N} Fun(\mathbb{R}^{2K+1})$. We want to emphasise that quasi-associativity is in no way inconsistent with the usual axioms of quantum mechanics, because the algebra $\mathcal{A}$ will not be interpreted as algebra of observables; it is only a tool which is useful to calculate correlation functions, just like Grassman variables are used to calculate fermionic correlation functions. In fact, it is possible to avoid the use of quasiassociative algebras alltogether, see [1]. Any lingering doubts can be eliminated by showing the equivalence of our models to ordinary QFT on the undeformed fuzzzy sphere, with slightly derformed interactions.

The chosen approach is rather general and is applicable in a more general context, such as for higher–dimensional theories.

The Quasiassociative Star Product

As discussed, we assume that the coefficients $a^{K,n}$ transform in the spin K representation of $\tilde{U}_q(su(2))$,

$$u\,\tilde{\triangleright}_q a^{K,n} = \pi^n_m(\tilde{S}u)\; a^{K,m}\,. \tag{5}$$

Let φ be the algebra (not coalgebra!)–isomorphism [9]

$$\varphi : \tilde{U}_q(su(2)) \to U(su(2))[[h]]\,,$$

where $q = e^h$. Moreover, let $\mathcal{F} = \mathcal{F}_1 \otimes \mathcal{F}_2 \in U(su(2))[[h]] \otimes U(su(2))[[h]]$ be the "Drinfeld–twist" [9] which relates the Hopf algebras $\tilde{U}_q(su(2))$ and $U(su(2))$, and satisfies among others

$$\begin{aligned}
\mathcal{F} &= \mathbf{1}\otimes\mathbf{1} + o(h)\,,\\
(\varepsilon\otimes \mathrm{id})\mathcal{F} &= \mathbf{1} = (\mathrm{id}\otimes\varepsilon)\mathcal{F}\,,\\
(\varphi\otimes\varphi)\tilde{\Delta}_q(u) &= \mathcal{F}\Delta(\varphi(u))\mathcal{F}^{-1}\,,\\
(\varphi\otimes\varphi)\mathcal{R} &= \mathcal{F}_{21} q^{\frac{t}{2}} \mathcal{F}^{-1}\,,
\end{aligned} \tag{6}$$

for any $u \in \tilde{U}_q(su(2))$. Using this twist, there is an action of $U(su(2))$ on the coefficients $a^{K,n}$, by $u \triangleright a^{K,n} = \varphi^{-1}(u)\,\tilde{\triangleright}_q a^{K,n}$. Hence we can consider the usual commutative algebra

$\mathcal{A}^K := Fun(\mathbb{R}^{2K+1})$ generated by the $a^{K,n}$, and view it as a $U(su(2))$–module algebra $(\mathcal{A}^K, \cdot, \triangleright)$. We now then define a new multiplication on the same space $\mathcal{A}$ by

$$a \star b := (\mathcal{F}_1^{-1} \triangleright a) \cdot (\mathcal{F}_2^{-1} \triangleright b) = \cdot(\mathcal{F}^{-1} \triangleright (a \otimes b)), \tag{7}$$

for any $a, b \in \mathcal{A}^K$. This is analogous to the Moyal product in deformation quantization. It is easy to verify that it satisfies

$$u \tilde{\triangleright}_q (a \star b) = \star \left(\tilde{\Delta}_q(u) \tilde{\triangleright}_q a \otimes b\right),$$

which means that $(\mathcal{A}^K, \star, \tilde{\triangleright}_q)$ is a $\tilde{U}_q(su(2))$–module algebra. It follows from (6) that if a is invariant under $\tilde{U}_q(su(2))$, then it is also central in $(\mathcal{A}^K, \star)$. Moreover, the following commutation relations are derived in [4]:

$$a_i \star a_j - a_k \star a_l \; \tilde{\mathcal{R}}^{lk}_{ij} = 0, \tag{8}$$

were $\tilde{\mathcal{R}}^{lk}_{ij}$ is obtained from the universal $\mathcal{R}$ matrix of $\tilde{U}_q(su(2))$. This new product is not associative, but quasiassocative:

$$(a \star b) \star c = (\tilde{\phi}_1 \triangleright a) \star \left((\tilde{\phi}_2 \triangleright b) \star (\tilde{\phi}_3 \triangleright c)\right).$$

Here

$$\tilde{\phi} := (\mathbf{1} \otimes \mathcal{F})[(\mathrm{id} \otimes \Delta)\mathcal{F}][(\Delta \otimes \mathrm{id})\mathcal{F}^{-1}](\mathcal{F}^{-1} \otimes \mathbf{1}),$$

is the coassociator, which is invariant under $U_q(su(2))$ and closely related to the KZ equation [9]. It is much easier to work with than the Drinfeld-twist $\mathcal{F}$, which in fact is never needed explicitly.

Finally, $(\mathcal{A}, \star, \tilde{\triangleright}_q)$ is defined as in (7), applied to any element of $\mathcal{A} = \otimes_K \mathrm{Fun}(\mathbb{R}^{2K+1})$. Polynomials $P_\star(a)$ must now be given including some "bracketing". Nevertheless, the Poincaré series of $\mathcal{A}$ is undeformed, because the vector space $\mathcal{A}$ is undeformed, and the new product preserves the grading. Different bracketings can always be related using the coassociator.

Invariant actions are now considered of the form

$$S_{int}[\Psi] = \int \Psi(x) \star (\Psi(x) \star \Psi(x)) = I^{(3)}_{K,K',K'';\,n,m,l} \; a^{K,n} \star (a^{K',m} \star a^{K'',l}), \tag{9}$$

which are invariant polynomials in $\mathcal{A}$.

Quantization

The path integral should be a "functional" on $\mathcal{A}$ which is invariant under $\tilde{U}_q(su(2))$. As in deformation quantization, we view $\mathcal{A}^K$ as the vector space of complex–valued functions on $\mathbb{R}^{2K+1}$, and consider the usual classical integral over $\mathbb{R}^{2K+1}$. Observe that it is also invariant under the action $\tilde{\triangleright}_q$ (5) of $\tilde{U}_q(su(2))$, because the algebra structure does not enter here at all. Explicitly, let $\int d^{2K+1}a^K\, f$ be the ordinary integral of $f \in \mathcal{A}^K$ over $\mathbb{R}^{2K+1}$. The path integral is then defined as

$$\int \mathcal{D}\Psi\, f[\Psi] := \int \prod_K d^{2K+1}a^K\, f[\Psi],$$

where $f[\Psi] \in \mathcal{A}$ denotes any integrable function (in the usual sense) of the variables $a^{K,m}$. It is by construction invariant under $\tilde{U}_q(su(2))$.

Correlation functions can now be defined as functionals of "bracketed polynomials" $P_\star(a) = a^{K_1,n_1} \star (a^{K_2,n_2} \star (\ldots \star a^{K_l,n_l}))$ in the field coefficients by

$$\langle P_\star(a)\rangle := \frac{\int \mathcal{D}\Psi\, e^{-S[\Psi]} P_\star(a)}{\int \mathcal{D}\Psi\, e^{-S[\Psi]}}. \tag{10}$$

This is natural, because all invariant actions $S[\Psi]$ commute with the generators $a^{K,n}$. Strictly speaking there should be a factor $\frac{1}{h}$ in front of the action, which we shall omit. Invariance of the action $S[\Psi] \in \mathcal{A}$ implies that

$$\langle u \,\tilde{\triangleright}_q P_\star(a)\rangle = \varepsilon_q(u)\, \langle P_\star(a)\rangle\, .$$

By construction, the number of independent modes of a polynomial $P_\star(a)$ with given degree is the same as for $q = 1$. One can in fact order them, using quasiassociativity together with the commutation relations (8). Therefore the symmetry requirement (4) above is satisfied. Using a suitable formalism, one can show that the requirements (2) and (3) are satisfied as well, see [1].

Finally, the field theories defined in this way are equivalent to ordinary QFT's on the undeformed fuzzy sphere, with slightly nonlocal interactions. Consider an interaction term of the form (9). If we write down explicitly the definition of the $\star$ product of the $a^{K,n}$ variables, then it can be viewed as an interaction term of $a^{K,n}$ variables with a tensor which is invariant under the *undeformed* $U(su(2))$, obtained from the $\tilde{U}_q(su(2))$–invariant tensor by multiplication with the twist $\mathcal{F} = \mathbf{1} + o(h)$. In other words, the above actions can also be viewed as actions on the undeformed fuzzy sphere $S^2_{q=1,N}$, with interactions which are slightly "nonlocal" in the sense of $S^2_{q=1,N}$, i.e. they are given by traces of products of matrices only to the lowest order in h. Upon spelling out the $\star$ product in the correlation functions (10) as well, they can be considered as ordinary correlation functions of a slightly nonlocal field theory on $S^2_{q=1,N}$, disguised by the transformation $\mathcal{F}$. Therefore q–deformation simply amounts to some kind of nonlocality of the interactions. A similar interpretation is well–known in the context of field theories on spaces with a Moyal product.

Finally, it is possible to calculate correlators in perturbation theory, and to derive an analog of Wicks theorem. For lack of space, the reader is referred to [1].

ACKNOWLEDGMENTS

The author thanks J. Lukierski and the organizers of the XXXVII Winter School of Theoretical Physics in Karpacz for hospitality and stimulating atmosphere during the school. It is also a pleasure to thank H.Grosse and J. Madore for collaboration on the two papers [4, 1] underlying this report, and G. Fiore and J. Pawelczyk for useful discussions.

REFERENCES

1. Grosse, H., Madore, J., and Steinacker, H., hep-th/0103164.
2. Madore, J., *Class. Quant. Grav.* **9**, 69 (1992).
3. Podleś, P., *Lett. Math. Phys.* **14**, 193 (1987).
4. Grosse, H., Madore, J., and Steinacker, H., hep-th/0005273, to appear in *J. Geom. Phys.*
5. Alekseev, A. Yu., Recknagel, A., and Schomerus, V., *JHEP* **9909**, 023 (1999), hep-th/9908040.
6. Alekseev, A. Yu., Recknagel, A., and Schomerus, V., *J. High-Energy Phys.* **0005**, 010 (2000), hep-th/0003187.
7. Grosse, H., Klimcik, C., and Presnajder, P., *Int. J. Theor. Phys.* **35**, 231 (1996), hep-th/9505175.
8. Oeckl, R., *Commun. Math. Phys.* **217**, 451–473 (2001);
   Chaichian, M., Demichev, A., and Presnajder, P., *J.Math.Phys.* **41**, 1647–1671 (2000).
9. Drinfel'd, V.G., *Leningrad Math. J.* **2**, No.4, 829 (1991);
   Drinfel'd, V.G., *Leningrad Math. J.* **1**, No.6, 1419 (1991).

Rational-Trigonometric Deformation

V.N. Tolstoy

Institute of Nuclear Physics, Moscow State University 119899 Moscow, Russia, email: tolstoy@nucl-th.sinp.msu.ru

Abstract. We discuss a rational-trigonometric deformation for two algebraic cases: for the universal enveloping algebra $U(g[u])$ of a polynomial loop algebra $g[u]$, where g is a finite–dimensional complex simple Lie algebra, and for the two-dimensional plane (x,y). In the both cases these deformations are obtained by a singular transformation (at $q=1$) of the q-deformation of $U(g[u])$ and (x,y). In the first case the quantum Hopf algebra called Drinfeldian $D_{q\eta}(g)$ is a quantization of $U(g[u])$ in the direction of a classical r-matrix which is a sum of the simple rational and trigonometric r-matrices. The Drinfeldian $D_{q\eta}(g)$ contains $U_q(g)$ as a Hopf subalgebra, moreover $U_q(g[u])$ and Yangian $Y_\eta(g)$ are its limit quantum algebras when the deformation parameters η goes to 0 and q goes to 1, respectively. Using the rational-trigonometric deformation of the plane (x,y) we introduce the (q,η)- and η-numbers, (q,η)- and η-exponentials, and (q,η)- and η-hypergeometric series.

INTRODUCTION

As it is well known, an universal enveloping algebra $U(g[u])$ of a polynomial current Lie algebra $g[u]$, where g is a finite–dimensional complex simple Lie algebra, admits two type deformations [2, 3]: a trigonometrical deformation $U_q(g[u])$ and a rational deformation or Yangian $Y_\eta(g)$. The algebras $U_q(g[u])$ and $Y_\eta(g)$ are quantizations of $U(g[u])$ in direction of the simple trigonometric and rational solutions of the classical Yang-Baxter (CYBE) over g, correspondingly.

It turns out that $U(g[u])$ also admits a two-parameter deformation [7, 8], which is a quantization of $U(g[u])$ in the direction of a classical r-matrix which is a sum of the simple rational and trigonometric r-martices. This Hopf algebra contains the quantum algebra $U_q(g)$ as a Hopf subalgebra, and the quantum (q-deformed) algebra $U_q(g[u])$ and the Yangian $Y_\eta(g)$ are its limit quantum algebras when the parameters of deformation η goes to 0 and q goes to 1, correspondingly. It should be noted that existence of such two-parameter deformation of $U(g[u])$ was observed by V.G. Drinfeld [4]. By virtue of this fact, this two-parameter deformation is called the Drinfeldian and it will be denoted by $D_{q\eta}(g)$.

In order to understand the connection between the quantum algebras $U_q(g[u])$, $Y_\eta(g)$ and $D_{q\eta}(g[u])$ let us consider their quasiclassical picture, i.e. bialgebras which correspond to $U_q(g[u])$, $Y_\eta(g)$ and $D_{q\eta}(g[u])$. For the sake of simplicity we take the case $g=sl_2$. The simple rational $r_{rt}(u,v)$ and trigonometric $r_{tr}(u,v)$ solutions of CYBE have the form

$$\begin{aligned} r_{rt}(u,v) &= \frac{\eta}{u-v}\Omega_2^{(sp)}, \\ r_{tr}(u,v) &= \hbar\Big(\frac{u+v}{u-v}\Omega_2^{(sp)} + e_{-\alpha}\otimes e_{\alpha} - e_{a}\otimes e_{-\alpha}\Big). \end{aligned} \qquad (1)$$

CP589, *Fundamental Interaction Theories*, edited by J. Lukierski and J. Rembieliński

Here $\{h_\alpha, e_{\pm\alpha}\}$ is a standard Cartan-Weyl basis of sl_2; $\Omega_2^{(sp)}$ is the "splitted" Casimir element: $\Omega_2^{(sp)} = \frac{1}{2}\big(\Delta_0(\Omega) - \Omega_2 \otimes 1 - 1 \otimes \Omega_2\big)$; $\Omega_2 = \frac{1}{2}(h_\alpha^2 + e_\alpha e_{-\alpha} + e_\alpha e_{-\alpha})$ is a second order Casimir element; Δ_0 is the primitive comultiplication. It is easy to see that the sum of rational and trigonometric solutions of CYBE is also a solution of CYBE. Indeed, we have

$$r_{tr}(u+\frac{a}{2}, v+\frac{a}{2}) = r_{tr}(u,v) + \frac{\hbar a}{\eta} r_{rt}(u,v) \qquad \text{for } a \in \mathbb{C}\,. \tag{2}$$

Since $r_{tr}(u,v)$ satisfies CYBE

$$[r_{tr}^{12}(u_1,u_2), r_{tr}^{13}(u_1,u_3) + r_{tr}^{23}(u_2,u_3)] + [r_{tr}^{13}(u_1,u_3), r_{tr}^{23}(u_2,u_3)] = 0\,, \tag{3}$$

the sum $r_{rtr}(u,v) := r_{rt}(u,v) + r_{tr}(u,v)$ also satisfies CYBE.

The classical r-matrices $r_{rt}(u,v)$, $r_{tr}(u,v)$ and $r_{rt-tr}(u,v)$ define the different Lie bialgebras $(sl_2[u], \delta_i)$, where δ_i is the cocommutator given by $\delta_i(x) = [x \otimes 1 + 1 \otimes x, r_i]$, ($x \in sl_2[u]$, r_i is one of our classical r-matrices). Using the comultiplication formulas of the quantum algebras $U_q(sl_2[u])$, $Y_\eta(sl_2)$ and $D_{q\eta}(sl_2[u])$ (see Sections 3) we can show that our bialgebras coincide with quasiclassical limits of $U_q(sl_2[u])$, $Y_\eta(sl_2)$ and $D_{q\eta}(sl_2[u])$, i.e., for example, $\delta_{rtr} = \Delta'_{q\eta} - \Delta_{q\eta}\ (\mathrm{mod}(\hbar^2, \eta^2, \hbar\eta))$. Thus the Drinfeldian $D_{q\eta}(sl_2[u])$ is a quantization of $U(sl_2[u])$ in direction of a classical r-matrix which is a sum of the simple rational and trigonometric r-matrices, and the quantum algebra $U_q(sl_2[u])$ and the Yangian $Y_\eta(sl_2)$ are its limit quantum algebras when the parameters of deformation η goes to 0 and q goes to 1, correspondingly.

In the case of the rational-triginometric deformation of the two-dimensional plane, (q,η)-plane (x,y), we discuss the (q,η)-binomial coefficients introduced by H.B. Benaoum [1] and via solutions of the functional equation $f_1(x+y) = f_2(y)f_3(x)$ we introduced the (q,η)-exponential, the (q,η)-numbers, and (q,η)-hypergeometric series. Since we can show that a quantum algebra, which is dual to a quantum group of linear transformations of the (q,η)-plane, is a rational-trigonometric deformation of $U(sl_2)$ (see [5]) we can assume the the (q,η)-plane is the rational-trigonometric deformation of the plane (x,y). In the case, when the parameter q goes to 1, we obtain the rational (Jordanian) η-deformation of the plane (x,y) (see [6]) and of the special functions.

QUANTUM ALGEBRA $U_q(\widetilde{g[u]})$

Let g be a finite-dimensional complex simple Lie algebra of a rank r with a standard Cartan matrix $A = (a_{ij})_{i,j=1}^r$, with a system of simple roots $\Pi := \{\alpha_1, \dots, a_r\}$, and with a Chevalley basis h_{α_i}, $e'_{\pm\alpha_i}$ $(i = 1,2,\dots,r)$. Let θ be a maximal positive root of g. The corresponding nontwisted affine algebra $\hat{g}$ is generated by g and the additional affine elements $e'_{\pm(\delta-\theta)}$ and $h_{\delta-\theta}$.

Let $\widetilde{g[u]}$ be a Lie algebra generated by g, the positive root vector $e'_{\delta-\theta}$ and the Cartan element $h_{\delta-\theta}$, i.e. $\widetilde{g[u]} \simeq g[u] \oplus \mathbb{C}h_\delta$, where h_δ is a central element. The standard defining relations of $\widetilde{g[u]}$ and of its universal enveloping algebra $U(\widetilde{g[u]})$ are given by the formulas:

$$[h_\delta, \text{everything}] = 0\,, \tag{4}$$

$$[h_{\alpha_i}, h_{\alpha_j}] = 0\,, \tag{5}$$

$$[h_{\alpha_i}, e'_{\pm\alpha_j}] = \pm(\alpha_i,\alpha_j)e'_{\pm\alpha_j}\,, \tag{6}$$

$$[e'_{\alpha_i}, e'_{-\alpha_j}] = \delta_{ij}h_{\alpha_i}\,, \tag{7}$$

$$(\mathrm{ad}\, e'_{\pm\alpha_i})^{n_{ij}} e'_{\pm\alpha_j} = 0 \qquad \text{for } i\neq j,\ n_{ij}{:=}1-a_{ij}\,, \tag{8}$$

$$[h_{\alpha_i}, e'_{\delta-\theta}] = -(\alpha_i,\theta)\, e'_{\delta-\theta}\,, \tag{9}$$

$$[e'_{-\alpha_i}, e'_{\delta-\theta}] = 0\,, \tag{10}$$

$$(\mathrm{ad}\, e'_{\alpha_i})^{n_{i0}} e'_{\delta-\theta} = 0 \qquad \text{for } n_{i0} = 1{+}2(\alpha_i,\theta)/(\alpha_i,\alpha_i), \tag{11}$$

$$[[e'_{\alpha_i}, e'_{\delta-\theta}], e'_{\delta-\theta}] = 0 \qquad \text{for } g\neq sl_2 \text{ and } (\alpha_i,\theta)\neq 0\,, \tag{12}$$

$$[[[e'_{\alpha}, e'_{\delta-\alpha}], e'_{\delta-\alpha}], e'_{\delta-\alpha}] = 0 \qquad \text{for } g = sl_2\,. \tag{13}$$

Here "ad" is the adjoint action of $\widetilde{g[u]}$ in $\widetilde{g[u]}$, i.e. $(\mathrm{ad}x)y = [x,y]$ for $x,y\in\widetilde{g[u]}$. The relations (12) relate to the case $g\neq sl_2$, and the relation (12) belongs to the case $g = sl_2$ (in this case $\theta = \alpha$).

The defining relations for the quantum algebra $U_q(\widetilde{g[u]})$ which is a q-deformation of $U(\widetilde{g[u]})$, can be obtained from the defining relations of the quantum algebra $U_q(\hat{g})$ by removing the relations with the negative affine root vector $e_{-\delta+\theta}$. If $k_\delta^{\pm1} := q^{\pm h_\delta}$, $k_{\alpha_i}^{\pm1} := q^{\pm h_{\alpha_i}}$, $e_{\pm\alpha_i}$ $(i = 1,2,\ldots,r)$, and $e_{\pm(\delta-\theta)}\}$ is the Chevalley basis of $U_q(\hat{g})$ then the standard defining relations of $U_q(\widetilde{g[u]})$ are given by the formulas:

$$[k_\delta^{\pm1}, \text{everything}] = 0\,, \tag{14}$$

$$k_{\alpha_i}^{\pm1}k_{\alpha_j}^{\pm1} = k_{\alpha_j}^{\pm1}k_{\alpha_i}^{\pm1}\,, \tag{15}$$

$$k_\delta k_\delta^{-1} = k_\delta^{-1}k_\delta = k_{\alpha_i}k_{\alpha_i}^{-1} = k_{\alpha_i}^{-1}k_{\alpha_i} = 1\,, \tag{16}$$

$$k_{\alpha_i}e_{\pm\alpha_j}k_{\alpha_i}^{-1} = q^{\pm(\alpha_i,\alpha_j)}e_{\pm\alpha_j}\,, \tag{17}$$

$$[e_{\alpha_i}, e_{-\alpha_i}] = \frac{k_{\alpha_i}-k_{\alpha_i}^{-1}}{q-q^{-1}}\,, \tag{18}$$

$$(\mathrm{ad}_q\, e_{\pm\alpha_i})^{n_{ij}} e_{\pm\alpha_j} = 0 \qquad \text{for } i\neq j,\ n_{ij}{:=}1-a_{ij}\,, \tag{19}$$

$$k_{\alpha_i}e_{\delta-\theta}k_{\alpha_i}^{-1} = q^{-(\alpha_i,\theta)}e_{\delta-\theta}\,, \tag{20}$$

$$
\begin{aligned}
[e_{-\alpha_i}, e_{\delta-\theta}] &= 0\,, && && (21)\\
(\mathrm{ad}_q e_{\alpha_i})^{n_{i0}} e_{\delta-\theta} &= 0 && \text{for } n_{i0} = 1+2(\alpha_i,\theta)/(\alpha_i,\alpha_i), && (22)\\
[[e_{\alpha_i}, e_{\delta-\theta}]_q, e_{\delta-\theta}]_q &= 0 && \text{for } g \neq sl_2 \text{ and } (\alpha_i,\theta)\neq 0\,, && (23)\\
[[[e_{\alpha}, e_{\delta-\alpha}]_q, e_{\delta-\alpha}]_q, e_{\delta-\alpha}]_q &= 0 && \text{for } g = sl_2\,, && (24)
\end{aligned}
$$

where $(\mathrm{ad}_q e_\beta)e_\gamma$ is the q-commutator:

$$
(\mathrm{ad}_q e_\beta)e_\gamma := [e_\beta, e_\gamma]_q := e_\beta e_\gamma - q^{(\beta,\gamma)} e_\gamma e_\beta\,. \tag{25}
$$

The comultiplication Δ_q, the antipode S_q, and the co-unite ε_q of $U_q(\widetilde{g[u]})$ are given by

$$
\begin{aligned}
\Delta_q(k_\gamma^{\pm 1}) &= k_\gamma^{\pm 1}\otimes k_\gamma^{\pm 1} \qquad (\gamma=\delta,\ \alpha_i)\,, && (26)\\
\Delta_q(e_{-\alpha_i}) &= e_{-\alpha_i}\otimes k_{\alpha_i} + 1\otimes e_{-\alpha_i}\,, && (27)\\
\Delta_q(e_{\alpha_i}) &= e_{\alpha_i}\otimes 1 + k_{\alpha_i}^{-1}\otimes e_{\alpha_i}\,, && (28)\\
\Delta_q(e_{\delta-\theta}) &= e_{\delta-\theta}\otimes 1 + k_{\delta-\theta}^{-1}\otimes e_{\delta-\theta}\,, && (29)\\
S_q(k_\gamma^{\pm 1}) &= k_\gamma^{\mp 1} \qquad (\gamma=\delta,\ \alpha_i)\,, && (30)\\
S_q(e_{-\alpha_i}) &= -e_{-\alpha_i}k_{\alpha_i}^{-1}\,, && (31)\\
S_q(e_{\alpha_i}) &= -k_{\alpha_i}e_{\alpha_i}\,, && (32)\\
S_q(e_{\delta-\theta}) &= -k_{\delta-\theta}e_{\delta-\theta}\,, && (33)\\
\varepsilon_q(e_{\pm\alpha_i}) &= \varepsilon_q(e_{\delta-\theta}) = 0\,, && (34)\\
\varepsilon_q(k_\delta^{\pm 1}) &= \varepsilon_q(k_{\alpha_i}^{\pm 1}) = \varepsilon_q(1) = 1\,. && (35)
\end{aligned}
$$

We set

$$
k_{\delta-\theta} = k_\delta k_{\alpha_1}^{-n_1} k_{\alpha_2}^{-n_2}\cdots k_{\alpha_r}^{-n_r}\,, \tag{36}
$$

if $\theta = n_1\alpha_1 + n_2\alpha_2 + \cdots + n_r\alpha_r$.

It is easy to check that the following proposition is valid.

Proposition 1 *There is a one-parameter group of Hopf algebra automorphisms $\mathcal{T}_a$ of $U_q(\widetilde{g[u]})$, $a\in\mathbb{C}$, given by*

$$
\begin{aligned}
\mathcal{T}_a(k_\gamma) &= k_\gamma\,,\\
\mathcal{T}_a(e_{\pm\alpha_i}) &= e_{\pm\alpha_i}\,, && (37)\\
\mathcal{T}_a(e_{\delta-\theta}) &= a\,e_{\delta-\theta}\,.
\end{aligned}
$$

Now starting from the quantum algebra $U_q(\widetilde{g[u]})$ we construct the two-parameter Hopf algebra which coincides with the Yangian $Y_\eta(g)$ at $q=1$.

DRINFELDIAN $D_{q\eta}(g)$

The quantum algebra $U_q(\widetilde{g[u]})$ does not contain any singular elements when q goes to 1 ($q=\exp(\hbar)$). This means that $\lim\limits_{q\to 1} x \in U(\widetilde{g[u]})$ for any $x\in U_q(\widetilde{g[u]})$. Now we extend $U_q(\widetilde{g[u]})$ by singular elements, i.e. we consider the algebra

$$\bar{U}_q(\widetilde{g[u]}) := U_q(\widetilde{g[u]}) \otimes_{\mathbb{C}[[\hbar]]} \mathbb{C}((\hbar))\ . \tag{38}$$

This algebra contains singular elements, i.e. elements which have not any limit when $q\to 1$. For example, if $x\in U_q(\widetilde{g[u]})$ and $\lim\limits_{q\to 1} x \neq 0$ then the element $x/(q-q^{-1})$ is singular.

Let $\tilde{e}_{-\theta}\in U_q(\widetilde{g[u]})$ be any element of the weight $-\theta$, i.e.

$$k_{\alpha_i}\tilde{e}_{-\theta}k_{\alpha_i}^{-1} = q^{-(\alpha_i,\theta)}\tilde{e}_{-\theta}\ , \tag{39}$$

and moreover

$$\lim_{q\to 1}\tilde{e}_{-\theta} = e'_{-\theta}\ , \tag{40}$$

where $e'_{-\theta}$ is a root vector of g with the minimal weight $-\theta$. Let us introduced the new affine generator

$$\xi_{\delta-\theta} = e_{\delta-\theta} + \frac{\eta}{q-q^{-1}}\tilde{e}_{-\theta}\ . \tag{41}$$

Then the defining relations (20)–(24) take the form

$$k_{\alpha_i}\xi_{\delta-\theta}k_{\alpha_i}^{-1} = q^{-(\alpha_i,\theta)}\xi_{\delta-\theta}\ , \tag{42}$$

$$[e_{-\alpha_i},\xi_{\delta-\theta}] = \tau[e_{-\alpha_i},\tilde{e}_{-\theta}]\ , \tag{43}$$

$$(\mathrm{ad}_q e_{\alpha_i})^{n_{i0}}\xi_{\delta-\theta} = \tau(\mathrm{ad}_q e_{\alpha_i})^{n_{i0}}\tilde{e}_{-\theta}\ , \tag{44}$$

for $n_{i0} = 1+2(\alpha_i,\theta)/(\alpha_i,\alpha_i)$, and

$$\begin{aligned}[][[e_{\alpha_i},\xi_{\delta-\theta}]_q,\xi_{\delta-\theta}]_q &= -\tau^2[[e_{\alpha_i},\tilde{e}_{-\theta}]_q,\tilde{e}_{-\theta}]_q \\ &\quad +\tau\Big([[e_{\alpha_i},\tilde{e}_{-\theta}]_q,\xi_{\delta-\theta}]_q+[[e_{\alpha_i},\xi_{\delta-\theta}]_q,\tilde{e}_{-\theta}]_q\Big)\ ,\end{aligned} \tag{45}$$

for $g\neq sl_2$ and $(\alpha_i,\theta)\neq 0$,

$$\begin{aligned}[][[[e_\alpha,\xi_{\delta-\alpha}]_q,\xi_{\delta-\alpha}]_q,\xi_{\delta-\alpha}]_q &= \tau^3[[[e_\alpha,\tilde{e}_{-\alpha}]_q,\tilde{e}_{-\alpha}]_q,\tilde{e}_{-\alpha}]_q \\ &\quad -\tau^2\Big([[[e_\alpha,\tilde{e}_{-\alpha}]_q,\tilde{e}_{-\alpha}]_q,\xi_{\delta-\alpha}]_q+[[[e_\alpha,\tilde{e}_{-\alpha}]_q,\xi_{\delta-\alpha}]_q,\tilde{e}_{-\alpha}]_q \\ &\quad +[[[e_\alpha,\xi_{\delta-\alpha}]_q,\tilde{e}_{-\alpha}]_q,\tilde{e}_{-\alpha}]_q\Big)+\tau\Big([[[e_\alpha,\tilde{e}_{-\alpha}]_q,\xi_{\delta-\alpha}]_q,\xi_{\delta-\alpha}]_q \\ &\quad +[[[e_\alpha,\xi_{\delta-\alpha}]_q,\tilde{e}_{-\alpha}]_q,\xi_{\delta-\alpha}]_q+[[[e_\alpha,\xi_{\delta-\alpha}]_q,\xi_{\delta-\alpha}]_q,\tilde{e}_{-\alpha}]_q\Big)\ ,\end{aligned} \tag{46}$$

for $g = sl_2$. Here and elsewhere $\tau := \eta/(q-q^{-1})$.

The formulas (26)–(34) for the comultiplication $\Delta_{q\eta}$ and the antipode $S_{q\eta}$ are rewritten as follows

$$\Delta_{q\eta}(x) = \Delta_q(x) \qquad (x \in U_q(g)\otimes k_\delta)\ , \tag{47}$$

$$\begin{aligned} \Delta_{q\eta}(\xi_{\delta-\theta}) &= \xi_{\delta-\theta}\otimes 1 + k_{\delta-\theta}^{-1}\otimes \xi_{\delta-\theta} \\ &\quad + \tau\Big(\Delta_q(\tilde{e}_{-\theta}) - \tilde{e}_{-\theta}\otimes 1 - k_{\delta-\theta}^{-1}\otimes \tilde{e}_{-\theta}\Big)\ , \end{aligned} \tag{48}$$

$$S_{q\eta}(x) = S_q(x) \qquad (x \in U_q(g)\otimes k_\delta)\ , \tag{49}$$

$$S_{q\eta}(\xi_{\delta-\theta}) = -k_{\delta-\theta}\xi_{\delta-\theta} + \tau\Big(S_q(\tilde{e}_{-\theta}) + k_{\delta-\theta}\tilde{e}_{-\theta}\Big)\ . \tag{50}$$

The following theorem is valid.

Theorem 1 *If the element $\tilde{e}_{-\theta} \in U_q(\widetilde{g[u]})$ satisfies the conditions (39) and (40) then the right-hand sides of the relations (43)–(46) and (48), (50) are nonsingular at $q = 1$.*

Thus, an associative subalgebra $D_{q\eta}(g) \subset \bar{U}_q(\widetilde{g[u]})$ generated (in the $\hbar$-adic sense [4]) by the quantum algebra $U_q(g)\otimes k_\delta$ and by the elements $k_\delta^{\pm 1}$, $\xi_{\delta-\theta}$ with the relations (42)–(46), where $\tilde{e}_{-\theta} \in U_q(\widetilde{g[u]})$ satisfies the conditions (39) and (40), and with the Hopf structure given by the formulas (47)–(50) is nonsingular at $q = 1$.

Theorem 2 *(i) The Hopf algebra $D_{q\eta}(g)$ is a two-parameter quantization of $U(\widetilde{g[u]})$, ($\widetilde{g[u]} := g[u]\oplus h_\delta$), in the direction of a classical r-matrix which is a sum of the simple rational and trigonometric r-matrices.*
(ii) The Hopf algebra $D_{q=1,\eta}(g) := D_{q\eta}(g)/\hbar D_{q\eta}(g)$ is isomorphic to the Yangian $Y'_\eta(g)$ (with the additional central element h_δ). Moreover, $D_{q\eta=0}(g) = U_q(\widetilde{g[u]})$.

By virtue of this theorem, the Hopf algebra $D_{q\eta}(g)$ can be called as *the rational-trigonometric quantum algebra* or *the q-deformed Yangian*. We also call this algebra as *the Drinfeldian* (in honor of V.G. Drinfeld, who first observed existence of such subalgebra in the extension (38) of the q-deformed affine algebra $U_q(\hat{g})$ (see [4])).

The relations between the Drinfeldian $D_{q\eta}(g)$ and the algebras $U_q(g[u])$, $Y_\eta(g)$, $U(g[u])$ (and also their subalgebras) are shown in the picture:

$$\begin{array}{ccccc} U_q(g)\subset & D_{q\eta}(g) & \xrightarrow{\ \eta\to 0\ } & U_q(g[u]) & \supset U_q(g) \\ & {\scriptstyle q\to 1}\big\downarrow & & \big\downarrow{\scriptstyle q\to 1} & \\ U(g)\subset & Y_\eta(g) & \xrightarrow{\ \eta\to 0\ } & U(g[u]) & \supset U(g) \end{array} \tag{51}$$

Fig.1. A diagram of the limit Hopf algebras of the Drinfeldian $D_{q\eta}(g)$ and their subalgebras. The arrows show passages to the limits.

Remark. Since the defining relations for $D_{q\eta}(g)$ and $U(\widetilde{g[u]})$ in terms of the Chevalley basis differ only in the right-hand sides of the relations (43)-(46), therefore the Dynkin diagram of $g[u]$ can be also used for classification of the Drinfeldian $D_{q\eta}(g)$ and the Yangian $Y_{\eta}(g)$.

As an immediate consequence of Proposition 1 we have the following result.

Proposition 2 *There is a one-parameter group of Hopf algebra automorphisms $\mathcal{T}_a$ of $D_{q\eta}(g)$, $a \in \mathbb{C}$, given by*

$$\begin{aligned} \mathcal{T}_a(x) &= x \qquad (x \in U_q(g) \otimes k_{\delta}) , \\ \mathcal{T}_a(\xi_{\delta-\theta}) &= \Big(1-(q-q^{-1})a\Big)\xi_{\delta-\theta} + \eta a \tilde{e}_{-\theta} . \end{aligned} \tag{52}$$

The results of this Section can be generalized to a supercase (see ([9])).

RATIONAL-TRIGONOMETRIC DEFORMATION OF FUNCTIONS AND OF NUMBERS

The q-, η-, and (q,η)-Analogs of the Binomial Formula

The usual Newton's binomial formula is defined as follows. Let x and y be two commuting coordinate variables, i.e.

$$xy - yx = 0 \qquad \text{(the usual plane)}, \tag{53}$$

then

$$(x+y)^n = \sum_{k=0}^{n} \binom{n}{k} y^k x^{n-k} , \tag{54}$$

where

$$\binom{n}{k} = \frac{n!}{k!(n-k)!} , \tag{55}$$

is the usual Newton's binomial coefficient.

If the coordinates x and y q-commute now,

$$xy - qyx = 0 \qquad \text{(the Manin's plane)}, \tag{56}$$

then a q-analog of the formula (55) is

$$(x+y)^n = \sum_{k=0}^{n} \binom{n}{k}_q y^k x^{n-k} , \tag{57}$$

where

$$\binom{n}{k}_q = \frac{(n)_q!}{(k)_q!(n-k)_q!} , \tag{58}$$

is called the q-analog of $\binom{n}{k}$ or the q-binomial coefficient with

$$(a)_q = \frac{1-q^a}{1-q}, \qquad (59)$$

$$(m)_q = (1)_q(2)_q\cdots(m)_q. \qquad (60)$$

The relation (59) is called the q-analog of the number a, and the relation (60) is the q-factorial of m.

For the Manin's plane (56) we take the linear singular transformation [6] (cp. (41)

$$\begin{aligned} x &\to x - \frac{\eta}{1-q}y, \\ y &\to y, \end{aligned} \qquad (61)$$

then the Manin's plane is transformed to

$$xy - qyx = \eta y^2 \qquad \text{(the } (q,\eta)\text{–plane)}. \qquad (62)$$

Again we obtained although the linear transformation is singular for $q = 1$ the resulting quantum plane is well-defined. If the coordinates x and y satisfy the relation (60) then we obtain the following (q,η)-binomial formula

$$(x+y)^n = \sum_{k=0}^{n} \binom{n}{k}_{q\eta} y^k x^{n-k}, \qquad (63)$$

where $\binom{n}{k}_{q\eta}$ is the (q,η)-binomial coefficient given as follows:

$$\binom{n}{k}_{q\eta} = \frac{(n)_q!}{(k)_{q\eta}!(n-k)_q!} = \binom{n}{k}_q \prod_{s=1}^{k}\Big(1+\eta(s-1)_q\Big), \qquad (64)$$

with

$$(k)_{q\eta} = \frac{(k)_q}{1+\eta(k-1)_q}, \qquad (65)$$

$$(k)_{q\eta}! = (1)_{q\eta}(2)_{q\eta}\cdots(k)_{q\eta}. \qquad (66)$$

It is evident that

$$\begin{aligned} (0)_{q\eta} &= 0, \qquad (1)_{q\eta} = 1, \\ \binom{n}{0}_{q\eta} &= 1, \qquad \binom{n}{1}_{q\eta} = \binom{n}{1}_q. \end{aligned} \qquad (67)$$

The (q,η)-binomial coefficints were introduced by H.B. Benaoum [1]. They obey the following properties [1]:

$$\binom{n+1}{k}_{q\eta} = q^k \binom{n}{k}_{q\eta} + (1+\eta(k-1)_q)\binom{n}{k-1}_{q\eta}, \qquad (68)$$

$$\binom{n+1}{k}_{q\eta} = (1+\eta(k-1)_q)\,\frac{(n+1)_q}{(k)_q}\binom{n}{k-1}_{q\eta}. \qquad (69)$$

Setting in the (q,η)-binomial formula (63) $q=1$ we obtain the η-analog of the Newton's binomial formula and the η-analog of the binomial coefficients:

$$\binom{n}{k}_\eta = \frac{n!}{(k)_\eta!(n-k)!}\,, \tag{70}$$

where

$$(k)_\eta = \frac{k}{1+\eta(k-1)}\,, \tag{71}$$

$$(k)_\eta! = (1)_\eta(2)_\eta\cdots(k)_\eta\,. \tag{72}$$

The q-, η- and (q,η)-Analogs of the Exponential and of Hypergeometric Series

Let us consider the functional equation

$$f_1(x+y) = f_2(y)f_3(x)\,, \tag{73}$$

where $f_i(z)$ are regular functions satisfying the initial conditions

$$f_1(0) = f_2(0) = f_3(0) = 1. \tag{74}$$

(i) If the variables x and y commute, $xy-yx=0$, then a general solution of (73) is the exponentials

$$\begin{aligned} f_1(z) &= f_2(z) = f_3(z) = exp(Cz) \\ &= 1+Cz+\tfrac{1}{2!}(Cz)^2+\cdots+\tfrac{1}{n!}(Cz)^n+\cdots\,. \end{aligned} \tag{75}$$

(ii) If the variables x and y q-commute, $xy-qyx=0$, then a general solution of (73) is the q-exponentials

$$\begin{aligned} f_1(z) &= f_2(z) = f_3(z) = exp_q(Cz) \\ &= 1+Cz+\tfrac{1}{(2)_q!}(Cz)^2+\cdots+\tfrac{1}{(n)_q!}(Cz)^n+\cdots\,. \end{aligned} \tag{76}$$

(iii) If $xy-qyx=\eta y^2$, then a general solution of (73) is the functions:

$$f_1(z) = f_2(z) = \exp_q(Cz)\,, \tag{77}$$

$$f_3(z) = 1+Cz+\tfrac{1}{(2)_{q\eta}!}(Cz)^2+\cdots+\tfrac{1}{(n)_{q\eta}!}(Cz)^n+\cdots\,, \tag{78}$$

where $(n)_{q\eta}$ and $(n)_{q\eta}!$ are given by (65) and (66). The function (78) is called the (q,η)-exponential, $\exp_{q\eta}(x)$.

(iv) If $xy-yx=\eta y^2$, then a general solution (77) and (78) of the equation (73) for $(q=1)$ is the functions:

$$f_1(z) = f_2(z) = \exp(Cz)\,, \tag{79}$$

$$f_3(z) = 1+Cz+\tfrac{1}{(2)_\eta!}(Cz)^2+\cdots+\tfrac{1}{(n)_\eta!}(Cz)^n+\cdots\,, \tag{80}$$

where $(n)_\eta$ and $(n)_\eta!$ are given by (71) and (72). The function (80) is called the η-exponential, $\exp_\eta(x)$.

According to (76), (78) and (80) the relations

$$(a)_{q\eta} = \frac{(a)_q}{1+\eta(a-1)_q}, \tag{81}$$

$$(a)_\eta = \frac{a}{1+\eta(a-1)}, \tag{82}$$

will be called the (q,η)- and η-analogs of the number a.

So we introduced the (q,η)- and η-numbers. Using these numbers we can construct another functions, e.g. (q,η)- and η-hypergeometric series. The standard hypergeometric series $F_{n,m}(x)$ is defined by

$$F_{n,m}(x) = \sum_{k=0}^{\infty} \frac{(a_1)_{(k)}(a_2)_{(k)}\cdots(a_n)_{(k)}}{k!(b_1)_{(k)}(b_2)_{(k)}\cdots(b_m)_{(k)}} x^k, \tag{83}$$

where

$$(a)_{(k)} = a(a+1)\cdots(a+k-1). \tag{84}$$

If we replace the parameters a_i, b_j, and $k!$ by the q-analogs $(a_i)_q$, $(b_j)_q$, and $(k)_q!$ we obtain the basic hypergeometric (q-hypergeometric) series $F_{n,m}^{(q)}(x)$. Analogously the replacement of a_i, b_j, and $k!$ by the (q,η)-analogs $(a_i)_{q\eta}$, $(b_j)_{q,\eta}$, and $(k)_{q\eta}!$ give us the (q,η)-hypergeometric series

$$F_{n,m}^{(q,\eta)}(x), = \sum_{k=0}^{\infty} \frac{((a_1)_{q\eta})_{(k)}((a_2)_{q\eta})_{(k)}\cdots((a_n)_{q\eta})_{(k)}}{(k)_{q\eta}!((b_1)_{q\eta})_{(k)}((b_2)_{q\eta})_{(k)}\cdots((b_m)_{q\eta})_{(k)}} x^k, \tag{85}$$

where

$$((a)_{q\eta})_{(k)} = (a)_{q\eta}(a+1)_{q\eta}\cdots(a+k-1)_{q\eta}. \tag{86}$$

Setting here $q=1$ we obtain the η-hypergeometric series $F_{n,m}^{(\eta)}(x)$. We can also introduce the (q,η)-, and η-analogs of another special functions.

In conclusion generalizing the picture (51) we present the picture of the deformations of the mathematics (math.) which includes Lie algebras, special functions and calculus:

$$\begin{array}{ccc} (q,\eta)\text{-math.} & \xrightarrow{\eta\to 0} & q\text{-math.} \\ {\scriptstyle q\to 1}\big\downarrow & & \big\downarrow{\scriptstyle q\to 1} \\ \eta\text{-math.} & \xrightarrow{\eta\to 0} & \text{math.} \end{array} \tag{87}$$

ACKNOWLEDGMENTS

I would like to thank Institute of Theoretical Physics, University of Wrocław, and the Organizing Committee of the Karpacz Winter School for the support of my visit on the

School. This work was supported by Russian Foundation for Fundamental Research, grant No. RFBR-99-01-01163, and the INTAS grant No. 00-00055.

REFERENCES

1. Benaoum, H.B., (q,η)-analogue of Newton's binomial formula, math-ph/9812028.
2. Drinfeld, V.G., Hopf algebras and quantum Yang-Baxter equation, *Soviet Math. Dokl.,* **283**, 1060–1064 (1985).
3. Drinfeld, V.G., A new realization of Yangians and quantized affine algebras, *Soviet Math. Dokl.,* **32**, 212–216 (1988).
4. Drinfeld, V.G., Quantum groups, in *Proc. ICM-86 (Berkeley USA)* vol. 1, 798-820. Amer. Math. Soc. Providence, RI (1987) 798-820.
5. Khoroshkin, S., Stolin, A., and Tolstoy, V., Q-power function over Q-commuting variables and deformed XXX, XXZ chains, math.QA/0012207
6. Ogievetsky, O., Hopf Structures on the Borel Subalgebra of $sl(2)$, *Supplemento ai Rendiconti del Circolo Mathematico di Palermo,* Serie II, No.37, (1993) 285–199.
7. Tolstoy, V.N., Connection between Yangians and Quantum Affine Algebras, in *Proceedings of the X-th Max Born Symposium, (Wroclav, 1996)*, edited by J. Lukierski, M. Mozrzymas, PWN - Polish Sci. Publishers, Warszawa, 1997, pp.99-117.
8. Tolstoy, V.N., Drinfeldians, in *Lie theory and its application in physics II, (Clausthal, 1997)*, 225–337, edited by H.-D. Doebner, V.K. Dobrev and J. Hilgert, World Sci. Publishing, River Edge, NJ, 1998, pp.225–337; math.QA/9803008.
9. Tolstoy, V.N., From Quantum Affine Superalgebras to super-Drinfeldians and super-Yangians, in *Supersymmetries and Quantum Symmetries, (Dubna, 1999)*, edited by E. Ivanov, S. Krivonos, Joint Institute for Nuclear Research, Dubna, 2000, pp. 431–439.

FIELD-THEORETIC FRAMEWORK: GENERAL FORMALISM AND MODELS

Finite Temperature and Large Gauge Invariance

A. Das

Department of Physics and Astronomy, University of Rochester, Rochester, New York, 14627, email: das@pas.rochester.edu

Abstract.
In this talk, I will summarize the status of our understanding of the puzzle of large gauge invariance at finite temperature.

INTRODUCTION

Gauge theories are beautiful theories which describe physical forces in a natural manner and because of their rich structure, the study of gauge theories at finite temperature [1] is quite interesting in itself. However, to avoid getting into technicalities, we will not discuss the intricacies of such theories either at zero temperature or at finite temperature. Rather, we would study a particular puzzle that arises when a fermion interacts with an external gauge field in odd space-time dimensions.

To motivate, let us note that gauge invariance is realized as an internal symmetry in quantum mechanical systems. Consequently, we do not expect a macroscopic external surrounding, such as a heat bath, to modify gauge invariance. This is more or less what is also found by explicit computations at finite temperature, namely, that gauge invariance and Ward identities continue to hold even at finite temperature [2]. This is certainly the case when one is talking about small gauge transformations for which the parameters of transformation vanish at infinity.

However, there is a second class of gauge transformations, commonly known as large gauge transformations, where the parameters do not vanish at infinity. In odd space-time dimensions, invariance under large gauge transformations leads to some interesting features in physical theories. Let us note that, in odd space-time dimensions, one can, in addition to the usual Maxwell (Yang-Mills) term, have a topological term in the gauge Lagrangian known as the Chern-Simons term. For example, in $2+1$ dimensions, a fermion interacting with a non-Abelian gauge field can be described by a Lagrangian density of the form

$$\begin{aligned}\mathcal{L} &= \mathcal{L}_{\text{gauge}} - m\mathcal{L}_{\text{CS}} + \mathcal{L}_{\text{fermion}} \\ &= \frac{1}{2}\operatorname{tr}F_{\mu\nu}F^{\mu\nu} - m\varepsilon^{\mu\nu\lambda}\operatorname{tr}A_\mu(\partial_\nu A_\lambda + \frac{2g}{3}A_\nu A_\lambda) + \overline{\psi}(\gamma^\mu(i\partial_\mu - gA_\mu) - M)\psi, \quad (1)\end{aligned}$$

where m is a mass parameter, A_μ a matrix valued non-Abelian gauge field in a given representation and "tr" stands for the matrix trace. The first term, on the right hand side, is the usual Yang-Mills term while the second is known as the Chern-Simons term which exists only in odd space-time dimensions. It is a topological term (since it does

CP589, *Fundamental Interaction Theories*, edited by J. Lukierski and J. Rembieliński

not involve the metric) and, in the presence of a Yang-Mills term, its effect is to provide a gauge invariant mass term to the gauge fields. Consequently, such a term is also known as a topological mass term [3]. (Such a term also breaks various discrete symmetries, but we will not get into that.)

Under a gauge transformation of the form

$$\begin{aligned} \psi &\rightarrow U^{-1}\psi, \\ A_\mu &\rightarrow U^{-1}A_\mu U - \frac{i}{g}U^{-1}\partial_\mu U, \end{aligned} \tag{2}$$

it is straightforward to check that both the Yang-Mills and the fermion terms are invariant, while the Chern-Simons term changes by a total divergence leading to

$$S = \int d^3x\,\mathcal{L} \rightarrow S + \frac{4\pi m}{g^2} 2i\pi W, \tag{3}$$

where

$$W = \frac{1}{24\pi^2}\int d^3x\,\varepsilon^{\mu\nu\lambda}\,\mathrm{tr}\partial_\mu UU^{-1}\partial_\nu UU^{-1}\partial_\lambda UU^{-1} \tag{4}$$

is known as the winding number for the gauge transformation. It is a topological quantity which is an integer and which groups all gauge transformations into topologically distinct classes. Basically, it counts how many times the gauge transformations wrap around the sphere. For small gauge transformations, the winding number vanishes since the gauge transformations vanish at infinity whereas non-vanishing winding numbers give rise to large gauge transformations.

Let us note from Eq. (3) that even though the action is not invariant under a large gauge transformation, if m is quantized in units of $\frac{g^2}{4\pi}$, the change in the action would be a multiple of $2i\pi$ and, consequently, the path integral would be invariant under a large gauge transformation. Thus, we have the constraint coming from the consistency of the theory that the coefficient of the Chern-Simons term must be quantized in units of $\frac{g^2}{4\pi}$. (From an operator point of view, such a condition arises from an analysis of the dimensionality of the Hilbert space.)

We have derived the quantization of the CS coefficient from an analysis of the large gauge invariance of the tree level action and we have to worry if the quantum corrections can change the behavior of the theory. At zero temperature, an analysis of the quantum corrections shows that the theory continues to be well defined with the tree level quantization of the Chern-Simons coefficient provided the number of fermion flavors is even. The even number of fermion flavors is also necessary for a global anomaly of the theory to vanish and so, everything is well understood at zero temperature. (Namely, the quantum corrections, with an even number of fermion flavors shift the tree level integer value to another, thereby maintaining large gauge invariance.)

At finite temperature, however, the situation appears to change drastically. Namely, the fermions induce a temperature dependent Chern-Simons term leading to [4]

$$m \rightarrow m - \frac{g^2}{4\pi}\frac{MN_f}{2|M|}\tanh\frac{\beta|M|}{2}. \tag{5}$$

Here, N_f is the number of fermion flavors and $\beta = \frac{1}{kT}$. This shows that, at zero temperature ($\beta \to \infty$), m changes by an integer (in units of $g^2/4\pi$) for an even number of flavors. However, at finite temperature, the CS coefficient becomes a continuous function of temperature and, consequently, it is clear that it can no longer be an integer for arbitrary values of the temperature, as is required for large gauge invariance. It seems, therefore, that temperature would lead to a breaking of large gauge invariance in such a system. This, on the other hand, is completely counter intuitive considering that temperature should have no direct influence on gauge invariance of the theory. As a result, we are left with a puzzle, whose resolution, as we will see, is quite interesting.

C-S THEORY IN 0+1 DIMENSION

The $2+1$ dimensional theory described in the previous section is quite complicated to carry out higher order calculations. On the other hand, as we have noted, Chern-Simons terms can exist in odd space-time dimensions. Consequently, let us try to understand this puzzle of large gauge invariance in a simple quantum mechanical theory. Let us consider a simple theory of an interacting massive fermion with an Abelian gauge field in $0+1$ dimension described by [5, 6]

$$L = \overline{\psi}_j(i\partial_t - A - M)\psi_j - \kappa A\,. \tag{6}$$

Here, $j = 1, 2, \cdots, N_f$ labels the fermion flavors. There are several things to note from this. First, we are considering an Abelian gauge field for simplicity. Second, in this simple model, the gauge field has no dynamics (in $0+1$ dimension the field strength is zero) and, therefore, we do not have to get into the intricacies of gauge theories. There is no Dirac matrix in $0+1$ dimension as well making the fermion part of the theory quite simple as well. And, finally, the Chern-Simons term, in this case, is a linear field so that we can, in fact, think of the gauge field as an auxiliary field. (Note that we have set the coupling constant to unity for simplicity.)

In spite of the simplicity of this theory, it displays a rich structure including all the properties of the $2+1$ dimensional theory that we have discussed earlier. For example, let us note that under a gauge transformation

$$\psi_j \to e^{-i\lambda(t)}\psi_j\,, \qquad A \to A + \partial_t\lambda(t)\,, \tag{7}$$

the fermion part of the Lagrangian is invariant, but the Chern-Simons term changes by a total derivative giving

$$S = \int dt\, L \to S - 2\pi\kappa N\,, \tag{8}$$

where

$$N = \frac{1}{2\pi}\int dt\, \partial_t\lambda(t) \tag{9}$$

is the winding number and is an integer which vanishes for small gauge transformations. Let us note that a large gauge transformation can have a parametric form of the form, say,

$$\lambda(t) = -iN \log\left(\frac{1+it}{1-it}\right). \tag{10}$$

The fact that N has to be an integer can be easily seen to arise from the requirement of single-valuedness for the fermion field. Once again, in light of our earlier discussion, it is clear from Eq. (8) that the theory is meaningful only if κ, the coefficient of the Chern-Simons term, is an integer.

Let us assume, for simplicity, that $M > 0$ and compute the correction to the photon one-point function arising from the fermion loop at zero temperature.

$$iI_1 = -(-i)N_f \int \frac{dk}{2\pi} \frac{i(k+M)}{k^2 - M^2 + i\varepsilon} = \frac{iN_f}{2}. \tag{11}$$

This shows that, as a result of the quantum correction, the coefficient of the Chern-Simons term would change as

$$\kappa \to \kappa - \frac{N_f}{2}.$$

As in $2+1$ dimensions, it is clear that the coefficient of the Chern-Simons term would continue to be quantized and large gauge invariance would hold if the number of fermion flavors is even. At zero temperature, we can also calculate the higher point functions due to the fermions in the theory and they all vanish. This has a simple explanation following from the small gauge invariance of the theory [7]. Namely, suppose we had a nonzero two point function, then, it would imply a quadratic term in the effective action of the form

$$\Gamma_2 = \frac{1}{2} \int dt_1\, dt_2\, A(t_1) F(t_1 - t_2) A(t_2). \tag{12}$$

Furthermore, invariance under a small gauge transformation would imply

$$\delta\Gamma_2 = -\int dt_1\, dt_2\, \lambda(t_1) \partial_{t_1} F(t_1 - t_2) A(t_2) = 0. \tag{13}$$

The solution to this equation is that $F = 0$ so that there cannot be a quadratic term in the effective action which would be local and yet be invariant under small gauge transformations. A similar analysis would show that small gauge invariance does not allow any higher point function to exist at zero temperature.

Let us also note that Eq. (13) has another solution, namely,

$$F(t_1 - t_2) = \text{constant}.$$

In such a case, however, the quadratic action becomes non-extensive, namely, it is the square of an action. We do not expect such terms to arise at zero temperature and hence the constant has to vanish for vanishing temperature. As we will see next, the constant does not have to vanish at finite temperature and we can have non-vanishing higher point functions implying a non-extensive structure of the effective action.

The fermion propagator at finite temperature (in the real time formalism) has the form [1]

$$\begin{aligned} S(p) &= (p+M)\left(\frac{i}{p^2-M^2+i\varepsilon}-2\pi n_F(|p|)\delta(p^2-M^2)\right) \\ &= \frac{i}{p-M+i\varepsilon}-2\pi n_F(M)\delta(p-M) \end{aligned} \tag{14}$$

and the structure of the effective action can be studied in the momentum space in a straightforward manner. However, in this simple model, it is much easier to analyze the amplitudes in the coordinate space. Let us note that the coordinate space structure of the fermion propagator is quite simple, namely,

$$S(t)=\int\frac{dp}{2\pi}e^{-ipt}\left(\frac{i}{p-M+i\varepsilon}-2\pi n_F(M)\delta(p-M)\right)=(\theta(t)-n_F(M))e^{-iMt}. \tag{15}$$

In fact, the calculation of the one point function is trivial now

$$iI_1=-(-i)N_fS(0)=\frac{iN_f}{2}\tanh\frac{\beta M}{2}. \tag{16}$$

This shows that the behavior of this theory is completely parallel to the $2+1$ dimensional theory in that, it would suggest

$$\kappa\to\kappa-\frac{N_f}{2}\tanh\frac{\beta M}{2}$$

and it would appear that large gauge invariance would not hold at finite temperature.

Let us next calculate the two point function at finite temperature.

$$\begin{aligned} iI_2 &= -(-i)^2\frac{N_f}{2!}S(t_1-t_2)S(t_2-t_1) \\ &= -\frac{N_f}{2}n_F(M)(1-n_F(M)) \\ &= -\frac{N_f}{8}\text{sech}^2\frac{\beta M}{2}=\frac{1}{2}\frac{1}{2!}\frac{i}{\beta}\frac{\partial(iI_1)}{\partial M}. \end{aligned} \tag{17}$$

This shows that the two point function is a constant as we had noted earlier implying that the quadratic term in the effective action would be non-extensive.

Similarly, we can also calculate the three point function trivially and it has the form

$$iI_3=\frac{iN_f}{24}\tanh\frac{\beta M}{2}\text{sech}^2\frac{\beta M}{2}=\frac{1}{2}\frac{1}{3!}\left(\frac{i}{\beta}\right)^2\frac{\partial^2(iI_1)}{\partial M^2}. \tag{18}$$

In fact, all the higher point functions can be worked out in a systematic manner. But, let us observe a simple method of computation for these. We note that because of the gauge invariance (Ward identity), the amplitudes cannot depend on the external time coordinates as is clear from the calculations of the lower point functions. Therefore, we can always simplify the calculation by choosing a particular time ordering convenient to us. Second, since we are evaluating a loop diagram (a fermion loop) the initial and the final time coordinates are the same and, consequently, the phase factors in the propagator (15) drop out. Therefore, let us define a simplified propagator without the phase factor as

$$\widetilde{S}(t) = \theta(t) - n_F(M)\,, \tag{19}$$

so that we have

$$\widetilde{S}(t>0) = 1 - n_F(M), \qquad S(t<0) = -n_F(M)\,. \tag{20}$$

Then, it is clear that with the choice of the time ordering, $t_1 > t_2$, we can write

$$\begin{aligned} \frac{\partial\widetilde{S}(t_1-t_2)}{\partial M} &= -\beta\widetilde{S}(t_1-t_3)\widetilde{S}(t_3-t_2)\,, \qquad t_1>t_2>t_3\,, \\ \frac{\partial\widetilde{S}(t_2-t_1)}{\partial M} &= -\beta\widetilde{S}(t_2-t_3)\widetilde{S}(t_3-t_1)\,, \qquad t_1>t_2>t_3\,. \end{aligned} \tag{21}$$

In other words, this shows that differentiation of a fermionic propagator with respect to the mass of the fermion is equivalent to introducing an external photon vertex (and, therefore, another fermion propagator as well) up to constants. This is the analogue of the Ward identity in QED in four dimensions except that it is much simpler. From this relation, it is clear that if we take a n-point function and differentiate this with respect to the fermion mass, then, that is equivalent to adding another external photon vertex in all possible positions. Namely, it should give us the $(n+1)$-point function up to constants. Working out the details, we have,

$$\frac{\partial I_n}{\partial M} = -i\beta(n+1)I_{n+1}\,. \tag{22}$$

Therefore, the $(n+1)$-point function is related to the n-point function recursively and, consequently, all the amplitudes are related to the one point function which we have already calculated. (Incidentally, this is already reflected in Eqs. (17,18)).

With this, we can now determine the full effective action of the theory at finite temperature to be

$$\begin{aligned} \Gamma &= -i\sum_n a^n\,(iI_n) \\ &= -\frac{i\beta N_f}{2}\sum_n \frac{(ia/\beta)^n}{n!}\left(\frac{\partial}{\partial M}\right)^{n-1}\tanh\frac{\beta M}{2} \\ &= -iN_f \log\left(\cos\frac{a}{2} + i\tanh\frac{\beta M}{2}\sin\frac{a}{2}\right), \end{aligned} \tag{23}$$

where we have defined

$$a = \int dt\, A(t)\,. \tag{24}$$

There are several things to note from this result. First of all, the higher point functions are no longer vanishing at finite temperature and give rise to a non-extensive structure of the effective action. More importantly, when we include all the higher point functions, the complete effective action is invariant under large gauge transformations, namely, under

$$a \to a + 2\pi N, \tag{25}$$

the effective action changes as

$$\Gamma \to \Gamma + N N_f \pi, \tag{26}$$

which leaves the path integral invariant for an even number of fermion flavors. This clarifies the puzzle of large gauge invariance at finite temperature in this model. Namely, when we are talking about large changes (large gauge transformations), we cannot ignore higher order terms if they exist. This may provide a resolution to the large gauge invariance puzzle in the $2+1$ dimensional theory as well.

EXACT RESULT

In the earlier section, we discussed a perturbative method of calculating the effective action at finite temperature which clarified the puzzle of large gauge invariance. However, this quantum mechanical model is simple enough that we can also evaluate the effective action directly and, therefore, it is worth asking how the perturbative calculations compare with the exact result.

The exact evaluation of the effective action can be done easily using the imaginary time formalism. But, first, let us note that the fermionic part of the Lagrangian in Eq. (6) has the form

$$L_f = \overline{\psi}(i\partial_t - A - M)\psi, \tag{27}$$

where we have suppressed the fermion flavor index for simplicity. Let us note that if we make a field redefinition of the form

$$\psi(t) = e^{-i\int_0^t dt' A(t')}\,\tilde{\psi}(t) \tag{28}$$

then, the fermionic part of the Lagrangian becomes free, namely,

$$L_f = \overline{\tilde{\psi}}(i\partial_t - M)\tilde{\psi}. \tag{29}$$

This is a free theory and, therefore, the path integral can be easily evaluated. However, we have to remember that the field redefinition in (28) changes the periodicity condition for the fermion fields. Since the original fermion field was expected to satisfy anti-periodicity

$$\psi(\beta) = -\psi(0)$$

it follows now that the new fields must satisfy

$$\tilde{\psi}(\beta) = -e^{-ia}\,\tilde{\psi}(0). \tag{30}$$

Consequently, the path integral for the free theory (29) has to be evaluated subject to the periodicity condition of (30).

Although the periodicity condition (30) appears to be complicated, it is well known that this can be absorbed by introducing a chemical potential [1], in the present case, of the form

$$\mu = \frac{ia}{\beta}. \tag{31}$$

With the addition of this chemical potential, the path integral can be evaluated subject to the usual anti-periodicity condition. The effective action can now be easily determined

$$\begin{aligned}
\Gamma &= -i\log\left(\frac{\det(i\partial_t - M + \frac{ia}{\beta})}{(i\partial_t - M)}\right)^{N_f} \\
&= -iN_f \log\left(\frac{\cosh\frac{\beta}{2}(M - \frac{ia}{\beta})}{\cosh\frac{\beta M}{2}}\right) \\
&= -iN_f \log\left(\cos\frac{a}{2} + i\tanh\frac{\beta M}{2}\sin\frac{a}{2}\right),
\end{aligned} \tag{32}$$

which coincides with the perturbative result of Eq. (24).

LARGE GAUGE WARD IDENTITY

It is clear from the above analysis that, to see if large gauge invariance is restored, we have to look at the complete effective action. In the $0+1$ dimensional model, it was tedious, but we can derive the effective action in closed form which allows us to analyze the question of large gauge invariance. On the other hand, in the theory of interest, namely, the $2+1$ dimensional Chern-Simons theory, we do not expect to be able to evaluate the effective action in a closed form. Consequently, we must look for an alternate way to analyze the question of large gauge invariance in a more realistic model. One such possible method may be to derive a Ward identity for large gauge invariance which will relate different amplitudes much like the Ward identity for small gauge invariance does. In such a case, even if we cannot obtain the effective action in a closed form, we can at least check if the large gauge Ward identity holds perturbatively.

It turns out that the large gauge Ward identities are highly nonlinear [8], as we would expect. Hence, looking for them within the context of the effective action is extremely hard (although it can be done). Rather, it is much simpler to look at the large gauge Ward identities in terms of the exponential of the effective action. Let us define

$$\Gamma(a) = -i\log W(a). \tag{33}$$

Namely, we are interested in looking at the exponential of the effective action (*i.e.* up to a factor of i, W is the basic determinant that would arise from integrating out the fermion field). We will restrict ourselves to a single flavor of massive fermions. The advantage of studying $W(a)$ as opposed to the effective action lies in the fact that, in order for $\Gamma(a)$ to have the right transformation properties under a large gauge transformation, $W(a)$ simply has to be quasi-periodic. Consequently, from the study of harmonic oscillator (as well as Floquet theory), we see that $W(a)$ has to satisfy a simple equation of the form

$$\frac{\partial^2 W(a)}{\partial a^2} + \nu^2 W(a) = g\,, \tag{34}$$

where ν and g are parameters to be determined from the theory. In particular, let us note that the constant g can depend on parameters of the theory such as temperature whereas we expect the parameter ν, also known as the characteristic exponent, to be independent of temperature and equal to an odd half integer for a fermionic mode. However, all these properties should automatically result from the structure of the theory. Let us also note here that the relation (34) is simply the equation for a forced oscillator whose solution has the general form

$$W(a) = \frac{g}{\nu^2} + A\cos(\nu a + \delta) = \frac{g}{\nu^2} + \alpha_1 \cos\nu a + \alpha_2 \sin\nu a\,. \tag{35}$$

The constants α_1 and α_2 appearing in the solution can again be determined from the theory. Namely, from the relation between $W(a)$ and $\Gamma(a)$, we recognize that we can identify

$$\begin{aligned} \nu^2\alpha_1 &= -\left.\frac{\partial^2 W}{\partial a^2}\right|_{a=0} = \left.\left(\left(\frac{\partial\Gamma}{\partial a}\right)^2 - i\frac{\partial^2\Gamma}{\partial a^2}\right)\right|_{a=0}\,, \\ \nu\alpha_2 &= \left.\frac{\partial W}{\partial a}\right|_{a=0} = i\left.\frac{\partial\Gamma}{\partial a}\right|_{a=0}\,. \end{aligned} \tag{36}$$

From the general properties of the fermion theories we have discussed, we intuitively expect $g = 0$. However, these should really follow from the structure of the theory and they do, as we will show shortly.

The identity (34) is a linear relation as opposed to the Ward identity in terms of the effective action. In fact, rewriting this in terms of the effective action (using Eq. (33)), we have

$$\frac{\partial^2\Gamma(a)}{\partial a^2} = i\left(\nu^2 - \left(\frac{\partial\Gamma(a)}{\partial a}\right)^2\right) - ig\,e^{-i\Gamma(a)}\,. \tag{37}$$

So, let us investigate this a little bit more in detail. We know that the fermion mass term breaks parity and, consequently, the radiative corrections would generate a Chern-Simons term, namely, in this theory, we expect the one-point function to be nonzero. Consequently, by taking derivative of Eq. (37) (as well as remembering that $\Gamma(a=0) = 0$), we determine (The superscript represents the number of flavors.)

$$(\nu^{(1)})^2 = \left[\left(\frac{\partial\Gamma^{(1)}}{\partial a}\right)^2 - 3i\frac{\partial^2\Gamma^{(1)}}{\partial a^2} - \left(\frac{\partial\Gamma^{(1)}}{\partial a}\right)^{-1}\left(\frac{\partial^3\Gamma^{(1)}}{\partial a^3}\right)\right]_{a=0}\,,$$

$$g^{(1)} = -\left[2i\frac{\partial^2\Gamma^{(1)}}{\partial a^2} + \left(\frac{\partial\Gamma^{(1)}}{\partial a}\right)^{-1}\left(\frac{\partial^3\Gamma^{(1)}}{\partial a^3}\right)\right]_{a=0}. \tag{38}$$

This is quite interesting, for it says that the two parameters in Eq. (34) or (37) can be determined from a perturbative calculation. Let us note here some of the perturbative results in this theory, namely,

$$\begin{aligned}
\left.\frac{\partial\Gamma^{(1)}}{\partial a}\right|_{a=0} &= \frac{1}{2}\tanh\frac{\beta M}{2}, \\
\left.\frac{\partial^2\Gamma^{(1)}}{\partial a^2}\right|_{a=0} &= \frac{i}{4}\operatorname{sech}^2\frac{\beta M}{2}, \\
\left.\frac{\partial^3\Gamma^{(1)}}{\partial a^3}\right|_{a=0} &= \frac{1}{4}\tanh\frac{\beta M}{2}\operatorname{sech}^2\frac{\beta M}{2}.
\end{aligned} \tag{39}$$

Using these, we immediately determine from Eq. (38) that

$$(\nu^{(1)})^2 = \frac{1}{4}, \qquad g^{(1)} = 0, \tag{40}$$

so that the equation (37) leads to the large gauge Ward identity for a single fermion theory of the form,

$$\frac{\partial^2\Gamma^{(1)}}{\partial a^2} = i\left(\frac{1}{4} - \left(\frac{\partial\Gamma^{(1)}}{\partial a}\right)^2\right). \tag{41}$$

Furthermore, we determine now from Eq. (36)

$$\alpha_1^{(1)} = 1, \qquad \alpha_2^{(1)} = \pm i\tanh\frac{\beta M}{2}. \tag{42}$$

The two signs in of $\alpha_2^{(1)}$ simply corresponds to the two possible signs of $\nu^{(1)}$. With this then, we can solve for $W(a)$ in the single flavor fermion theory and we have (independent of the sign of $\nu^{(1)}$)

$$W_f^{(1)}(a) = \cos\frac{a}{2} + i\tanh\frac{\beta M}{2}\sin\frac{a}{2}, \tag{43}$$

which can be compared with Eq. (32). For N_f flavors, similarly, we can determine the Ward identity to be

$$\frac{\partial\Gamma^{(N_f)}}{\partial a^2} = iN_f\left(\frac{1}{4} - \frac{1}{N_f^2}\left(\frac{\partial\Gamma^{(N_f)}}{\partial a^2}\right)^2\right), \tag{44}$$

where the nonlinearity of the Ward identity is manifest.

Similarly, we can determine the large gauge Ward identity for scalar theories as well as supersymmetric theories, but we will not go into the details of this.

BACK TO 2+1 DIMENSIONS

The analysis of various 0+1 dimensional models shows [9] how the large gauge invariance puzzle gets resolved. A crucial feature in this was the existence of non-extensive higher order terms in the effective action. To further understand this feature, we have also studied the effective action for a fermion interacting with an external gauge field in 1+1 dimensions [10]. In that case, the effective action is non-local, as it should be at finite temperature, but extensive. Furthermore, the effective action shows non-analyticity, as one would expect at finite temperature, unlike the 0+1 dimensional model, where one does not expect any non-analyticity.

Following the results of the 0+1 dimensional model, it was shown [11] that, for the special choice of the gauge field backgrounds where $A_0 = A_0(t)$ and $\vec{A} = \vec{A}(\vec{x})$, the parity violating part of the effective action of a fermion interacting with an Abelian gauge field takes the form

$$\Gamma^{PV} = \frac{ie}{2\pi}\int d^2x \arctan\left(\tanh\frac{\beta M}{2}\tan(\frac{ea}{2})\right)B(\vec{x})\,. \tag{45}$$

However, because the choice of the background is very special, it would seem that this may not represent the complete effective action in a general background. In fact, in higher dimensions, such as 2+1, one also has to tackle with the question of the non-analyticity of the thermal amplitudes which leads to a nonuniqueness of the effective action [1, 12].

With these issues in mind, we have studied the parity violating part of the four point function in 2+1 dimensions at finite temperature. The calculations are clearly extremely difficult and we have evaluated the amplitudes at finite temperature by using the method of forward scattering amplitudes [13]. Without going into details, let me summarize the results here [14]. First, the parity violating part of the box diagram is nontrivial at zero temperature and comes from an effective action of the form

$$\Gamma^4_{T=0} = -\frac{e^4}{64\pi M^6}\int d^3x \varepsilon^{\mu\nu\lambda}F_{\mu\nu}(\partial^\tau F_{\tau\lambda})F^{\rho\sigma}F_{\rho\sigma}\,. \tag{46}$$

This is Lorentz invariant and is invariant under both small and large gauge transformations and is compatible with the Coleman-Hill theorem [15].

At finite temperature, however, the amplitude is not manifestly Lorentz invariant (because of the heat bath) and is non-analytic. We have investigated the amplitude in two interesting limits. Namely, in the long wave limit (all spatial momenta vanishing), the leading term of the amplitude, at high temperature, can be seen to come from an effective action of the form

$$\Gamma^4_{LW} = \frac{e^4}{512MT}\int d^3x \varepsilon_{0ij}E_i(\partial_t^{-1}E_j)(\partial_t^{-1}E_k)(\partial_t^{-1}E_k)\,, \tag{47}$$

where $\vec{E}$ represents the electric field. There are several things to note here. First, this is an extensive action, be it non-local. Second, it is manifestly large gauge invariant and finally, the leading behavior at high temperature goes as $\frac{1}{T}$.

In contrast, we can evaluate the amplitude in the static limit (all energies vanishing) where we find the presence of both extensive as well as non-extensive terms at high temperature. However, the extensive terms are suppressed by powers of T and the leading term seems to come from an effective action of the form

$$\Gamma_S^4 = \frac{e^4}{4\pi T^2}\left(\tanh\frac{\beta M}{2} - \tanh^3\frac{\beta M}{2}\right)\int d^3x a^3 B\,. \tag{48}$$

This coincides with the amplitude that will come from Eq. (45) and has the leading behavior of $\frac{1}{T^3}$ at high temperature. Such a term is not invariant under a large gauge transformation. However, we can now derive a large gauge Ward identity for the leading part of the static action and the solution of the Ward identity coincides with the form given in Eq. (45). This, therefore, clarifies the meaning of the effective action in the special background, namely, it represents the leading term in the effective action in the static limit.

HIGHER ORDER CORRECTIONS

Since we have calculated the box diagram at finite temperature, we can also ask about possible higher loop corrections to the CS coefficient. Let us note that, at zero temperature, there is a result due to Coleman and Hill [15], which says that in an Abelian theory, there cannot be any correction to the CS coefficient beyond one loop. Their result basically uses two simple assumptions, i) small gauge invariance and ii) analyticity of the amplitudes in the momentum space. Small gauge invariance is, of course, known to be true at finite temperature. However, as we have pointed out earlier, amplitudes become non-analytic in the momentum space at finite temperature. Therefore, the second assumption of Coleman-Hill breaks down at finite temperature and one may expect higher loop correction to the CS coefficient at finite temperature.

We have explicitly computed the two loop correction to the CS coefficient at finite temperature [16]. Parameterizing the self-energy in a covariant gauge as

$$\Pi^{\mu\nu}(p,u) = \Pi_1^{\mu\nu}(p,u) + i\varepsilon^{\mu\nu\lambda}p_\lambda\Pi_2(p,u)\,, \tag{49}$$

where u^μ represents the velocity of the heat bath, we find that, in the static limit, the two loop correction to the CS coefficient at high temperatures takes the form

$$\Pi_2^{(2)}(0) = (2m - 3M)\frac{e^4}{192\pi^2 T^2}\ln\frac{T}{m}\,. \tag{50}$$

This result explicitly shows that the Coleman-Hill result breaks down at finite temperatures. Furthermore, the form of the correction is interesting in that it diverges as $m \to 0$. This is the usual manifestation of infrared divergence and shows that, although the zero temperature theory is well defined in the limit of vanishing m, there are infrared divergences at finite temperature. In addition, since there is a two loop correction to the CS coefficient, one may ask the structure of the effective action when higher order corrections are taken into account. This can be determined from the large gauge Ward identity that we have derived and it determines the form of the parity violating effective action, satisfying large gauge invariance, at any order to be

$$\Gamma^{PV} = \tan^{-1}\left(2\Gamma'(0)\tan\frac{ea}{2}\right)\int d^2xB\,, \tag{51}$$

where $\Gamma'(0)$ represents the correction to the CS coefficient to that order.

NON-ABELIAN THEORIES

Our main interest was in the study of the question of large gauge invariance at finite temperature in a non-Abelian theory. As we have shown, the question of large gauge invariance is well understood in an Abelian theory. However, not much is known about the non-Abelian theory yet. To that extent, let us note that even the Coleman-Hill result was derived for an Abelian theory. In this section, let me describe briefly how the Coleman-Hill result can be generalized to non-Abelian theories (at zero temperature).

There are several qualitative differences between the Abelian and the non-Abelian theories. First, while the Abelian theory is well defined even when the tree level CS coefficient vanishes, the infrared divergences in the non-Abelian theory are severe if there is no tree level CS term. Second, while the Abelian theory can be defined in any gauge, the non-Abelian theory is well defined only in a select class of infrared safe gauges such as the Landau gauge, the axial gauge etc. Finally, in the Abelian theory, the CS coefficient is a gauge independent quantity while in the non-Abelian theory, the CS coefficient is gauge dependent. As a result, it is not clear how to generalize the Coleman-Hill result in the non-Abelian case.

On the other hand, it is known from various arguments that, in a non-Abelian theory, it is the ratio $\frac{4\pi m}{g^2}$ which has a physical meaning and, therefore, must be gauge independent. In fact, a one loop calculation verifies that in all infrared safe gauges the one loop correction to this ratio is given by

$$\frac{4\pi m}{g^2} \to \frac{4\pi m}{g^2} + N\,. \tag{52}$$

It is, therefore, meaningful to generalize the Coleman-Hill result for this ratio. (Recall that it is also this ratio that needs to be quantized for large gauge invariance to hold.)

This can indeed be done as follows [17]. First, let us note that, although the CS coefficient is gauge dependent in general, it takes on a physical meaning in the axial gauge. This is because in the axial gauge,

$$\frac{4\pi m}{g^2} \to \frac{4\pi m}{g^2}\left(1+\Pi_2(0)\right). \tag{53}$$

Since this ratio is physical, in this gauge $\Pi_2(0)$ does carry a physical meaning. Using the Ward identities in the axial gauge, it is straight forward to show that the CS coefficient does not receive any corrections beyond one loop provided i) small gauge invariance holds and ii) amplitudes are analytic in the momentum space. A consequence of this is that the ratio $\frac{4\pi m}{g^2}$ does not have any correction beyond one loop in this gauge. However, this is a gauge independent quantity and hence it holds in any gauge that this ratio does not receive any correction beyond one loop. Thus, we understand some features of the zero temperature non-Abelian theory which are parallel to the Abelian theory and what remains is to understand systematically if the issue of large gauge invariance also gets resolved in a parallel manner.

CONCLUSION

In this talk, we have discussed the question of large gauge invariance at finite temperature. We have discussed the resolution of the problem in a simple $0+1$ dimensional model. We have derived the Ward identity for large gauge invariance in this model. We have analyzed the box diagram in $2+1$ dimensions and have obtained the form of the effective action at zero temperature. We have also obtained the amplitude as well as the quartic effective actions in the long wave as well as static limits, at finite temperature. The LW limit has only extensive terms in the action which goes as $\frac{1}{T}$ at high temperature and is invariant under large gauge transformations. The leading term in the static action, however, is non-extensive, goes as $\frac{1}{T^3}$ at high temperature and coincides with the effective action proposed earlier for a restrictive gauge background. This action is not invariant under large gauge transformations. However, using a large gauge Ward identity, we can determine the full leading order action in the static limit which coincides with the effective action obtained in a restrictive gauge background. We have shown explicitly that higher loop corrections to the CS coefficient do not vanish at finite temperature. This violation of the Coleman-Hill result is a consequence of the fact that one of their assumptions (namely, analyticity of the amplitudes) breaks down at finite temperature. We have also extended the result of Coleman-Hill to the case of non-Abelian gauge theories.

ACKNOWLEDGMENTS

It is a pleasure to thank the organizers for their hospitality. I have learnt a lot from all of my collaborators in the field, namely, K. Babu, J. Barcelos-Neto, F. Brandt, A. J. da Silva, G. Dunne, J. Frenkel, P. Panigrahi, K. Rao and J. C. Taylor, and I am grateful to all of them. This work was supported in part by the U.S. Dept. of Energy Grant DE-FG 02-91ER40685.

REFERENCES

1. See for example, Das, A., *Finite Temperature Field Theory*, World Scientific, 1997.
2. Das, A., and Hott, M., *Mod. Phys. Lett.*, **A9**, 3383 (1994).
3. Deser, S., Jackiw, R., and Templeton, S., *Ann. Phys.*, **140**, 372 (1982).
4. Babu, K. S., Das, A., and Panigrahi, P., *Phys. Rev.*, **D36**, 3725 (1987); I. Aitchison and J. Zuk, *Ann. Phys.*, **242**, 77 (1995);

Bralić, N., Fosco, C., and Schaposnik, F., *Phys. Lett.*, **B383**, 199 (1996);
Cabra, D., Fradkin, E., Rossini, G., and Schaposnik, F., *Phys. Lett.*, **B383**, 434 (1996).
5. Dunne, G., Jackiw, R., and Trugenberger, C., *Phys. Rev.*, **D41**, 661 (1990).
6. Dunne, G., Lee, K., and Lu, C., *Phys. Rev. Lett.*, **78**, 3434 (1997).
7. Das, A., and Dunne, G., *Phys. Rev.*, **D57**, 5023 (1998).
8. Das, A., Dunne, G., and Frenkel, J., *Phys. Lett.*, **B472**, 332 (2000).
9. Barcelos-Neto, J., and Das, A., *Phys. Rev.*, **D58**, 085022 (1998);
Barcelos-Neto, J., and Das, A., *Phys. Rev.*, **D59**, 087701 (1999);
Das, A., and Dunne, G., *Phys. Rev.*, **D60**, 085010 (1999).
10. Das, A., and da Silva, A. J., *Phys. Rev.*, **D59**, 105011 (1999).
11. Deser, S., Griguolo, L., and Seminara, D., *Phys. Rev. Lett.*, **79**, 1976 (1997);
Fosco, C., Rossini, G., and Schaposnik, F., *Phys. Rev.*, **79**, 1980 (1997);
Deser, S., Griguolo, L., and Seminara, D., *Phys. Rev.*, **D57**, 7444 (1998);
Fosco, C., Rossini, G., and Schaposnik, F., *Phys. Rev.*, **D56**, 6547 (1997).
12. Das, A., and Hott, M., *Phys. Rev.*, **D50**, 6655 (1994).
13. Frenkel, J., and Taylor, J. C., *Nuc. Phys.*, **B374**, 156 (1992);
Brandt, F. T., and Frenkel, J., *Phys. Rev.*, **D56**, 2453 (1997).
14. Brandt, F. T., Das, A., and Frenkel, J., *Phys. Rev.*, **D62**, 085012 (2000);
Brandt, F. T., Das, A., and Frenkel, J., `hep-ph/0004195`.
15. Coleman, S., and Hill, B., *Phys. Lett.*, **B159**, 184 (1985).
16. Brandt, F. T., Das, A., Frenkel, J., and Rao, K., *Phys. Lett.*, **B492**, 393 (2000).
17. Brandt, F. T., Das, A., and Frenkel, J., *Phys. Lett.*, **B494**, 339 (2000);
Brandt, F. T., Das, A., and Frenkel, J., `hep-th/0012087`.

Aspects of Born-Infeld Theory and String/M-Theory

G.W. Gibbons

D.A.M.T.P., Cambridge University, Wilberforce Road, Cambridge CB3 0WA, U.K., email: g.w.gibbons@damtp.cam.ac.uk

INTRODUCTION

M/String Theory is the currently most popular approach to a unified quantum theory of gravity and the other interactions. We still lack a complete formulation of the theory, but there is a general consensus that whatever finally emerges it will involve in some way or to some degree of approximation, $p-$ branes, i.e. $p+1$-dimensional Lorentzian submanifolds Σ of a Lorentzian spacetime manifold M. In M-theory one supposes that M is eleven dimensional. In string theory it is usually taken to be ten dimensional. Branes may crudely be sub-divided into two types Heavy and Light. In the former case one is usually thinking of many coincident branes whose gravitational field and hence the ambient spacetime metric is non-trivial. Semi-classically these may be studied using supergravity techniques. The other extreme is to study a single isolated brane moving in flat Minkowski spacetime as a solution of the Dirac-Born-Infeld equations of motion. This will be the approach taken in these lectures. It is well suited to newcomers to the subject because, as I will try to show, considerable insights into string theory can be gained by asking some of the simplest physical questions. There is little need for the full heavy technical machinery of supergravity or superstring theories. Thus the material is well suited for presentation at a School. I have deliberately tried to keep things simple. This runs the risk that experts may feel that I have not done full justice to the subject or indeed their contributions to it. If so, I apologize but I repeat my aim was to provide the beginner with a rapid survey of the subject. I will mainly assume that the brane is flat. It is fairly straightforward to extend the present circle of ideas to the case of a curved background. In the case of Born-Infeld theory the reader is referred to [22]

The detailed material to be covered is given below.

CLASSICAL CAUSALITY AND THE DOMINANT ENERGY CONDITION

A major pre-occuppation of at least some of the lectures at this school are various issues concering causality and locality in quantum and classical and in commutative and non-commutative field theories. Indeed they have been part of the motivation of much of the work reported in in what follows. It therefore seems appropriate to begin

CP589, *Fundamental Interaction Theories*, edited by J. Lukierski and J. Rembieliński

by recapitulating the role of the dominant energy condition, particularly in the light of some recent papers either reporting or theoretically analsying experimental results on the speed of light and which will be described elsewhere in these proceedings.

The Language of Cones

The appropriate formal language for the discussion is that of convex cones. Since this seems to be playing an increasingly important role in M-theory [2, 3] we shall pause to develop it a little. In fact the theory may be developed for general convex cones, but the most interesting case is that of homogeneous self-dual cones. The discussion below will be assuming that the cones are quadratic but is couched in such a way that it extends in a straightforwrd way to a more general setting.

We suppose that an n-dimensional vector space X, ultimately the tangent space T_xM of a spacetime M at some point x, is equipped with a Lorentzian metric g. In this section we use the mainly minus signature $+,-,-,\dots,-$. Picking a time orientation allows us to define the (solid) cone C_g of future directed causal vectors. In a a time oriented orthonormal basis or Lorentz frame such that $V = (V^0, \mathbf{V})$, this consists of vectors satisfying $V^0 \geq |\mathbf{V}|$. Conversly a vector $V \in C_g$ iff $V^0 \geq |\mathbf{V}|$ in all Lorentz frames. If W is another member of of C_g then $g(V,W) \geq 0$ and conversely if $g(V,W) \geq 0 \quad \forall V \in C_g$ then $W \in C_g$. One deduces that C_g is a convex cone homogeneous with respect to the Causal group Caus(n − 1, 1) , i.e. the semi-direct product of Lorentz group $SO(n-1,1)$ with $\mathbb{R}_+$ acting as dilations. Thus $C_g = \mathbb{R}_\times SO(n-1,1)/SO(n-1)$. The set Y of Lorentzian metrics g on a fixed vector space X is the homogeneous space $GL(n,\mathbb{R})/SO(n-1,1)$ and it admits a partial order $<$ (in fact an interesting type of causal structure) which corresponds to inclusion of cones $g < g'$ iff $C_g \subset C_{g'}$. The inclusion need not be strict, i.e. the two cones may touch.

Associated with any convex cone $C \in X$ is the dual cone-cone or co-cone $C^\star$ in the dual space $X^\star$. This is defined as the set of covectors $\omega \in X^\star$ such that $\omega.V \geq 0 \quad \forall \quad V \in C$. It is a simple exercise to convince one'self that duality reverses inclusion, $C \subset C'$ iff $C'^\star \subset C^\star$. For a Lorentzian cone C_g the dual cone is given by the inverse metric $C^\star = G_{g^{-1}}$, and because one may use the metric to set up an isosomorphism between X and $X^\star$, the cone is said to be self-dual and one does not normally distingush between C_g and $C_g{}^\star$. However with more than one metric in the game it is essential to make the distinction.

The idea duality provides a duality between *ray* (or particle) and *wave* in all areas of physics. The basic observation of De Broglie's Ph D Thesis, [1] may be summarized by saying was the in Relativity the *unique* Einstein light cone demanded by the Equivalence Principle permits an identification of these two dual concepts and hence leads to Quantum Mechanics.

The Energy Conditions

Given a metric may regard the energy momentum tensor as a billinear form $T_{\mu\nu}$ or an endomorphism $T^{\mu}{}_{\nu}$, according to taste. The various energy conditions [5] depend on the metric. Thus

The Weak Energy Condition

One regards the energy momentum tensor as a quadratic form and demands that $T_{\mu\nu}V^{\mu}V^{\nu} \geq 0 \quad \forall \quad V \in C_g$. In other words the quadratic form is non-negative on the cone C_g.

Because it is equivalent to $T_{00} \geq 0$ in all Lorentz frames, the weak energy condition is regarded as a fairly mininmal requrement but it is violated in gauged supergravity theories. This is because they may contain scalar filelds with *negative* potentials. Note that the sign of T_{00} is independent of spacetime signature.

The Strong Energy Condition

This is similar and captures the idea that gravity is attractive. It is used to prove the singularity theorems but it also implies that any cosmological term must be *negative* and is inconsistent with inflation. It is satisified by all supergravity models in all dimensions. Essentially it is incompatible with potentials for scalar fields which are *positive*.

One again regards the energy momentum tensor as a quadratic form and demands now that $(T_{\mu\nu} - \frac{1}{n-2}g_{\mu\nu}T^{\sigma}_{\sigma})V^{\mu}V^{\nu} \geq 0 \quad \forall \quad V \in C_g$.

By contrast

The Dominant Energy Condition

is most easilly expressed by regarding the energy momentum tensor as an endomorphism and demanding that it maps C_g into itself. That is if V^{μ} is causal and future directed then so is $T^{\mu}{}_{\nu}V^{\nu}$. Thus $T_{\mu\nu}V^{\mu}W^{\mu} \geq 0 \quad \forall V, W \in C_g$. An equivalent requirent is that $T_{00} \geq T_{\mu\nu} \quad \forall \quad \mu\nu$, hence the name.

Note that the set $C_{\text{condition},g}$ of energy momentum tensors satsifying any one of these conditions with respect to a fixed metric g is itself a convex cone inside the $\frac{1}{2}n(n+1)$ -dimensional vector space T of symmetric tensors. This accords with one's general prejudice that the state spaces of physical systems or substances are often convex cones. The structure of these cones, their boundaries and extreme points and mutual dispositions and their dependence on g is an interesting topic which time does not permit us to pursue in detail here. We merely remark that one may classify the possible energy momentum tensors by bringing them to canonical form (see e.g. [5] in the case of four dimensions). Generically one may diagonalize $T_{\mu\nu}$ with respect to the metric $g_{\mu\nu}$ and we get simple conditions in terms of the energy density and principal pressures. Because the metric $g_{\mu\nu}$ is not positive definite there are also some exceptional cases. In this way one classifies the orbits of $SO(n-1-,1)$ on the space of symmetric tensors. Now one may identify the extreme points, faces etc of the relevant cone.

An important application of the dominant energy condition is to the Positive Energy Theorem of Classical General Relativity which states that if locally the stress tensor lies everywhere in the dominant energy cone then the the ADM energy momentum vector $P_{ADM}{}^{\mu}$ of a regular asymptotically flat spacetime lies in the cone of future directed causal vectors.

Ex Nihilo Nihil Fit

We now outline an elegant argument of Hawking [4, 5] which shows that even if the background metric is time dependent the dominat energy condition implies causal propagation. If the metric is time independent this follows form energy conservation but energy conservation fails if the metric is time dependent and one might worry that that classically matter might appear "out of nowhere" that is it might travel at super-luminal speeds. The point of Hawking's argument is that this cannot happen classically. Of course quantum mechnically things are different, a point made strenuously by Zel'dovich [6, 7]. Pair-creation processes in external fields often give the appearance of a-causality because, thought of as a tunnelling process, especially usng the semi-classical or instanton approximation the particles suddenly materialize at spacelike separations. In this context the instanton is often called a bounce. Think of an electron positron pair in an external electric for example. The instanton is a closed circle in Euclidean spacetime which analytcally continues to a pair of causally disjoint timelike hyperbolae in Minkowski spacetime [6]. A related point is that the quantum mechnaical Feynmam propagator in constrat to the classical retarded or advanced propgators has support both inside and *outside* the light cone. As Feynman has pointed out, this is bacause while one may not be able to join two spacelike points by a smooth timelike curve one may join them by one which is piecewise smooth and consisting of some past directed and some future directed intervals. Where the past directed and future directed intervals join is the site of a pair-annihilation of or pair-creation event.

Now one might try to describe the pair creation process using a regularized expectation value of energy momentum tensor operator, that is

$$T^{\mu\nu} = \langle \hat{\mathbf{T}}^{\mu\nu} \rangle . \tag{1}$$

Zel'dovich and Paitaevsky [7] pointed out that the dominant energy property cannot and does not survive the regularization process.

Here is Hawking's argument. Let U be a compact region of a spacetime M admitting a time function t whose gradient $\partial_\mu t = V_\mu$. Let the level surfaces of the time function be called Σ_t and the part of U earlier than Σ_t , i.e. that part containing events at which the time function is less than t is called U_t. The boundary ∂U decomposes into three components ∂U_1 and ∂U_2 on which the normal is non- spacelike and time function is decreasing or increasing along the outward normal , and ∂U_3 with spacelike normal. We note that

$$J^{\mu}{}_{;\mu} = T^{\mu\nu} V_{\mu;\nu} \,, \tag{2}$$

where, by the dominat nergy condition, $J^{\mu} = T^{\mu}_{\nu} V^{\nu}$ is a future directed timelike vector field. Byy the dominant energy condition and the compactness of of U that there exists a positive constant P such that

$$T^{\mu\nu} V_{\mu;\nu} \leq P T^{\mu\nu} V_{\mu} V_{\nu} . \tag{3}$$

Let

$$E(t) = \int_{\Sigma_t} J_{\mu} \Sigma^{\mu} \,. \tag{4}$$

Clearly $E(t) \geq 0$ and $E(t) = 0$ implies that $T^{\mu\nu}$ vanishes on Σ_t.
Integration of (2) over $U(t)$ gives

$$E(t) \leq -\int_{U(t)\cap\partial U_1} J_{\mu} d\Sigma^{\mu} + \int_{U(t)\cap\partial U_3} J_{\mu} d\Sigma^{\mu} + P \int^{t} dt' E(t') \,. \tag{5}$$

We have used the fact that by the dominant energy condition

$$\int_{U(t)\cap\partial U_2} J_{\mu} d\Sigma^{\mu} \geq 0 \,. \tag{6}$$

Now suppose that $T^{\mu\nu}$ vanishes on ∂U_3, the timelike component of the boundary ∂U and so nothing flows into the region U. We deduce that

$$\frac{dE}{dt} \leq PE(t) \,. \tag{7}$$

Integration of this simple this simple differential identity implies that

$$E(t) \leq E(t') \exp P(t - t') \,, \tag{8}$$

and hence that if $E(t)$ vanishes at some time t' , then it must vanish for all times. This despite the fact that we have allowed for the possibility of a time dependent metric doing work on the matter. one cannot get somethig from nothing. To get a statement about causality we apply this result to the case when $U = D^{+}(S)$ the future Cauchy development of some set S. This is the set of all points p such that every past directed causal curve through p intersects S. If $T^{\mu\nu}$ vanishes on S then it vanishes everywhere in S.

The results just given, and obvious generalizations show clearly that according to Maxwell's equations, electromagnetic waves can never, in the sense defined above, travel faster than light.

OPEN STRINGS AND D-BRANES

Branes may be incorporated in string theory if one contemplates opens strings whose ends are constrained (by Dirichlet boundary conditions) to lie on a $(p+1)$-dimensional submanifold Σ_{p+1}. Now open strings can couple minimally to vector A_μ at the ends of the strings. In the Polyakov approach one has an action of the form

$$-\frac{1}{2}\int_{\Sigma_1} d^2\sigma(G_{ab}+B_{ab})\partial y^a\partial y^b + \int_{\partial\Sigma_1} A_a dy^a\,, \tag{9}$$

where the embedding of the string world sheet $\Sigma_1 \to M$ is given by $y^a = y^a(\sigma^A)$, $A = 1,2$ and $a = 1,2,\ldots,n = \dim M$, and G_{ab} and B_{ab} are the spacetime metric and Neveu-Schwarz two-form respectively.

One obtains an effective action for a D-brane if one "integrates out" all possible string motions subject to the Dirichlet boundary condition.

The resulting action depends on the position of the D-brane and the pullback to the D-brane of the metric and Neveu-Schwarz two-form. It also contains the vector field A_μ.

DIRAC-BORN-INFELD ACTIONS

This is governs the embedding $y : \Sigma_{p+1} \to M$ given in local coordinates by $y^a = y^a(x^\mu)$, where $a = 1,2,\ldots,n = \dim M$ and $\mu = 0,1,2,\ldots,p$. It is

$$-T_p \int dx^{p+1}\sqrt{\det(g_{\mu\nu} + (2\pi\alpha')F_{\mu\nu} + B_{\mu\nu})}\,, \tag{10}$$

where

$$g_{\mu\nu} = \eta_{ab}\partial_\mu y^a \partial_\nu y^b\,, \tag{11}$$

and

$$B_{\mu\nu} = B_{ab}\partial_\mu y^a \partial_\nu y^b\,, \tag{12}$$

are the pull-backs of the metric η_{ab} and Neveu-Schwarz two-form B_{ab} to the world volume Σ_{p+1} of the p-brane.

The the world-volume field $F_{\mu\nu}$ is given by

$$F_{\mu\nu} = \partial_\mu A_\nu - \partial_\nu A_\mu\,. \tag{13}$$

One often defines

$$\mathcal{F}_{\mu\nu} = F_{\mu\nu} + B_{\mu\nu}\,. \tag{14}$$

This is invariant under a Neveu-Schwarz gauge transformation $B \to B - dC$ where C is a one-form, if we transform $F \to F + dC$. One may check that this is consistent with the behaviour of the open string metric.

Monge Gauge

To proceed we fix some of the gauge-invariance associated with world sheet diffeomorphisms of the coordinates X^{μ} by using what is usually, and misleadingly called static gauge (since it applies in non-static situations) and which is more accurately and with more justice called Monge gauge. In effect we project onto a $p+1$ plane by setting $y^a = x^{\mu}, y^i$ and use the $n-p-1$ height functions y^i, $i = 1,2,\dots,n-p-1$ as scalar fields on the world volume. In the theory of minimal surfaces this is called a non-parametric representation. For Monge's work see [8]. Of course there may not be a global Monge gauge, and we shall encounter this situation later.

The determinant then becomes (we use units in which $2\pi\alpha' = 1$),

$$\det(\eta_{\mu\nu} + \partial_\mu y^i \partial_\nu y^i + F_{\mu\nu}) . \tag{15}$$

It is evidently consistent to set the scalars to zero $y^i = 0$ and we then obtain the Lagrangian of Born and Infeld which is a special form of Non-Linear Electrodynamics.

Dimensional Reduction

The previous section result has a sort of converse. We could start with a pure Born-Infeld action in n flat dimensions and dimensionally reduce to $p+1$ dimensions. We begin with

$$-\int d^n x \sqrt{-\det(\eta_{ab} + F_{ab})} . \tag{16}$$

We make the ansatz $A_a = (A_\mu(x^\lambda), y^i(x^\lambda))$ and obtain the Monge-gauge-fixed Dirac-Born-Infeld action

$$-\int d^{p+1} \sqrt{-\det(\eta_{\mu\nu} + F_{\mu\nu} + \partial_\mu y^i \partial_\nu y^i)} . \tag{17}$$

Thus all solutions of the Dirac Born-Infeld action are solutions of the Born-Infeld action. Interestingly in the case $p = 1$ we get a string action from the pure Born-Infeld action.

NON-LINEAR ELECTRODYNAMICS

There are advantages in viewing the theory in this context. An excellent account of the theory is given in [10]. The general theory in four-spacetime dimensions (p=3) has equations

$$\text{curl}\mathbf{E} = -\frac{\partial \mathbf{B}}{\partial t} : \qquad \text{div}\mathbf{B} = 0, \tag{18}$$

$$\text{curl}\mathbf{H} = -\frac{\partial \mathbf{D}}{\partial t} : \qquad \text{div}\mathbf{D} = 0. \tag{19}$$

Constitutive Relations

To close the system one needs constitutive relations $\mathbf{H} = \mathbf{H}(\mathbf{E},\mathbf{B})$ and $\mathbf{D} = \mathbf{D}(\mathbf{E},\mathbf{B})$ which, if one has a Lagrangian $L = L(\mathbf{E},\mathbf{B})$, take the form

$$\mathbf{H} = -\frac{\partial L}{\partial \mathbf{B}}, \qquad \mathbf{D} = \frac{\partial L}{\partial \mathbf{E}}. \tag{20}$$

Because

$$\mathbf{D} = \frac{\partial L}{\partial \dot{\mathbf{A}}}, \tag{21}$$

D is the canonical momentum density. Note also that the conserved electric charge is given by the flux of **D** and *not* as is often assumed, the flux of **E**.

In what follows we shall denote by $K_{\mu\nu}$ the Ampère 2-form with components $(\mathbf{D},\mathbf{H})$ and refer to $F_{\mu\nu}$ as the Faraday 2-form. Thus the equations of motion without sources are

$$dF = 0 \qquad d\star K = 0. \tag{22}$$

Lorentz-Invariance

The symmetry of the energy momentum tensor $T_{0i} = T_{i0}$ and hence the uniqueness of the Poynting vector requires that the latter be given by

$$\mathbf{E}\times\mathbf{H} = \mathbf{D}\times\mathbf{B}. \tag{23}$$

This will follow if L is constructed from the two Lorentz invariants

$$x = \frac{1}{2}(\mathbf{B}^2 - \mathbf{E}^2), \tag{24}$$

$$y = \mathbf{E}\cdot\mathbf{B}. \tag{25}$$

Duality Invariance

The constitutive relations will permit the obvious rotation needed to rotate the two sets of equations (18, 19) into themselves

$$\mathbf{E} + i\mathbf{H} \to e^{i\theta}(\mathbf{E} + i\mathbf{H}), \tag{26}$$

$$\mathbf{D} + i\mathbf{B} \to e^{i\theta}(\mathbf{D} + i\mathbf{B}), \tag{27}$$

with θ constant if

$$\mathbf{E}\cdot\mathbf{B} = \mathbf{D}\cdot\mathbf{H}. \tag{28}$$

Note that what we are encountering here is a *non-linear form of the familiar linear Hodge duality* This gives a constraint on possible theories. For example if the Lagrangian depends arbitrarily on the invariants x and y it gives rise to a Lorentz-invariant theory. Imposing duality invariance reduces this freedom to that of a function of a single variable. For more details on duality invariance see [18, 19] and [10] which was not known to the authors of [18, 19] when they were written.

Hamiltonian Density

One has

$$\mathcal{H} = T_{00} = \mathbf{E} \cdot \mathbf{D} - L, \tag{29}$$

one may think of $\mathcal{H} = \mathcal{H}(\mathbf{B}, \mathbf{D})$ as the Legendre transform of the Lagrangian and is thus expressed in terms of the canonical variables $\mathbf{B}$ and $\mathbf{D}$ whose Poisson Brackets are

$$\{B_i(\mathbf{x}), D_j(\mathbf{y})\} = -\varepsilon_{ijk}\partial_k \delta(\mathbf{x} - \mathbf{y}). \tag{30}$$

Born-Infeld

We have

$$L = 1 - \sqrt{1 - \mathbf{E}^2 + \mathbf{B}^2 - (\mathbf{E} \cdot \mathbf{B})^2} \tag{31}$$

and

$$\mathcal{H} = \sqrt{1 + \mathbf{B}^2 + \mathbf{D}^2 + (\mathbf{B} \times \mathbf{D})^2} - 1. \tag{32}$$

A constant has been added to make the zero field have zero energy. This is not strictly necessary in the theory of banes since the notion of world volume energy is not well defined because there are no privileged coordinates on the brane. However it is convenient when making comparisons with standard flat space field theory. To do so we must however use Monge gauge.

Lorentz and Duality invariance are clear. Before the advent of String/M-theory the latter was rather mysterious. Nowadays it may be thought of as a manifestation of S-duality. In this way we see how Born-Infeld theory considered *sui generis* has important lessons for M/String theory. Conversely M/String theory throws light on Born-Infeld theory. We shall see more examples of this mutually symbiotic behaviour later.

THE MAXIMAL ELECTRIC FIELD STRENGTH

If $\mathbf{B} = 0$, the Born-Infeld Lagrangian is

$$L = 1 - \sqrt{1 - \mathbf{E}^2}. \tag{33}$$

If we use a gauge in which $A_0 = 0$, we have

$$L = 1 - \sqrt{1 - \dot{\mathbf{A}}^2}. \tag{34}$$

The analogy with special relativity is clear. There will be an upper bound to the electric field strength. The special relativistic analogy may also be understood from the point of T-duality.

In string theory the existence of a maximal electric field strength may be understood dynamically as follows. A stretched open string of length L has, in our units, elastic energy L. If it has charges $+1$ at one end and -1 at the other it will, in an electric field have energy $-EL$. This if $E > 1$ one may gain energy from the background electric field by creating open strings. This an electric field with strength greater than 1 will quickly breakdown and the electric field will be reduced to a value less than one.

Note that if one restores dimensions and units the critical field strength E_c is given by

$$E_c = \frac{1}{2\pi\alpha'} . \tag{35}$$

In the zero slope limit $\alpha' \to 0$ there is no upper bound and in the strong coupling limit $\alpha' \to \infty$ the critical field goes down to zero. Later we will investigate the behaviour of the theory in this limit.

BIONS

The maximal electric field was originally invoked to ensure the existence of a classical solution representing a charged object with finite total energy

$$\int_{\mathbb{E}^3} d^3x T_{00} < \infty . \tag{36}$$

This can be achieved by setting

$$\mathbf{D} = \frac{q}{r^2}\hat{\mathbf{r}} . \tag{37}$$

Because

$$\mathbf{E} = \frac{\mathbf{D}}{\sqrt{1+\mathbf{D}^2}} , \tag{38}$$

the electric field achieves its maximal value at the centre. Note that

$$\mathbf{D} = \frac{\mathbf{D}}{\sqrt{1-\mathbf{D}^2}} , \tag{39}$$

the electric induction $\mathbf{D}$ diverges at the origin and so does the energy density

$$T_{00} = \mathcal{H} = \mathbf{E}\cdot\mathbf{D} - L . \tag{40}$$

Thus this solution is **not** *a smooth soliton solution without sources.* In fact there is a distributional source

$$\text{div}\mathbf{D} = 4\pi q\delta(\mathbf{r}) . \tag{41}$$

Finite energy but singular solutions like this of non-linear theories with distributional sources are a sufficiently distinct phenomenon from the familiar finite energy non-singular lump solutions without sources as to deserve a different name. The suggestion has been made [14] that they be called BIons. From the string point of view the source has a natural interpretation as being associated with a string ending on a three-brane. In fact one returns in this way to a picture very close to late nineteenth century speculations in which an electron is regarded as an "ether-squirt" on a 3-surface embedded in four dimensional space [9]. The application to strings is contained in [14] and [21]. The present account is largely based on [14].

Maximal Spacelike Hypersurfaces

Another interpretation of the static solutions may be obtained as follows. One introduces the electrostatic potential $\phi = A_0$ and finds the Lagrangian density to be given by

$$1-\sqrt{1-(\nabla\phi)^2}\,. \tag{42}$$

The Euler-Lagrange equation

$$\mathrm{div}\Big(\frac{\nabla\phi}{\sqrt{1-(\nabla\phi)^2}}\Big)=0\,, \tag{43}$$

is just that which would be obtained if one sought a maximal spacelike hypersurface of minkowski spacetime where ϕ is now thought of as a time function

$$x^0(\mathbf{x})=\phi(\mathbf{x})\,. \tag{44}$$

The maximal hypersurface becomes null at the critical field strength.

Catenoids and $D-\bar{D}$ Solutions

Rather than exciting the electric field we can excite a single scalar y. We get as Lagrangian density

$$1-\sqrt{1+(\nabla y)^2}\,. \tag{45}$$

The Euler-Lagrange equation

$$\mathrm{div}\Big(\frac{\nabla y}{\sqrt{1+(\nabla y)^2}}\Big)\,, \tag{46}$$

is that governing the height function of a minimal surface in four space-like dimensions. One readily checks that Monge gauge is not global. In the spherically symmetric case There is a branch 2-surface at a finite radius. One needs two Monge patches. The resulting two sheeted worm-hole or better Einstein-Rosen bridge type surface looks like two parallel three planes a with finite separation joined by a neck. The solution is not stable and therefore one thinks of it as a Brane-Anti-Brane pair.

Charged Catenoids: $O(1,1)$ Symmetry Relating Catenoids and Bions

Including both electric and scalar fields gives a Lagrangian

$$1-\sqrt{1+(\nabla y)^2-(\nabla\phi)^2}. \tag{47}$$

It and the Euler-Lagrange equations

$$\operatorname{div}\Big(\frac{\nabla y}{\sqrt{1+(\nabla y)^2-(\nabla\phi)^2}}\Big), \tag{48}$$

and

$$\operatorname{div}\Big(\frac{\nabla\phi}{\sqrt{1+(\nabla y)^2-(\nabla\phi)^2}}\Big)=0, \tag{49}$$

which are manifestly invariant under an obvious $O(1,1)$ action analogous to the well-known Harrison transformation of static Einstein Maxwell theory. Using this action one may construct everywhere smooth charged catenoids, the electric field lines passing through the neck or throat in a way similar to that discussed by Wheeler in the case of Einstein-Maxwell theory. This family I call under-extreme. They are obviously analogous to under extreme Reissner-Nordstrom solutions. One may also excite the scalar field of the BIon solution. The original flat three-brane acquires a cusp as if it were being pulled. All of these solutions are singular. I call them over-extreme. They are obviously analogous to over extreme Reissner-Nordstrom solutions.

The BPS Solution: S-Duality

In the limit of infinite $O(1,1)$ parameter one obtains an extreme solution analogous to extreme Reissner-Nordstrom. This solution is in fact supersymmetric. It may be interpreted as a fundamental (F-) or $(1,0)$ string ending on a three-brane. Using the electric-magnetic duality one may easily obtain a magnetic monopole solution which represents a D-string or $(0,1)$ string ending on a three-brane. In fact using $SL(2,\mathbb{Z})$ and the Dirac quantization condition we can get dyon or (p,q) strings ending on a three-brane.

OPEN STRING CAUSALITY

In string theory, open string states propagating in a background $\mathcal{F}_{\mu\nu}$ field do so according to a different metric from the Einstein metric $g_{\mu\nu}$ felt by closed strong states.

One has

$$\Big(\frac{1}{g+\mathcal{F}}\Big)^{\mu\nu}=G^{\mu\nu}+\theta^{\mu\nu}, \tag{50}$$

where $G^{\mu\nu}=G^{(\mu\nu)}$ and $\theta^{\mu\nu}=\theta^{[\mu\nu]}$. If

$$G_{\mu\lambda}G^{\lambda\mu}=\delta^{\nu}_{\mu}, \tag{51}$$

then

$$G = g_{\mu\nu} - \mathcal{F}_{\mu\lambda} g^{\lambda\rho} \mathcal{F}_{\rho\nu} . \tag{52}$$

Note that even if $B_{\mu\nu} = 0$ so that $\mathcal{F}_{\mu\nu} = F_{\mu\nu}$, the metric $G_{\mu\nu}$ is not invariant under electric-magnetic duality.

Boillat Metrics

One may investigate the propagation of small disturbances of vectors, A_μ scalars y and spinors ψ around a Born-Infeld background using the method of characteristics. This was done in great detail by Boillat for a general non-linear electrodynamic theory. He found that in general, because of bi-refringence, there are a pair of characteristic surfaces $S =$ constant satisfying

$$\left(T^{\mu\nu}_{\text{Maxwell}} + \mu g^{\mu\nu}\right) \partial_\mu S \partial_\nu S = 0 , \tag{53}$$

where $T^{\mu\nu}_{\text{Maxwell}}$ is the Maxwell stress tensor constructed form $F_{\mu\nu}$. Of course the stress tensor $T^{\mu\nu}$ of the non-linear electrodynamic theory is different from $T^{\mu\nu}_{\text{Maxwell}}$. The quantity $\mu = \mu(x,y)$ satisfies a quadratic equation whose coefficients depend upon first and second derivatives of the Lagrangian $L(x,y)$ with respect to x and y. Boillat finds it convenient to fix the arbitrary conformal rescaling freedom in the characteristic co-metric by setting

$$C^{\mu\nu} = \frac{1}{\sqrt{\mu^2 - x^2 - y^2}} \left(\mu g^{\mu} + T^{\mu\nu}_{\text{Maxwell}}\right) , \tag{54}$$

with inverse or metric

$$C^{-1}_{\mu\nu} = \frac{1}{\sqrt{\mu^2 - x^2 - y^2}} \left(\mu g^{\mu\nu} - T^{\mu\nu}_{\text{Maxwell}}\right) . \tag{55}$$

In general the boundaries of the two Boillat cones $C_{\text{Boillat}} : C^{-1}_{\mu\nu} v^\mu v^\nu \geq 0, v^0 > 0$ and the Einstein cone $C_{\text{Boillat}} : g_{\mu\nu} v^\mu v^\nu \geq , v^0 > 0$ will touch along the two principle null directions of $F_{\mu\nu}$. One sometimes find that one at least of the Boillat cones lies outside the Einstein cone. In other words small fluctuations can travel faster than gravitational waves whose speed is governed by $g_{\mu\nu}$.

To check causality we examine the Boillat co-cones $C^\star_{\text{Boillat}} : C^{\mu\nu} p_\mu p_\nu \geq 0, p_0 > 0$ and the Einstein co-cone $C^\star_{\text{Einstein}} : C^{\mu\nu} p_\mu p_\nu \geq 0, p_0 > 0$ in the cotangent space $T^\star \Sigma_{p+1}$. Suppose that l_μ is the co-normal the Einstein co-cone $C^\star_{\text{Einstein}}$

$$g^{\mu\nu} l_\mu l_\nu = 0 . \tag{56}$$

The weak energy condition implies

$$T^{\mu\nu}_{\text{Maxwell}} l_\mu l_\nu \geq 0 . \tag{57}$$

Thus

$$C^{\mu\nu} l_\mu l_\nu \geq 0 . \tag{58}$$

This means that if μ is positive then $C^\star_{\text{Einstein}}$ lies inside or touches $C^\star_{Boillat}$. Remembering that duality reveres inclusions one finds then that the Einstein cone C_{Einstein} lies outside or touches the Boillat cone C_{Boillat}. Note that what we are calling a cone here is the solid cone. The light cone is the boundary of this solid cone.

Hooke's' Law

Born-Infeld is exceptional in that there is just one solution for μ:

$$\mu = 1 + x. \tag{59}$$

Thus there is no bi-refringence. Moreover one finds that the Boillat co-metric satisfies the remarkable identity

$$C^{\mu\nu}_{\text{BI}} = g^{\mu\nu} + T^{\mu\nu}. \tag{60}$$

I call this identity Hooke's Law for reasons which will be explained below. Another striking identity is

$$\det\left(\delta^\mu_\nu + T^\mu_\nu\right) = 1. \tag{61}$$

This follows form another useful identity is

$$\det C^{\mu\nu} = \det g^{\mu\nu}. \tag{62}$$

From Hooke's Law it is easy to see, since $T^{\mu\nu}$ for Born-Infeld theory satisfies the Weak Energy Condition, that the Boillat cone lies inside or touches the Einstein cone. In other words small fluctuations travel with a speed no greater than gravitational waves. Because the Born-Infeld energy momentum tensor is invariant under electric-magnetic duality rotations, the Boillat metric, unlike the open string metric $G_{\mu\nu}$ is also invariant. One has

$$C^{-1}_{\mu\nu} = \frac{1}{\sqrt{1+2x-y^2}} G_{\mu\nu}. \tag{63}$$

The conformal factor is related to the Lagrangian:

$$\sqrt{-\det(\eta_{\mu\nu} + F_{\mu\nu})} = \sqrt{1+2x-y^2}. \tag{64}$$

The reason for the name Hooke's law is that Hooke asserted, in the days when the archive was in Latin that *eiiiouucnssstt*. In this way he hoped to make both a priority claim and preserve his discovery for his own later use. Bearing in mind that u v are not distinguished in Latin, the earth shattering discovery that he wished to hide was that *ut tensio sic vis* . In other words stress is proportional to strain. A standard measure of strain in non-linear elasticity theory is the difference of two metrics. More precisely, the configuration space of an elastic medium is a map from an elastic manifold to an embedding space. There is usually a rest or un-deformed configuration and one takes as a measure of stress the difference between the pullbacks from the embedding space to the elastic manifold in the strained and unstrained configuration.

What we have is an expression involving the difference of two co-metrics but the idea is similar. One is the co-metric induced on the brane from the Einstein co-metric and the other is a measure of the vector field excitations.

Hooke's Law, the Monge-Ampère Equation and Pulse Interactions

The striking determinantal identity has an interesting application to the propagation of pulses in Born-Infeld theory.

In flat two dimensional spacetime, the conservation law for the stress tensor implies that it is given by a single free function, the Airy stress function ψ, such that

$$T_{tt} = \psi_{zz}\,, \quad T_{zz} = \psi_{tt}\,, \quad T_{tz} = \psi_{zt}\,. \tag{65}$$

Written in terms of the Airy stress function, the determinantal identity becomes the Monge-Ampère equation

$$\psi_{zz}\psi_{tt} - \psi_{zt}^2 = \psi_{zz} - \psi_{tt}\,. \tag{66}$$

This can be solved exactly (see [15] and references therein) by a Legendre transform under which it becomes D'Alembert's equation with respect to a new set of variables T and Z.

One has

$$T^{tt} = \frac{A+B+2AB}{1-AB}\,, \tag{67}$$

$$T^{zz} = \frac{A+B-2AB}{1-AB}\,, \tag{68}$$

$$T^{zt} = \frac{B-A}{1-AB}\,, \tag{69}$$

where $A = A(T+Z)$ and $B = B(Z-T)$ are arbitrary functions of their arguments. The relation between the new coordinates (T,Z) and usual coordinates (t,z) is most conveniently expressed using null coordinates. Let $v+t+z, u = t-z, \xi = Z-T, \eta = Z-T$. The asymmetrical definition of η is so as to agree with previous work cited in [15], One has

$$dv = d\xi - Bd\eta, \qquad du = -d\eta + Ad\xi. \tag{70}$$

Thus

$$(1-AB)d\eta = Adv - du\,, \qquad (1-AB)d\xi = dv - Bdu\,, \tag{71}$$

one checks that

$$\begin{aligned} dT^2 - dZ^2 &= dt^2(1-A-B+AB) - dz^2(1+A+B+AB) - 2dtdz(A-B) \\ &= C^{-1}_{\mu\nu}dx^\mu dx^n u\,, \end{aligned} \tag{72}$$

where

$$C^{\mu\nu} = \eta^{\mu\nu} + T^{\mu\nu}. \tag{73}$$

Thus we see that the Legendre transformation to the new coordinates (T,Z) used to solve the Monge-Ampère equation in effect passes to flat inertial coordinates with respect to the Boillat metric. It should be noted that one does not expect the Boillat metric to be flat in general.

The general solution consists of two pulses, one right-moving and one left moving which pass through-another without distortion. In terms of the usual coordinates (t,z)

they two pulses experience a *delay* That is measured with respect to the closed string metric. However with respect to the Boillat coordinates, that is measured with respect to the Boillat metric, there is no delay.

Scalars and Fermions: Open String Equivalence Principle

The coupling of scalars has already been given above. It is easy to check that the Boillat co-metric determines their fluctuations around a background, they are in fact governed by the D'Alembert equation constructed from the co-metric $C^{\mu\nu}$. One may also consider fermion fields ψ. Omitting four-fermion terms, they couple in a typical Volkov-Akulov fashion.

$$-\int dx^{p+1}\sqrt{\det(g_{\mu\nu}+i\bar{\psi}\gamma_{\mu}\nabla_{\nu}\psi+F_{\mu\nu}+B_{\mu\nu})}\,, \tag{74}$$

where

$$\gamma_{\mu}\gamma_{\nu}+\gamma_{\nu}\gamma_{\mu}=2g_{\mu\nu}\,. \tag{75}$$

Let's define Boillat gamma matrices by

$$a^{\mu\nu}=G^{\mu\nu}+\theta^{\mu\nu}\,, \tag{76}$$

and

$$\tilde{\gamma}^{\mu}=a^{\mu\nu}\gamma_{\nu}\,. \tag{77}$$

One has

$$\tilde{\gamma}^{\mu}\tilde{\gamma}^{\nu}+\tilde{\gamma}^{\nu}\tilde{\gamma}^{\mu}=2G^{\mu\nu}\,. \tag{78}$$

Because the leading derivative term in the action is

$$i\bar{\psi}\tilde{\gamma}^{\mu}\partial_{\mu}\psi\,. \tag{79}$$

It is clear that the characteristics of the fermions are also given by the Open String metric or equivalently the Boillat metric.

Thus we have a sort of world sheet equivalence principle or universality holding: all open string fields have the same characteristics and hence the same maximum speed.

TOLMAN REDSHIFTING OF THE HAGEDORN TEMPERATURE

As an application of the equivalence principle it is interesting to consider open strings at finite temperature in a background electromagnetic field. . This was done for the neutral bosonic string in [12]. If the free energy density in the absence of a background is $F=F(\beta)$ where β is the inverse temperature, then the free energy in a background is obtained by the replacement

$$F\rightarrow\sqrt{G_{00}}\sqrt{-G_{ij}}F(\beta\sqrt{G_{00}})\,, \tag{80}$$

where $G_{\mu\nu}$ is the open strong metric. The first factor may be thought of as a redshift and volume contraction factor. The rescaling of the argument is essentially the Tolman effect whereby in order to retain local equilibrium in an external static or stationary metric $G_{\mu\nu}$, the local temperature must vary as $\frac{1}{\sqrt{G_{00}}}$. Note that $G_{00} = 1 - \mathbf{E}^2$ and so the redshifting is indeed *red* shifting and it depends only on the electric field, the effect diverging at the critical electric field strength.

Alternatively, one may regard the effect as being due to the fact that finite temperature physics corresponds to working in imaginary time with a period given by the inverse temperature. If the global time variable is identified with period β, the local period will be $\beta\sqrt{G_{00}}$. Thus the locally measured temperature will be higher. If more than one metric is involved, then the temperature of states in local equilibrium may differ, since each will be redshifted by the appropriate Tolman factor. In the present case one has closed string states at temperature $\frac{1}{\beta}$ and open strong states at temperature $\frac{1}{\beta\sqrt{G_{00}}}$. The redshifting of open string states is universal, was confirmed in [13].

In the absence of a background field the open string has, in perturbation theory, the free energy has a singularity at the Hagedorn temperature $T_{\text{Hagedorn}} = \frac{1}{\beta_{\text{Hagedorn}}}$. This is represents a maximum possible temperature because above it there are so many massive string states that thermal equilibrium becomes impossible. In a background electric field the maximum temperature is reduced to

$$T_{\text{Hagedorn}}\sqrt{1-\mathbf{E}^2}\,. \tag{81}$$

This effect has been interpreted as being due to a reduction in the effective string tension in an electric field. This is certainly true but one cannot derive the exact formulae from that assumption alone whereas everything follows rather naturally by an application of the equivalence principle, as long as one uses the open string metric.

Shocks and Exceptionality

Loosely speaking, shocks can occur of the speed of waves depends on the phase or amplitude in such a way that different waves surfaces $S =$ comstant can catch up and form caustics. More precisely one assumes the ansatz

$$F_{\mu\nu} = F^0{}_{\mu\nu}\big(f(S)\big)\,, \tag{82}$$

with

$$S = \mathbf{n}\cdot\mathbf{x} - v(\mathbf{n},S)t\,, \tag{83}$$

and $f(S)$ an arbitrary function. The surfaces $S =$ constant are hyperplanes and are to be thought of as surfaces of constant phase. If the phase speed v depends non-trivially on the phase S there will be shocks along the envelope of the hyperplanes. Theories without shocks for which v is independent of S are called exceptional.

Boillat has shown that the only form of non-linear electrodynamics with a sensible weak field limit is that of Born-Infeld.

Theories with shocks are essentially incomplete. In a sense, like General Relativity they predict their own demise. By contrast Born-Infeld, like classical Non-Abelian Yang-Mills theory seems to be a perfect example of a classical theory. As far as one can tell it appears to possess the property which is known to be true for Yang-Mills theory,that regular Born-Infeld initial data with finite energy may evolved for all time to give everywhere non-singular solutions of the field equations. For a more detailed discussion and references to the original literature see [17]

STRONG COUPLING BEHAVIOUR OF BORN-INFELD

There are (at least) two interesting strong coupling limits of Born-Infeld theory.

- A Weyl-invariant duality invariant theory which appears to be related to a fluid of massless magnetic Schild type strings and may describe string theory near critical electric field strengths.
- A massive theory which is related to a fluid of massive strings and may be related to current ideas about $D-\bar{D}$ annihilation and tachyon condensates.

In both cases the key to understanding these limiting theories is passing to the Hamiltonian formulation. It also helps to bear in mind some facts about:

Simple 2-Forms, Distributions and String Fluids

A 2-form Ω is simple iff

$$\Omega=\alpha\wedge\beta, \tag{84}$$

equivalently

$$\Omega\wedge\Omega=0. \tag{85}$$

In particular since the matrix of components has $\Omega_{\mu\nu}$ has rank two:

$$\det\Omega_{\mu\nu}=0. \tag{86}$$

In four spacetime dimensions Ω is simple iff

$$\Omega_{\mu\nu}\star\Omega^{\mu\nu}=0. \tag{87}$$

Of course α and β are not unique but a field of simple 2-forms defines the unique two-dimensional sub-space which they span in the cotangent space T_xM at every point of spacetime. Hence, given a metric, a simple two form is equivalent to a simple bi-vector $\Omega_{\mu\nu}$ which defines a distribution D of 2-planes in the tangent space TM. Raising indices with the metric, the simplicity condition becomes in terms of the bi-vector

$$\Omega^{[\mu\nu}\Omega^{\alpha]\beta}=0. \tag{88}$$

One may think of the distribution D as a sub-bundle of the tangent bundle with two-dimensional fibres. The 2-planes will be timelike, null or spacelike depending upon whether $\Omega_{\mu\nu}\Omega^{\mu\nu}$ is negative, zero or positive respectively. (Note that this statement is signature indepbedent.) In the timelike case one may chose α to be timelike and β to be spacelike. In the null case one may choose α to be null and β to be spacelike.

In general the distribution D will not be integrable. That is neighbouring 2-planes will not mesh together to form the tangent spaces of a co-dimension two family of 2-dimensional surfaces. If it is, then if two vector fields X and Y belong to D then their Lie bracket $[X,Y]$ must belong to D. Such an integrable distribution may be identified as a gas or soup, perhaps more accurately a spaghetti of strings. The condition for integrability may be expressed in various ways. For us the simplest condition is in terms of the bi-vector and is

$$\Omega^{[\alpha\beta}\partial_\kappa \Omega^{\mu]\kappa} = 0. \tag{89}$$

Note that if f is a smooth function, then Ω and $f\Omega$ define the same distribution and if the first is integrable then so is the second. Moreover, the partial derivative in (89) may be replaced by a torsion free covariant derivative. In four spacetime dimensions we may re-express the integrability condition as

$$\star\Omega_{\mu\nu}\nabla_\kappa \Omega^{\nu\kappa} = 0. \tag{90}$$

We may re-write this as

$$\Omega \wedge \delta\Omega = 0, \tag{91}$$

where $\delta\Omega = \star d \star \Omega$.

Now if we take for Ω the Ampère tensor $K_{\mu\nu}$ of any non-linear electrodynamic theory. We see that any simple solution of the equations of motion

$$\nabla K^{\mu\nu} = 0, \tag{92}$$

automatically defines an integrable distribution. In other words non-linear electrodynamic theory supplemented with the constraint

$$F \wedge K = 0, \tag{93}$$

may be re-interpreted as a (vorticity free) string fluid. Different Lagrangians correspond to different equations of state.

0-Brane Fluids

This section is based on part on [15] The situation described above should be compared with the familiar case of a non-linear scalar field theory with a Lagrangian $L(\partial\phi)$ containing no explicit dependence on the scalar field ϕ. The equations of motion may be cast in the form

$$\nabla_\mu(sU^\mu) = 0, \tag{94}$$

where U^μ is a normalized timelike vector given by

$$U_\mu = \frac{\partial_\mu \phi}{\sqrt{(\partial\phi)^2}}, \tag{95}$$

and

$$sU^{\mu} = \frac{\partial L}{\partial(\partial_{\mu}\phi)}, \tag{96}$$

may be interpreted as a conserved entropy current. The quantity s corresponds to the entropy density and ρ to the local energy density. The energy momentum tensor takes the perfect fluid form

$$T^{\mu\nu} = (\rho + P)U^{\mu}U^{\nu} - Pg^{\mu\nu}. \tag{97}$$

One has

$$P = L. \tag{98}$$

If one defines

$$T^2 = (\partial\phi)^2, \tag{99}$$

it is natural to regard the pressure as a function of the temperature T but the energy density as a function of the entropy density s. In fact they are related by a Legendre transform. One finds that

$$\rho + P = sT, \tag{100}$$

and

$$s = \frac{\partial P}{\partial T}, \qquad T = \frac{\partial \rho}{\partial s}. \tag{101}$$

It is an illuminating exercise to convince oneself that finding the speed of small fluctuations by the calculating the sound speed

$$c_s = \sqrt{\frac{\partial P}{\partial \rho}}, \tag{102}$$

is equivalent to calculating the characteristics , that is the Boillat metric.

The most interesting case from the present point of view arises when one takes the scalar Born-Infeld Lagrangian

$$L = 1 - \sqrt{1 - (\partial\phi)^2}. \tag{103}$$

One has

$$P = 1 - \sqrt{1 - T^2}, \tag{104}$$

and

$$\rho = \frac{1}{\sqrt{1 + s^2}} - 1, \tag{105}$$

which has a maximum temperature reminiscent of the Hagedorn temperature. However the detailed equation of state is different. One has the equation of state

$$P = \frac{\rho}{1 + \rho}, \tag{106}$$

and hence

$$c_s = \frac{1}{1 + \rho}. \tag{107}$$

Note that one need not regard the conserved current as an entropy current if one does not wish to. One could regard it as a conserved particle number.

The Weyl-Invariant Bialynicki-Birula Limit

The Hamiltonian density , with units restored is

$$\mathcal{H} = T^2 \sqrt{1 + \frac{\mathbf{B}^2 + \mathbf{D}^2}{T^2} + \frac{(\mathbf{D} \times \mathbf{B})^2}{T^4}} - T^2 . \tag{108}$$

One can take the limit $T \downarrow 0$ to get

$$\mathcal{H} = |\mathbf{D} \times \mathbf{B}| . \tag{109}$$

This gives the constitutive relations

$$\mathbf{E} = -\mathbf{n} \times \mathbf{B} , \qquad \mathbf{H} = \mathbf{n}\mathbf{D} , \tag{110}$$

where we have defined a unit vector in the direction of the Poynting vector $\mathbf{D} \times \mathbf{B}$

$$\mathbf{n} = \frac{\mathbf{D} \times \mathbf{B}}{|\mathbf{D} \times \mathbf{B}|} . \tag{111}$$

Remarkably these constitutive relations (which arise as the limiting form of the constitutive relations of the full theory) imply the constraints

$$\mathbf{E}^2 - \mathbf{B}^2 = 0 , \qquad \mathbf{E}.\mathbf{B} = 0 . \tag{112}$$

Defining a null vector $l^\mu = (1, \mathbf{n})$, the energy momentum tensor becomes

$$T^{\mu\nu} = \mathcal{H} l^\mu l^\nu . \tag{113}$$

It follows that the trace vanishes

$$T^\mu_\mu = 0 , \tag{114}$$

and hence the limiting theory is Weyl-invariant. It may be checked that is Lorentz-invariant and invariant under electric-magnetic duality rotations. One may also check from the equation of motion that there are infinitely many conserved symmetric tensors

$$T^{\mu_1 \mu_2 \dots \mu_k} = \mathcal{H} l^{\mu_1} l^{\mu_2} \dots l^{\mu_k} . \tag{115}$$

The constraints (112) tell us that the Faraday tensor $F_{\mu\nu}$ is simple

$$\det F_{\mu\nu} = 0 , \tag{116}$$

and null,

$$F_{\mu\nu} F^{\mu\nu} = 0 . \tag{117}$$

Thus $F_{\mu\nu}$ defines a two plane which is null, that is, the two-plane is tangent to the light cone along the lightlike vector l^μ and

$$F_{\mu\nu} l^\mu = 0 . \tag{118}$$

The equations of motion tell us that the two-plane distribution in the tangent space defined by the Faraday two-form $F_{\mu\nu}$ is integrable, that is surface forming, and hence that spacetime is foliated by two-dimensional lightlike surfaces which may be interpreted as the world sheets of magnetic null or Schild strings. In other words, in this critical limit which may be interpreted as describing Born-Infeld theory near critical field strength, the system dissolves into a gas or fluid of Schild strings.

Since electric-magnetic duality is maintained in the limit, one can of course pass to a dual description in terms of $K_{\mu\nu}$. This amounts to the observation that $\mathbf{H}^2 = \mathbf{D}^2$ and $\mathbf{H}.\mathbf{D} = 0$, i.e. $K_{\mu\nu}K^{\mu\mu} = 0$ and $K_{mu\nu} \star K^{\mu\nu} = 0$.

Covariant Formulation of UBI Using Auxiliary Fields

The Weyl-invariant limit was called by Bialynicki-Birula [10, 11], Ultra-Born-Infeld. Let us follow him and consider

$$L = -\frac{\mu}{4}F_{\mu\nu}F^{\mu\nu} + \frac{\nu}{4}F_{\mu\nu} \star F^{\mu\nu}, \tag{119}$$

where μ and ν are dimensionless auxiliary fields, variation with respect to which gives the constraints

$$F_{\mu\nu}F^{\mu\nu} = 0 = F_{\mu\nu} \star F^{\mu\nu}. \tag{120}$$

Variation with respect to A_μ gives the field equation.

Note that in axion-dilaton Maxwell theory, the auxiliary fields could be functions of the dimensionless dilaton Φ and axion χ, $\mu = \mu(\Phi,\chi)$, $\nu(\Phi,\chi)$ chosen in such a way that the system was $SL(2,\mathbb{R})$ invariant. The dilaton and axion provide a map from spacetime into $SL(2,\mathbb{R})/SO(2)$ and one would, in general, have a non-linear sigma model type kinetic term for them (see e.g. [19]). For dimensional reasons it must be multiplied by T. In the limit we are considering the kinetic term vanishes and the axion and dilaton become auxiliary fields.

Tachyon Condensation

This subsection is based on [20]where references to the string literature may be found. The basic idea goes back to Ashoke Sen. In the presence of a tachyon field the Born-Infeld Lagrangian density is believed to be modified by the tachyon potential V to take the form

$$L = V - V\sqrt{1 - \mathbf{E}^2 + \mathbf{B}^2 - (\mathbf{E}\cdot\mathbf{B})^2}. \tag{121}$$

We are now keeping α' fixed and using units in which $2\pi\alpha' = 1$. Now it is believed that V has a critical point away from zero a which V vanishes. It is also believed that dynamically the system will relax to the state with $V = 0$, a so-called tachyon condensate. One may thus ask, what happens to the Born-Infeld vector in this limit. Again the Lagrangian density causes confusion: it vanishes identically in the limit.

However the Hamiltonian density is

$$\mathcal{H} = \sqrt{V^2(1+\mathbf{B}^2)+\mathbf{D}^2+(\mathbf{D}\times\mathbf{B})^2} - V\,, \tag{122}$$

and the limiting form is

$$\mathcal{H} = \sqrt{\mathbf{D}^2+(\mathbf{D}\times\mathbf{B})^2}\,. \tag{123}$$

The resulting constitutive relations are

$$\mathbf{H} = \frac{\mathbf{B}\mathbf{D}^2 - \mathbf{D}(\mathbf{B}.\mathbf{D})}{\sqrt{\mathbf{D}^2+(\mathbf{B}\times\mathbf{D})^2}}\,, \tag{124}$$

$$\mathbf{E} = \frac{\mathbf{D}+\mathbf{D}\mathbf{B}^2 - \mathbf{B}(\mathbf{B}.\mathbf{D})}{\sqrt{\mathbf{D}^2+(\mathbf{B}\times\mathbf{D})^2}}\,. \tag{125}$$

They tell us that $\mathbf{E}\times\mathbf{H} = \mathbf{D}\times\mathbf{B}$, and therefore the theory is Lorentz-invariant. One may check that electric-magnetic duality invariance is lost in this limit. The constitutive relations also imply that

$$\mathbf{D}.\mathbf{H} = 0\,, \tag{126}$$

but

$$\mathbf{D}^2 - \mathbf{H}^2 > 0\,. \tag{127}$$

It follows that the Ampère tensor $K_{\mu\nu}$ with components $\mathbf{D},\mathbf{H}$, is simple but timelike. Thus the two-form $K_{\mu\nu}$ it defines a 2-plane distribution in the tangent space. As discussed above the equation of motion for K implies that the distribution is integrable.

The limiting theory maybe expressed in terms of the Ampère tensor $K_{\mu\nu}$. One way to proceed is to consider a dual Lagrangian. We define $G = \star K$. The field equation $d\star K = 0$ becomes the Bianchi-Identity $dG = 0$. We now set $G = dC$ and consider the Lagrangian

$$\hat{L} = \sqrt{\frac{1}{2}G_{\mu\nu}G^{\mu\nu}} = \sqrt{-\frac{1}{2}K_{\mu\nu}K^{\mu\nu}}\,. \tag{128}$$

The action is now varied with respect to C but *subject to the constraint* that

$$K_{\mu\nu}\star K^{\mu\nu} = 0\,. \tag{129}$$

The resulting energy momentum tensor is given by

$$T_{\mu\nu} = -\frac{K_{\mu\lambda}K_{\nu}{}^{\lambda}}{\sqrt{-\frac{1}{2}K_{\sigma\tau}K^{\sigma\tau}}}\,. \tag{130}$$

The trace is given by

$$T^{\mu}_{\mu} = -\sqrt{-2K_{\sigma\tau}K^{\sigma\tau}}\,, \tag{131}$$

and therefore this is certainly not a conformally invariant theory.

Locally one may pass to a rest frame in which $\mathbf{B} = 0$. Then

$$\mathcal{H} = |\mathbf{D}|. \tag{132}$$

This is precisely what one expects of electric flux tubes with an energy proportional to the length and to the total flux carried by the tube.

In this rest frame one finds that

$$T_{\mu\nu} = \begin{pmatrix} \tau & 0 & 0 & 0 \\ 0 & -\tau & 0 & 0 \\ 0 & 0 & 0 & 0 \\ 0 & 0 & 0 & 0 \end{pmatrix}. \tag{133}$$

This is just what one expects for a string fluid.

THE M5-BRANE

In this concluding section I will indicate how many of the ideas described above extend to the theory of the M5-brane. To paraphrase Hooke *ut D3-brane sic M5-brane*. Indeed from the M-Theory point of view one should perhaps have reversed the logic, since one may regard the equations of Born-Infeld theory as the dimensional reduction of the M5-brane equations. The theory and it's equation have a reputation for complexity and so I will try to present them in as direct a way as possible. The interested reader may find references to the original papers and the statements made below the paper on which section is based [16].

One is of course considering a 6-dimensional non-linear theory involving scalar and spinor fields and in addition and closed 3-form $H_{\alpha\beta\gamma}$. In what follows I shall follow the original papers except that $\mu = 0, 1, \ldots, 5$. In particular in this section I shall follow their lead in this section be using the mainly positive signature convention.

Bianchi Identity

Consider the simplest situation: just the 3-form in a fixed background Einstein-metric $g_{\mu\nu}$. Thus

$$dH = 0. \tag{134}$$

Locally therefore one has $H = dA$, for some 2-form A.

Non-Linear Self-Duality

Now in six-dimensional Minkowski spacetime and acting on three-forms the standard linear Hodge duality is an involution of order two: $\star\star = 1$ and in linear theory self-duality is a consistent field equation. In other words closure of H and the self-duality condition give the *complete* set of equations of motion. For the M5-brane a very remarkable *non-linear self-duality* condition is possible which fulfills the same purpose. This was first discovered by Perry and Schwarz and its covariant form written down by Howe Sezgin and West. One does not seem to be able to construct a covariant Lagrangian just using the 2-form A. Non-covariant variational principles exist and a covariant action principles has been written down using an additional scalar field which acts as a time function. For the time being we need in these lectures only the equations of motion.

This remarkable condition is perhaps most expeditiously written as

$$\star H_{\alpha\beta\gamma} = \frac{1}{\sqrt{1+\frac{2}{3}H^2}}\Big[(1+\frac{4}{3}H^2)\delta^{\varepsilon}_{\alpha} - 4(H^2)^{\varepsilon}_{\alpha}\Big]H_{\varepsilon\beta\gamma}. \tag{135}$$

Of course in the limit of small H we have $H \approx \star H$. As stated above, if one reduces to five spacetime dimensions the equations reduce to the standard Born-Infeld equations.

Boillat Cone and Hooke's Law

One may introduce the analogue of the Boillat co-metric:

$$C^{\alpha\gamma} = \frac{Q}{(2-Q)}\Big(g^{\alpha\gamma}(1+\frac{4}{3}H^2) - 4(H^2)^{\alpha\gamma}\Big), \tag{136}$$

where

$$Q = -\frac{3}{H^2}(1+\frac{2}{3}H^2). \tag{137}$$

The characteristics of the scalar, spinor and 3-form equations of motion are determined by the Boillat metric $C^{\mu\nu}$. Moreover one may introduce an energy momentum tensor $T^{\mu\nu}$ which satisfies Hooke's Law:

$$T^{\alpha\beta} = g^{\alpha\beta} - C^{\alpha\beta}, \tag{138}$$

and is conserved

$$T^{\mu\nu}{}_{;\nu} = 0. \tag{139}$$

Note the sign change in Hooke's law because of the signature change. (However $T^{00} \geq 0$ in both conventions.)

One may prove that $T^{\mu\nu}$ satisfies the Dominant Energy Condition and hence, as with Born-Infeld theory, that the Einstein cone never lies inside the Boillat cone. In general the two cones touch along a circle of directions.

Weyl-Invariant Strong Coupling Limit

In general the trace of the energy momentum tensor T^{μ}_{μ} does not vanish. The theory is not Weyl-invariant except at vanishing field strength. However it becomes Weyl invariant in the limit of strong coupling. As with Born-Infeld, the most direct route to this result is the non-covariant (in our case $SO(5) \subset SO(5,1)$ symmetric) form of the equations. One defines a pair of two-forms $E_{ij} = H_{0ij}$ and $B_{ij} = -\frac{1}{6}\varepsilon_{ijpqr}H^{pqr}$. The Bianchi identity may be written in an obvious notation as

$$\frac{\partial \mathbf{B}}{\partial t} + \mathrm{curl}\mathbf{E} = 0, \qquad \mathrm{div}\mathbf{B} = 0. \tag{140}$$

To close the system one need a constitutive relation. To this end one defines

$$\mathcal{H} = \sqrt{\det(\delta_{ij} + 4B_{ij})} - 1. \tag{141}$$

The full non-linear self duality constraint has as solution

$$E_{ij} = \frac{1}{16}\frac{\partial \mathcal{H}}{\partial B_{ij}}. \tag{142}$$

The quantity $\mathcal{H}$ is the energy density T_{00}.

One may now restore dimensions by setting

$$\mathcal{H} = T^2\sqrt{\det(\delta_{ij} + 4\frac{B_{ij}}{T})} - T^2, \tag{143}$$

where T has dimension mass cubed. The limit $T \downarrow$ is now easily taken. More interesting than the general formulae for E_{ij} are the results for the energy momentum tensor. It takes the null matter form

$$T^{\mu\nu} = \mathcal{H} l^\mu l^\nu, \tag{144}$$

where, l^μ is again a null vector in the direction of the Poynting flux. Thus, just as is the case with Born-Infeld in fours spacetime dimensions, we attain Weyl-invariance in this limit and the theory has infinitely many conservation laws.

Point wise, one may skew diagonalize B_{ij}. In general it has rank four and two distinct skew eigenvalues B_1 and B_2 respectively. Of course the basis in which B_{ij} is skew diagonalized will in general vary with position. Pointwise one finds that if $l^\mu = (1,0,0,0,0,1)$

$$H = (dt - dx^5) \wedge (B_2 dx^2 \wedge dx^2 + B_1 dx^4 \wedge dx^5). \tag{145}$$

The three form H is in general not self-dual and is the sum of two totally simple three-forms. One factor is the null one form $L_\mu dx^\mu$, and one has $H_{\mu\nu\sigma}l^\mu = 0$, $\star H_{\mu\nu\sigma}l^\mu = 0$.

The quantum mechanical nature of this mysterious conformally invariant theory is an interesting challenge for the future.

REFERENCES

1. de Broglie, L., Un Itinéraire Scientifique, *Éditions Scientifique* (1987).

2. Gauntlett, J.P., Gibbons, G.W., Hull, C.M., and Townsend, P., *Comm. Math. Phys.*, **216**, 431–459 (2001).
3. Gibbons, G.W., Convex Cones in Physics, unpublished notes of lectures delivered at the Yuakawa Insitute.
4. Hawking, S.W., *Comm. Math. Phys.*, **18**, 303–306 (1970).
5. Hawking, S.W., and Ellis, G.F.R., *The Large Scale Structure of Spacetime* (1973), Cambridge University Press.
6. Zeldovich, Ya. B., The Creation of Particles and Anti-Particles in Electric and Gravitational Fields, *Magic Without Magic*, Ed. J. R., Klauder (277-288).
7. Zeldovich, Ya. B., and Pataevsky, L.P., *Comm. Math. Phys.*, **23**, 185 (1971).
8. Monge, G., Application de l'analyse al̀a geométrie, (1807).
9. Rouse Ball, W., *Messenger of Mathematics*, **21**, 20–24 (1891).
10. Białynicki-Birula, I., Non-linear Electrodynamics: Variations on a Theme of Born and Infeld, in *Quantum Theory of Fields and Particles*, Eds. B. Jancerwicz and J. Lukierski, World Scientific, Singapore (1983).
11. Białynicki-Birula, I., *Acta Physica Polonica*, **B23**, 553–559 (1992).
12. Ferrer, E.J., Fradkin, E.S., and de la Incarra, V., *Phys. Lett.*, **B248**, 281–287 (1990).
13. Tseytlin, A.A., *Nucl. Phys.*, **B460**, 69 (1998).
14. Gibbons, G.W., *Nucl. Phys.*, hep=th/9709027.
15. Gibbons, G.W., Pulse Propagation in Born-Infeld Theory, hep-th/0104015.
16. Gibbons, G.W., and West, P.C., Subm. to: *J. Math. Phys.*, hep-th/0011149.
17. Gibbons, G.W., and Herdeiro, C. A. R., *Phys. Rev.*, **D63**, 064006-1–064006-17 (2001); hep-th/0008052.
18. Gibbons, G.W., and Rasheed, D.A., *Nucl. Phys.*, **B454**, 185–206 (1995).
19. Gibbons, G.W., and Rashheed, D.A., *Phys. Lett.*, **B365**, 46–50 (1996).
20. Gibbons, G.W., Hori, K., and Yi, P., String Fluid from Unstable D-brane, hep-th/0007019.
21. Callan, C.G., and Maldacena, J., *Nucl. Phys.*, **B513**, 198 (1998); hep-th/9708147.
22. Gibbons, G.W., and Hashimoto, K., Nonlinear Electrodynamics in Curved Backgrounds, hep-th/0007019.

Regularizing Effect of the Graviton Background

Z. Haba

Institute of Theoretical Physics, University of Wrocław, pl. M. Borna 9, 50-205 Wrocław, Poland,
emai:zhab@ift.uni.wroc.pl

Abstract. We consider field theoretic propagators in a quantum gravitational field. It is shown that the propagators are less singular than the ones with a fixed gravitational background.

INTRODUCTION

A long time ago it has been suggested by Pauli and other authors (see [1]-[3]) that quantization of the gravitational field can remove divergencies of the conventional quantum field theory. It could be imagined that as a result of the fluctuations of the metric the light-cone singularities of quantum propagators would disappear.

In this paper we consider a non-perturbative approach to the model. We do not treat the complete quantum gravity . Such a model is too complicated for a non-perturbative rigorous study. In Ref. [4] we assumed that only some components of the metric are non-zero and moreover that they depend only on one coordinate. In this paper we set up another scheme which starts from multidimensional quantum gravity. The metric is assumed to depend only on the unphysical coordinates. Subsequently, a dimensional reduction to the physical dimension is performed. This simplifies the analysis of singularities and enables a rigorous treatment of the model.

We split the coordinates $x = (\mathbf{x}_G, \mathbf{x}_F)$ in $2N$ dimensions as $2n$ physical coordinates and $2N-2n$ unphysical ones. The metric tensor g_{BC} splits [5]-[6] into a block form with $g_{\mu\nu} + g_{jl}A^j_\mu A^l_\nu$ and g_{ij} on the diagonal and $A^l_\mu g_{jl}$ off the diagonal. Now, the volume element is $V_{2N} = V_{2N-2n}(\det g_{\mu\nu})^{\frac{1}{2}}$ and the scalar curvature

$$R_{2N} = R_{2N-2n} - \frac{1}{4}g_{\mu\nu}F^{\mu}_{jk}F^{\nu jk} + \frac{1}{4}g^{jk}(g^{\alpha\beta}g^{\sigma\rho} - g^{\alpha\sigma}g^{\beta\rho})\partial_j g_{\alpha\beta}\partial_k g_{\sigma\rho}\,, \tag{1}$$

where F^{μ}_{jk} is the field strength for $O(2N-2n-1,1)$ Yang-Mills field and the Latin indices index the unphysical coordinates. We consider also a scalar field ϕ with the propagator which is the kernel of the inverse of the operator

$$\Box_g = g^{BC}\partial_B\partial_C + \frac{1}{2}\Gamma^B\partial_B\,. \tag{2}$$

We average the propagators over the metric. In order to reduce the task to a menageable problem we consider a simplified model where only $g^{\mu\mu} = \alpha^2_\mu$ is different from a constant. The eventual corrections could be treated by means of a perturbation method.

CP589, *Fundamental Interaction Theories*, edited by J. Lukierski and J. Rembieliński

We consider the Schrödinger equation in R^{2N}

$$i\partial_\tau\psi = -\mathcal{A}\psi, \tag{3}$$

and its Euclidean version

$$\partial_\tau\psi = \mathcal{A}\psi, \tag{4}$$

where

$$\mathcal{A} = \frac{1}{2}\sum_{\mu=1}^{2n}\alpha_\mu^2(\mathbf{x}_F)\partial_\mu^2 + \frac{1}{2}\sum_{k=2n+1}^{2N}\partial_k^2, \tag{5}$$

where we denote $\mathbf{x}_F = (x_{2n+1}, ..., x_{2N})$ and α_μ are independent homogeneous Gaussian random fields with the same covariance

$$\langle\alpha_\mu(x)\alpha_\mu(y)\rangle = G(\mathbf{x}_F - \mathbf{y}_F). \tag{6}$$

We solve the Schrödinger equation (3) with the initial condition ψ by means of a probabilistic version of the Feynman integral [8]-[9]

$$(\exp(i\mathcal{A}\tau)\psi)(x) = E[\psi(q_\tau(x))], \tag{7}$$

where $E[..]$ is an expectation value with respect to the Brownian motion defined as the Gaussian process on R^{2N} with the covariance

$$E[b_j(t)b_k(s)] = \delta_{jk}\min(s,t).$$

In order to write the solution of the Schrödinger equation (3) in the form (7) we assume that ψ is an analytic function. Then, we begin with regularized random fields which are analytic functions as well.

The stochastic process q_τ consists of two vectors $(\mathbf{q}_G, \mathbf{q}_F)$ where

$$\mathbf{q}_F(\tau, \mathbf{x}_F) = \mathbf{x}_F + \lambda\mathbf{b}_F(\tau) \tag{8}$$

and $\mathbf{q}_G$ has the components for $\mu = 1, ..., 2n$

$$q_\mu(\tau, \mathbf{x}) = x_\mu + \lambda\int_0^\tau \alpha_\mu(\mathbf{q}_F(s, \mathbf{x}_F))\,db_\mu(s), \tag{9}$$

where $\tau \geq 0$ and

$$\lambda = \sqrt{i} = \frac{1}{\sqrt{2}}(1+i).$$

We consider also Euclidean (imaginary time) version of Eq. (7) described by the semi-group $\exp\tau\mathcal{A}$. The Euclidean version corresponds to $\lambda = 1$ in Eqs. (8)-(9).

The process $\mathbf{q}_\tau(\mathbf{x})$ starts from $\mathbf{x}$ at $\tau = 0$. It is a solution of a set of stochastic equations [7]. It can be checked by direct differentiation using the Ito calculus [7] that ψ_τ solves the Schrödinger equation (3) for each (regularized) α.

The time evolution (3) is determined by the propagator

$$\psi_\tau(\mathbf{x}) = \int K_\tau(\mathbf{x}, \mathbf{y})\psi(\mathbf{y})d\mathbf{y}.$$

We further specify the random fields α_μ in Eq. (6) by assuming that their covariance is of the form

$$G(\mathbf{x}_F) = \int_{-\infty}^{\infty} d\nu\rho(\nu)\cos(\nu|\mathbf{x}_F|^4). \tag{10}$$

Then, $G(\sqrt{i}x) = G(x)$. We represent K by means of of the Fourier transform and the stochastic process (8)-(9)

$$K_\tau(\mathbf{x},\mathbf{y}) = \int d\mathbf{p}E[\exp(i\mathbf{p}(\mathbf{y}-\mathbf{q}_\tau(\mathbf{x})))]. \tag{11}$$

We compute the Gaussian integral over α resulting from the Lagrangian (1) as in the Euclidean quantum gravity. Then, we obtain for the expectation value of the kernel K

$$\begin{aligned}\langle K_\tau(\mathbf{x},\mathbf{y})\rangle &= \int d\mathbf{p}\exp(i\mathbf{p}(\mathbf{y}-\mathbf{x}))E[\prod_{k>2n}\exp(-i\lambda p_k b_k(\tau)) \\ &\prod_{\mu\leq 2n}\exp(-\frac{i}{2}p_\mu^2\int_0^\tau\int_0^\tau db_\mu(s)db_\mu(t)G(\mathbf{q}_F(t)-\mathbf{q}_F(s)))]. \end{aligned}\tag{12}$$

If the gravitational field was treated in the formalism of the Feynman integral in the Minkowski space then $p_\mu^2 \to -ip_\mu^2$ in Eq. (12). Conclusions concerning the short distance behavior of the propagator do not depend on whether we work in Minkowski or Euclidean space.

The Euclidean version of the path integral can be expressed in a more compact form

$$\begin{aligned}\langle K_\tau(\mathbf{x},\mathbf{y})\rangle &= (2\pi\tau)^{-N+n}\exp\left(-\frac{1}{2\tau}(\mathbf{x}_F-\mathbf{y}_F)^2\right)\int\prod_k dp_k\exp(ip_k(y_k-x_k)) \\ &E[\prod_\mu\exp(-\frac{1}{2}p_\mu^2\int_0^\tau\int_0^\tau db_\mu(s)db_\mu(t)G(\mathbf{Q}_F(t)-\mathbf{Q}_F(s)))], \end{aligned}\tag{13}$$

where

$$\mathbf{Q}_F(s) = \mathbf{x}_F + \frac{s}{\tau}(\mathbf{y}_F-\mathbf{x}_F) + \sqrt{\tau}(1-\frac{s}{\tau})\mathbf{b}_F(\frac{s}{\tau-s}). \tag{14}$$

Let us note that

$$\int_0^\tau\int_0^\tau db_\mu(s)db_\mu(t)G(\mathbf{q}_F(t)-\mathbf{q}_F(s)) = \Gamma_\mu(\tau)+G(0)\tau, \tag{15}$$

where

$$\Gamma_\mu(\tau) = 2\int_0^\tau db_\mu(t)\int_0^t db_\mu(s)G(\mathbf{b}_F(t)-\mathbf{b}_F(s)). \tag{16}$$

So, we can remove the infinite term $G(0)\tau$ defining the renormalized semigroups as

$$\exp\left(-\frac{i}{2}G(0)\,\tau\triangle\right)\exp(i\mathcal{A}\tau), \tag{17}$$

in the Minkowski case and

$$\exp\left(-\frac{1}{2}G(0)\,\tau\triangle\right)\exp(\mathcal{A}\tau), \tag{18}$$

in the Euclidean case. Now, the momentum integrals p_k can be calculated. As a result, after the above mentioned renormalization, we obtain the following formula

$$\begin{aligned}\langle K_\tau(\mathbf{x},\mathbf{y})\rangle &= (2\pi)^{-n}E\Big[\int d\mathbf{p}_F \exp\left(i\mathbf{p}_F\left(\mathbf{x}_F-\mathbf{y}_F\right)+i\lambda\mathbf{p}_F\mathbf{b}_F\left(\tau\right)\right) \\ &\quad \prod_{\mu=1}^{2n}\left(2\pi\Gamma_\mu\left(\tau\right)\right)^{-\frac{1}{2}}\exp\left(\frac{i}{2}\left(x_\mu-y_\mu\right)^2/\Gamma_\mu(\tau)\right)\Big].\end{aligned} \tag{19}$$

The Euclidean version is

$$\begin{aligned}\langle K_\tau(\mathbf{x},\mathbf{y})\rangle &= (2\pi\tau)^{-N+n}\exp\left(-\frac{1}{2\tau}\left(\mathbf{x}_F-\mathbf{y}_F\right)^2\right) \\ &\quad E\Big[\prod_{\mu=1}^{2n}\left(2\pi\Gamma_\mu\left(\tau\right)\right)^{-\frac{1}{2}}\exp\left(-\frac{1}{2}\left(x_\mu-y_\mu\right)^2/\Gamma_\mu(\tau)\right)\Big],\end{aligned} \tag{20}$$

where

$$\Gamma_\mu(\tau)=2\int_0^\tau db_\mu(t)\int_0^t db_\mu(s)G\left(\mathbf{Q}_F\left(t\right)-\mathbf{Q}_F\left(s\right)\right). \tag{21}$$

We choose $\rho(\nu)$ in G (Eq. (10)) in a scale invariant form

$$\rho(\nu)=const|\nu|^{\frac{\gamma}{2}-1},$$

where $\gamma<\frac{1}{2}$ if the stochastic integrals in Eq. (21) are to make sense. Then

$$G(\mathbf{x})=|\mathbf{x}|^{-2\gamma}. \tag{22}$$

THE PROPAGATOR

Let us consider the wave operator in the harmonic gauge (propagators of other operators [10] entering quantum gravity can be treated in a similar way via the Feynman formula, but their analysis is more complicated)

$$\mathcal{A}=\frac{1}{2}g^{BC}\partial_B\partial_C. \tag{23}$$

We can obtain the causal Feynman propagator of $\mathcal{A}$ (the Green's function) by means of the proper time method [11]

$$\mathcal{A}^{-1}(x,y)=-i\int_0^\infty d\tau\left(\exp\left(i\tau\mathcal{A}\right)\right)(x,y). \tag{24}$$

We consider a complex scalar field and define

$$\phi(x_1,\ldots,x_d)=\int dx_{d+1}\ldots.dx_{d+n}\phi(x), \tag{25}$$

and

$$\phi^*(x_1,...,x_d) = \int dx_{d+n+1}....dx_{d+2N}\bar{\phi}(x)\,. \qquad (26)$$

If $\alpha = const$ then the correlation functions of ϕ and ϕ^* coincide with the ones of a complex free scalar field in d dimensions. If α is a regular deterministic or random function then the correlations will have the same singularity as for the free field. If however α is a singular Gaussian field then we show that the correlations are more regular than the free ones. The Green's function depends only on $x-y$. The propagator of dimensionally reduced model is now equal to

$$\begin{aligned}\langle \mathcal{A}^{-1}(\mathbf{x}_G,\mathbf{y}_G)\rangle &= \int_0^\infty d\tau \prod_{k=1}^{n} dx_{2n+k}dy_{3n+k}\langle K_\tau(\mathbf{x},\mathbf{y})\rangle = \\ &\approx \int_0^\infty d\tau E\left[\prod_\mu \left(2\pi\Gamma_\mu(\tau)\right)^{-\frac{1}{2}}\exp\left(-\frac{1}{2}\left(x_\mu-y_\mu\right)^2/\Gamma_\mu(\tau)\right)\right], \qquad (27)\end{aligned}$$

where the approximation concerns the neglect of the $\mathbf{x}_F$ dependence of the Green function G.

$\Gamma_\mu(\tau)$ behaves as $\tau^{1-\gamma}$ for a small τ. Then, by a change of variables

$$\tau^{1-\gamma} = \sum_{\mu=1}^{\mu=2n}(x_\mu-y_\mu)^2\,\tilde{\tau}\,,$$

the dependence on $x-y$ in the integral (27) can be extracted. In this way we obtain (similarly as in [4]) the bound

$$\langle \mathcal{A}^{-1}(\mathbf{x}_G,\mathbf{y}_G)\rangle \simeq \left((\mathbf{x}_G-\mathbf{y}_G)^2\right)^{-n+1+\frac{\gamma}{1-\gamma}}\,. \qquad (28)$$

We would obtain the same behavior in a complete perturbative quantum gravity if we calculated the propagator via the Feynman formula (7) where the stochastic process q_τ is derived to the lowest order in the perturbative expansion around Minkowski metric.

We can generalize this method to Wick powers taking the Wick powers of the operators (25)-(26). So, for the Wick square

$$\begin{aligned}\langle \phi^2(\mathbf{x}_G)(\phi^2)^*(\mathbf{y}_G)\rangle &= \int d\tau d\tilde{\tau}\prod dx_{2n+k}dy_{3n+k}d\tilde{x}_{2n+k}d\tilde{y}_{3n+k} \\ &\quad \langle K_\tau(\mathbf{x}_G,x_{2n+1},...;\mathbf{y}_G,y_{2n+1},...) \\ &\quad K_{\tilde{\tau}}(\mathbf{x}_G,\tilde{x}_{2n+1}...;\mathbf{y}_G,\tilde{y}_{2n+1},....)\rangle\,. \qquad (29)\end{aligned}$$

If we neglect the $\mathbf{x}_F-\mathbf{y}_F$ dependence of the propagators G then the x_{2n+k} and y_{3n+k} integrals can be calculated with the result (in the Euclidean version)

$$\begin{aligned}\langle \phi^2(\mathbf{x}_G)(\phi^2)^*(\mathbf{y}_G)\rangle &= \int d\tau d\tilde{\tau}E[\prod_\mu(\det \mathcal{M}^{(\mu)})^{\frac{1}{2}} \\ &\quad \exp(-\frac{1}{2}(X-Y)_\mu\mathcal{M}_\mu^{-1}(X-Y)_\mu)]\,, \qquad (30)\end{aligned}$$

where

$$(X-Y)_\mu = (x_\mu - y_\mu, x_\mu - y_\mu),$$

is the two component vector and $\mathcal{M}$ is 2×2 matrix with the matrix elements (where $\mathbf{x}_F - \mathbf{y}_F \simeq 0$ and $\tilde{\mathbf{x}}_F - \tilde{\mathbf{y}}_F \simeq 0$ in Eq. (14) defining $\mathbf{Q}$)

$$\mathcal{M}_{11}^{(\mu)} = 2\int_0^\tau db_\mu(t)\int_0^t db_\mu(s) G\left(\mathbf{Q}_F(t) - \mathbf{Q}_F(s)\right), \tag{31}$$

$$\mathcal{M}_{22}^{(\mu)} = 2\int_0^{\tilde{\tau}} d\tilde{b}_\mu(t)\int_0^t d\tilde{b}_\mu(s) G\left(\tilde{\mathbf{Q}}_F(t) - \tilde{\mathbf{Q}}_F(s)\right), \tag{32}$$

$$\mathcal{M}_{12}^{(\mu)} = \mathcal{M}_{12}^{(\mu)} = \int_0^\tau \int_0^{\tilde{\tau}} db_\mu(t) d\tilde{b}_\mu(s) G\left(\mathbf{Q}_F(t) - \tilde{\mathbf{Q}}_F(s)\right). \tag{33}$$

A detailed analysis of the behavior of the integral in Eq. (30) for small τ and $\tilde{\tau}$ shows that the correlation functions of Wick powers are also less singular than the ones for a fixed gravitational background. There is still a gap between these results and a complete perturbative scheme where some components of the metric tensor (e.g. the diagonal part) are treated non-perturbatively whereas the remaining non-quadratic terms in the Lagrangian (including, e.g., the ϕ^4 interaction) are expanded in a perturbation series. However, we believe that our results are encouraging for a research in this direction.

REFERENCES

1. Pauli, W., *Helv. Phys. Acta*, Suppl., **4**, 69 (1956).
2. Deser, S., *Rev. Mod. Phys.*, **29**, 417 (1957).
3. Isham, C.J., Salam, A., and Strathdee, J., *Phys. Rev.*, **D3**, 1805 (1971).
4. Brzezniak, Z., and Haba, Z., *Journ. Phys.*, **A34**, L139 (2001).
5. Kerner, R., *Ann. Inst. Henri Poincaré*, **9**, 143 (1968).
6. Cremmer, E., and Julia, B., *Nucl. Phys.*, **B159**, 141 (1979).
7. Ikeda, N., and Watanabe, S., *Stochastic Differential Equations and Diffusion Processes*, North Holland, 1981.
8. Haba, Z., *Journ. Phys.*, **A27**, 6457 (1994).
9. Haba, Z., *Feynman Integral and Random Dynamics in Quantum Physics*, Kluwer, 1999.
10. Christensen, S.M., and Duff, M.J., *Nucl. Phys.*, **B154**, 301 (1979).
11. Feynman, R.P., *Phys. Rev.*, **80**, 440 (1950).

Localization of Particles, Spreading and the Notion of Einstein Causality

G.C. Hegerfeldt

Institut für Theoretische Physik, Universität Göttingen, Bunsenstr. 9, 37073 Göttingen, Germany, email: hegerf@theorie.physik.uni-goettingen.de

Abstract. The notion of Einstein causality, i.e. the limiting role of the velocity of light in the transmission of signals, is discussed. We show that under quite general assumptions instantaneous spreading of particle localization occurs in quantum theory, relativistic or not, with fields or without. We discuss if this affects Einstein causality. It is also pointed out that the controversy about recent experiments of Nimtz and coworkers reporting superluminal signal velocities in micro-wave tunneling arise from an indiscriminate use of terminology and that the experimental results are in full agreement with Einstein causality in its ordinary sense.

INTRODUCTION

Einstein's principle of finite signal velocity is of fundamental importance for the foundations of physics, both classical and quantum. If signal velocities could be arbitrarily high this would either lead to the possibility of absolute clock synchronization and to a change of special relativity or to the possible existence of superluminal tachyons with their associated acausal effects [1].

Are there superluminal phenomena in the quantum realm? For a free nonrelativistic particle instantaneous spreading of the wave function is well known. If, at time $t = 0$, the wave function vanishes outside some finite region V then the particle is localized in V with probability 1. Instantaneous spreading implies that the probability of finding the particle arbitrarily far away from the initial region is nonzero for any $t > 0$. In a nonrelativistic theory, however, this superluminal propagation is of no great concern.

If one describes the localization of a free relativistic particle by the Newton-Wigner position operator then instantaneous spreading also occurs, as noted in Refs. [2] and [3] (cf. also Ref. [4]). In 1974 the present author [5] showed that the Newton-Wigner position operator is extraneous to this phenomenon. Instantaneous spreading occurs for any notion of free particle localization, be it in the sense of Newton-Wigner or others. Later an alternative proof of this result was given [6] and the result was extended to the center-of-mass motion of relativistic systems with possibly more than one particle [7]. Ruijsenaars and the author [8] then showed that instantaneous spreading occurs for quite general, relativistic or nonrelativistic, interactions. The main result of Ref. [8] was that this instantaneous spreading is mainly due to positivity of the energy plus translation invariance. More recently it was shown by the author [9] that translation invariance is also not needed. Hilbert space and positivity of the Hamiltonian (energy) suffices to ensure either instantaneous spreading or, alternatively, confinement in a fixed

CP589, *Fundamental Interaction Theories,* edited by J. Lukierski and J. Rembieliński

region for all times. Another extension was given by the author [10] for free relativistic particles and for relativistic systems which have exponentially bounded tails in their localization outside some region V. It was shown that the state spreads out to infinity faster than allowed by a probability flow with finite propagation speed. Probably the most astonishing part of our results is the fact that so little is needed to derive them. They hold with and without field theory and with and without relativity. Only Hilbert space and positivity of the energy is needed.

What do these results mean for Einstein causality? This will be discussed in the following where we concentrate on the role played by positivity of the energy for instantaneous spreading. We also briefly discuss Fermi's two-atom model [11, 12]. In the last section the micro-wave experiments of Nimtz and coworkers [13] in which superluminal tunneling was reported will be briefly discussed and clarified. Related material appeared in Ref. [14].

POSITIVITY OF THE ENERGY AND FERMI'S MODEL

In this section we will prove a simple mathematical result on the temporal behavior of certain expectation values. If the time-development operator is positive one might expect analyticity properties, but for expectation values of the form

$$\langle e^{-iHt}\psi, Ae^{-iHt}\psi\rangle ,$$

this is not true since the first operator involves a complex conjugate. One has, however, the following result.

Theorem: Let H be a selfadjoint operator, positive or bounded from below, in a Hilbert space $\mathcal{H}$. For given $\psi_0 \in \mathcal{H}$ let $\psi_t, t \in \mathbb{R}$, be defined as

$$\psi_t = e^{-iHt}\psi_0 \,. \tag{1}$$

Let A be a positive operator in $\mathcal{H}, A \geq 0$, and let $p_A(t)$ be defined as

$$p_A(t) = \langle \psi_t, A\psi_t\rangle \,. \tag{2}$$

Then either

$$p_A(t) \neq 0 \quad \text{for almost all} \quad t \tag{3}$$

and the set of such t's is dense and open, or

$$p_A(t) \equiv 0 \quad \text{for all} \quad t \,. \tag{4}$$

Proof. The proof is based on an analyticity argument for which, however, a little care – and the positivity of A – is needed. Evidently, since $H \geq -c$, one can define $\exp\{-iH(t+iy)\}$ for $y \leq 0$, and $\exp\{-iHz\}$ is analytic in z for Im $z < 0$, and hence ψ_t can be analytically continued to the lower half-plane, with continuous boundary values on the real axis. However, as noted above the r.h.s. of Eq. (2) can in general not be analytically continued since it equals

$$\langle \psi, e^{iHt}Ae^{-iHt}\psi\rangle$$

and since $\exp\{i(H+iy)t\}$ is in general unbounded for $y < 0$. To by-pass this the positivity of A can be used. We write

$$p_A(t) = \langle A^{1/2}\psi_t, A^{1/2}\psi_t \rangle \,, \tag{5}$$

where $A^{1/2}$ is the positive square root of A, and denote by $\mathcal{N}_0$ the set of t's for which $p_A(t) = 0$. By continuity of $p_A(t)$, $\mathcal{N}_0$ is closed and its complement $\mathcal{N}_0^c$ is open. Eq. (5) now implies

$$A^{1/2}\psi_t = 0 \text{ for } t \in \mathcal{N}_0 \,. \tag{6}$$

For fixed $\phi \in \mathcal{H}$ we define the function $F_\phi(z)$ for Im $z \leq 0$ by

$$F_\phi(z) = \langle \phi, A^{1/2} e^{-iHz}\psi_0 \rangle \,. \tag{7}$$

By the above remark on $\exp\{-iHz\}$, $F_\phi(z)$ is a continuous function for Im $z \leq 0$ and is analytic for Im $z < 0$. By Eq. (6) one has

$$F_\phi(t) = 0 \quad \text{for } t \in \mathcal{N}_0 \,. \tag{8}$$

Now let us assume that the complement $\mathcal{N}_0^c$ is not dense. Then $\mathcal{N}_0$ contains some interval I of nonzero length, and $F_\phi(z)$ vanishes on I. One can now directly employ the Schwarz reflexion principle [15] to conclude that $F_\phi(z) \equiv 0$ or proceed in a more pedestrian way as follows. One defines an extension of F_ϕ to the upper half plane by putting

$$F_\phi(z) = F_\phi(z^*)^* \text{ for Im } z > 0 \,. \tag{9}$$

Since $F_\phi(t) = 0$ for $t \in I$ and thus, *a fortiori*, real for $t \in I$, the extension $F_\phi(z)$ is continuous for $z \in I$. From this one easily shows [15] that $F_\phi(z)$ is analytic for $z \notin \mathbb{R}\backslash I$, and thus I is contained in the domain of analyticity. Since $F_\phi(z)$ vanishes on I it must therefore vanish on the analyticity domain, i.e. for $z \notin \mathbb{R}\backslash I$. By continuity $F_\phi(z)$ then vanishes everywhere. Since ϕ was arbitrary, we obtain $A^{1/2}\psi_t = 0$ for all t. Hence,

$$A\psi_t = 0 \quad \text{for all } t \tag{10}$$

and thus $p_A(t) \equiv 0$ if $\mathcal{N}_0^c$ is not dense, i.e. alternative (ii) holds in this case.

Since a dense open set need not have full Lebesgue measure, it remains to show that $\mathcal{N}_0$ is a null set if alternative (ii) does not hold. To prove this we use the fact that, as a boundary value of a bounded analytic function, $F_\phi(t)$ satisfies the inequality [16]

$$\int_{-\infty}^{\infty} dt \, \frac{\ln|F_\phi(t)|}{1+t^2} > -\infty, \tag{11}$$

unless it vanishes identically. If $\mathcal{N}_0$ had positive measure the integral would be $-\infty$, and thus $F_\phi(t)$ would vanish for all t, for each ϕ. This would again imply alternative (ii). This proves the theorem.

The theorem is a more abstract version of a result in [12] on Fermi's two-atom problem. To check the speed of light in quantum electrodynamics, Fermi had considered two atoms, separated by a distance R and with no photons present initially. One of the atoms was assumed to be in its ground state, the other in an excited state. The latter could then decay with the emission of a photon. Fermi calculated the excitation probability of the atom which had initially been in its ground state. Using standard approximations he found the excitation probability to be zero for $t < R/c$.

Now, if one takes for ψ_0 in the theorem the initial state considered by Fermi and for A the operator describing the excitation probability, e.g. the projector onto the excited states, then $p_A(t)$ becomes the excitation probability, and the theorem states that this probability is immediately nonzero. Already in [12] it was discussed how to avoid a possible conflict with causality, and this was continued in more detail for example in [17, 18, 19, 20], as explained further below.

PARTICLE LOCALIZATION AND SPREADING

Let us suppose that it makes sense to speak of the probability to find a particle at a given time inside a space region V. This is a highly nontrivial assumption. In a quantum theory the probability to find a particle or system inside V should be given by the expectation of an operator, $N(V)$ say. Since probabilities lie between 0 and 1, one must have

$$0 \leq N(V) \leq 1 \,. \tag{12}$$

Now let us assume that the system, with state ψ_0 at $t = 0$, is strictly localized in a region V_0, i.e. with probability 1, so that $\langle \psi_0, N(V_0)\psi_0 \rangle = 1$ or, equivalently,

$$\langle \psi_0, (1 - N(V_0))\psi_0 \rangle = 0 \,. \tag{13}$$

From Eq. (12) one has $1 - N(V_0) \geq 0$ and hence the theorem can be applied, with

$$A \equiv 1 - N(V_0) \,. \tag{14}$$

As a consequence one either has

$$\langle \psi_t, N(V_0)\psi_t \rangle \equiv 1 \quad \text{for all } t \,, \tag{15}$$

or

$$\langle \psi_t, N(V_0)\psi_t \rangle < 1 \quad \text{for almost all } t \,. \tag{16}$$

The alternative in Eq. (15) means that the particle or system stays in V_0 for all times, as might happen for a bound state in an external potential.

Now, if the particle or system is strictly localized in V_0 at $t = 0$ it is, *a fortiori*, also strictly localized in any larger region V containing V_0. If the boundaries of V and V_0 have a finite distance and if finite propagation speed would hold then the probability to find the system in V would also have to be 1 for sufficiently small times, e.g. $0 \leq t < \varepsilon$. But then the theorem, with $A \equiv 1 - N(V)$, states that the system stays in V for *all* times. Now we can make V smaller and smaller and let it approach V_0. Thus we conclude that if a particle or system is strictly localized in a region V_0 at time $t = 0$, then finite propagation speed implies that it stays in V_0 for all times and therefore prohibits motion to infinity. Or put conversely, if there exist particle states which are strictly localized in some finite region at $t = 0$ and later move towards infinity, then finite propagation speed cannot hold for localization of particles.

This can be formulated somewhat more strongly as follows. If at $t = 0$ a particle is strictly localized in a bounded region V_0 then, unless it remains in V_0 for all times, it cannot be strictly localized in a bounded region V, however large, for any finite time interval thereafter, and the particle localization immediately develops infinite "‘tails"’. The spreading is over all space except possibly for "‘holes"’ which, if any, will persist for all times, by the same arguments as before. If the theory is translation invariant then there can be no holes, as shown in Ref. [8] under some mild spectrum conditions.

At first sight the Dirac equation might seem to be a counterexample to our results on instantaneous spreading. Indeed, this wave equation is hyperbolic, implying finite propagation speed. For the localization operator $N(V)$ one might take the characteristic function $\chi_V(\mathbf{x})$, just as in the nonrelativistic case and in contrast to the Newton-Wigner operator. Then, for a wave function initially vanishing outside a finite region, i.e. of finite support, the localization does evolve with finite propagation speed. Doesn't this contradict the results of the preceding section?

This example is instructive since it shows the importance of the positive-energy condition. The Dirac equation contains positive and negative energy states. Now, consider a solution of the Dirac equation which vanishes outside some finite region and make the additional assumption that it is composed of positive-energy solutions only. Then one gets a contradiction to our results and therefore the additional assumption must be wrong, i.e. a solution with finite support at some time must contain negative-energy contributions. This means that positive-energy solutions of the Dirac equation always have *infinite* support to begin with. This is phrased as a mathematical result for instance in the book of Thaller [21].

Thus the results of the preceding section do not apply if there are no strictly localized states in the theory! Strict localization of a state ψ in a region V means that $\langle\psi, N(V)\psi\rangle = 1$, and this gives

$$0 = \langle\psi, (\mathbf{1} - N(V))\psi\rangle = \|(\mathbf{1} - N(V))^{1/2}\psi\|^2,$$

where the root exists by positivity of $N(V)$. This implies

$$N(V)\psi = \psi. \tag{17}$$

Hence ψ is an eigenvector of $N(V)$ for the eigenvalue 1 if ψ is strictly localized in V, and vice versa. The eigenvalue 0 means strict localization outside V.

The existence or nonexistence of strictly localized states depends on the form of $N(V)$. For example, if one has a self-adjoint position operator $\hat{\mathbf{X}}$ with commuting components, then $N(V)$ is a projection operator from the spectral decomposition of $\hat{\mathbf{X}}$ and thus has eigenvalues 1 and 0. Hence in this case there are strictly localized states for any region V, and the result of the previous section implies instantaneous spreading.

This instantaneous spreading also occurs for position operators with self-adjoint but *non*-commuting components $\hat{X}_i$. Each $\hat{X}_i$ has a spectral decomposition whose projection operators give the localization operators for infinite slabs. Eigenvectors for the eigenvalue 1 represent states strictly localized in these slabs, and there is instantaneous spreading in this case, too.

To avoid instantaneous spreading one therefore has to consider localization operators $N(V)$ which are not projectors, for example positive operator-valued measures. However, if one insists on arbitrary good localization, i.e. on tails which drop off arbitrarily fast, then one runs into our results in Ref. [10] .

FIELD-THEORETIC ASPECTS AND DISCUSSION

Localization of particles. In field theory difficulties with particle localization have been known for a long time [22]. However, our results are more general since no fields explicitly needed and none of the particular assumptions or axioms of field theory, except for positivity of the energy, are used.

In a field-theoretic context permanent infinite tails can be understood intuitively through clouds of virtual particles due to renormalization ('dressed states'). It is also conceivable that whenever one tries to prepare a localized particle or system one automatically creates particle-antiparticle pairs outside the desired localization region, and this would have the same effect as tails.

Instead of speaking about infinite tails one may also envisage that all particle detectors exhibit, for example, inherent noise and that therefore localization with probability 1 or zero can never be recorded. This would essentially lead to the same conclusions as permanent infinite tails.

Fermi's model. From the foregoing discussion it is evident that, in a field-theoretic context, excitation of atom B need not be due to absorption of a photon emitted by atom A. The excitation could rather be due to vacuum fluctuations, photon clouds etc. Or the excitation maybe just spontaneous, whatever that means. At a more fundamental level, the notion of bounds states has its intricacies in field theory, and the corresponding observables in the Fermi model might be hard to put on a mathematically rigorous footing.

If infinite tails always exist, or if all counters are influenced by vacuum fluctuation, then how can finite propagation speed or Einstein causality be checked at all? Here it is useful for differentiate between two notions of causality.

Strong causality. By this we mean that for *each* individual experiment in which two systems, separated by a distance R, are prepared at time $t = 0$ *no* disturbance or excitation of the second system occurs for $t < R/c$ [17]. This notion is analogous to energy momentum and baryon conservation in each individual scattering process in particle physics. Strong causality would hold in the Fermi model if the transition probability were strictly zero for $t < R/c$. It seems that Fermi had this causality notion in mind. Our results show that strong causality cannot be *checked* (unless a possible way out via cut-off theories holds [17]), or it may fail in a theory.

Weak causality. This notion was introduced by Schlieder [4], and it means Einstein causality for expectation values or ensemble averages only, not for each individual process. Thus for the above two systems, expectation values for the second system need not vanish for $t < R/c$, but it would take at least a time $t = R/c$ to produce an effect on them. To exhibit this effect it is convenient to subtract possible fluctuations of the second system alone. The above theorem does not apply to this situation since this difference is not the expectation value of a positive operator.

The weak assumptions of the above theorems (just Hilbert space and positive energy) will not be enough to prove weak causality. In Refs. [19, 20] the Fermi model was been studied using the methods and approximations of quantum optics. Vacuum fluctuations and virtual photons contribute to the excitation of atom B, and once the expectation value of this contribution has been subtracted, the remainder behaves causally, at least within the approximations employed. This just corresponds to the notion of weak causality. In how far this can be measured will be discussed below.

In the framework of quantum field theory it is sometimes simply argued that local commutation or anti-commutation relations must clearly ensure causal behavior, for instance of localized particles. This implicitly presupposes, however, that the relevant operators, e.g. $N(V)$ or O_{e_B} –if they exist – are local functions of the fields. It should be recalled that fields do not enter in our formulation, and it should be noted that if $N(V)$ is a local function of the fields then the Reeh-Schlieder theorem [23] implies that its vacuum expectation is nonzero.

Within local quantum field theory a rigorous proof of *weak* causality for *local* observables has been given by Schlieder [4] and by Buchholz and Yngvason [18], as well as by Neumann and Werner [24] in an alternative algebraic framework. In Ref. [18] it was moreover shown that restrictions of states to local algebras cannot be tested by means of transition probabilities.

How can one check weak causality experimentally? In the Fermi model one would not use a single pair of atoms but rather an ensemble, either by repetition or by simultaneous realization. If in the Fermi case one has N pairs of atoms A and B one would measure at time t how many B atoms are excited. For $N \to \infty$ their fraction would be given by $\langle \psi_t, O_{eB} \psi_t \rangle$, while for finite N this would hold only approximately, due to statistical fluctuations. Then one would subtract the – either calculated or measured – excitation probability of B without A present. Weak causality asserts that the difference should be zero for $t < R/c$ – but only for $N \to \infty$, while for finite N there are always fluctuations. Hence in a strict sense, weak Einstein causality can only be checked experimentally for infinite ensembles, and this suggests a macroscopic context.

SUPERLUMINAL TUNNELING?

In the experiments of Nimtz and coworkers [13], typically, a rapid sequence of microwave pulses is generated. Each pulse is split into two and sent over different paths of the same length to a receiver. Calibration of the path length is achieved by displaying the two pulse sequences stroboscopically as still pictures on an oscillograph. Then a photonic tunnel barrier is inserted into one of the paths which attenuates the corresponding pulses and reshapes them. To compare tunneled and non-tunneled pulses the former are

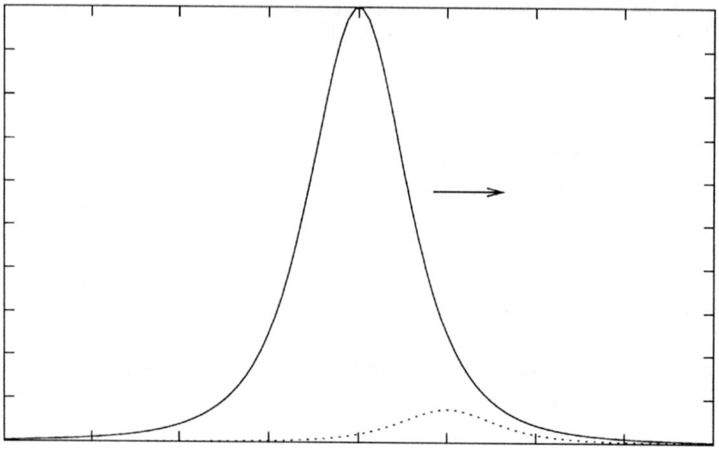

FIGURE 1. Typical behavior of airborne pulse (solid line) and tunneled pulse (dashed line), traveling from left to right (not to scale). In the experiments, the amplitude of the latter is always smaller than that of the non-tunneled pulse, although its maximum arrives at an earlier time.

re-amplified to their original amplitude height at the receiving end and again displayed stroboscopically on the oscillograph. The effect is dramatic. Upon insertion of the tunnel barrier the still picture of the tunneled pulses makes a jump to earlier times, seemingly indicating that they are arriving earlier than the non-tunneled pulses. With the definition of signal velocity and arrival time used by Nimtz and coworkers this is indeed true.

To see, however, whether this has anything to do with superluminal signal velocities in the Einstein sense it is an eye opener to look at the tunneled pulses without amplification. Experimentally it has been verified by Nimtz and coworkers that the amplitudes of the tunneled pulses are always *below* the amplitude of the non-tunneled pulses [25]. In these experiments, the maxima as well as the half widths of the tunneled pulses are ahead of those of the non-tunneled pulses and therefore arrive earlier. This is graphically depicted in Fig. 1 by the pulses traveling from left to right. The figure is not to scale and and does not represent experimental curves, but is just for illustration.

For the signal velocity in the Einstein sense, however, the arrival time of the pulse maximum or the read-out time of the half width is not relevant since they are not used for clock synchronization. Relevant, rather, is the first possible response time of the measuring device, as explained in the Introduction. Now, since experimentally the tunneled pulses are always below the non-tunneled pulses in amplitude, any measuring device will respond first to the non-tunneled pulses and then to the tunneled ones, or at most simultaneously to both. Thus the limiting role of the speed of light as signal velocity in the sense of Einstein is not violated in the experiments.

What then is superluminal here? Let us consider the group and the phase velocity of light. Both are mathematical constructs useful for the description of electromagnetic phenomena. It is well known [26] that both can be larger than c, but this cannot be used for superluminal signals in the Einstein sense. Similarly, it has been shown in Ref. [27] that in a somewhat idealized situation the tunneling pulse can be fully described within

Maxwell theory by means of another mathematically introduced auxiliary phase-time velocity notion. Again, this auxiliary velocity cannot be used for superluminal signal transmission in the Einstein sense.

So it seems that the controversy about the interpretation of Nimtz's experiments arises from an indiscriminate use of terminology. Terms like signal velocity and arrival time are used by Nimtz and coworkers in a sense different from that of Einstein. Using the notions in the original sense the experiments are fully compatible with Einstein causality as ordinarily understood.

REFERENCES

1. Cf., e.g., Rindler, W., *Introduction to Special Relativity*, 2nd Edition, Clarendon, Oxford 1991, p. 17 f. But see also
   Bilaniuk, O. M., Deshpande, V. K., and Sudarshan, E. C. G., *Am. J. Phys.*, **30**, 718 (1962).
2. Weidlich, W., and Mitra, K., *Nuovo Cim.*, **30**, 385 (1963).
3. Fleming, G. N., *Phys. Rev.*, **139**, B963 (1965).
4. Schlieder, S., in: *Quanten und Felder*, H.P. Dürr Ed., Vieweg, Braunschweig (1971), p. 145.
5. Hegerfeldt, G. C., *Phys. Rev.*, **D10**, 3320 (1974).
6. Skagerstam, B., *Int. J. Theor. Phys.*, **15**, 213 (1976).
7. Perez, J. F., and Wilde, I. F., *Phys. Rev.*, **D16**, 315 (1977).
8. Hegerfeldt, G. C., and Ruijsenaars, S. N. M. , *Phys. Rev.*, **D22**, 337 (1980).
9. Hegerfeldt, G. C., in: *Irreversibility and Causality in Quantum Theory*, edited by A. Bohm, H.-D. Doebner and P. Kielanowski, Lecture Notes in Physics **504**, p. 238, Springer, Berlin (1998).
10. Hegerfeldt, G. C., *Phys. Rev. Lett.*, **54**, 2395 (1985).
11. Fermi, E., *Rev. Mod. Phys.*, **4**, 87 (1932);
    Shirokov, M. I. , *Yad. Fiz.*, **4**, 1077 (1966), [*Sov. J. Nucl. Phys.*, **4**, 774 (1967)];
    Heitler, W., and Ma, S. T. , *Proc. R. Ir. Acs.*, **52**, 123 (1949);
    Hamilton, J., *+Proc. Phys. Soc.*, **A62**, 12 (1949);
    Fierz, M., *Helv. Phys. Acta*, **23**, 731 (1950);
    Ferretti, B., In: *Old and new Problems in Elementary Particles*, edited by G. Puppi, Academic Press, New York (1968), p. 108;
    Milonni, P. W. , and Knight, P. L , *Phys. Rev.*, **A10**, 1096 (1974);
    Shirokov, M. I. , *Sov. Phys. Usp.*, **21**, 345 (1978);
    Rubin, M. H. , *Phys. Rev.*, **D35**, 3836 (1987);
    Biswas, A. K. Compagno, G., Palma, G. M. , Passante, R., and Persico, R., *Phys. Rev.*, **A42**, 4291 (1990);
    Valentini, A., *Phys. Lett.*, **A153**, 321 (1991).
12. Hegerfeldt, G. C., *Phys. Rev. Lett.*, **72**, 596 (1994).
13. Aichmann, H., Haibel, A., Lennartz, W., Nimtz, G., and Spanoudaki, A., Proc. of the International Symposium on Quantum Theory and Symmetries, Goslar, 18-22 July 1999 (ISBN 9-8102-4237-9), 605-611 (2000);
    Nimtz, G., *Ann. Phys.*, (Leipzig) **7**, 618 (1998);
    Enders, A., and Nimtz, G., *Phys. Rev.*, **E48**, 632 (1994), and references therein. For related experiments with photons see
    Ciao, R. Y. , and Steinberg, A. M. , *Progress in Optics* vol. XXXVII, edited by E. Wolf, Elsevier, Amsterdam (1997), p. 345;
    Wang, L. J., Kuzmich, A., and Dogariu, A., *Nature*, **406**, 277 (2000).
14. Hegerfeldt, G. C., in: *Extensions of Quantum Theory*, Eds. A. Horzela and E. Kapuscik, published by Apeiron, Montreal, 2001, p. 9 16.
15. Levinson, N., and Redheffer, R.M., *Complex Variables*, Holden-Day, San Francisco 1970.
16. Garnett, J.B., *Bounded Analytic Functions*, Academic Press, New York 1981; p. 64.

17. Hegerfeldt, G.C., in *Nonlinear, deformed, and irreversible quantum systems*, edited by H.-D. Doebner, V.K. Dobrev, and P. Nattermann, World Scientific, Singapore (1995), p. 253.
18. Buchholz, D., and Yngvason, J., *Phys. Rev. Lett.*, **73**, 613 (1994).
19. Labarbara, A., and Passante, R., *Phys. Lett.*, **206**, 1 (1994).
20. Milonni, P.W., James, D.F.V., and Fearn, H., *Phys. Rev.*, **A52**, 1525 (1995).
21. Thaller, B., *The Dirac Equation*, Springer, Berlin (1992).
22. Wightman, A.S., and Schweber, S.S., *Phys. Rev.*, **98**, 812 (1955) and references therein.
23. Reeh, H., and Schlieder, S., *Nuovo Cim.*, **22**, 1051 (1961). This theorem also exploits analyticity but uses, in addition to positivity of the energy, stronger assumptions of field theory, in particular locality.
24. Neumannn, H., and Werner, R., *Int. J. Theor. Phys.*, **22**, 781 (1983).
25. Aichmann, H., private communication.
26. Brillouin, L., *Wave Propagation and Group Velocity,* Academic Press, New York (1960).
27. Emig, T., Diplomarbeit, Universität Köln (1995); Phys. Rev **E54**, 5780 (1996).

Composite Fields, Generalized Hypergeometric Functions and the $U(1)_Y$ Symmetry in the AdS/CFT Correspondence

L. Hoffmann*, T. Leonhardt*, L. Mesref* and W. Rühl[1]*

*Department of Physics, Theoretical Physics, University of Kaiserslautern, Postfach 3049, 67653 Kaiserslautern, Germany

Abstract. We discuss the concept of composite fields in flat CFT as well as in the context of AdS/CFT. Furthermore we show how to represent Green functions using generalized hypergeometric functions and apply these techniques to four-point functions. Finally we prove an identity of $U(1)_Y$ symmetry for four-point functions.

THE CONCEPT OF COMPOSITE FIELDS

Consider a renormalizable CFT in D spacetime dimensions defined by k fundamental fields $\phi_1, \phi_2, \ldots, \phi_k$ of respective conformal dimensions $\delta_1, \delta_2, \ldots, \delta_k$ and polynomial interactions. We assume that this CFT is "perturbative": any n-point function can be expanded into "skeleton graphs" with renormalized propagators

$$\langle \phi_i(x)\phi_j(0)\rangle = \delta_{i,j} A_i (x^2)^{-\delta_i}. \tag{1}$$

These propagator normalizations A_i can be cast on the vertex where

$$\phi_{i_1}\phi_{i_2}\ldots\phi_{i_f} \tag{2}$$

meet. Then this vertex obtains a coupling constant

$$z(i_1, i_2, \ldots, i_f)^{\frac{1}{2}} = \left(\prod_{j=i_1}^{i_f} A_j\right)^{\frac{1}{2}}. \tag{3}$$

Assume that all vertices are also "dressed". Then the whole theory is characterized by the dimensions δ_i and the couplings z_i. For each propagator and each type of vertex there is a bootstrap equation which replace the dynamical equations of the theory. They can be evaluated perturbatively. In the limit of vanishing coupling, the CFT degenerates into a *generalized* free field theory which admits normal products. The perturbative corrections of these normal product fields are the composite fields. The composite fields must be conformal (or quasi primary); this eliminates ambiguities in this perturbative definition.

[1] email:ruehl@physik.uni-kl.de

CP589, *Fundamental Interaction Theories*, edited by J. Lukierski and J. Rembieliński

CFTs possess operator product expansions (OPEs): Each pair of conformal fields can be expanded into ("blocks" of) conformal fields and this expansion converges if inserted into any n-point function in the maximal sphere excluding the locations of the other $n-2$ fields. All conformal fields appearing in such an OPE which are not fundamental are composite. A CFT with k fundamental fields may possess a sub-CFT (closed under OPE) consisting solely of composite fields. E.g. a CFT with bosonic and fermionic fields possesses a sub-CFT consisting of only the bosonic fields of the CFT. If all fundamental fields are fermionic the sub-CFT consists of bosonic composite fields. In a 1-particle reducible graph with exchange of a fundamental field ϕ_i integration over the vertices attached to the ϕ_i-propagator gives two terms

$$\sim F(\delta_i)(x^2)^{-\delta_i} + F(D-\delta_i)(x^2)^{-(D-\delta_i)}, \tag{4}$$

which is the original propagator plus a "shadow propagator" . This symmetrization is the consequence of conformal invariant integration. This shadow propagator can be attributed to a shadow field $\phi_i^{(s)}$ of dimension $\delta_i^{(s)} = D - \delta_i$, if Wightman positivity allows this, i.e. for a scalar field besides the positivity condition for ϕ_i

$$\delta_i \geq \frac{D}{2} - 1 \tag{5}$$

also the positivity condition for $\phi_i^{(s)}$ has to be fulfilled:

$$D - \delta_i \geq \frac{D}{2} - 1. \tag{6}$$

This means that δ_i must obey

$$\frac{D}{2} + 1 \geq \delta_i \geq \frac{D}{2} - 1. \tag{7}$$

Let us assume that this is fulfilled. Then we can formulate

Theorem 1. *The CFT admits two different but equivalent representations. In the first we have ϕ_i external legs, but no $\phi_i^{(s)}$ legs; in the second we have $\phi_i^{(s)}$ external legs but no ϕ_i legs.*

Proof. The internal propagators are symmetric in ϕ_i and $\phi_i^{(s)}$. An external ϕ_i-leg goes into a $\phi_i^{(s)}$-leg by amputation (their propagators are inverses as integral kernels), and vice versa. □

Remark. An artificial "symmetrization" by adding a ϕ_i-leg and a $\phi_i^{(s)}$-leg is of no meaning.

Turning our attention to OPEs we discover

Theorem 2. *Fundamental fields appear with a direct and a shadow term, composite fields only with a direct term, if OPEs are applied to n-point functions.*

This is just everybody's experience with applications of OPEs. There is a complementary theorem:

Theorem 3. *The shadow field of a fundamental field cannot appear as a composite field. The bootstrap equation for the propagator of a fundamental field is equivalent to the absence of its shadow field as composite field.*

Theorem 3 is not so popular, therefore let us illustrate how it works in the case of the critical non-linear $O(N)$ sigma model [1] with the two fundamental fields

$$S_a(x): \quad O(N) \text{ vector, spacetime scalar} \tag{8}$$
$$\alpha(x): \quad \text{the auxiliary field, } O(N) \text{ and spacetime scalar} \tag{9}$$

with propagators

$$\langle S_a(x)S_b(0)\rangle = A\,\delta_{ab}(x^2)^{-\delta_1} = \underline{\qquad} \tag{10}$$
$$\langle \alpha(x)\alpha(0)\rangle = B(x^2)^{-\delta_2} = G(x) = \text{----} \tag{11}$$

with dimensions

$$\delta_1 = \frac{1}{2}D - 1 + O(\frac{1}{N}), \quad \delta_2 = 2 + O(\frac{1}{N}) \tag{12}$$

and interaction term

$$z^{\frac{1}{2}} \int dx S_a(x) S_a(x) \alpha(x), \quad \text{with } z^{\frac{1}{2}} = AB^{\frac{1}{2}}. \tag{13}$$

The bootstrap equation for G is

$$-G^{-1} = z\,\bigcirc\bigcirc\, + z^2\,\Diamond\, + \cdots \tag{14}$$

Consider the normal product with corrections

$$T^{(0)}(x) = \sum_{a=1}^{N} :S_a(x)S_a(x): . \tag{15}$$

If this field exists, it should be the shadow field of α. Its dimension in $2 < D < 4$ is

$$\dim T^{(0)} = D - \dim \alpha = D - 2 + O(\frac{1}{N}). \tag{16}$$

Since the dimension of $T^{(0)}$ fulfills the positivity condition (6) one would expect by theorem 1 that we could reformulate the $O(N)$ sigma model in terms of $T^{(0)}$. The definition of $T^{(0)}$ implies that it should appear in the OPE of $S_a(x)S_a(y)$. Calculation of the 4-point function

$$\langle S_a(x+\varepsilon_1)S_a(x-\varepsilon_1)S_b(y+\varepsilon_2)S_b(y-\varepsilon_2)\rangle \tag{17}$$

in the limit $\varepsilon \to 0$ and subsequent decomposition into 1-P-reducible and -irreducible graphs gives

$$\langle T^{(0)}(x)T^{(0)}(y)\rangle \sim \bigcirc\!F\!\bigcirc + \bigcirc\!F\!\bigcirc\text{--}\bigcirc\!F\!\bigcirc \tag{18}$$

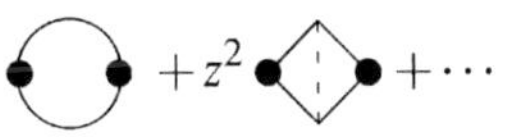

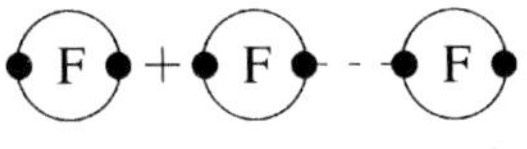

where the circles with the Fs are (up to renormalization) the r.h.s. of the α-bootstrap equation (14). Call it $-\tilde{G}^{-1}$. Then

$$\langle T^{(0)}(x)T^{(0)}(y)\rangle \sim -\tilde{G}^{-1}+\tilde{G}^{-1}G\,\tilde{G}^{-1}. \tag{19}$$

The bootstrap equation says $G^{-1}=\tilde{G}^{-1}$, therefore

$$\langle T^{(0)}(x)T^{(0)}(y)\rangle = 0. \tag{20}$$

On the other hand

$$\langle T^{(0)}(x)T^{(0)}(y)\rangle = 0 \Longrightarrow G^{-1}=\tilde{G}^{-1}. \tag{21}$$

Now we turn to the AdS/CFT correspondence and apply our theorems. Any field ϕ on AdS with mass $m^2 \geq -\frac{D^2}{4}$ is mapped holographically on a CFT field $\tilde{\phi}$ with dimension

$$\delta = \frac{D}{2} \pm \left(\frac{D^2}{4}+m^2\right)^{\frac{1}{2}}, \tag{22}$$

where the upper/lower sign corresponds to $\tilde{\phi}/\tilde{\phi}^{(s)}$ respectively. In this context Aharony et al. [2] considered square integrability and boundary conditions. But we have

Theorem 4. *Square integrability of an AdS wave function and Wightman positivity of the corresponding CFT field are the same thing.*

Thus we arrive at the conclusion: Taking for a field always the upper sign or always the lower sign (if Wightman positivity allows this) leads to two different but equivalent CFTs.

By evaluating AdS exchange graphs and analyzing the result one finds that there are no shadow terms [3]. This leads us to

Theorem 5. *The holographic image of an AdS field theory is a CFT without fundamental fields.*

In fact in the standard example of N= 4 SYM_4, which is dual to type *IIB* superstring theory on $\mathrm{AdS}_5 \times S^5$, the relevant coupling constant is ‘t Hooft’s

$$\lambda = g_{YM}^2 N. \tag{23}$$

It is a perturbative CFT if $\lambda \to 0$ with (say) the gauge supermultiplett as fundamental fields. They belong to the adjoint representation of the gauge group SU(N). The holographic image possesses only gauge invariant fields which are therefore composed of at least two fundamental fields with the adjoint representation matrices “traced” away.

But the content of Theorem 5 derives from the geometry of AdS_5 and not from gauge invariance. In other examples compositeness and gauge invariance may not have the same effect.

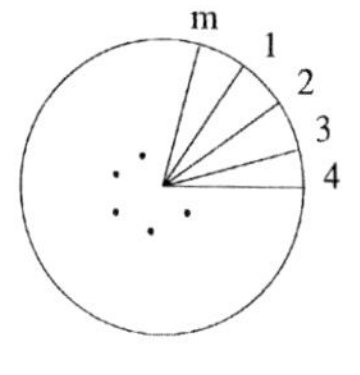

FIGURE 1. AdS_{D+1} "star graph" (vertex)

GENERALIZED HYPERGEOMETRIC FUNCTIONS

A "star graph" (vertex) of m scalar fields in AdS_{D+1} conformal field theory defines a Green function ($x_i \in \mathbf{R}_D = \partial AdS_{D+1}$)

$$\Gamma(x_1,...,x_m) = \pi^{-\frac{D}{2}} \int_{y_0>0} \frac{d^{D+1}y}{y_0^{D+1}} \prod_{i=1}^{m} (\frac{y_0}{y_0^2 + (\vec{y}-\vec{x}_i)^2})^{\alpha_i}, \tag{24}$$

where$\{\alpha_i\}$ are field dimensions constrained only by

$$\alpha_i \geq \frac{1}{2}D - 1. \tag{25}$$

In conformal field theory on $\mathbf{R_D}$ such a vertex is given by

$$G_m(x_1,...,x_m) = \pi^{-\frac{D}{2}} \int_{\mathbf{R_D}} d^D y \prod_{i=i}^{m} ((y-x_i)^2)^{-\alpha_i}, \tag{26}$$

with the conformal "uniqueness" condition

$$\sum_{i=1}^{m} \alpha_i = D. \tag{27}$$

The usual parametric representation of G_m

$$G_m(x_1,...,x_m) = [\prod_{i=1}^{m} \Gamma(\alpha_i)]^{-1} (\prod_{i=1}^{m} \int_0^{\infty} dt_i t_i^{\alpha_i - 1}) T^{-\frac{D}{2}} e^{-\frac{1}{T}\sum_{i<j} t_i t_j x_{ij}^2}, \tag{28}$$

with

$$T = \sum_{i=1}^{m} t_i. \tag{29}$$

allows to continue analytically in $\{\alpha_i\}$ and D independently, such that one has the Green function

$$G_m(D;\alpha_1,...,\alpha_m). \tag{30}$$

The AdS_{D+1} function Γ_m is [6]

$$\Gamma_m(D;\alpha_1,...,\alpha_m) = \frac{1}{2}\Gamma(\frac{1}{2}(\sum_{i=1}^{m} \alpha_i - D))\, G_m(\sum_{i=1}^{m} \alpha_i; \alpha_1,...,\alpha_m). \tag{31}$$

All the Green functions G_m have been represented in the form of a Mellin-Barnes integral and by expansion in generalized hypergeometric form by K. Symanzik [7]. For $m=4$ and with the biharmonic ratios

$$u=\frac{x_{13}^2 x_{24}^2}{x_{12}^2 x_{34}^2}, \quad v=\frac{x_{14}^2 x_{23}^2}{x_{12}^2 x_{34}^2}, \tag{32}$$

we get

$$G_4(x_1,x_2,x_3,x_4)=[\prod_{i=1}^{4}\Gamma(\alpha_i)]^{-1}(x_{12}^2)^{\alpha_4-\frac{D}{2}}(x_{14}^2)^{\frac{D}{2}-\alpha_1-\alpha_4}(x_{24}^2)^{\frac{D}{2}-\alpha_2-\alpha_4}(x_{34}^2)^{-\alpha_3}F(u,v), \tag{33}$$

with

$$\begin{aligned} F(u,v)=&u^{-\frac{1}{2}(\alpha_1+\alpha_3)}\sum_{n,m=0}^{\infty}\frac{u^n(1-v)^m}{n!m!}\{u^{\frac{1}{2}(\alpha_1+\alpha_3)}\Gamma(\frac{1}{2}(\alpha_2+\alpha_4-\alpha_1-\alpha_3)) \\ &\times\frac{\Gamma(\alpha_1+n)\Gamma(\frac{1}{2}D-\alpha_2+n)\Gamma(\alpha_3+n+m)\Gamma(\frac{1}{2}D-\alpha_4+n+m)}{\Gamma(\alpha_1+\alpha_3+2n+m)(\frac{1}{2}(\alpha_1+\alpha_3-\alpha_2-\alpha_4)+1)_n} \\ &+u^{\frac{1}{2}(\alpha_2+\alpha_4)}\Gamma(\frac{1}{2}(\alpha_1+\alpha_3-\alpha_2-\alpha_4)) \\ &\times\frac{\Gamma(\alpha_4+n)\Gamma(\frac{1}{2}D-\alpha_3+n)\Gamma(\alpha_2+n+m)\Gamma(\frac{1}{2}D-\alpha_1+n+m)}{\Gamma(\alpha_2+\alpha_4+2n+m)(\frac{1}{2}(\alpha_2+\alpha_4-\alpha_1-\alpha_3)+1)_n}. \end{aligned} \tag{34}$$

This form converges in a complex neighborhood of

$$u=0, \quad v=1, \tag{35}$$

and is therefore suited for an operator product expansion (OPE) in the limit, "t-channel"

$$x_{13}\longrightarrow 0, \quad x_{24}\longrightarrow 0. \tag{36}$$

The analytic continuation to the "s-channel"

$$\begin{aligned} &x_{12}\longrightarrow 0, \quad x_{34}\longrightarrow 0, \\ &u\underset{g_{t\to s}}{\longrightarrow} u'=\frac{1}{u}, \\ &v\underset{g_{t\to s}}{\longrightarrow} v'=\frac{v}{u} \end{aligned} \tag{37}$$

and to the "u-channel"

$$\begin{aligned} &x_{14}\longrightarrow 0, \quad x_{23}\longrightarrow 0, \\ &u\underset{g_{t\to u}}{\longrightarrow} u''=v, \\ &v\underset{g_{t\to u}}{\longrightarrow} v''=u, \end{aligned} \tag{38}$$

can be obtained by two methods

1. using the symmetry group of the graphs [6];

2. using the Kummer formulae for ${}_2F_1$-functions [8].

The AdS graphs at order $O(\frac{1}{N^2})$ for dilaton-axion four-point functions with

$$m^2 = 0 \longrightarrow \Delta = 4, \quad (D = 4), \tag{39}$$

can all be represented by

$$\Gamma_4(x_1,x_2,x_3,x_4) = \pi^{-\frac{D}{2}}(x_{12}^2)^{-\Delta}(x_{24}^2)^{\Delta-\Delta'}(x_{34}^2)^{-\Delta}G_{\Delta\Delta'}(u,v) \tag{40}$$

evaluated at $\alpha_1 = \alpha_3 = \Delta$, $\alpha_2 = \alpha_4 = \Delta'$; $\Delta' \geq \Delta$; $\Delta, \Delta' \in \mathbf{N}$, with

$$\begin{aligned} G_{\Delta\Delta'}(u,v) = & \frac{\pi^2}{2}\frac{\Gamma(\Delta+\Delta'-2)}{\Gamma(\Delta)^2\Gamma(\Delta')^2}\sum_{m=0}^{\infty}\frac{(1-v)^m}{m!}\{\sum_{n=0}^{\Delta'-\Delta-1}\frac{(-1)^n u^n}{n!}(\Delta'-\Delta-n-1)! \\ & \times\frac{\Gamma(\Delta+n)^2\Gamma(\Delta+n+m)^2}{\Gamma(2\Delta+2n+m)} + \sum_{n=\Delta'-\Delta}^{\infty}\frac{(-1)^{\Delta'-\Delta}u^n}{n!(n-\Delta'+\Delta)!}\frac{\Gamma(\Delta+n)^2\Gamma(\Delta+n+m)^2}{\Gamma(2\Delta+2n+m)} \\ & \times[-log\,u+\Psi(n-\Delta'+\Delta+1) \\ & +\Psi(n+1)-2\Psi(\Delta+n)+2\Psi(2\Delta+2n+m)-2\Psi(\Delta+n+m)]. \end{aligned} \tag{41}$$

In the other channels one has

$$\begin{aligned} G_{\Delta\Delta'}(u',v') = & \frac{\pi^2}{2}\frac{\Gamma(\Delta+\Delta'-2)}{\Gamma(\Delta)^2\Gamma(\Delta')^2}u^{\Delta}\sum_{n,m=0}^{\infty}\frac{u^n(1-v)^m}{(n!)^2 m!}\frac{\Gamma(\Delta+n)\Gamma(\Delta'+n)\Gamma(\Delta+n+m)}{\Gamma(\Delta+\Delta'+2n+m)} \\ & \times\Gamma(\Delta'+n+m)\{-log\,u+2\Psi(n+1)-\Psi(\Delta+n)-\Psi(\Delta'+n) \\ & -\Psi(\Delta+n+m)-\Psi(\Delta'+n+m)+2\Psi(\Delta+\Delta'+2n+m)\} \end{aligned} \tag{42}$$

and

$$\begin{aligned} G_{\Delta\Delta'}(u'',v'') = & \frac{\pi^2}{2}\frac{\Gamma(\Delta+\Delta'-2)}{\Gamma(\Delta)^2\Gamma(\Delta')^2}v^{\Delta'-\Delta}\sum_{n,m=0}^{\infty}\frac{u^n(1-v)^m}{(n!)^2 m!}\frac{\Gamma(\Delta'+n)^2\Gamma(\Delta+n+m)^2}{\Gamma(\Delta+\Delta'+2n+m)} \\ & \times\{-log\,u+2\Psi(n+1)-2\Psi(\Delta+n)-2\Psi(\Delta'+n+m)+2\Psi(\Delta+\Delta'+2n+m)\}. \end{aligned} \tag{43}$$

$U(1)_Y$ SYMMETRY

In the standard model of AdS/CFT correspondence

$$AdS_5 \times S_5 \longrightarrow \mathrm{SYM}_4 \tag{44}$$

the $U(1)_Y$ symmetry carries over to the strong coupling domain of SYM_4 and makes predictions on three and four-point functions [4]. We will discuss here an identity of two four-point functions [6]. Let

$$\begin{aligned} \text{Dilaton}: & \quad \Phi(y) \longrightarrow \tilde{\Phi}(x) \sim Tr(F_{\mu\nu}(x)F_{\mu\nu}(x)), \\ \text{Axion}: & \quad C(y) \longrightarrow \tilde{C}(x) \sim Tr(F_{\mu\nu}(x)\tilde{F}_{\mu\nu}(x)), \end{aligned} \tag{45}$$

and consider

$$O_\tau(x) = \tilde{\Phi}(x) + i\tilde{C}(x)\,. \tag{46}$$

This field has $U(1)_Y$ charge $q = -4$. Therefore

$$\langle \prod_{i=1}^{4} O_\tau(x_i) \rangle = 0\,, \tag{47}$$

due to charge non-conservation ($\sum_i q_i \neq 0$). Introducing the bilocal operators

$$\Psi_\mp(x_1,x_3) = \frac{1}{\sqrt{2}}[\mp\tilde{\Phi}(x_1)\tilde{\Phi}(x_3) + \tilde{C}(x_1)\tilde{C}(x_3)]\,, \tag{48}$$

$$\Psi_0(x_1,x_3) = \frac{1}{\sqrt{2}}[\tilde{\Phi}(x_1)\tilde{C}(x_3) + \tilde{C}(x_1)\tilde{\Phi}(x_3)]\,, \tag{49}$$

which by OPE produce tensor fields of even rank with opposite parities, we obtain from (47)

$$\langle \Psi_-(x_1,x_3)\Psi_-(x_2,x_4)\rangle - \langle \Psi_0(x_1,x_3)\Psi_0(x_2,x_4)\rangle = 0\,. \tag{50}$$

This relation can be worked out up to order $O(\frac{1}{N^2})$ at present. At leading order $O(1)$, both four-point functions are

$$(x_{12}^2 x_{34}^2)^{-4}(1+v^{-4}) \quad \text{with} \quad v = \frac{x_{14}^2 x_{23}^2}{x_{12}^2 x_{34}^2}\,. \tag{51}$$

At order $O(\frac{1}{N^2})$ the relevant graphs are given by Figure 2, while the dilaton and graviton

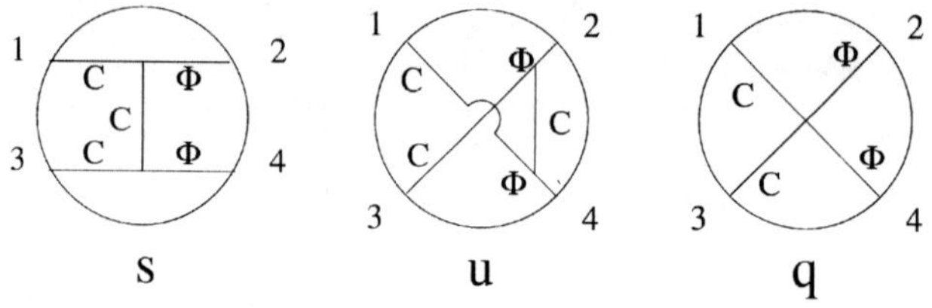

FIGURE 2. Relevant graphs at order $O(\frac{1}{N^2})$

exchange graphs cancel in (50). The sum is denoted

$$I_{(1,3)}^{(s+u+q)}\,, \tag{52}$$

where we indicate the position of the C-fields. However,

$$I_{(1,3)}^{(s+u+q)} = I_{(2,4)}^{(s+u+q)}\,. \tag{53}$$

Then the constraint (50) is expressible as

$$I_{(1,3)}^{(s+u+q)} + I_{(1,2)}^{(s+u+q)} + I_{(1,4)}^{(s+u+q)} = 0\,. \tag{54}$$

This is typically a constraint for a crossing invariant function.

Now we take from d'Hoker et al. [5]

$$I^{(s+u+q)}_{(1,3)} = 32(\frac{6}{\pi^2})^4(x_{12}^2x_{34}^2)^{-4}[2G_{45}(u,v) - G_{44}(u,v)] \tag{55}$$

so that, with the analytic continuations in all channels, we get from (54)

$$\begin{aligned} G_{45}(u,v)+u^{-4}G_{45}(u',v')+G_{45}(u'',v'') &= \frac{1}{2}[G_{44}(u,v)+u^{-4}G_{44}(u',v')+G_{44}(u'',v'')] \\ &= \frac{3}{2}G_{44}(u,v) \end{aligned} \tag{56}$$

since G_{44} is a star function with equal legs and therefore itself crossing invariant.

In fact the following identity can be proved

$$G_{\Delta,\Delta+1}(u,v)+u^{-\Delta}G_{\Delta,\Delta+1}(u',v')+G_{\Delta,\Delta+1}(u'',v'') = \frac{2\Delta-2}{\Delta}G_{\Delta,\Delta}(u,v). \tag{57}$$

It is a recursion for the generalized hypergeometric function G.

For the proof of (57) we do not use the explicit form of $G_{\Delta,\Delta'}$ given earlier, but make use of an "ε-trick"

$$\begin{aligned} G_{\Delta,\Delta'}(u,v) = \lim_{\varepsilon\to 0}\frac{\partial}{\partial\varepsilon}[\frac{\pi^2}{2}\frac{\Gamma(\Delta+\Delta'-2)}{\Gamma(\Delta)^2\Gamma(\Delta')^2}(-1)^{\Delta'-\Delta}\prod_{k=1}^{\Delta'-\Delta}(n-\Delta'+\Delta+k-\varepsilon) \\ \times\sum_{m,n=0}^{\infty}\frac{u^{n-\varepsilon}(1-v)^m}{\Gamma(n+1-\varepsilon)^2 m!}\frac{\Gamma(\Delta+n-\varepsilon)^2\Gamma(\Delta+n+m-\varepsilon)^2}{\Gamma(2\Delta+2n+m-2\varepsilon)}], \end{aligned} \tag{58}$$

$$\begin{aligned} u^{-\Delta}G_{\Delta,\Delta'}(u',v') = \lim_{\varepsilon\to 0}\frac{\partial}{\partial\varepsilon}[\frac{\pi^2}{2}\frac{\Gamma(\Delta+\Delta'-2)}{\Gamma(\Delta)^2\Gamma(\Delta')^2}\sum_{m,n=0}^{\infty}\frac{u^{n-\varepsilon}(1-v)^m}{\Gamma(n+1-\varepsilon)^2 m!} \\ \times\frac{\Gamma(\Delta+n-\varepsilon)\Gamma(\Delta'+n-\varepsilon)\Gamma(\Delta+n+m-\varepsilon)\Gamma(\Delta'+n+m-\varepsilon)}{\Gamma(\Delta+\Delta'+2n+m-2\varepsilon)}], \end{aligned} \tag{59}$$

$$\begin{aligned} G_{\Delta,\Delta'}(u'',v'') = \lim_{\varepsilon\to 0}\frac{\partial}{\partial\varepsilon}[\frac{\pi^2}{2}\frac{\Gamma(\Delta+\Delta'-2)}{\Gamma(\Delta)^2\Gamma(\Delta')^2} \\ \times\sum_{m,n=0}^{\infty}\frac{u^{n-\varepsilon}(1-v)^m}{\Gamma(n+1-\varepsilon)^2 m!}\frac{\Gamma(\Delta'+n-\varepsilon)^2\Gamma(\Delta+n+m-\varepsilon)^2}{\Gamma(\Delta+\Delta'+2n+m-2\varepsilon)}]. \end{aligned} \tag{60}$$

Since $\Delta'-\Delta\in\mathbf{N_0}$, we can sum the coefficients as

$$\begin{aligned} \frac{\Gamma(\Delta+n-\varepsilon)^2\Gamma(\Delta+n+m-\varepsilon)^2}{\Gamma(\Delta+\Delta'+2n+m-2\varepsilon)}\{(2\Delta+2n+m-2\varepsilon)_{\Delta'-\Delta}\prod_{k=1}^{\Delta'-\Delta}(\Delta'-\Delta-n-k+\varepsilon) \\ +(\Delta+n-\varepsilon)^2_{\Delta'-\Delta}+(\Delta+n-\varepsilon)_{\Delta'-\Delta}(\Delta+n+m-\varepsilon)_{\Delta'-\Delta}\} \end{aligned} \tag{61}$$

For $\Delta' = \Delta + 1$ we have

$$\{\ldots\} = \Delta(2\Delta + 2n + m - 2\varepsilon), \tag{62}$$

which leads to $G_{\Delta\Delta}$ by the formula given above. This completes the proof.

Let us quote some detailed results on the OPE of the bilocal fields $\Psi_{+,-,0}$ [6].

1. $\Psi_+(x_1,x_3)$ expands into the stress-energy tensor plus the towers of conformal fields: tensor rank $l \in 2\mathbf{N_0}$, parity $(+1)$ and dimension

$$\delta_+(l,t) = 8 + l + 2t + \eta_+(l,t); \quad t \in \mathbf{N_0}(\text{tower label, twist}); \eta_+ = O(\frac{1}{N^2}), \tag{63}$$

2. $\Psi_-(x_1,x_3)$ expands into conformal fields: tensor rank $l \in 2\mathbf{N_0}$, parity $(+1)$ and dimension

$$\delta_-(l,t) = 8 + l + 2t + \eta_-(l,t); \quad t \in \mathbf{N_0}, \eta_- = O(\frac{1}{N^2}), \tag{64}$$

3. $\Psi_0(x_1,x_3)$ expands into conformal fields: tensor rank $l \in 2\mathbf{N_0}$, parity (-1) and dimension

$$\delta_0(l,t) = 8 + l + 2t + \eta_0(l,t); \quad t \in \mathbf{N_0}, \eta_0 = O(\frac{1}{N^2}). \tag{65}$$

For $t = 0$ we obtain

$$\eta_-(l,0) = \eta_0(l,0) = -\frac{96}{N^2(l+1)(l+6)} \tag{66}$$

and the corresponding fusion constants squared

$$\varepsilon_-(l,0) = \varepsilon_0(l,0) = \frac{1}{18}(2l+7)(l+1)_6. \tag{67}$$

The remaining quantities η_+, ε_+ and $\eta_- = \eta_0(t \neq 0)$, $\varepsilon_- = \varepsilon_0(t \neq 0)$ are accessible by solving infinite linear equations with triangular matrices, where t counts the off-diagonality of the matrix elements. The main labor resides in calculating the matrix elements in a generic form (no numbers).

REFERENCES

1. Lang, K., and Rühl, W., *Phys. Lett.*, 93 **B275**, (1992).
2. Aharony, O., Gubser, S. S., Maldacena, J., Ooguri, H., and Oz, Y., *Phys.Rep.*, **323**, 183 (2000), hep-th/9905111.
3. Liu, H., *Phys.Rev.*, **D60**, 106005 (1999), hep-th/9811152.
4. Intriligator, K., *Nucl. Phys.*, **B551**, 575 (1999), hep-th/9811047;
   Intriligator, K., and Skiba, W., *Nucl. Phys.*, **B559**, 165 (1999), hep-th/9905020.
5. D'Hoker, E., Freedman, D. Z., Mathur, S. D., Matusis, A., and Rastelli, L., *Nucl. Phys.*, **B562**, 353 (1999), hep-th/9903196.
6. Hoffmann, L., Mesref, L., and Rühl, W., hep-th/0012153.
7. Symanzik, K., *Lett. Nuovo Cim.*, **3**, 734 (1972).
8. Gradshteyn, I. S., and Ryzhik, I. M., *Table of Integrals, Series and Products*, 4th Ed., Academic Press (1965).

Topics in Born-Infeld Electrodynamics

R. Kerner*, A.L. Barbosa[1*] and D.V. Gal'tsov[2*]

*L.P.T.L. - Tour 22, 4-ème étage, Boite 142, Université Paris-VI, 4, Place Jussieu, 75005 Paris

INTRODUCTION - A SHORT GLANCE AT THE HISTORY

Since the discovery of the electron by J.J.Thomson in 1899, physicists tried to develop models of finite energy charge concentrations that could describe the elementary electric charge. One of the ideas was to use non-linear generalizations of Maxwell's theory, deviating from it only at very short distances and very strong fields in order to ensure a cut-off and to avoid singularity at $r \to 0$.

G. Mie ([1]) was first to introduce such a model, based on the assumption that the electric field $\mathbf{E}$ can not exceed the limiting value $\mathbf{E_0}$, and that the repulsive force should be proportional to the expression

$$\mathbf{F} \sim \frac{\mathbf{E}}{\sqrt{1 - \frac{\mathbf{E}^2}{\mathbf{E}_0^2}}}. \tag{1}$$

In this model it was possible to find a nonsingular solution with finite energy and charge, and with the field $\mathbf{E}$ falling off as r^{-2} at great distances, but this solution was not covariant with respect to the Lorentz transformations.

In 1932 and in 1934 Born and Infeld have published by now celebrated version of non-linear electrodynamics, in which they proposed the following Lorentz-invariant Lagrangian:

$$\mathcal{L} = \beta^2 \left[\sqrt{\det\left(\delta^{\mu}_{\lambda} + \beta^{-1} F^{\mu}_{\lambda} \right)} \right]. \tag{2}$$

The constant β appears for dimensional reasons, and plays the same rôle here as the limiting value of the electric field in G. Mie's non-linear electrodynamics.

When expressed in terms of two invariants of Maxwell's tensor,

$$P = \frac{1}{4} F_{\mu\nu} F^{\mu\nu} \quad \text{and} \quad S = \frac{1}{4} F_{\mu\nu} \tilde{F}^{\mu\nu}, \quad \text{with} \quad \tilde{F}^{\mu\nu} = \frac{1}{2} \varepsilon^{\mu\nu\lambda\rho} F_{\lambda\rho}$$

---

[1] Permanent address: IFT, Universidade Estadual Paulista, Rua Pamplona 145, 01405 São Paulo, Brazil.
[2] Permanent address: Department of Theoretical Physics, Moscow State University, 119899 Moscow, Russia.

CP589, *Fundamental Interaction Theories*, edited by J. Lukierski and J. Rembieliński

this Lagrangian can be written explicitly as

$$\mathcal{L}_{BI} = \beta^2 \left[1 - \sqrt{1 + 2P - S^2} \right], \tag{3}$$

or as $$\mathcal{L}_{BI} = \beta^2 \left[1 - \sqrt{1 + \frac{1}{2\beta^2}(\mathbf{B}^2 - \mathbf{E}^2) - \frac{1}{16\beta^4}(\mathbf{E}\cdot\mathbf{B})^2} \right]. \tag{4}$$

With the advent of Quantum Mechanics and Dirac's equation for the electron, the interest in classical models of charged particles has considerably faded. However, in 1970 G. Boillat ([3] considered the Born-Infeld electrodynamics as an example of non-linear theory in order to study its propagation properties. Starting with the most general non-linear theory derived from an arbitrary Lagrangian depending on two Lorentz invariants of Maxwell's tensor, $\mathcal{L}(P,S)$, he discovered that among all such non-linear theories, the Born-Infeld electrodynamics is the only one ensuring the absence of birefringence, i.e. propagation along a single light-cone, and the absence of shock waves. In this respect the Born-Infeld theory is unique (except for another singular and unphysical Lagrangian $\mathcal{L} = P/S$). A beautiful discussion of these properties can be found in I. Bialynicki-Birula's paper ([4]). Let us remind very shortly how the birefringence phenomenon may occur in non-linear theories.

From the mathematical point of view, these theories are based on systems of second order partial differential equations, linear in highest derivatives, with coefficients which depend only on the fields (but not on their derivatives). The systems of this type can be reduced to a set of differential equations of first order via introduction of auxiliary fields, which are the independent linear combinations of the first partial derivatives of functions corresponding to the degrees of freedom of our system.

The differential system can be represented by means of a matrix whose entries contain the operators of partial derivation or multiplicative coefficients, acting on a vector-column representing auxiliary fields. If the vector-column $\mathbf{u}$ with the fields ψ, χ_i, E_i and B_i contains N elements, then let us denote by $\mathcal{A}$ the NxN matrix containing the partial derivatives and by $\mathcal{B}$ the NxN matrix containing the multiplicative factors. Then the field equations can be written as:

$$\mathcal{A}^{\mu}(\mathbf{u})\partial_{\mu}\mathbf{u} + \mathcal{B}(\mathbf{u})\mathbf{u} = 0. \tag{5}$$

If the hypersurface defined by the implicit equation

$$\Sigma(x^{\mu}) = 0 \tag{6}$$

is a surface of discontinuity, then the first derivatives of fields are discontinuous across this surface, whereas the fields themselves are continuous. So, when applied to the *discontinuities* across the hypersurface (6), the equation (5) reduces to

$$\left(\mathcal{A}^{\mu}\Sigma_{\mu}\right)\delta_1\mathbf{u} = 0, \tag{7}$$

where $\Sigma_{\mu} \equiv \partial_{\mu}\Sigma$, and $\delta_1\mathbf{u}$ denotes the discontinuity of the first derivative across Σ, $\delta_1\mathbf{u} \equiv \partial\mathbf{u}/\partial\Sigma|_{+} - \partial\mathbf{u}/\partial\Sigma|_{-}$. By definition, for a characteristic surface one has $\delta_1\mathbf{u} \neq 0$, therefore, in order for (7) to hold, one must have

$$\det\left(\mathcal{A}^{\mu}\,\Sigma_{\mu}\right) = 0\,, \tag{8}$$

on the surface of discontinuity. The characteristic equation (8) determines the surface whose generic equation is $H(x,\Sigma_\mu) = 0$, with H a homogeneous function of order N in Σ_μ. The Born-Infeld theory turns out to be *completely exceptional* since it obeys the corresponding condition of [3], namely $\delta_0 H \equiv H|_+ - H|_- = 0$.

Let us give an illustration of this principle on the simplest case: the scalar field wave equation in a two-dimensional space-time (t,x):

$$\partial_0^2\phi - \partial_x^2\phi = 0\,. \tag{9}$$

(Partial derivatives in Lorentz indices, $(0,x,y,z)$ or $(0,1,2,3)$, will be denoted by ∂_0, ∂_x), etc.. According to the prescription, we can use as auxiliary fields ψ and χ the first derivatives of the scalar field ϕ, $\partial_0\phi = \psi$ and $\partial_x\phi = \chi$. Then by definition, the first derivatives of auxiliary fields are not independent, because we have, as $\partial_0(\partial_x\phi) = \partial_x(\partial_0\phi)$, automatically $\partial_0\chi - \partial_x\psi = 0$. On the other hand the dynamical equation (9) can be written as $\partial_0\psi - \partial_x\chi = 0$. In the matrix notation of (7) these two equations can be combined to yield

$$\begin{pmatrix} 0 & 1 \\ 1 & 0 \end{pmatrix} \partial_0 \begin{pmatrix} \psi \\ \chi \end{pmatrix} + \begin{pmatrix} -1 & 0 \\ 0 & -1 \end{pmatrix} \partial_x \begin{pmatrix} \psi \\ \chi \end{pmatrix} = \begin{pmatrix} 0 \\ 0 \end{pmatrix}. \tag{10}$$

We then find

$$\mathcal{A}^{\mu}\Sigma_{\mu} = \begin{pmatrix} -\Sigma_x & \Sigma_0 \\ \Sigma_0 & -\Sigma_x \end{pmatrix}, \tag{11}$$

and the characteristic equation $\det(\mathcal{A}^{\mu}\Sigma_{\mu}) = 0$ can be written as

$$\Sigma_0{}^2 - \Sigma_x{}^2 = 0\,, \tag{12}$$

where $\Sigma_0 \equiv \partial_0\Sigma$ and $\Sigma_x \equiv \partial_x\Sigma$. The last equation defines the characteristic surfaces $\Sigma(t,x)$, which in this case are the light-cones in two space-time dimensions.

The same technique can be easily applied to the electromagnetic Maxwellian field. We have to solve a 6×6 matrix, because we have now six independent combinations of its first derivatives (the fields $\mathbf{E}$ and $\mathbf{B}$) appearing in the first-order Maxwell's equations. As we know, the characteristic surfaces in four dimensions are given by $\Sigma_{,\mu}\Sigma^{,\mu} = \Sigma_0{}^2 - \Sigma_x{}^2 - \Sigma_y{}^2 - \Sigma_z{}^2 = 0$.

The same is true for the Born-Infeld non-linear electrodynamics. A more exhaustive discussion of the propagation properties of various non-linear generalizations of the electromagnetism can be found in recent papers ([5], [6]).

An entirely new and unexpected impulse for the revival of interest in the Born-Infeld electrodynamics, and in its non-abelian generalizations, came from recent developments of the string and brane theories. The string Lagrangian in (4 + D) dimensions, which defines a minimal surface in a (4+D)-dimensional Minkowskian space-time, is in fact a generalization of geodesic equation for a point-like particle.

Consider a two-dimensional surface with cylindrical topology, parametrized with one time-like and one space-like parameter, τ and σ, respectively, and embedded in a (4+D)-dimensional space-time. The embedding functions will be denoted by $X^{\mu}(\tau,\sigma)$, or equivalently, as $X^{\mu}(\xi^{a})$, with $\mu,\nu = 0,1,2,...3+D$, and $A,B,.. = 1,2$ so that $\xi^1 = \tau$, $\xi^2 = \sigma$.

The exterior space-time metric $g_{\mu\lambda}$ induces the internal metric of the world-sheet spanned by the string,

$$G_{AB} = g_{\mu\lambda}\,\partial_a X^{\mu}\,\partial_b X^{\lambda}. \tag{13}$$

Let $h^{AB}(\xi^c)$ be an arbitrary metric on the world-sheet; the variational principle introduced first by A. Polyakov reads then

$$\delta S = -\frac{1}{4\pi\alpha'}\,\delta\int\!\!\int \sqrt{-h}\,h^{AB}\,G_{AB}\,d\tau d\sigma = 0. \tag{14}$$

Under the independent variations δx^{μ} and δh^{AB} one gets the following equations:

$$G_{AB} - \frac{1}{2}h_{AB}\left(h^{CD}\,G_{CD}\right) = 0, \tag{15}$$

$$h^{AB}\left[\nabla_A\nabla_B x^{\mu} + \Gamma^{\mu}_{\lambda\rho}\,\partial_A x^{\lambda}\partial_B x^{\rho}\right] = 0. \tag{16}$$

After dimensional reduction from 11 to 10 dimensions, auxiliary fields A_{μ} and ϕ do appear, and the total Lagrangian takes on the form that contains the Born-Infeld Lagrangian ([14, 15, 16]).

$$\mathcal{L} = \frac{1}{2}\,\mathcal{D}_{\mu}\Phi\mathcal{D}^{\mu}\Phi^{*} + \beta^2\,(1-\mathcal{R}) - \frac{\lambda}{2}\,(\Phi^{*}\,\Phi - v^2)^2 + \frac{1}{16\pi G}R \tag{17}$$

with Φ denoting scalar field, R the Riemann curvature scalar, and $\mathcal{R}$ given by

$$\mathcal{R} = \sqrt{1 + \frac{1}{2\beta^2}F^a_{\mu\nu}F^{\mu\nu}_a - \frac{1}{16\beta^4}(F^a_{\mu\nu}\tilde{F}^{\mu\nu}_a)^2}\,. \tag{18}$$

For dimensional reductions onto lower dimensions, the non-abelian generalizations of this Lagrangian are naturally produced.

In a pure Yang-Mills theory in flat space-time, with the usual Lagrangian density $\mathcal{L}_{YM} = -\frac{1}{4}g_{AB}F^{A}_{\mu\nu}F^{B\mu\nu}$ there are no finite energy static non-singular solutions describing a charged soliton. This fact can be explained by the conformal invariance of the theory, and the tracelessness of the energy-momentum tensor,

$$T^{\mu}{}_{\mu} = -T_{00} + \sum_{i=1}^{3} T_{ii} = 0. \tag{19}$$

Given the positivity of energy, i.e.$T_{00} > 0$, this means that the sum of principal pressures is positive, too, $\sum T_{ii} > 0$, which leads to the conclusion that the Yang-Mills "matter" is naturally subjected to repulsive forces only.

The presence of the Higgs field breaks the conformal invariance, which leads to the existence of 't Hooft and Prasad-Sommerfield magnetic monopoles. In what follows, we are interested in soliton-like solutions arising in other non-linear theories, including non-abelian versions of Yang-Mills theories, which are no more conformally invariant.

NON-LINEAR ELECTRODYNAMICS FROM THE KALUZA-KLEIN THEORY

An interesting non-linear generalization of electrodynamics derived from the Kaluza-Klein theory in five dimensions has been proposed in ([8], [9]). It is based on the addition of the Gauss-Bonnet term, $R_{ABCD}R^{ABCD} - 4R_{AB}R^{AB} + R^2$, which in five dimensions is not a topological invariant, leading to non-trivial equations of motion of second order when added to the Einstein-Hilbert Lagrangian.

In a flat space-time and without the scalar field the Kaluza-Klein metric is

$$g_{AB} = \begin{pmatrix} g_{\mu\nu} + A_\mu A_\nu & A_\mu \\ A_\nu & 1 \end{pmatrix}, \tag{20}$$

where $A, B = 0, 1, 2, 3, 5$ and $\mu, \nu = 0, 1, 2, 3$ (or $\mu, \nu = 0, x, y, z$ following the convention we have been using). The full Lagrangian is taken to be (see [8, 9]):

$$\mathcal{L} = R + \gamma (R_{ABCD}R^{ABCD} - 4R_{AB}R^{AB} + R^2), \tag{21}$$

with γ being a certain dimensional parameter characterizing the strength of the non-linearity. When expressed in four dimensions in terms of the Maxwell tensor, it becomes

$$\mathcal{L} = -\frac{1}{4} F_{\mu\nu} F^{\mu\nu} - \frac{3\gamma}{16} \left[(F_{\mu\nu}F^{\mu\nu})^2 - 2(F_{\mu\lambda}F_{\nu\rho}F^{\mu\nu}F^{\lambda\rho}) \right]. \tag{22}$$

In terms of the invariants P and S this Lagrangian is given by $\mathcal{L} = 2P + \frac{3\gamma}{2} S^2$, which for the choice $\gamma = -\frac{2}{3}$ yields essentially the square of the Born-Infeld Lagrangian. The equations of motion are:

$$F_{\lambda\rho,\mu} + F_{\rho\mu,\lambda} + F_{\mu\lambda,\rho} = 0, \tag{23}$$

which correspond to the Bianchi identities and are geometrical equations valid independently of the Lagrangian chosen, and the dynamical equations resulting from the variational principle,

$$[F^{\lambda\rho} - \frac{3\gamma}{2}(F_{\mu\nu}F^{\mu\nu})F^{\lambda\rho} + \frac{3\gamma}{2} F_{\mu\nu}F^{\lambda\mu}F^{\rho\nu}]_{,\lambda} = 0. \tag{24}$$

The Lagrangian (22) is particularly simple when expressed in more familiar terms with the fields $\mathbf{E}$ and $\mathbf{B}$:

$$\mathcal{L} = \frac{1}{2}(\mathbf{B}^2 - \mathbf{E}^2) + \frac{3\gamma}{2}(\mathbf{E} \cdot \mathbf{B})^2. \tag{25}$$

The equations of motion also display a clear physical meaning when expressed in terms of **E** and **B**. The equation (24) becomes

$$\mathbf{div}\,\mathbf{B} = 0, \qquad \mathbf{rot}\,\mathbf{E} = -\partial_0 \mathbf{B}, \tag{26}$$

whereas the equations (23) become

$$\begin{aligned} \mathbf{div}\,\mathbf{E} &= -3\gamma \mathbf{B}\cdot \mathbf{grad}\,(\mathbf{E}\cdot\mathbf{B}) \\ \mathbf{rot}\,\mathbf{B} &= \partial_0 \mathbf{E} + 3\gamma \left[\mathbf{B}\partial_0(\mathbf{E}\cdot\mathbf{B}) - \mathbf{E}\times \mathbf{grad}(\mathbf{E}\cdot\mathbf{B}) \right], \end{aligned} \tag{27}$$

which show how the density of charge and the current are created by the non-linearity of the field: indeed, we can introduce

$$\rho = -3\gamma \mathbf{B}\cdot \mathbf{grad}\,(\mathbf{E}\cdot\mathbf{B}) \quad \text{and} \quad \mathbf{j} = 3\gamma \left[\mathbf{B}\partial_0(\mathbf{E}\cdot\mathbf{B}) - \mathbf{E}\times \mathbf{grad}(\mathbf{E}\cdot\mathbf{B}) \right] \tag{28}$$

which satisfy the continuity equation

$$\partial_0 \rho + \mathbf{div}\,\mathbf{j} = 0. \tag{29}$$

The Poynting vector conserves its form known from the Maxwellian theory, but the energy density is modified:

$$\mathbf{S} = \mathbf{E}\times\mathbf{B}, \qquad \mathcal{E} = \frac{1}{2}(\mathbf{E}^2 + \mathbf{B}^2) + \frac{3\gamma}{2}(\mathbf{E}\cdot\mathbf{B})^2, \tag{30}$$

with the continuity equation resuming the energy conservation satisfied by virtue of the equations of motion:

$$\partial_0 \mathcal{E} + \mathbf{div}\mathbf{S} = 0. \tag{31}$$

It can be easily proved that there is birefringence in this theory. One wave propagates in a Maxwellian way, the other possible wave solution propagates differently; in fact, it is delayed (see [5] for details).

The properties of possible stationary axisymmetric solutions, endowed with non-vanishing charge, intrinsic kinetic and magnetic moments, have been discussed in [8, 9]. In the theory based on the Gauss-Bonnet term in 5 dimensions, one can try to find axially-symmetric configurations displaying both finite electric charge and finite magnetic moment; also, a kinetic momentum can be expected, parallel to the magnetic moment.

In cylindric coordinates ρ, φ, z we expect the induced current density to be aligned on the $\mathbf{e}_\varphi$-vector of the local frame, giving a current density circulating around the z-axis; the fields **E** and **B** should be contained in the $\rho - z$ planes orthogonal to $\mathbf{e}_\varphi$. Recalling the fact that the lines of strength of **B** must be closed, the best description of this configuration can be obtained using the toroidal curvilinear coordinates (μ, η, φ) defined as follows in terms of cylindric coordinates:

$$\rho = \frac{a\cosh\mu}{\cosh\mu - \cos\eta}, \qquad z = \frac{a\sin\eta}{\cosh\mu - \cos\eta}, \qquad \varphi.$$

with $0 \leq \varphi < 2\pi$, $0 \leq \eta < 2\pi$, $0 \leq \mu \leq \infty$. The coordinate lines of φ are concentric circles in the $(z = 0)$-plane, while the coordinate lines of the variable η are excentric tori concentrating around the circle $\rho = a$. We shall suppose that the lines of force of the magnetic field coincide with the coordinate curves given by $\varphi =$ Const. and $\mu =$ Const. The configuration we seek can be written as:

$$\mathbf{E} = E_\rho \, \mathbf{e}_\rho + E_z \, \mathbf{e}_z = E_\mu \mathbf{e}_\mu + E_\eta \, \mathbf{e}_\eta; \tag{32}$$

$$\mathbf{B} = B_\rho \, \mathbf{e}_\rho + B_z \, \mathbf{e}_z = B_\eta \, \mathbf{e}_\eta; \quad \mathbf{j}_{ind} = j_{ind} \, \mathbf{e}_\varphi. \tag{33}$$

It can be also shown that the whole problem can be reduced to determining just two unknown functions of the variables (μ, η), because $\mathbf{B} = rot\mathbf{A}$ and $B_\mu = 0$, and because here $\mathbf{E} = -gradV$, we have $\mathbf{A} = A(\mu, \eta)\, \mathbf{e}_\varphi$, and $V = V(\mu, \eta)$.

Approximate solutions of this form have been found in ([8], [9]); here we shall only remind their essential features. At great distances, the fields $\mathbf{E}$ and $\mathbf{B}$ behave as if they were generated by a finite charge Q and a finite magnetic dipole $\mathbf{m}$:

$$\mathbf{E}_\infty \simeq \frac{Q\,\mathbf{r}}{4\pi r^3}, \quad \mathbf{B}_\infty \simeq \frac{\mathbf{m} \wedge \mathbf{r}}{4\pi r^3}. \tag{34}$$

The charge is concentrated around the circle $\rho = a$ and "smeared" in its vicinity; if it is chosen to be positive, there is a little "halo" of negative charge density farther away, imitating the vacuum polarization effect. The charge density's fall-off is vary rapid, behaving at short distances as r^{-9}, and then falling off exponentially; the same concerns the density of induced current $\mathbf{j}_{ind}$ which falls off as r^{-8}. The induced current behaves as if it were produced by the charge density rotating around the z-axis with the speed of light.

Another interesting feature of this solution is its $Z_2 \times Z_2$ symmetry. Indeed, any such solution displaying the total energy (mass) $\mathcal{E}$, the total charge Q, magnetic momentum $\mathbf{m}$ and the total spin $\mathbf{s}$ is followed by three similar solutions with the same energy, but either with the same charge, but with the spin and magnetic momentum in the opposite direction (both "down"), or another couple of solutions having the opposite charge, and spin and magnetic moment up or down, but always opposite to each other - just like with what we know about the electron and the positron. The following table shows the properties of the four solutions:

Fields	*Energy*	*Charge*	$\mathbf{m}$	*Spin*
$\mathbf{E}, \ \mathbf{B}$	$\mathcal{E}$	Q	$\mathbf{m}$	$\mathbf{s}$
$\mathbf{E}, -\mathbf{B}$	$\mathcal{E}$	Q	$-\mathbf{m}$	$-\mathbf{s}$
$-\mathbf{E}, \ \mathbf{B}$	$\mathcal{E}$	$-Q$	$\mathbf{m}$	$-\mathbf{s}$
$-\mathbf{E}, -\mathbf{B}$	$\mathcal{E}$	$-Q$	$-\mathbf{m}$	$\mathbf{s}$

Tab 1. The symmetry properties of four solutions.

Unfortunately, these solutions present a mild singularity on the circle $\rho = a$, which can not be avoided. Its presence can be proved by using Poincaré's lemma; the details can be found in ([8], [9]).

AN SU(2)-BASED NON-ABELIAN GENERALIZATION OF BORN-INFELD THEORY

The superstring theory gives rise to one important modification of the standard Yang-Mills quadratic Lagrangian suggesting the action of the Born-Infeld (BI) type [14, 15, 16]. Because this modification breaks the scale invariance, the natural question arises whether in the Born–Infeld–Yang–Mills (BIYM) theory the non-existence of classical particle-like solutions can be overruled. Although a mere scale invariance breaking, being a necessary condition, by no means guarantees the existence of particle-like solutions, a detailed study ([10] has shown that the $SU(2)$ BIYM classical glueballs indeed do exist.

Non–Abelian generalization of the Born–Infeld action presents an ambiguity in specifying how the trace over the the matrix–valued fields is performed in order to define the Lagrangian [14, 17]. Here we adopt the version with the ordinary trace which leads to a simple closed form for the action. The BIYM action with the ordinary trace looks like a straightforward generalisation of the corresponding $U(1)$ action in the "square root" form

$$S = \frac{\beta^2}{4\pi} \int (1 - \mathcal{R})\, d^4x, \tag{35}$$

where

$$\mathcal{R} = \sqrt{1 + \frac{1}{2\beta^2} F^a_{\mu\nu} F^{\mu\nu}_a - \frac{1}{16\beta^4} (F^a_{\mu\nu} \tilde{F}^{\mu\nu}_a)^2}\,. \tag{36}$$

It is easy to see that the BI non-linearity breaks the conformal symmetry ensuring the non-zero trace of the stress–energy tensor

$$T^\mu_\mu = \mathcal{R}^{-1} \left[4\beta^2(1 - \mathcal{R}) - F^a_{\mu\nu} F^{\mu\nu}_a\right] \neq 0. \tag{37}$$

This quantity vanishes in the limit $\beta \to \infty$ when the theory reduces to the standard one. For the Yang-Mills field we assume the usual monopole ansatz

$$A^a_0 = 0, \quad A^a_i = \varepsilon_{aik} \frac{n^k}{r} (1 - w(r)), \tag{38}$$

where $n^k = x^k/r$, $r = (x^2 + y^2 + z^2)^{1/2}$, and $w(r)$ is the real-valued function. After the integration over the sphere in (35) one obtains a two-dimensional action from which β can be eliminated by the coordinate rescaling $\sqrt{\beta} t \to t$, $\sqrt{\beta} r \to r$. The following static action results then:

$$S = \int L dr, \quad L = r^2 (1 - \mathcal{R}), \tag{39}$$

with

$$\mathcal{R} = \sqrt{1 + 2\frac{w'^2}{r^2} + \frac{(1 - w^2)^2}{r^4}}, \tag{40}$$

where prime denotes the derivative with respect to r. It is worth noticing that the non-linearity arises here because of the non-linear dependence of the tensor $F^a_{\mu\nu}$ on the potentials A_μ^b. The corresponding equation of motion reads

$$\left(\frac{w'}{\mathcal{R}}\right)' = \frac{w(w^2-1)}{r^2\mathcal{R}}. \tag{41}$$

A trivial solution $w \equiv 0$ corresponds to the point-like magnetic BI-monopole with the unit magnetic charge (embedded $U(1)$ solution). In the Born–Infeld theory it has a finite self-energy [18]. For time-independent configurations the energy density is equal to minus the Lagrangian, so the total energy (mass) is given by the integral

$$M = \int_0^\infty (\mathcal{R}-1)r^2 dr. \tag{42}$$

For $w \equiv 0$ one finds

$$M = \int \left(\sqrt{r^4+1} - r^2\right) dr = \frac{\pi^{3/2}}{3\Gamma(3/4)^2} \approx 1.23604978. \tag{43}$$

Looking now for the essentially non–Abelian solutions of finite mass, we observe that in order to assure the convergence of the integral (42) the quantity $\mathcal{R}-1$ must fall down faster than r^{-3} as $r \to \infty$. Thus, far from the core the BI corrections have to vanish and the Eq.(41) should reduce to the ordinary Yang-Mills equation, equivalent to the following two-dimensional autonomous system [19, 20, 21, 22]:

$$\dot{w} = u, \quad \dot{u} = u + (w^2-1)w, \tag{44}$$

where a dot denotes the derivative with respect to $\tau = \ln r$. This dynamical system has three non-degenerate stationary points $(u = 0, w = 0, \pm 1)$, from which $u = w = 0$ is a focus, while two others $u = 0$, $w = \pm 1$ are saddle points with eigenvalues $\lambda = -1$ and $\lambda = 2$. The separatices along the directions $\lambda = -1$ start at infinity and after passing through the saddle points go to the focus with the eigenvalues $\lambda = (1 \pm i\sqrt{3})/2$.

It has been proved in ([10]) that *the only finite-energy configurations with non vanishing magnetic charge are the embedded U(1) BI-monopoles*. Indeed, such solutions should have asymptotically $w = 0$, which does not correspond to bounded solutions unless $w \equiv 0$. The remaining possibility is $w = \pm 1, \dot{w} = 0$ asymptotically, which corresponds to zero magnetic charge. Coming back to r-variable one finds from (41)

$$w = \pm 1 + \frac{c}{r} + O(r^{-2}), \tag{45}$$

where c is a free parameter. This gives a convergent integral (42) as $r \to \infty$. The two values $w = \pm 1$ correspond to two neighboring topologically distinct Yang-Mills vacua.

Now consider local solutions near the origin $r = 0$. For convergence of the total energy (42), w should tend to a finite limit as $r \to 0$. Then using the Eq.(41) one finds that the only allowed limiting values are $w = \pm 1$ again. In view of the symmetry of (44) under reflection $w \to \pm w$, one can take without loss of generality $w(0) = 1$. Then the following Taylor expansion can be checked to satisfy the Eq.(44):

$$w = 1 - br^2 + \frac{b^2(44b^2+3)}{10(4b^2+1)}r^4 + O(r^6), \tag{46}$$

with b being (the only) free parameter.

As $r \to 0$, the function $\mathcal{R}$ tends to a finite value

$$\mathcal{R} = \mathcal{R}_0 + O(r^2), \qquad \mathcal{R}_0 = 1 + 12b^2, \tag{47}$$

therefore it is not a solution of the initial system (42). What remains to be done is to find appropriate values of constant b leading to smooth finite-energy solutions by gluing together the two asymptotic solutions between 0 and ∞.

It has been proved in ([10] that *any regular solution of the* Eq.(41) *belongs to the one-parameter family of local solutions* (46) *near the origin.*

It follows that the global finite energy solution starting with (46) should meet some solution from the family (45) at infinity. Since both these local solutions are non–generic, one can at best match them for some discrete values of parameters. This technique has been used first in ([20])

For some precisely tuned value of b the solution will remain a monotonous function of τ reaching the value -1 at infinity (Fig.1). This happens for $b_1 = 12.7463$.

By a similar reasoning one can show that for another fine-tuned value $b_2 > b_1$ the integral curve $w(\tau)$ which has a minimum in the lower part of the strip and then becomes positive will be stabilized by the friction term in the upper half of the strip $[-1, 1]$ and tend to $w = 1$. This solution will have two nodes. Continuing this process we obtain the increasing sequence of parameter values b_n for which the solutions remain entirely within the strip $[-1, 1]$ tending asymptotically to $(-1)^n$. The lower values b_n found numerically are given in Tab. 2.

n	b	M
1	1.27463×10^1	$1.13559 =$
2	8.87397×10^2	1.21424
3	1.87079×10^4	1.23281
4	1.27455×10^6	1.23547
5	2.65030×10^7	1.23595

Tab 2. Parameters b, M for first five solutions.

AN SU(2) $\times$ U(1) GENERALIZATION OF BORN-INFELD LAGRANGIAN AND ITS EMBEDDING IN THE STANDARD ELECTROWEAK MODEL

The Born-Infeld Lagrangian generalizes the usual Maxwell theory; however, since we know that this theory is a part of the non-abelian field theory which accounts for electromagnetic and weak interactions, a natural question can be asked: is the original abelian version of Born-Infeld theory just a "shadow" of a more complicated non-abelian analog of the Born-Infeld Lagrangian ? If so, we should be able to compare the pure electromagnetic (abelian) BI-Lagrangian with what can be extracted from its non-abelian version based on the symmetry group $SU(2) \times U(1)$ after defining physical fields as linear combinations of the $U(1)$ and $SU(2)$ gauge fields with the coefficients defined by a rotation with the Weinberg angle. The ultimate comparison is beyond the scope of this paper; we will show how the first few terms in the Taylor expansions of these two Lagrangians can be compared. Performing the expansion of the BI Lagrangian and we obtain the following first few terms in the series up to the fourth order.

$$\begin{aligned} L_{BI} &= -\frac{1}{4}F_{\mu\nu}F^{\mu\nu}+\frac{1}{32}\beta^{-2}(F_{\mu\nu}F^{\mu\nu})^2+\frac{1}{32}\beta^{-2}(F_{\mu\nu}\widetilde{F}^{\mu\nu})^2 \\ &- \frac{1}{128}\beta^{-4}(F_{\mu\nu}F^{\mu\nu})^3-\frac{1}{128}\beta^{-4}F_{\mu\nu}F^{\mu\nu}(F_{\mu\nu}\widetilde{F}^{\mu\nu})^2+\frac{5}{2048}\beta^{-6}(F_{\mu\nu}F^{\mu\nu})^4 \\ &+ \frac{3}{1024}\beta^{-6}(F_{\mu\nu}F^{\mu\nu})^2(F_{\mu\nu}\widetilde{F}^{\mu\nu})^2+\frac{1}{2048}\beta^{-6}(F_{\mu\nu}\widetilde{F}^{\mu\nu})^4. \end{aligned} \tag{48}$$

For non-abelian groups we shall use the same generalization of the Born-Infeld Lagrangian as in the previous section, (35). Here we will construct the non-abelian Lagrangian for $SU(2)$ and $U(1)$, to be compared with (48). We expand the series in powers of β^{-2} in terms of Lorentz invariants of the fields, the abelian ones, P and S, and their non-abelian generalizations P' and S':

$$P' \equiv F_{\mu\nu}F^{\mu\nu}, \qquad S' \equiv F_{\mu\nu}\widetilde{F}^{\mu\nu} = \frac{1}{2}\varepsilon^{\mu\nu\rho\sigma}F_{\mu\nu}F_{\rho\sigma}, \tag{49}$$

with $F_{\mu\nu} = F^a_{\mu\nu}J_a$, with $a = 0 =$ for $U(1)$ and $a = 1..3$ for $SU(2)$. With the invariants (49) replacing the abelian ones, and taking all the traces in the Lagrangian, which in the non-abelian case takes value in the matrix algebra of the fundamental representation of $SU(2)\times U(1)$ chosen here, we obtain

$$\begin{aligned} L_{SU(2)U(1)} &= 2a\,F^0_{\mu\nu}F^{0\mu\nu}+\tfrac{1}{2}a\,F^a_{\mu\nu}F^{a\mu\nu} \\ &+\beta^{-2}M^2\{b\,[2F^0_{\mu\nu}F^{0\mu\nu}F^0_{\rho\sigma}F^{0\rho\sigma}+\tfrac{1}{8}F^a_{\mu\nu}F^{a\mu\nu}F^c_{\rho\sigma}F^{c\rho\sigma} \\ &+F^0_{\mu\nu}F^{0\mu\nu}F^a_{\rho\sigma}F^{a\rho\sigma}+2F^0_{\mu\nu}F^{a\mu\nu}F^0_{\rho\sigma}F^{a\rho\sigma}]+c\,[2F^0_{\mu\nu}\widetilde{F}^{0\mu\nu}F^0_{\rho\sigma}\widetilde{F}^{0\rho\sigma} \\ &+\tfrac{1}{8}F^a_{\mu\nu}\widetilde{F}^{a\mu\nu}F^c_{\rho\sigma}\widetilde{F}^{c\rho\sigma}+F^0_{\mu\nu}\widetilde{F}^{0\mu\nu}F^a_{\rho\sigma}\widetilde{F}^{a\rho\sigma}+2F^0_{\mu\nu}\widetilde{F}^{a\mu\nu}F^0_{\rho\sigma}\widetilde{F}^{a\rho\sigma}]\}+\dots, \end{aligned} \tag{50}$$

where β is the BI-Lagrangian parameter, M the mass scale of the unified theory, and $a, b, c, \ldots$ are complicated numerical coefficients coming from traces and representation-dependent. Introducing physical fields with linear combinations of the $U(1)$ and the $SU(2)$ gauge fields

$$F^0_{\mu\nu} = F_{\mu\nu}\cos\theta-(\partial_\mu Z_\nu-\partial_\nu Z_\mu)\sin\theta, \tag{51}$$

$$F^3_{\mu\nu} = F_{\mu\nu}\sin\theta+(\partial_\mu Z_\nu-\partial_\nu Z_\mu)\cos\theta = +ig(W^\dagger_\mu W^+_\nu - W^+_\mu W^\dagger_\nu), \tag{52}$$

$$\begin{aligned} F^1_{\mu\nu} &= \frac{1}{\sqrt{2}}[(\partial_\mu W_\nu-\partial_\nu W_\mu) + (\partial_\mu W^\dagger_\nu-\partial_\nu W^\dagger_\mu)] \\ &- ig[(W_\nu-W^\dagger_\nu)(A_\mu\sin\theta+Z_\mu\cos\theta) \\ &+ (W_\mu-W^\dagger_\mu)(A_\nu\sin\theta+Z_\nu\cos\theta)], \end{aligned} \tag{53}$$

$$
\begin{aligned}
F^2_{\mu\nu} &= \frac{i}{\sqrt{2}}[(\partial_\mu W_\nu - \partial_\nu W_\mu) - (\partial_\mu W^\dagger_\nu - \partial_\nu W^\dagger_\mu)] \\
&+ g[(W_\nu + W^\dagger_\nu)(A_\mu \sin\theta + Z_\mu \cos\theta) \\
&- (W_\mu + W^\dagger_\mu)(A_\nu \sin\theta + Z_\nu \cos\theta)], \qquad (54)
\end{aligned}
$$

where A_μ is the pure electromagnetic field, Z_μ is the neutral boson, W^+_μ and W^-_μ are the charged W-bosons, we can now compare the two series, term by term, trying to fix the coefficients in order to make coincide as many terms as possible. With the Weinberg angle θ we can identify the pure electromagnetic sector in (50), then evaluate the difference. Because of the lack of space, we show here only first terms of this expression:

$$
\begin{aligned}
L_{SU(2)U(1)} - L_{EM} &= a(2\cos^2\theta + \tfrac{1}{2}\sin^2\theta)F_{\mu\nu}F^{\mu\nu} \\
&+\beta^{-2}M^2\{b[2\cos^4\theta + \tfrac{1}{8}\sin^4\theta + 3\sin^2\theta\cos^2\theta]F_{\mu\nu}F^{\mu\nu}F_{\rho\sigma}F^{\rho\sigma} \\
&+c[2\cos^4\theta + \tfrac{1}{8}\sin^4\theta + 3\sin^2\theta\cos^2\theta]F_{\mu\nu}\widetilde{F}^{\mu\nu}F_{\rho\sigma}\widetilde{F}^{\rho\sigma}]\} \qquad (55) \\
&+\beta^{-4}M^3\{g\,[\tfrac{1}{32}\sin^6\theta + \tfrac{15}{8}\cos^2\theta\sin^4\theta + \tfrac{15}{2}\cos^4\theta\sin^2\theta + 2\cos^6\theta] + \ldots\}.
\end{aligned}
$$

It is possible to show that the coefficient for the n-th order of β^{-2} is given by

$$
C_n(\theta) = \frac{1}{4^n}[(1-2\cot\theta)^{2n} + (1+2\cot\theta)^{2n}](\sin\theta)^{2n}, \qquad (56)
$$

and its derivative is given by

$$
\begin{aligned}
C'_n(\theta) &= \frac{n}{2^{2n-1}}\{\cot\theta[(1-2\cot\theta)^{2n} + (1+2\cot\theta)^{2n}] \\
&+2[(1-2\cot\theta)^{2n-1} + (1+2\cot\theta)^{2n-1}](\csc\theta)^2\}(\sin\theta)^{2n}. \qquad (57)
\end{aligned}
$$

It is interesting to examine the behaviour of these coefficients. Surprisingly enough, starting from $n = 3$ they display a maximum, whose position converges to a certain value with growing n; moreover, this position is very close to the established value of the Weinberg angle (satisfying $\sin^2\theta_W = 0.227 \pm 0.014$, corresponding to $\theta_W = 28^o, 45$ or to 0.497 radians). The maxima were found solving $C'_n(\theta) = 0$ for a given value of n. We show below examples of the coefficients C_n starting from $n = 3$, and the value of the angle (in radians) for the first maximum of C_n

$$
C_3 = \frac{1}{32}\sin^6\theta + \frac{15}{8}\sin^4\theta\cos^2\theta + \frac{15}{2}\sin^2\theta\cos^4\theta + 2\cos^6\theta. \qquad (58)
$$

The first maximum corresponds to $C_3 = 2.0838$ for $\theta = 0.34682$.

Later on, we have:

First maximum for $C_4 = 2.4886$ at $\theta = 0.43522$.

First maximum at $C_5 = 3.0713$ for $\theta = 0.45474$.

First maximum at $C_6 = 3.8232$ for $\theta = 0.4606$.

For $n = 8$ we obtain

First maximum at $C_8 = 5.9622$ for $\theta = 0.46327$.

The value of the angle corresponding to the first maximum for higher order tends to $\theta = 0.463648$ and remains constant for $n = 50$ and higher.

The fact that the value of the mixing angle obtained for the first maximum of the of the so defined coefficients approaches the value of the Weinberg angle seems to be rather accidental; nevertheless, it is worth noticing.

ACKNOWLEDGMENTS

A. L. Barbosa thanks FAPESP (São Paulo, Brazil) for financial support. D. V. Gal'tsov acknowledges support from RFBR under grant 00-02-16306.

REFERENCES

1. Mie, G., *Annalen der Physik*, **37**, 511 (1912).
2. Born, M,. and Infeld, L., *Nature* **132**, 970 (1932); also, *Proc. Roy. Soc.*, **A 144**, 425 (1934).
3. Boillat, G., *J. Math. Phys.*, **11**, 941 (1970).
4. Bialynicki-Birula, I., in J. Lopuszanski's *Festschrift, Quantum Theory of Particles and Fields"*, Eds. B. Jancewicz and J. Lukierski, p. 31 - 42, *World Scientific*, Singapore (1983).
5. Lemos, J.P.S., and Kerner, R., *Gravitation and Cosmology*, **6**, 49-58 (2000).
6. Gibbons, G.W., and Herdeiro, C., *Phys. Rev.*, **D63**, 064006 (2001).
7. Müller-Hoissen, F., *Phys. Lett.*, **B156**, 315 - 320 (1985).
8. Kerner, R., *Comptes Rendus Acad. Sci. Paris*, **304**, série 2, 621-624 (1987).
9. Kerner, R., in *Infinite dimensional Lie algebras and Quantum Field Theory*, proceedings of the Varna Summer School (Bulgaria) 1987, H.D. Doebner, J.D. Hennig and T.D. Palev eds., p.53 - 72, *World Scientific*, (1988).
10. Gal'tsov, D.V., and Kerner, R., *Phys. Rev. Lett.*, **84**, (26), 5955-5959, (2000).
11. Gal'tsov, D.V., and Volkov, M.S., *Phys. Lett.*, **B273**, 255–259 (1991).
12. Sudarsky, D., and Wald, R.M., *Phys. Rev.*, **D 46**, 1453–1447 (1992).
13. Volkov, M.S., and Gal'tsov, D.V., *Phys. Rep.*, **C319**, 1–83 (1999), `hep-th/9810070`.
14. Tseytlin, A., *Nucl. Phys.*, **B501** , 41 (1997).
15. Gauntlett, J.P., Gomis, J., and Townsend, P.K., *JHEP*, **01**, 003 (1998).
16. Brecher, D., and Perry, M.J., *Nucl. Phys.*, **B527**, 121 (1998).
17. Grandi, N., Moreno, E.F., and Shaposhnik, F.A., Monopoles in non-Abelian Dirac-Born-Infeld theory, `hep-th/9901073`.
18. Gibbons, G.W., Wormholes on the World Volume: Born-Infeld particles and Dirichlet p-branes, `hep-th/9801106`.
19. Chernavskii, D.S., and Kerner, R., *J. Math. Phys.*, **9**, (1), 287 (1978).
20. Kerner, R., *Phys. Rev.*, **D19**, (4), 1243 (1979).
21. Protogenov, A.P., *Phys. Lett.*, **B87**, 80 (1979).
22. Breitenlohner, P., Forgacs, P., and Maison, D., *Comm. Math. Phys.*, **163**, 141–172 (1994).

RELATED TOPICS

Domain Wall as a 2-Brane

H. Arodź

Institute of Physics, Jagellonian University, Reymonta 4, 30-059 Cracow, Poland, email: ufarodz@th.if.uj.edu.pl

Abstract. We give an overview of research work on the problem of relation betwen domain walls and membranes.

INTRODUCTION

Quantum field theory (QFT) has had many stunning successes. Nevertheless, at the moment it is far from becoming a completely understood standard formalism. At such a stage QFT would become a tool which could be readily applied, and we would have perfect understanding of its workings. The present day QFT has several intrinsic problems. One of them is posed by existence of special classical solutions called topological defects. They belong to nontrivial topological sectors in the space of fields. To illustrate the point, let us have a look at Abelian Higgs model with spontaneous breakdown of U(1) symmetry. In the vacuum sector Higgs mechanism leads, upon standard quantisation, to the presence of massive scalar and vector particles. However, on classical level the model possesses the well-known Nielsen-Olesen static vortex solutions which are enumerated by integer topological charge. Such vortex solutions have infinite energy. It is not known how to construct the corresponding QFT model which would incorporate topological defects in a scheme useful for practical calculations. One can easily ask questions about the model which seem to be beyond the present day calculational capability of QFT. For example, what is the probability of creating intermediate vortex loops in a scattering of the vector or scalar particles? The vortex loop has finite energy and vanishing topological charge, hence it can appear in the vacuum, but in order to take it into account one has to go beyond the standard perturbative expansion with respect to coupling constants of the model. Another sample question: what are amplitudes for scattering of the vector and scalar particles in presence of the vortex? One of difficulties in the latter problem is presence of very soft excitations of the vortex, related to existence of so called translational zero modes. The vortex can not be regarded as merely an infinitely heavy spectator to the scattering, equivalent to a static background field.

Existence of topological defects is predicted by fundamental unifying theories such as, e. g., superstring theories. They are also present in condensed matter, for instance, in quantum superfluids or magnetic materials. Especially beautiful examples of topological defects can be found in liquid crystals. There is a whole variety of them, including domain walls, disclination lines and hedgehogs.

Apart from the indicated above direct physical relevance, the topological defects, such as domain walls or vortices, provide physical models of membranes or strings — more

CP589, *Fundamental Interaction Theories*, edited by J. Lukierski and J. Rembieliński

generally, physical models of p-branes [1]. It is clear that the relation of a p-brane to the topological defect is analogous to the relation of material point to a real material particle. However, it turns out that the task to formulate that correspondence rigorously is not quite simple. In the following part of this article we sketch certain attempts in this direction. Such mathematically controlled correspondence could be used in at least two ways. First, even simplest p-branes can have very intricate dynamics. One can find unexpected counter-intuitive motions already for Nambu-Goto string, which is rather simple example of 1-brane. Therefore, it is quite enlightening to compare the dynamics of the p-branes, which are mathematical idealisations, with behaviour of real objects, such as domain walls or vortices.

Second way of utilizing that correspondence goes in the direction opposite to the previous one — one can use results about dynamics of the p-branes in order to approximately describe behaviour of the topological defects, similarly as the material point idealisation is exploited in classical mechanics.

Apart from those applications, the investigations of the relationship between the p-branes and the topological defects have revealed many details of the intricate and beautiful non-linear dynamics of the defects. We believe that these results bring us closer to the ultimate goal of constructing QFT with all topological sectors accessible for practical calculations.

HISTORICAL REMARKS

Nowadays, the correspondence between domain walls and membranes, or vortices and strings, is discussed in the framework of expansion with respect to dimensionless ratios l_0/R, where l_0 denotes a transverse width of the domain wall or the vortex at rest, and R denotes main curvature radia of the corresponding membrane or string in a local rest frame. The fact that the vortex can be regarded as a physical model of the relativistic string was pointed out by Nielsen and Olesen in the seminal paper [1]. Their argument was strengthened by Förster [2]. Gervais and Sakita [3] tried to reformulate Abelian Higgs model in terms of Nambu-Goto strings using the ideas of Nielsen, Olesen and Förster. From the present day perspective, in those papers the correspondence between the vortex and the string was considered in a leading order in l_0/R.

The subject stayed dormant until mid-eighties. The new impetus came from the cosmic string hypothesis [4]. It became clear that the Nambu-Goto string sometimes is too crude an approximation to the vortex. It particular, it can have arbitrary number of spikes, that is points at which direction of the vector tangent to the string changes abruptly. Such spikes can travel along the string and they exist forever. On the other hand, the vortex is always smooth due to its nonzero thickness. Therefore, the spikes seem to be artifacts of the approximation. It turned out that going beyond the leading approximation is not a simple task. The problem was known as the question of rigidity corrections to the Nambu-Goto action — majority of authors tried to approximate the

---

[1] p+1 gives the dimension of the corresponding world-volume.

vortex by a string with the corresponding action functional of more complicated form than just the Nambu-Goto action. The expectation was that such corrections would introduce certain rigidity against bending, and thus prevent the appearence of spikes. The Nambu-Goto string is perfectly flexible — it resist only stretching. At present it is clear that that hope is not fulfilled. The first papers in which corrections to the Nambu-Goto action were computed were [5], [6]. These papers were followed by several others. Let us mention here paper [7] in which so called classical effective action (CEA) method was explicitly formulated. That method was used in most papers on the problem. An exception is the paper [8] in which effective equations of motion for the string were derived directly from pertinent field equations as evolution equations for certain averaged quantities, without resorting to action functionals. The CEA method is used till nowadays, see, e.g., [9].

More or less at the same time it became clear that there is an analogous, and slightly simpler, problem of formulating the correspondence between domain walls and membranes (2-branes). This problem was considered by Gregory and collaborators [10], again in the framework of CEA method.

The classical effective action method has several shortcomings. It leads to p-brane models with time derivatives of the order higher than second in equations of motion. As a rule, such models are pathological. They possess "run-away" solutions, and energy is not bounded from below. Moreover, the CEA method can be used only if field equations governing the motion of the extended object are obtained as Euler-Lagrange equations from an action functional. This requirement excludes dissipative type of evolution which is quite common in condensed matter systems.

Evolution of domain walls and vortices was investigated also in condensed matter physics. The corresponding models are not relativistically invariant of course. From our viewpoint, these investigations brought in one very important element: projections on translational zero-modes, [11], [12].

It has turned out that combining the projections on zero-modes with so called co-moving coordinate frame from the CEA method, one can construct a systematic perturbative expansion which seems to clarify the relation of domain walls to membranes [13]. This approach, called the improved expansion in width, avoids the problems of the CEA method. It also works in the case of vortices [14]. (Comment: due to an unfortunate mistake in transformation of field equations to the co-moving coordinates, the model actually considered in [14] is not the Abelian Higgs model, contrary to the claim made in the paper. The correct application of the improved expansion in width to the vortex in the Abelian Higgs model has been worked out in collaboration with A. L. Larsen [15] who pointed out the mistake.)

Recently it has been found that the improved expansion in width can be applied to interfaces between stable and metastable phases in condensed matter systems [16], [17]. With its help one can compute, for example, critical radius of nucleating bubbles of the stable phase. In this case pertinent evolution equations for order parameters are provided by Ginzburg-Landau effective models. Such interfaces are related to nonrelativistic membranes. They obey Allen-Cahn equations, which are nonrelativistic counterparts of Nambu-Goto equations.

RELATIVISTIC DOMAIN WALL AS A 2-BRANE

In this Section we outline the improved expansion in width. The presentation is based on the paper [18]. We explain how dynamics of a curved domain wall in a (3+1)-dimensional relativistic scalar field model can be represented by the Nambu-Goto membrane and certain (2+1)-dimensional scalar fields living on the worldsheet of the membrane. The expansion is constructed along the lines of Hilbert-Chapman-Enskog approach to singularly perturbed evolution equations [11], in which certain consistency conditions play the crucial role. Another important ingredient is a special co-moving coordinate system [2]. It is introduced in order to maintain the Lorentz invariance order by order in the expansion.

Let us consider a domain wall in the model defined by the following Lagrangian

$$\mathcal{L} = -\frac{1}{2}\eta_{\mu\nu}\partial^\mu\Phi\partial^\nu\Phi - \frac{\lambda}{2}(\Phi^2 - \frac{M^2}{4\lambda})^2, \tag{1}$$

where Φ is a single real scalar field, $(\eta_{\mu\nu})$=diag(-1,1,1,1) is the space-time metric, and λ, M are positive constants. There are two vacuum values of Φ: $\pm M/2\sqrt{\lambda}$. The mass of the scalar particle is equal to M. Instead of Φ we shall use the dimensionless field ϕ,

$$\Phi(x^\mu) = \frac{M}{2\sqrt{\lambda}}\phi(s, u^a).$$

The domain wall is a solution Φ of the corresponding Euler-Lagrange equation which smoothly interpolates between the two vacuum values. Then, at each instant of time the field Φ vanishes somewhere in the interior of the domain wall. The locus of these zeros is assumed to be a smooth connected surface $\tilde{S}$ in the space. It is called the core of the domain wall. The transverse width l_0 of such a domain wall is of the order M^{-1}, and energy density is exponentially localised around the core. The world-volume $\tilde{\Sigma}$ of the core is a 3-dimensional manifold embedded in Minkowski space-time. The core plays an essential role in the CEA method mentioned in the previous Section.

In our approach we use another smooth connected surface S attached to the domain wall. In general it differs from the core, except at the initial instant of time when it coincides with the core by assumption. We shall obtain Nambu-Goto equation for S, hence this surface can be regarded as the Nambu-Goto type relativistic membrane co-moving with the domain wall. The world-volume of S is denoted by Σ. We shall parametrise it as follows

$$\Sigma \ni (Y^\mu)(u^a) = (\tau, Y^i(u^a)).$$

We use here the notation $(u^a)_{a=0,1,2} = (\tau, \sigma^1, \sigma^2)$, where τ coincides with the laboratory frame time x^0, while (σ^1, σ^2) parametrise the co-moving membrane S at each instant of time. The index $i = 1,2,3$ refers to the spatial components of the four-vector in Minkowski space-time. The points of the co-moving membrane S at the instant τ are given by $(Y^i)(\tau, \sigma^1, \sigma^2)$. The coordinate system $(\tau, \sigma^1, \sigma^2, \xi)$ co-moving with the domain wall is defined by the formula

$$x^\mu = Y^\mu(u^a) + \xi n^\mu(u^a), \tag{2}$$

where x^μ are Cartesian, laboratory frame coordinates in Minkowski space-time, and (n^μ) is a normalised space-like four-vector orthogonal to Σ in the covariant sense,

$$n_\mu(u^a)Y^\mu_{,a}(u^a) = 0, \quad n_\mu n^\mu = 1,$$

where $Y^\mu_{,a} \equiv \partial Y^\mu/\partial u^a$. The three four-vectors $Y_{,a}$ are tangent to Σ. The definition (2) implies that ξ and u^a are Lorentz scalars. In the co-moving coordinates the co-moving membrane S is given by the simple Lorentz invariant condition $\xi = 0$. For points of S the parameter τ coincides with the laboratory time x^0, but for $\xi \neq 0$ τ is not equal to x^0 in general. Instead of the coordinate ξ we use the rescaled dimensionless coordinate s,

$$\xi = \frac{2}{M}s.$$

From now on we use the co-moving coordinates $(\tau, \sigma^1, \sigma^2, \xi)$ instead of the Cartesian ones.

In general, the core is shifted with respect to the co-moving membrane. In other words, the scalar field has non-zero values on the membrane. It is convenient to extract from the scalar field its component living on the co-moving membrane, and to treat it independently from the remaining part of the scalar field. To achieve this we use the identity

$$\phi(s,u^a) = B(u^a)\psi_0(s) + \chi(s,u^a), \tag{3}$$

where

$$B(u^a) \stackrel{df}{=} \phi(0,u^a) \tag{4}$$

is the component of the scalar field living on the co-moving membrane, and

$$\chi \stackrel{df}{=} \phi(s,u^a) - B(u^a)\psi_0(s) \tag{5}$$

is the remaining part of the field. The auxiliary, fixed function $\psi_0(s)$ depends on the variable s only. It is smooth, concentrated around $s = 0$, and $\psi_0(0) = 1$. It follows that

$$\chi(0,u^a) = 0. \tag{6}$$

It turns out that the best choice for $\psi_0(s)$ is given by formula

$$\psi_0(s) = \frac{1}{\cosh^2 s}. \tag{7}$$

Formulas (3-5) can be regarded as an invertible change of variables

$$\phi(s,u^a) \to (B(u^a), \chi(s,u^a))$$

in the configuration space of the scalar field. Therefore, it is legitimate to use formula (3) in Lagrangian (1) and to derive Euler-Lagrange equations by taking independent variations of $B(u^a)$ and χ.

The variation $\delta\chi$ has to respect the condition (6), hence

$$\delta\chi(0,u^a) = 0.$$

Because of this condition, variation of the action functional with respect to χ gives Euler-Lagrange equation in the regions $s < 0$ and $s > 0$. It has the following form

$$\frac{2}{M^2}\frac{1}{\sqrt{-g}}\partial_a[\sqrt{-g}hG^{ab}\partial_b(B\psi_0+\chi)] \tag{8}$$
$$+\frac{1}{2}\partial_s[h\partial_s(B\psi_0+\chi)]+h(B\psi_0+\chi)[1-(B\psi_0+\chi)^2]=0,$$

where h, g and G^{ab} are defined in the Appendix, and $\partial_s = \partial/\partial s, \partial_a = \partial/\partial u^a$. At $s = 0$ there is no Euler-Lagrange equation corresponding to the variation $\delta\chi$. Instead, we have the condition (6). Equation (8) should be solved in the both regions separately, with (6) regarded as a part of boundary conditions for χ. To complete the boundary conditions we also specify the behaviour of χ for $|\xi|$ much larger than the characteristic length $1/M$, that is for $|s| \gg 1$. In the model (1) the expected behaviour of the domain wall field Φ for large $|\xi|$ is given by an exponential approach to the vacuum values. Therefore, we shall seek the solution such that χ is exponentially close to $+1$ for $s \gg 1$, while for $s \ll -1$ it is exponentially close to -1.

At this stage of our considerations Eq. (8) should not be extrapolated to $s = 0$. For example, the r.h.s. of it could have a $\delta(s)$-type singularity which would occur if χ was smooth for $s > 0$ and for $s < 0$ but had a spike at $s = 0$.

In addition to Eq. (8) we also have the Euler-Lagrange equation corresponding to variations of $B(u^a)$. This equation has the following form

$$\frac{2}{M^2}\int ds\,\frac{1}{\sqrt{-g}}\partial_a\left[\sqrt{-g}hG^{ab}\partial_b(B\psi_0+\chi)\right]\psi_0 \tag{9}$$
$$-\frac{1}{2}\int ds\, h\partial_s\psi_0\partial_s(B\psi_0+\chi)$$
$$+\int ds\, h\psi_0(B\psi_0+\chi)\left[1-(B\psi_0+\chi)^2\right]=0.$$

Here and in the following we use $\int ds$ as a shorthand for the definite integral $\int_{-\infty}^{+\infty} ds$.

Comparing equations (8) and (9) one might think that they are not independent — it seems that multiplying Eq. (8) by ψ_0 and integrating the result over s (with the help of integration by parts) we obtain Eq. (9). This argument is false, namely it ignores the above mentioned possibility that the r.h.s. of Eq. (8) might have the δ-type singularity at $s = 0$. On the other hand, if the singularity at $s = 0$ is absent then indeed Eq. (9) does follow from Eq. (8). Actually, just because Eqs. (8) and (9) are independent we can prove [18] that the solution χ does not have the spike at $s = 0$ after all, and one may use Eq. (8) for all s. Then, Eq. (9) can be regarded as a generating equation for the consistency conditions described below.

Let us now solve Eq. (8) in the leading approximation which is obtained by putting $1/M = 0$. The equation is then reduced to

$$\frac{1}{2}\partial_s^2\phi^{(0)}+\phi^{(0)}[1-(\phi^{(0)})^2]=0, \tag{10}$$

where

$$\phi^{(0)} = B^{(0)}\psi_0 + \chi^{(0)}.$$

Mathematically, Eq. (10) coincides with equation for a static planar domain wall in our scalar field model. It has the following well-known solution

$$B^{(0)} = 0, \quad \chi^{(0)} = \tanh s. \tag{11}$$

The fact that the zeroth order term is given by the well-known static planar domain wall we regard as one of very attractive features of the expansion in width. Also in other models such planar solutions are rather simple, and there are many methods to obtain them.

The solution (11) in the co-moving coordinates does not determine the field ϕ in the laboratory frame because we do not know yet the position of the co-moving membrane with respect to the laboratory frame. Equations (8), (9) should also yield an equation for the co-moving membrane, otherwise they would not form the complete set of evolution equations for the domain wall. In fact, we shall see that they imply the Nambu-Goto equation for the membrane.

The expansion in the width has the form

$$\chi(s,u^a) = \tanh s + \frac{1}{M}\chi^{(1)}(s,u^a) + \frac{1}{M^2}\chi^{(2)}(s,u^a) + \ldots,$$

$$B(u^a) = \frac{1}{M}B^{(1)}(u^a) + \frac{1}{M^2}B^{(2)}(u^a) + \ldots,$$

where we have taken into account the zeroth order results (11). The expansion parameter is $1/M$ and not $1/M^2$ because $1/M$ in the first power appears in the h and G^{ab} functions after passing to the s variable, see formulas in the Appendix. In order to obey the condition (6) and to ensure the proper asymptotics of χ at large $|s|$ we assume that for $n \geq 1$

$$\chi^{(n)}(0,u^a) = 0, \quad \lim_{s\to\pm\infty} \chi^{(n)} = 0. \tag{12}$$

Expanding the r.h.s.'s of Eqs. (8), (9) in powers of $1/M$, and equating to zero coefficients in front of the powers of 1/M, we obtain a sequence of linear, inhomogeneous equations for $\chi^{(n)}(s,u^a), B^{(n)}(u^a)$ with $n \geq 1$.

The first order terms in Eq. (8) give the following equation

$$\hat{L}\chi^{(1)} = K^a_a \partial_s \chi^{(0)}, \tag{13}$$

where

$$\hat{L} \stackrel{df}{=} \frac{1}{2}\partial_s^2 + 1 - 3(\chi^{(0)})^2,$$

and $\chi^{(0)}$ is given by the second of formulas (11). We have also used the identity (14) given below. General solution of the equation of the form (13) is can easily be found with the help of standard methods [19].

The most important point in our derivation of the expansion in the width is the observation that the operator $\hat{L}$ has the normalizable zero-mode ψ_0,

$$\hat{L}\psi_0 = 0. \tag{14}$$

Notice that $\psi_0 = \partial_s \chi^{(0)}$ — this means that the zero-mode ψ_0 is related to the translational invariance of Eq. (10) under $s \to s$ + const. The presence of the zero-mode implies the consistency condition: we multiply Eq. (13) by ψ_0 and integrate over s. With the help of integration by parts we find that $\int \psi_0 \hat{L}\chi^{(1)}$ vanishes because of (14). Therefore, we obtain the following condition

$$K_a^a \int ds\, \psi_0^2(s) = 0,$$

which is equivalent to

$$K_a^a = 0. \tag{15}$$

It can be shown that this condition coincides with the well-known Nambu-Goto equation. It determines the motion of the co-moving membrane, that is the functions $Y^i(u^a)$, i=1,2,3, once initial data have been fixed. When we know these functions we can calculate the extrinsic curvature coefficients K_{ab} and the metric g_{ab}. Review of properties of relativistic Nambu-Goto membranes can be found in, e.g., [20].

Due to Nambu-Goto equation (15) the r.h.s. of Eq. (13) vanishes, and

$$\chi^{(1)} = 0,$$

in accordance with the boundary conditions (12).

Equation (8) expanded in powers of $1/M$ gives equations of the type

$$\hat{L}\chi^{(n)} = f^{(n)}, \tag{16}$$

where the source term $f^{(n)}$ is determined by the lower order terms. For example,

$$f^{(2)} = 2s\psi_0 K_b^a K_a^b + 3\chi^{(0)}\psi_0^2 (B^{(1)})^2. \tag{17}$$

Solution of Eq. (17) is given by the formula

$$\chi^{(n)}(s) = \int dx\, G(s,x) f^{(n)}(x), \tag{18}$$

where the Green's function $G(s,x)$ can be found explicitly [18]. This solution obeys the boundary conditions (12). From formula (18) we obtain the explicit dependence of $\chi^{(n)}$ on s. The integrals over x which appear on the r.h.s. of that formula are rather simple. Most of them can be evaluated analytically and the remaining ones numerically.

Equations of motion for the (2+1)-dimensional fields $B^{(n)}(u^a)$ can be obtained by expanding Eq. (9). The resulting equations coincide with the consistency conditions $\int \psi_0 f^{(n)} = 0$ for Eqs. (16). The first nontrivial equation appears in the order $1/M^3$. It has the form of non-linear, inhomogeneous (2+1)-dimensional wave equation for $B^{(1)}(u^a)$ regarded as a field defined on Σ

$$\Delta_3 B^{(1)} + K^a_b K^b_a B^{(1)} = d_0 K^a_b K^b_c K^c_a, \tag{19}$$

where

$$\Delta_3 \stackrel{df}{=} \frac{1}{\sqrt{-g}} \partial_a (\sqrt{-g} g^{ab} \partial_b)$$

is the three-dimensional d'Alembertian on the world-volume Σ of the co-moving membrane. The constant d_0 is given by

$$d_0 = 2 \int ds\, s^2 \psi_0^2 = \frac{2}{9}\pi^2 - \frac{4}{3}.$$

Before solving Eq. (19) one should first solve Nambu-Goto equation (15) for the co-moving membrane. This is necessary in order to determine g_{ab} and K_{ab}, and also in order to find the explicit form of transformation (2) which relates the co-moving coordinates to the Cartesian laboratory coordinates.

Each of Eqs. (8), (9), (15) describes a different aspect of the dynamics of the curved domain wall. Expanded in the powers of $1/M$ Eq. (8) determines dependence of χ on s, that is on the distance from the worldsheet Σ of the co-moving membrane along the perpendicular direction $n(u^a)$ at each point $Y(u^a)$ of the worldsheet. Because the term $B\psi_0$ in formula (3) has explicit dependence on s, we may say that Eq. (8) for χ fixes the transverse profile of the domain wall.

Equation (9) determines the $B^{(n)}(u^a)$ functions, which can be regarded as a (2+1)-dimensional scalar fields defined on the worldsheet Σ and having nontrivial nonlinear dynamics. The extrinsic curvature K_{ab} of Σ acts as an external source for these fields. The fields $B^{(n)}$ can propagate along Σ. One may regard this effect as causal propagation of deformations which are introduced by the extrinsic curvature.

Finally, Nambu-Goto equation (15) for the co-moving membrane determines the evolution of the shape of the domain wall.

Equations (9) and (15) are of the evolution type in the $1/M$ expansion — we have to specify initial data for them, otherwise their solutions are not unique. Equations for the perturbative contributions $\chi^{(n)}$ are of different type. In order to ensure uniqueness of their solutions it is sufficient to adopt the boundary conditions (12). The initial data for $B(u^a)$ and $Y^i(u^a)$ follow from initial data for the original field ϕ. Assuming that at the initial instant τ_0 the co-moving membrane and the core have the same position and velocity one can show that

$$B^{(n)}(\tau_0, \sigma^1, \sigma^2) = 0, \quad \partial_\tau B^{(n)}(\tau_0, \sigma^1, \sigma^2) = 0. \tag{20}$$

In order to find the domain wall solution one should first solve the collective dynamics, that is to compute evolution of the co-moving membrane and of the B field. The profile χ of the domain wall is found in the next step. In our perturbative scheme the profile of the domain wall can not be chosen arbitrarily even at the initial time — it is fixed uniquely once the initial data for the membrane and for the B field are given.

REMARKS

1. In the description of the collective dynamics of the domain wall we could use the core instead of the co-moving membrane. This might even seem a better choice because the definition of the core as the locus of zeros of the scalar field ϕ is very simple and it directly refers to the domain wall solution. Actually, it turns out that such equation is more complicated than the Nambu-Goto equation for the co-moving membrane and that it involves the $B^{(n)}$ fields.

In the literature one can find many attempts to represent the collective dynamics of the domain wall by the core only, without the $B^{(n)}$ fields. In our approach this would amount to expressing the $B^{(n)}$ fields by the extrinsic curvatures. In principle this could be done by solving Eq. (19) for $B^{(1)}$ and similar equations for the other $B^{(n)}$'s , but the resulting self-contained equation for the core would be nonlocal.

2. There is an important assumption we have tacitly made: that the derivatives $\chi^{(n)}_{,a}, B^{(n)}_{,a}$ are of the order $1/M^n$. It is not satisfied, for example, if χ and B contain modes oscillating with a frequency $\sim M$ which give positive powers of M upon differentiation with respect to u^a. If such oscillating components were present the counting of powers of $1/M$ would no longer be so straightforward as presented above. The assumption excludes radiation modes, as well as massive excitations of the domain wall. Therefore, the approximate solution we obtain gives what we may call the basic curved domain wall. To obtain more general domain wall solutions one would have to change appropriately the approximation scheme. Actually, the fact that such particular radiationless unexcited curved domain wall exists is a prediction coming from the $1/M$ expansion. The expansion yields domain walls of concrete transverse profile — the dependence on s is explicit in the approximate solution we construct even at the initial instant of time. Once we choose the initial position and velocity of points of the membrane the dependence of the scalar field on the variable s at the initial time is given by formula (18). This unique profile is characteristic for the basic curved domain wall.

3. The accuracy of the expansion in width has been checked by comparing the perturbative results with purely numerical simulation of evolution of cylindrical and spherical domain walls in the model (1) [21]. Very good agreement was found.

4. There is a limitation to the applicability of the improved expansion in width. It comes from the requirement that the zero-mode has to vanish for large $|s|$ in order to secure convergence of the integrals present in the consistency conditions. It turns out that in models with massless physical fields there exist translational zero-modes which do not obey that condition. In such cases the expansion has to be significantly modified. Work in this direction is in progress.

ACKNOWLEDGEMENTS

I would like to thank the organisers for pleasant and stimulating atmosphere during the 37th Karpacz School, and for the opportunity to present this work.

APPENDIX. THE CO-MOVING COORDINATES

The extrinsic curvature coefficients K_{ab} and induced metrics g_{ab} on Σ are defined by the following formulas:

$$K_{ab}(u) \stackrel{df}{=} n_\mu Y^\mu_{,ab}, \quad g_{ab}(u) \stackrel{df}{=} Y^\mu_{,a} Y_{\mu,b},$$

where $a, b = 0, 1, 2$. The covariant metric tensor in the co-moving coordinates has the following form

$$[G_{\alpha\beta}] = \begin{bmatrix} G_{ab} & 0 \\ 0 & 1 \end{bmatrix},$$

where $\alpha, \beta = 0, 1, 2, 3$ ($\alpha = 3$ corresponds to the ξ coordinate) and

$$G_{ab} = N_{ac} g^{cd} N_{db}, \quad N_{ac} \stackrel{df}{=} g_{ac} - \xi K_{ac}.$$

Thus, $G_{\xi\xi} = 1$, $G_{\xi a} = 0$. It follows that

$$\sqrt{-G} = \sqrt{-g}\, h,$$

where $g \stackrel{df}{=} det[g_{ab}]$, $G \stackrel{df}{=} det[G_{\alpha\beta}]$, and

$$h = 1 - \xi K^a_a + \frac{1}{2}\xi^2 (K^a_a K^b_b - K^b_a K^a_b) - \frac{1}{3}\xi^3 K^a_b K^b_c K^c_a.$$

For raising and lowering the Latin indices of the extrinsic curvature coefficients we use the induced metric tensors g^{ab}, g_{ab}.

The inverse metric tensor $G^{\alpha\beta}$ is given by the formula

$$[G^{\alpha\beta}] = \begin{bmatrix} G^{ab} & 0 \\ 0 & 1 \end{bmatrix},$$

where

$$G^{ab} = (N^{-1})^{ac} g_{cd} (N^{-1})^{db},$$

and

$$(N^{-1})^{ac} = \frac{1}{h}\Big\{ g^{ac}[1 - \xi K^b_b + \frac{1}{2}\xi^2 (K^b_b K^d_d - K^d_b K^b_d)] + \xi(1 - \xi K^b_b) K^{ac} + \xi^2 K^a_d K^{dc} \Big\}.$$

In general, the coordinates (u^a, ξ) are defined locally, in a vicinity of the world-volume Σ of the co-moving membrane. Roughly speaking, the allowed range of the ξ coordinate is determined by the smaller of the two main curvature radia of the membrane in the local rest frame. We assume that this curvature radius is sufficiently large so that on the outside of the region of validity of the co-moving coordinates there are only exponential tails of the domain wall, that is the field ϕ is exponentially close to one of the two constant vacuum solutions.

REFERENCES

1. Nielsen, H.B., and Olesen, P., *Nucl.Phys.* **B61**, 45 (1973).
2. Förster, D., *Nucl.Phys.* **B81**, 84 (1974).
3. Gervais, J.-L., and Sakita, B., *Nucl.Phys.* **B91**, 301 (1975).
4. Kibble, T.W.B., *J.Phys.* **A9**, 1387 (1976).
5. Maeda, K., and Turok, N., *Phys.Lett.* **B202**, 376 (1988).
6. Gregory, R., *Phys.Lett.* **B206**, 199 (1988).
7. Barr, S.M., and Hochberg, D., *Phys.Rev.* **D39**, 2308 (1989).
8. Silveira, V., *Phys. Rev.* **D41**, 1914 (1990).
9. Anderson, M., Bonjour, F., Gregory, R., and Stewart, J., *Phys. Rev.* **D56**, 8014 (1997).
10. Gregory, R., Haws, D., and Garfinkle, D., *Phys. Rev.* **D42**, 343 (1990).
11. van Kampen, N.G., *Stochastic Processes in Physics and Chemistry*, North-Holland Publ.Comp., Amsterdam, 1987. Chapt.8,§7.
12. Dorsey, A.T., *Ann. Phys.* **233**, 248 (1994).
13. Arodź, H., *Nucl. Phys.* **B450**, 174 (1995).
14. Arodź, H., *Nucl.Phys.* **B450**, 189 (1995).
15. H. Arodź and A. L. Larsen, in preparation.
16. Arodź, H., *Acta Phys. Pol.* **B29**, 3725 (1998).
17. Arodź, H., and Pełka, R., *Phys. Rev.* **E62**, 6749 (2000).
18. Arodź, H., *Nucl. Phys.* **B509**, 273 (1998).
19. Korn, G.A., and Korn, T.M., *Mathematical Handbook*, IInd Edition. McGraw-Hill Book Comp., New York, 1968. Chapt.9.3-3.
20. Carter, B., in *"Formation and Interactions of Topological Defects"*, p.303. Eds.: A.-Ch. Davis and R. Brandenberger, Plenum Press, New York and London, 1995.
21. Karkowski, J., and Świerczyński, Z., *Acta Phys. Pol.* **B30**, 234 (1996).

Localised Electron States and a Modified Nonlinear Schrödinger Equation

L. Brizhik*, B. Piette† and W.J. Zakrzewski†

*Bogolyubov Institute for Theoretical Physics, 252143 Kyiv, Ukraine
email:brizhik@nonlin.bitp.kiev.ua
†Department of Mathematical Sciences, University of Durham, Durham DH1 3LE, UK emails:
B.M.A.G.Piette@durham.ac.uk, W.J.Zakrzewski@durham.ac.uk

Abstract. We present and discus the results of [1], [2] and [3]. There we have shown that, for a suitable range of parameters, the two dimensional discrete equations describing a quasiparticle interacting with the displacements of a lattice of atoms possess solitonic solutions.

INTRODUCTION

There are many organic and inorganic substances which exhibit localised modes of excitations which resemble solitons [4]-[9]. These substances posses anisotropic properties and are usually classified as low-dimensional systems (LDS), including one- and two-dimensional. Many of them have wide applications in opto- and nano-electronics, others are natural biological macromolecules or synthesised biopolymers. Therefore, there is an ever increasing interest in studying their properties.

It is well established that the electron-phonon interactions manifest themselves profoundly in the case of the reduced dimensionality and can lead to some very particular phenomena. One of them is a quasiparticle self-trapping which is also referred to as a spontaneous or auto-localisation of a quasiparticle. Thus a comparative study of systems with similar chemical composition but belonging to classes of different dimensionalities would be of great interest. Here one can mention [9] where it was shown that the life-time of Amide-I excitation in myoglobin is much higher than the life-time of the excitations of the photoactive yellow protein.

We would like to add here that the idea of the self-trapping of Amide-I excitations in a soliton state in one-dimensional proteins was first suggested by Davydov[10, 8]. It is known that in 1DS a quasi-particle (electron, hole, exciton, etc) self-trapping takes place only for a particular range of the parameters characterising the system [11]. Qualitatively a similar, although more complex, situation takes place for 2DS. Thus, it has been shown in [1, 2] that the self-trapped solitonic states of a quasiparticle exist and are stable in isotropic and anisotropic 2D crystals within some intervals of numerical values of parameters.

From the general point of view, these low-dimensional substances are characterised by the existence of a regular, often anisotropic, lattice of atoms which may vibrate around their positions of equilibrium. The electrical or optical properties of these substances are governed by the collective excitations of electron states which are coupled to the vibra-

CP589, *Fundamental Interaction Theories*, edited by J. Lukierski and J. Rembieliński

tions of the lattice through the effective electron-phonon coupling constants. Depending on the strength of these coupling constants the ground electron states of a quasiparticle can be classified as almost free quasiparticle, small polaron, or soliton, which possess qualitatively different properties and correspond to different approximations.

When we look at one-dimensional systems, such as polypetides etc, we find that they possess acoustic and optical phonons and admit the existence of Davydov or molecular solitons, provided adiabatic approximation conditions are fulfilled. In the continuum limit the collective excitations of such systems are described by a one-dimensional nonlinear Schrödinger equation with an attractive interaction. Such systems have attracted many investigations. So far, however, relatively little study has been performed of higher dimensional systems where, of course, the spectrum of possibilities is much richer as the systems can be anisotropic.

Here we report some results of our attempts to perform such investigations (of two-dimensional systems) [1, 2, 3].

So we look at systems which are described by the Fröhlich Hamiltonian which is a sum of Hamiltonians describing electron, phonons and electron-phonon interactions:

$$\hat{H} = \hat{H}_e + \hat{H}_{ph} + \hat{H}_{int} \tag{1}$$

i.e., in the site representation,

$$\begin{aligned}\hat{H}_e = & \sum_{m,n}[\mathcal{E}_0 A^+_{m,n}A_{m,n} - j_x(A^+_{m,n}A_{m+1,n} + A^+_{m+1,n}A_{m,n}) \\ & - j_y(A^+_{m,n}A_{m,n+1} + A^+_{m,n+1}A_{m,n})],\end{aligned} \tag{2}$$

$$\begin{aligned}\hat{H}_{ph} = & \frac{1}{2}\sum_{m,n}\Big(\frac{\hat{p}^2_{m,n}}{M} + \frac{\hat{q}^2_{m,n}}{M} + k_x[(\hat{u}_{m,n} - \hat{u}_{m+1,n})^2 + (\hat{v}_{m,n} - \hat{v}_{m+1,n})^2] \\ & + k_y[(\hat{u}_{m,n} - \hat{u}_{m,n+1})^2 + (\hat{v}_{m,n} - \hat{v}_{m,n+1})^2]\Big),\end{aligned} \tag{3}$$

$$\hat{H}_{int} = \sum_{m,n} A^+_{m,n}A_{m,n}[\chi_x(\hat{u}_{m+1,n} - \hat{u}_{m-1,n}) + \chi_y(\hat{v}_{m,n+1} - \hat{v}_{m,n-1})]. \tag{4}$$

Here $A^+_{m,n}$ ($A_{m,n}$) are the creation (annihilation) operators of the electron on the site (m,n), $\hat{u}_{m,n}$, $\hat{v}_{m,n}$ and $\hat{p}_{m,n}$, $\hat{q}_{m,n}$ are the longitudinal and transverse components of the vector operator of molecule displacements and their respective conjugated momenta. The energy $\mathcal{E}_0 - 2j_x - 2j_y$ corresponds to the bottom of the electron energy band, while j_x, j_y are the exchange interaction energies, and χ_x, χ_y are the electron-phonon coupling constants in the x and y directions, respectively. Finally, k_x, k_y are the corresponding elasticity coefficients.

Next we follow the standard semi-classical analysis and derive an effective classical Hamiltonian H, with $\varphi_{m,n}$ - the probability amplitude for the electron, and $u_{m,n}$, $v_{m,n}$ - the variables describing molecular displacements, from their positions of equilibrium, in the x and y directions.

This Hamiltonian is given by

$$H = \sum_{m,n}((\mathcal{E}_I + \mathcal{W})\,|\varphi_{m,n}|^2 - j_x\varphi^*_{m,n}(\varphi_{m+1,n}+\varphi_{m-1,n}) - j_y\varphi^*_{m,n}(\varphi_{m,n+1}+\varphi_{m,n-1}) + |\varphi_{m,n}|^2\left[b\chi_x(u_{m+1,n} - u_{m-1,n}) + a\chi_y(v_{m,n+1} - v_{m,n-1})\right], \quad (5)$$

where W describes the phonon energy and is given by

$$\mathcal{W} = \frac{1}{2}\sum_{m,n}\Big(\frac{p^2_{m,n}}{M} + \frac{q^2_{m,n}}{M} + k_x\left[(u_{m,n}-u_{m+1,n})^2 + (v_{m,n}-v_{m+1,n})^2\right] + k_y\left[(u_{m,n}-u_{m,n+1})^2 + (v_{m,n}-v_{m,n+1})^2\right]\Big). \quad (6)$$

Furthermore, we have a constraint: the electron wave function must satisfy the normalisation condition

$$\sum_{m,n}|\varphi_{m,n}|^2 = 1. \quad (7)$$

Let us note that this normalisation condition is less important in one dimension for one extra electron. However, the condition becomes essential in the many-electron problems in 1-dimensional systems and it leads to nontrivial effects in the 2-dimensional lattice cases even for just one extra electron.

To proceed further it is convenient to introduce dimensionless units:

$$\tau = \frac{j_x t}{\hbar}, \quad U = C_x\frac{u}{b}, \quad V = C_x\frac{v}{b}, \quad E_s = \frac{b^2 M j_x}{\hbar^2}, \quad E_0 = \frac{\mathcal{E}_I}{j_x},$$
$$C_x = \frac{\chi_x b^2}{j_x}, \quad C_y = \frac{\chi_y a b}{j_x}, \quad K_x = \frac{k_x\hbar^2}{M j_x^2}, \quad K_y = \frac{k_y\hbar^2}{M j_x^2}, \quad g = \frac{2C_x^2}{K_x E_s} \quad (8)$$

and to introduce the anisotropy parameters

$$A_j = \frac{j_y}{j_x}, \qquad A_c = \frac{C_y}{C_x}, \qquad A_k = \frac{K_y}{K_x}. \quad (9)$$

The Euler-Lagrange equations describing our system are then given by:

$$i\frac{d\varphi_{m,n}}{d\tau} = (E_0 + W)\,\varphi_{m,n} - (\varphi_{m+1,n} + \varphi_{m-1,n}) - A_j\,(\varphi_{m,n+1} + \varphi_{m,n-1}) + \left[(U_{m+1,n} - U_{m-1,n}) + A_c\,(V_{m,n+1} - V_{m,n-1})\right]\varphi_{m,n}, \quad (10)$$

$$\frac{d^2U_{m,n}}{d\tau^2} = -\frac{K_x}{C_x}\Big[(2U_{m,n} - U_{m+1,n} - U_{m-1,n}) + A_k\,(2U_{m,n} - U_{m,n+1} - U_{m,n-1}) - \frac{g}{2}\left(|\varphi_{m+1,n}|^2 - |\varphi_{m-1,n}|^2\right)\Big], \quad (11)$$

$$\frac{d^2V_{m,n}}{d\tau^2} = -\frac{K_x}{C_x}\Big[(2V_{m,n} - V_{m+1,n} - V_{m-1,n}) + A_k\,(2V_{m,n} - V_{m,n+1} - V_{m,n-1}) - \frac{gA_c}{2}\left(|\varphi_{m,n+1}|^2 - |\varphi_{m,n-1}|^2\right)\Big], \quad (12)$$

while the phonon energy is given by:

$$W = \frac{1}{2}E_s \sum_{m,n}\Big(P_{m,n}^2 + Q_{m,n}^2 + \frac{K_x}{C_x}\,([(U_{m,n}-U_{m+1,n})^2 + (V_{m,n}-V_{m+1,n})^2]$$
$$+A_k[(U_{m,n}-U_{m,n+1})^2 + (V_{m,n}-V_{m,n+1})^2])\Big), \tag{13}$$

where

$$P_{m,n} = \frac{dU_{m,n}}{d\tau}, \quad Q_{m,n} = \frac{dV_{m,n}}{d\tau}. \tag{14}$$

STATIONARY CASE

If we restrict ourselves to stationary fields the equations (10-12) reduces to

$$\lambda\varphi_{m,n} + (2\varphi_{m,n} - \varphi_{m+1,n} - \varphi_{m-1,n}) + A_j\,(2\varphi_{m,n} - \varphi_{m,n+1} - \varphi_{m,n-1})$$
$$+\left[(U_{m+1,n} - U_{m-1,n}) + A_c\,(V_{m,n+1} - V_{m,n-1})\right]\varphi_{m,n} = 0, \tag{15}$$
$$(2U_{m,n} - U_{m+1,n} - U_{m-1,n}) + A_k\,(2U_{m,n} - U_{m,n+1} - U_{m,n-1}) =$$
$$\frac{g}{2}\,(|\varphi_{m+1,n}|^2 - |\varphi_{m-1,n}|^2), \tag{16}$$
$$(2V_{m,n} - V_{m+1,n} - V_{m-1,n}) + A_k\,(2V_{m,n} - V_{m,n+1} - V_{m,n-1}) =$$
$$\frac{gA_c}{2}\,(|\varphi_{m,n+1}|^2 - |\varphi_{m,n-1}|^2), \tag{17}$$

where $\lambda = E_0 + W - 2(1+A_j)$ and where A_k, A_c, A_j are the anisotropy parameters. We have analysed these equations, numerically, for various choices of parameters. The results are presented in [1] and [2]. The phase diagrams given in [1] clearly exhibit the role of the anisotropy. In Figure 1 we exhibit the range of parameters for which we have a solution of the equation as a function of g and of $A_j = A_k = A_c$. The solitons exists in the middle region marked "S" In Figure 1. The region "D" corresponds to the "almost free" (delocalised) electrons while the region "L" is the region of extremely localised solitons - "small polarons" (essentially localised on one lattice point).

In Figure 2 we present the plots of $|\varphi|$, and U for the static solution of equations (15-17) corresponding to $g = 6.8$.

In what follows we restrict our attention to the isotropic case.

ISOTROPIC CASE

The isotropic system corresponds to the choice $A_j = A_k = A_c = 1$. We now define

$$\Delta(1)U_{m,n} = 4U_{m,n} - U_{m+1,n} - U_{m-1,n} - U_{m,n+1} - U_{m,n-1}, \tag{18}$$

and

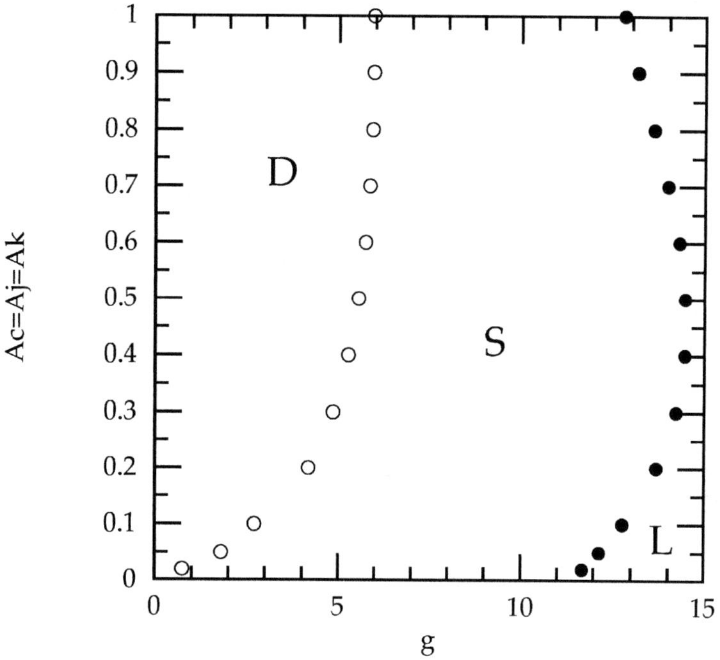

FIGURE 1. A typical phase diagram for the case $A_j = A_k = A_c$

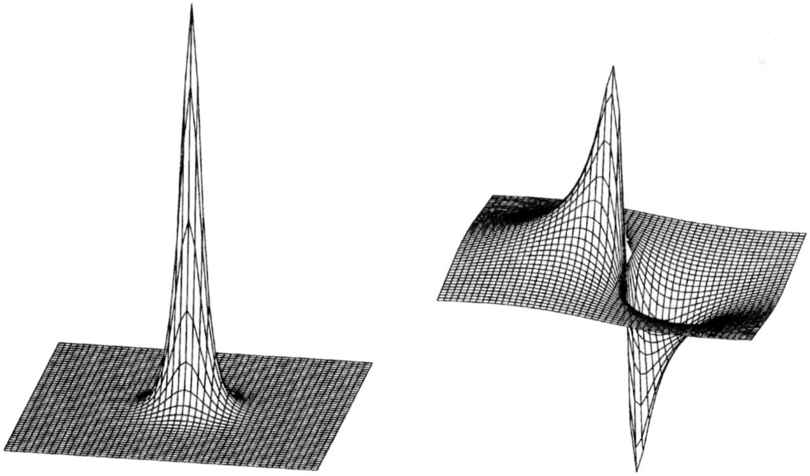

FIGURE 2. Soliton for $g = 6.8$. (a) $|\varphi|$ and (b) U.

$$\Delta(2)|\varphi_{m,n}|^2 = 4|\varphi_{m,n}|^2 - |\varphi_{m+2,n}|^2 - |\varphi_{m-2,n}|^2 - |\varphi_{m,n+2}|^2 - |\varphi_{m,n-2}|^2. \tag{19}$$

Noting that, as $A_k = 1$, (16) and (17) can be written as

$$\Delta(1)(U_{m+1,n} - U_{m-1,n}) = \frac{g}{2}(|\varphi_{m+2,n}|^2 - |\varphi_{m,n}|^2 - |\varphi_{m,n}|^2 + |\varphi_{m-2,n}|^2), \tag{20}$$

$$\Delta(1)(V_{m,n+1} - V_{m,n-1}) = \frac{g}{2}(|\varphi_{m,n+2}|^2 - |\varphi_{m,n}|^2 - |\varphi_{m,n}|^2 + |\varphi_{m,n-2}|^2), \tag{21}$$

we have

$$\Delta(1)\left((U_{m+1,n} - U_{m-1,n}) + (V_{m,n+1} - V_{m,n-1})\right) = -\frac{g}{2}\Delta(2)|\varphi_{m,n}|^2. \tag{22}$$

Thus we have obtained a very **interesting equation**

$$\Delta(1)Z_{m,n} = -\frac{g}{2}\Delta(2)|\varphi_{m,n}|^2, \tag{23}$$

where

$$Z_{m,n} = (U_{m+1,n} - U_{m-1,n}) + (V_{m,n+1} - V_{m,n-1}). \tag{24}$$

In this expression $\Delta(1)$ describes a 5 point Laplacian and $\Delta(2)$ is the same operator but, effectively, missing out the nearest points of the lattice (*ie* an operator on a lattice twice as large). Unfortunately, this discrete equation cannot be solved exactly except in 1 dim - (*ie* for a 3 point Laplacian). In this case

$$P_{i+1} + P_{i-1} - 2P_i = \kappa(2R_i - R_{i+2} - R_{i-2}). \tag{25}$$

is solved by

$$P_n = -\kappa(R_{n+1} + R_{n-1} + 2R_n) \tag{26}$$

where the boundary terms have been neglected. Equation (23) will nevertheless turn out to be useful later.

CONTINUUM LIMIT

To look at the continuum limit of equations (15-17) we define $\varphi(x,y) = \varphi_{m,n}$ as functions of the continuous variables x and y instead of the discrete variables m and n. Then using the Taylor expansion

$$\varphi_{m\pm1,n} = \varphi_{m,n} \pm \frac{\partial\varphi(x,y)}{\partial x} + \frac{1}{2}\frac{\partial^2\varphi(x,y)}{\partial x^2} \pm \frac{1}{6}\frac{\partial^3\varphi(x,y)}{\partial x^3} + \frac{1}{24}\frac{\partial^4\varphi(x,y)}{\partial x^4} + \ldots \tag{27}$$

$$U_{m\pm1,n} = U_{m,n} \pm \frac{\partial U}{\partial x} + \frac{1}{2}\frac{\partial^2 U}{\partial x^2} \pm \frac{1}{6}\frac{\partial^3 U}{\partial x^3} + \frac{1}{24}\frac{\partial^4 U}{\partial x^4} + \ldots \tag{28}$$

$$V_{m,n\pm1} = V_{m,n} \pm \frac{\partial V}{\partial y} + \frac{1}{2}\frac{\partial^2 V}{\partial x^2} \pm \frac{1}{6}\frac{\partial^3 V}{\partial x^3} + \frac{1}{24}\frac{\partial^4 V}{\partial x^4} + \ldots \tag{29}$$

we can write

$$\frac{\partial^2 U}{\partial x^2} = -g\left(\frac{\partial|\varphi|^2}{\partial x} + \frac{1}{6}\frac{\partial^3}{\partial x^3}|\varphi|^2\right), \tag{30}$$

$$\frac{\partial^2 V}{\partial x^2} = -g\left(\frac{\partial|\varphi|^2}{\partial y} + \frac{1}{6}\frac{\partial^3}{\partial y^3}|\varphi|^2\right), \tag{31}$$

where we have neglected $\frac{\partial^2 U}{\partial y^2}$ and $\frac{\partial^2 V}{\partial x^2}$. Integrating (30) and (31) we get

$$\frac{\partial U}{\partial x} = -g\left(|\varphi|^2 + \frac{1}{6}\frac{\partial^2}{\partial x^2}|\varphi|^2\right), \tag{32}$$

$$\frac{\partial V}{\partial y} = -g\left(|\varphi|^2 + \frac{1}{6}\frac{\partial^2}{\partial y^2}|\varphi|^2\right), \tag{33}$$

and so

$$i\frac{d\varphi}{d\tau} + \Delta\varphi + 2g\left(|\varphi|^2 + \frac{1}{12}\Delta|\varphi|^2\right)\varphi = 0, \tag{34}$$

ie a **Nonlinear Schrödinger Equation** but with an extra term.

This equation, without the extra term, has been studied before. It is well known that, in one dimension, it possesses a soliton solution for an arbitrary value of g. However, in two dimensions, the soliton solution is stable only if the extra term is present.

EXISTENCE OF SOLITONS

Here we present some arguments from [3] which allows us to "understand" our numerical results. We shall consider two limits of our equations, when we can perform some approximations; namely, we shall consider

- the broad solitons
- the narrow solitons.

Let us look first at **broad** solitons. Now we expect the continuum limit approximation to be valid and so we base our discussion on our Nonlinear Schrödinger Equation (34). This equation, clearly, possesses a conserved energy which is given by:

$$\mathcal{E} = \int \left[|\vec{\nabla}\varphi|^2 - g|\varphi|^4 + \frac{g}{12}\left(\Delta|\varphi|^2\right)^2\right] dxdy. \tag{35}$$

Clearly, had we neglected the last term and so used only the Schrödinger equation

$$i\frac{\partial\varphi}{\partial\tau} + \Delta\varphi + 2g|\varphi|^2\varphi = 0 \tag{36}$$

its solutions would have been unstable but, here we have an extra term which comes from the lattice and which stabilises the solitons. To see this consider first the square of the size of any localised, soliton-like, configuration

$$R^2 = \int |\varphi(x,y)|^2 (x^2+y^2) dxdy. \tag{37}$$

Differentiating with respect to τ and using (34) we get:

$$\frac{dR^2}{d\tau} = -\int (x^2+y^2)(\varphi\Delta\varphi^* - \varphi^*\Delta\varphi) dxdy. \tag{38}$$

and so

$$\frac{d^2R^2}{d\tau^2} = 8(\mathcal{E}+\delta), \tag{39}$$

where

$$\delta = \frac{g}{12}\int \left(\Delta|\varphi|^2\right)^2 dxdy. \tag{40}$$

Note that as $\mathcal{E} < 0$ and $\delta > 0$, the interplay between these two terms produces a behaviour which alternates between shrinking and expanding and leads thus to the stabilisation of the soliton.

Let us see how this works on the following example (which, in fact, is not a bad approximation for a soliton). Inserting the Gaussian function

$$\varphi(x,y) = \frac{\kappa}{\sqrt{\pi}} exp\left(-\frac{\kappa^2}{2}(x^2+y^2)\right). \tag{41}$$

into (35) we get

$$\mathcal{E} = \kappa^2(1-\frac{g}{2\pi}) + g\frac{\kappa^4}{12\pi}, \tag{42}$$

and so we see that the value of κ minimising $\mathcal{E}$ is given by

$$\kappa_0^2 = 3(1-\frac{2\pi}{g}). \tag{43}$$

Looking at this result we note the existence of a critical value $g_c = 2\pi$ below which there is no stable solution. After solving (15-17) numerically we have found $g_{cr} \approx 5.85$.

To prove the stability of the soliton we consider a small deviation from the static configuration, *ie* we take

$$R^2 = 1/\kappa_0^2 + f(\tau), \tag{44}$$

where f is small. Then

$$\frac{d^2 f}{d\tau^2}(\tau) = -\omega^2 f(\tau), \tag{45}$$

where

$$\omega^2 = \frac{6}{\sqrt{\pi}}(g-2\pi)^{3/2}. \tag{46}$$

We thus have

$$R^2 = 1/\kappa_0^2 + \varepsilon\cos(\omega\tau) \tag{47}$$

and we see that the soliton oscillates in size with an amplitude of oscillation ε which is determined by the initial condition. More results can be found in [3].

Next we look at the **narrow** solitons. Based on the experience gained by performing numerical simulations we assume that the solitons are restricted to a very few lattice points. This is meant to be true not only for the φ field but also for U and V fields (although they are less localised than the field φ).

To do this we look at the eigenvalue equation for φ

$$\lambda\varphi_{m,n} = -(\varphi_{m+1,n} + \varphi_{m-1,n} + \varphi_{m,n+1} + \varphi_{m,n-1} - 4\varphi_{m,n}) + Z_{m,n}\varphi_{m,n}, \tag{48}$$

where $Z_{m,n}$ defined by (24) satisfies (23).

We then attempt to solve (48) iteratively. First we assume that $|\varphi_{0,0}|^2 = 1$ and all other $|\varphi_{m,n}|^2$ zero. We find

$$Z_{0,0} = -\frac{8}{11}g, \qquad Z_{1,0} = -\frac{5}{22}g, \qquad Z_{1,1} = -\frac{4}{33}g, \qquad Z_{2,0} = \frac{2}{33}g, \qquad Z_{2,1} = -\frac{1}{66}g. \tag{49}$$

Then we use this result to determine a "new-modified" φ. So we assume

$$\begin{aligned} \varphi_{0,0} &= 1 - \frac{F}{g^2}, \\ \varphi_{m,n} &= f_{m,n}\, g^{-(|m|+|n|)}, \end{aligned} \tag{50}$$

and find

$$\lambda = 4 - \tfrac{8}{11}g, \qquad f_{1,0} = 2, \qquad f_{1,0} = \tfrac{33}{5}, \qquad f_{2,0} = \frac{33}{13}, \tag{51}$$

which is, in fact, in good agreement with the solutions determined numerically! In Figure 3 we show $F = g^2(1-\varphi_{0,0})$, $f_{1,0} = g\varphi_{1,0}$ and the energy $E = (4-\lambda)/g$ determined numerically. We see that our result are in fact, valid for a large range of g, and not only in the limit $g \to \infty$.

MOVING SOLITONS

We have also studied moving solitons. To do this we introduced periodic boundary conditions and boosted the field φ by multiplying it by the phase factor $\exp(ikx)$. Our studies have shown that such solutions exist and that the U and V fields get "dragged" by the φ field. Of course there is a relation between k and v, the velocity of the soliton. Moreover, the lattice coarseness effects lead to the existence of a critical velocity (which depends on g) below which the soliton is trapped close to a lattice site. This subject is currently under active study and we expect to have more to say on it soon.

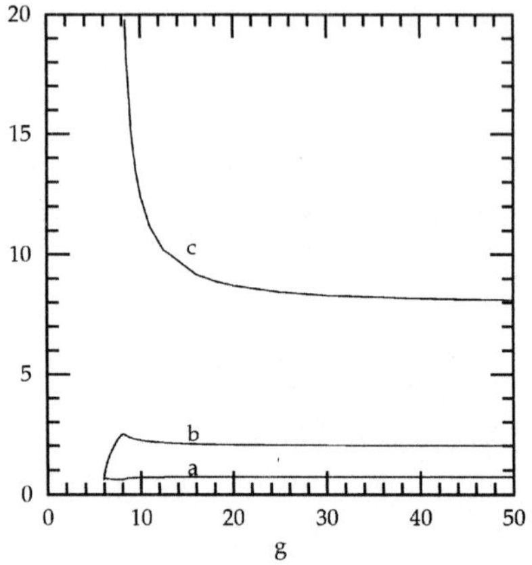

FIGURE 3. Numerically determined g dependence of (a) the energy $(4-\lambda)/g$ (b) $f_{1,0} = g\varphi_{1,0}$ and (c) $F = g^2(1-\varphi_{0,0})$,

CONCLUSIONS AND GENERAL COMMENTS

We have given an overview of our recent work [1, 2, 3]. We have shown that our discrete model, for a range of parameters, possesses solutions which are solitonic in nature. The model, in its continuum limit, becomes a **modified** nonlinear Schrödinger model - so our work establishes the existence of solitonic solutions for this model. We have also mentioned the role played by the anisotropies.

We have also proved the existence of moving solitons in this model and for slow velocities found that the solitons can get trapped on lattice sites. Other, more general, configurations were found to spread out. Now we are looking at some interesting anisotropic cases, concentrating our attention on the investigation of various properties of moving solitons. We are also looking at possible applications of our results.

ACKNOWLEDGMENTS

WJZ wants to thank J. Lukierski and Z. Popowicz for their invitation to give this lecture at the XXXVII Karpacz Winter School and for their hospitality.

REFERENCES

1. Brizhik, L., Piette, B.M.A.G., and Zakrzewski, W.J., *Physica*, **D 146**, 275-288 (2000).
2. Brizhik, L., Piette, B.M.A.G., and Zakrzewski, W.J., *Ukr. Fiz. Journal*, **46**, 503-511 (2001).
3. Brizhik, L., Eremko, A., Piette, B.M.A.G. and Zakrzewski, W.J., in preparation.
4. Nagamatsu, J., Nakagawa, N., Muranaka, T., Zenitani, Y., and Akimitsu, J., *Nature*, **410**, 63-64 (2001).

5. Schön, J.H., Dodapalapur, A., Bao, Z., Kloc, Ch., Schenker, O., and Battlogg, B., *Nature*, **410**, 189-192 (2001).
6. Devreese, J.T., Evrard. R.P., eds., *Highly conducting one-dimensional solids* (New York: Plenum Press) 1979.
7. Monceau, P., ed., *Electronic properties of inorganic quasi-one-dimensional compounds*. 2 parts. (Dordrecht, Boston, Lancaster: Reidel) 1985.
8. Davydov, A.S., *Solitons in Molecular Systems* (Dordrecht: Reidel) 1985.
9. Xie, A., van der Meer, L., Hoff, W., and Austin, R.H. *Phys. Rev. Lett.*, **84**, 5435-5438 (2000).
10. Davydov, A.S., and Kislukha, N.I., *Phys. Stat. Sol.*, **59**, 465-470 (1973).
11. Brizhik, L.S., and Eremko, A.A., *Z. Phys.*, **B104**, 771-775 (1997).

The Fedosov Class of the Wick Type Star-Product

V. A. Dolgushev [1]

Belinsky 86-28, 634034 Tomsk, Russia, email: vald@thsun1.jinr.ru

Abstract. The closed two-form is presented whose cohomological class coincides with the Fedosov class of the Wick-type star-product. For the case of the Kähler manifold this class is shown to be related to the first Chern class. In particular, we prove that the Wick type star-product on a Kähler manifold is equivalent to that of the Weyl type iff the manifold is the Calabi-Yau one.

INTRODUCTION

This talk is based on the paper [1], which has been done in collaboration with S. L. Lyakhovich and A. A. Sharapov.

The classification of the star-products on a symplectic manifold up to the equivalence relation has been carried out in several works [2, 3, 4]. It has been shown that each equivalence class of the star-products on a symplectic manifold corresponds to formal series in the deformation parameter taking values in second De Rham cohomologies.

This remarkable result seems to be most transparent in the context of the Fedosov method [5], where the space of series in second De Rham cohomologies naturally appears as a moduli space of the flat Fedosov connections of the auxiliary symbol bundle. Each equivalence class of the flat connections is associated to some set of the equivalent Fedosov star-products, and the star-products corresponding to different equivalence classes of the connections are easily shown to be inequivalent. A nontrivial result obtained by P. Xu [6] that allows to include his name in the list of people, who have classified the star-products is that each star-product on a symplectic manifold turns out to be equivalent to some Fedosov star-products. Hence, the moduli space of the flat Fedosov connections is identified with the set of equivalence classes of all the star-products on a symplectic manifold.

The aim of this note is to apply this consideration to the case of the Fedosov star-product of Wick type, obtained in our recent paper [1]. We calculate the Fedosov class of the star-product and show that for the case of the Kähler manifold the crucial order zero term of the class is proportional to the respective first Chern class, and in particular, it means that the Wick type star-product on a Kähler manifold is equivalent to that of the Weyl type iff the manifold is the Calabi-Yau one.

In the first section of the paper we recall the main steps of the original Fedosov

---

[1] This work is supported by the RFBR grant 00-02-17-956, Grant for Support of Young Scientists 01-02-06418, Grant for Support of Scientific Schools 00-15-96557, and INTAS grant 00-262.

CP589, *Fundamental Interaction Theories*, edited by J. Lukierski and J. Rembieliński

construction and present the modification of the construction that yields the star-product of the Wick type. In the second section we show how the moduli space of the flat Fedosov connections is parameterized by the series in second De Rham cohomologies and how it is identified with the set of equivalence classes of the Fedosov star-products. As an application of the consideration we calculate the Fedosov class of the Wick type star-product. The concluding section is devoted to some open questions.

FEDOSOV STAR-PRODUCT OF THE WICK TYPE

Suppose we are given a real manifold M, $dim M = 2n$, equipped with a real symplectic structure $\omega = \omega_{ij}(x)dx^i \wedge dx^j$. Then we say that the star-product $*$, deforming the pointwise multiplication in the algebra $C^\infty(M)$ is of *the Wick type* if it is a generally invariant star-product, satisfying the following "boundary condition"

$$a * b = ab + \frac{i\hbar}{2}\Lambda^{ij}\partial_i a \partial_j b + \dots . \tag{1}$$

Here $\hbar$ is the formal deformation parameter, dots mean the terms of higher orders in $\hbar$, and

$$\Lambda^{ij}(x) = \omega^{ij}(x) + g^{ij}(x), \tag{2}$$

where ω^{ij} is the Poisson tensor associated to $\omega_{ij}(x)$, $\omega^{ik}(x)\omega_{kj}(x) = \delta^i_j$ and g^{ij} is a complex-valued symmetrical tensor. In what follows we require that the matrix of the tensor $\Lambda^{ij}(x)$ is of the half rank at each point of the manifold

$$rank||\Lambda^{ij}(x)|| = n. \tag{3}$$

The above definition is motivated by the simple observation that the first terms in the expansion of the canonical Wick-type star-product takes the form (1) with the constant matrix Λ^{ij}, which is, in turn, of the half rank. In fact this definition generalizes along with the canonical Wick type star-product the so-called qp-type star-product, which satisfies the condition (1) with a real constant matrix Λ^{ij}.

There are two exceptional cases of the geometrical structure $\Lambda^{ij}(x)$ that are worthy of notice. The first case is that the symmetrical tensor has imaginary components in local real coordinates. In this case the symplectic manifold turns out to be an almost Kähler one [7]. The second case in which the symmetrical tenor is real corresponds to the symplectic manifold, equipped with the so-called para-Kähler structure [8].

In the paper [1] it has been shown that the star-product of the above type can be constructed in the frame of the Fedosov approach if there exists a torsion free Levi-Civita connection preserving the tensor Λ^{ij} (2), that is preserving both the symplectic structure and the symmetrical tensor. This condition is satisfied for a wide range of symplectic manifolds (see [1] for more detail). For example, all the Kähler manifolds as well as the para-Kähler ones [8] are suitable for our consideration. It is also worth noting that while the star-product on a Kähler manifold may be regarded as that of the genuine Wick type, the respective star-product on a para-Kähler manifold furnishes a generalizations of the canonical qp-type star-product to the case of the curved manifold.

Let us firstly recall the main steps of the original Fedosov method, which allows to construct a generally invariant star-product on an arbitrary symplectic manifold. The only ingredient that is required for this construction is a torsion free Levi-Civita connection, preserving the symplectic structure. This connection is shown to exist for an arbitrary symplectic manifold.

Following the Fedosov method [5], we start with *the symbol bundle* $\mathcal{A}$, whose sections are the following formal series

$$
\begin{aligned}
a(x,y,dx,\hbar) \quad &= \sum_{2k+p,q\geq 0} \hbar^k a_{k,i_1\ldots i_p j_1\ldots j_q}(x) \\
&\quad y^{i_1}\ldots y^{i_p} dx^{j_1}\wedge\ldots\wedge dx^{j_q}\,.
\end{aligned} \tag{4}
$$

Here the expansion coefficients $a_{k,i_1\ldots i_p j_1\ldots j_q}(x)$ are the complex-valued components of covariant tensor fields on M symmetric with respect to $i_1,\ldots,i_p$ and antisymmetric in $j_1,\ldots,j_q$, and $\{y^i\}$ are variables transforming as components of tangent vector.

We can endow the set of sections $\Gamma\mathcal{A}$ with a structure of an associative algebra over $\mathbf{C}$ with a unit if we define the product of two sections $a,b\in\Gamma\mathcal{A}$ as

$$
\begin{aligned}
a\circ b \quad &= \quad exp\left(\frac{i\hbar}{2}\omega^{ij}(x)\frac{\partial}{\partial y^i}\frac{\partial}{\partial z^j}\right) a(x,y,dx,\hbar)\wedge \\
&\quad \wedge\, b(x,z,dx,\hbar)|_{z=y}\ .
\end{aligned} \tag{5}
$$

In what follows it will cause no confusion if we refer to $\Gamma\mathcal{A}$ as $\mathcal{A}$ and call the sections of the bundle as the elements of the algebra $\mathcal{A}$.

The product (5) is obviously graded with respect to the standard exterior degree q. Moreover, it is easy to check that it is also graded with respect to the sum $2k+p$, where k is the order in $\hbar$ and p is the order in y.

Given the associative algebra $\mathcal{A}$, we can easily turn it into the differential graded Lie algebra if we define the commutator together with the nilpotent differential as

$$[a,b] = a\circ b - (-1)^{q_a q_b} b\circ a\,, \tag{6}$$

$$\delta a = dx^k\wedge\frac{\partial a}{\partial y^k}\,,\quad \delta(\delta a) = 0\,,\quad \forall a\in\mathcal{A}\,, \tag{7}$$

where a and b are monomial elements of the algebra, q_a and q_b are the respective exterior degrees.

It is easy to find another covariant superderivation of the algebra $\mathcal{A}$, along with (7). It is constructed by using the Christoffel symbols Γ^k_{ij} of the Levi-Civita connection, preserving the symplectic structure

$$\nabla = dx^i\wedge\left(\frac{\partial}{\partial x^i} - y^j\Gamma^k_{ij}(x)\frac{\partial}{\partial y^k}\right)\,. \tag{8}$$

In general (8) is no longer nilpotent, but the square of ∇ turns out to be a non-trivial interior derivative, which is expressed in terms of the curvature tensor $R^m{}_{jkl}$ of the Levi-Civita connection, namely

$$\nabla^2 a \quad = \quad \nabla(\nabla a) = \frac{1}{i\hbar}[R,a]\,,$$

$$R = \frac{1}{4} R_{ijkl} y^i y^j dx^k \wedge dx^l \,, \qquad (9)$$

where $R_{ijkl} = \omega_{im} R^m{}_{jkl}$.

One can also easily check that the derivations (7) and (8) are anticommuting.

Noting that all the nontrivial cocycles of (7) belongs to the commutative subalgebra $C^\infty(M)[[\hbar]] \in \mathcal{A}$ we construct the respective partial homotopy operator

$$\delta^{-1} a_{p,q} = \begin{cases} \frac{1}{p+q} y^k i(\frac{\partial}{\partial x^k}) a_{p,q}\,, & \text{if } p+q \neq 0 \\ 0\,, & \text{otherwise}\,, \end{cases} \qquad (10)$$

where $a_{p,q}$ denotes a monomial element with p being the order in y and q being the exterior degree, $i(\dfrac{\partial}{\partial x^k})$ denotes the interior derivative.

It is easy to check that the operators (7) and (10) obey the "Hodge-De Rham decomposition", holding for an arbitrary element $a \in \mathcal{A}$,

$$a = \sigma(a) + \delta\delta^{-1} a + \delta^{-1}\delta a\,, \qquad (11)$$

where $\sigma(a) = a(x,0,0,\hbar)$ denotes the projection of a onto $C^\infty(M)[[\hbar]]$.

According to the Fedosov method we look for a flat connection of the symbol bundle $\mathcal{A}$ in the framework of the following anzatz

$$D = \nabla - \delta + \frac{1}{i\hbar}[r, \cdot]\,, \qquad (12)$$

where $r = r_i(x, y, \hbar) dx^i$. A simple calculation yields that

$$D^2 a = \frac{1}{i\hbar}[\Omega, a]\,, \quad \forall a \in \mathcal{A}\,, \qquad (13)$$

$$\Omega = R - \delta r + \nabla r + \frac{1}{i\hbar} r \circ r\,. \qquad (14)$$

Thus we conclude that whenever the equation

$$R - \delta r + \nabla r + \frac{1}{i\hbar} r \circ r = 0 \qquad (15)$$

is satisfied the connection (12) turns out to be flat.

The connection (12) obviously defines a superderivative of the algebra $\mathcal{A}$ and we usually refer to the flatness condition of the connection as to the nilpotency condition for the derivative, and vice-versa.

Using the "Hodge-De Rham decomposition" (11), the Bianchi identity $\nabla R = 0$ and the symmetry property of the curvature tensor $\delta R = 0$ one may prove the following theorem

Theorem 1. (Fedosov,[5]) The solution r of the equation (15) obeying an additional condition $\delta^{-1} r = 0$ can be found by iterating the following equation

$$r = \delta^{-1}(R + \nabla r + \frac{1}{i\hbar} r \circ r)\,. \qquad (16)$$

Given the nilpotent derivative (12) (i.e. the flat connection) we define the subalgebra of flat(parallel) sections $\mathcal{A}^D \in \mathcal{A}$ as a subset consisting of the elements $a \in \mathcal{A}$, which are of the zero exterior degree, and such that $Da = 0$. It turns out that the subalgebra $\mathcal{A}^D$ is isomorphic as a linear space to the space of functions $C^\infty(M)[[\hbar]]$, and this isomorphism naturally induces the associative multiplication from the subalgebra $\mathcal{A}^D$ to the set of functions $C^\infty(M)[[\hbar]]$. The following statement gives us an explicit iteration procedure defining the isomorphism

Theorem 2. (Fedosov,[5]) For any function $a \in C^\infty(M)[[\hbar]]$ the iteration of the equation

$$\sigma^{-1}a = a + \delta^{-1}(\nabla\sigma^{-1}a + \frac{1}{i\hbar}[r, \sigma^{-1}a]), \tag{17}$$

yields a unique element $\sigma^{-1}a \in \mathcal{A}^D$ such that $\sigma(\sigma^{-1}a) = a$. Therefore, σ establishes an isomorphism between $\mathcal{A}_D$ and $C^\infty(M)[[\hbar]]$.

Note that the iteration of the equations (16) and (17) is performed with respect to the degree $m = 2k + p$ mentioned above, where k is the order in $\hbar$ and p is the order in y

Evaluating the lowest orders of the induced multiplication in $C^\infty(M)[[\hbar]]$

$$a * b(x, \hbar) = ab(x) + \frac{i\hbar}{2}\omega^{ij}\partial_i a(x)\partial_j b(x) + \ldots, \tag{18}$$

we see that the desired star-product is obtained. We refer to the star-product (18) as that of the Weyl type.

The Wick-type star-product (1) can be constructed in the framework of the analogous method, which is a minimal modification of the presented Fedosov approach. Namely, starting with the symbol product of the Wick type in the algebra $\mathcal{A}$

$$\begin{aligned} a \circ_g b \;\; &= \;\; exp\left(\frac{i\hbar}{2}\Lambda^{ij}(x)\frac{\partial}{\partial y^i}\frac{\partial}{\partial z^j}\right) a(x, y, dx, \hbar)\wedge \\ &\wedge b(x, z, dx, \hbar)|_{z=y} \end{aligned} \tag{19}$$

instead of (5) and using the Levi-Civita connection, which preserves both the symplectic structure and the symmetrical tensor, we can develop the analogue of the Fedosov method proceeding similarly at each stage and thus we obtain the star-product, obeying the above "boundary condition" (1).

THE FEDOSOV CLASS OF THE STAR-PRODUCTS

In fact the problem of constructing the flat Fedosov connection (12) leads to a more general equation then (15). Namely, the equation for the element r, defining the flat connection (12) looks as follows

$$R - \delta r + \nabla r + \frac{1}{i\hbar} r \circ r = \Omega, \tag{20}$$

where $\Omega = \hbar\Omega^{(1)}_{ij}(x)dx^i \wedge dx^j + \hbar^2\Omega^{(2)}_{ij}(x)dx^i \wedge dx^j + \ldots$ is formal series in $\hbar$ taking values in closed two-forms on M.

The solutions of the equation (20) can be found by iterating the analogous relation

$$r = \delta^{-1}(-\Omega + R + \nabla r + \frac{1}{i\hbar} r \circ r), \tag{21}$$

and the star-product corresponding to the respective flat connection is constructed quite similarly.

Thus, each flat Fedosov connection corresponds to some Fedosov star-product, or in other words each formal series $\Omega = \hbar\Omega^{(1)}_{ij}(x)dx^i \wedge dx^j + \hbar^2\Omega^{(2)}_{ij}(x)dx^i \wedge dx^j + \ldots$ taking values in closed two-forms defines a star-product on the given symplectic manifold.

It turns out that the Fedosov star-products, corresponding to the series of different cohomological classes are not equivalent. Let us illustrate the fact for the Weyl-type star-product $*$, presented in the previous section and the Fedosov star-product $\tilde{*}$, corresponding to the form $\Omega = \hbar\Omega^{(1)}_{ij}(x)dx^i \wedge dx^j$ with $\Omega^{(1)}_{ij}(x)dx^i \wedge dx^j$ being a nontrivial cocycle.

Evaluating lowest orders in $\hbar$ we get:

$$\begin{aligned} a * b &= ab + \frac{i\hbar}{2}\omega^{ij}\nabla_i a \nabla_j b - \frac{\hbar^2}{2}\omega^{ik}\omega^{jl}\nabla_i\nabla_j a \nabla_k \nabla_l b + O(\hbar^3), \\ a \tilde{*} b &= ab + \frac{i\hbar}{2}\tilde{\omega}^{ij}\nabla_i a \nabla_j b - \frac{\hbar^2}{2}\omega^{ik}\omega^{jl}\nabla_i\nabla_j a \nabla_k \nabla_l b + O(\hbar^3), \end{aligned} \tag{22}$$

where

$$\tilde{\omega}^{ij} = \omega^{ij} + \hbar\omega^{ij}_1, \quad \omega^{ij}_1 = \omega^{ik}\Omega^{(1)}_{kl}\omega^{lj}. \tag{23}$$

The equations (23), (23) mean that the second star product $\tilde{*}$ is called *the 1-differentiable deformation* of the first one $*$ [9]. Then, since the deformation is known to be trivial iff the 2-form Ω is exact [9] we conclude that the star-products $*$ and $\tilde{*}$ are not equivalent.

As the Fedosov star-products corresponding to the series Ω and Ω' of different cohomological classes are not equivalent, the star-products corresponding to the series of the same class turn out to be equivalent. Surprisingly, this statement turns out to be in full agreement with the fact that the respective flat Fedosov connections are equivalent iff the corresponding series Ω and Ω' are of same cohomological class.

We mean that the Fedosov connections $D_{(1)}$ and $D_{(2)}$ are equivalent if there exists an invertible element $U \in \mathcal{A}$ being a 0-form, such that

$$D_{(2)} = D_{(1)} + [U^{-1} \circ D_{(1)}U, \cdot]. \tag{24}$$

For example, given the flat connection D with r obeying the equation (15) and the flat connection $\tilde{D} = \nabla - \delta + \frac{1}{i\hbar}[\tilde{r}, \cdot]$, such that

$$R - \delta\tilde{r} + \nabla\tilde{r} + \frac{1}{i\hbar}\tilde{r} \circ \tilde{r} = \Omega = \hbar\Omega^{(1)}_{ij}(x)dx^i \wedge dx^j + \hbar^2\Omega^{(2)}_{ij}(x)dx^i \wedge dx^j + \ldots, \tag{25}$$

the equation for the element U, which would establish the equivalence between D and $\tilde{D}$ takes the following form

$$U^{-1} \circ DU = \frac{1}{i\hbar}\Delta r + \frac{1}{i\hbar}\psi, \tag{26}$$

where $\Delta r = \tilde{r} - r$, and $\psi = \hbar\psi_i^{(1)}(x)dx^i + \dots$ is formal series in $\hbar$ taking values in 1-forms on M.

Then applying the derivation D to both sides of the equation (26) and using the identity $D^2 = 0$ we get the compatibility condition for (26)

$$D\Delta r + \frac{1}{i\hbar}\Delta r \circ \Delta r + d\psi = 0. \tag{27}$$

The expression $D\Delta r + \frac{1}{i\hbar}\Delta r \circ \Delta r$ in (27) can be deduced to $-\Omega$ by using the equations (15), (25) and the properties of the derivatives (7) and (8) mentioned in the previous section. Hence the compatibility condition (27) is satisfied iff the form Ω is exact, and the flat connections D and $\tilde{D}$ are equivalent iff the series Ω is a trivial cocycle.

On the other hand, provided the condition (27) is fulfilled one advocates that the solution of (26) can be obtained by iterating the following relation

$$U = 1 + \delta^{-1}(\nabla U + \frac{1}{i\hbar}[r, U] - \frac{1}{i\hbar}U \circ (\Delta r + \psi)). \tag{28}$$

This statement is proved quite similarly to the that in [5,Theorem4.3].

By using the element U (28) one can establish the isomorphism between the subalgebras of the flat sections $\mathcal{A}^D$ and $\mathcal{A}^{\tilde{D}}$. Namely, this isomorphism is defined as the following conjugation

$$i_U a = U \circ a \circ U^{-1} \, : \, \mathcal{A}^{\tilde{D}} \mapsto \mathcal{A}^D. \tag{29}$$

Given the isomorphism (29), we automatically get the intertwining operator, establishing the equivalence between the respective star-products

$$Ta(x) = \sigma(U^{-1} \circ \sigma^{-1}(a) \circ U), \tag{30}$$

and

$$T(a * b) = (Ta \tilde{*} Tb), \tag{31}$$

where $*$ and $\tilde{*}$ are the star-products corresponding to the flat connections D and $\tilde{D}$ respectively, and σ^{-1} is the same as in the Theorem 2.

More generally, one concludes that the equivalence classes of the Fedosov star-products are in one-to-one correspondence with the points of *the moduli space of flat Fedosov connections*, and the space is, in turn, parameterized by the formal series in $\hbar$ taking values in the second De Rham cohomologies of the symplectic manifold. It is also worth noting that in view of the result of P. Xu [6], who has proved that each star-product on a symplectic manifold is equivalent to some Fedosov star-product, we conclude that the same statement holds for all the star-products on a symplectic manifold. This justifies the definition of the Fedosov class of the star-products on a symplectic manifold as the cohomological class of the respective formal series Ω. In fact for historical reasons the Fedosov class is defined as the cohomological class of the following two-form

$$F = \frac{1}{i\hbar}(\omega + \Omega), \tag{32}$$

where ω is the initial symplectic structure on M. Hereafter we are going to follow this definition.

As an application of the above considerations we calculate the Fedosov class of the Wick-type star-product (1). For this, we first note that the circle-products of the Weyl (5) and of the Wick types (19) are equivalent in the following sense

$$G(a \circ_g b) = (Ga \circ Gb), \tag{33}$$

where the formally invertible operator G is given by

$$G = \exp(-\frac{i\hbar}{4} g^{ij} \frac{\partial}{\partial y^i} \frac{\partial}{\partial y^j}). \tag{34}$$

In view of the simple identities

$$\nabla G = G\nabla, \qquad \delta G = G\delta, \tag{35}$$

the operator G establishes the isomorphism between the subalgebra of flat sections $\mathcal{A}^{D_g}$ and $\mathcal{A}^{\tilde{D}}$, corresponding to the flat connections D_g and $\tilde{D}$ respectively. Here $D_g a = \nabla a - \delta a + \frac{1}{i\hbar}(r_g \circ_g a - a \circ_g r_g)$ is the flat connection used in constructing the Wick-type star-product, and $\tilde{D} = \nabla - \delta + \frac{1}{i\hbar}[\tilde{r}, \cdot]$ with $\tilde{r} = Gr_g$ is a flat connection corresponding to some Fedosov star-product.

It is easy to check that the element $\tilde{r} = Gr_g$ satisfies the following equation

$$R - \delta\tilde{r} + \nabla\tilde{r} + \frac{1}{i\hbar}\tilde{r} \circ \tilde{r} = \Omega, \tag{36}$$

where $\Omega = -\frac{i\hbar}{8} g^{ij} R_{ijkl} dx^k \wedge dx^l$.

Thus we conclude that the Fedosov class of the Wick-type star-product is represented by the following two-form

$$F_{Wick} = \frac{1}{i\hbar}\omega - \frac{1}{8} g^{ij} R_{ijkl} dx^k \wedge dx^l. \tag{37}$$

Comparing the class with the one of the Weyl-type star-product we see that the Wick-type star-product is equivalent to that of the Weyl type iff the order zero term of the class (37) is trivial. Then, since the for the case of the Kähler manifold the two-form $F^{(0)} = -\frac{1}{8} g^{ij} R_{ijkl} dx^k \wedge dx^l$ represents the respective first Chern class we conclude that the Wick-type star-product on a Kähler manifold is equivalent to that of the Weyl type iff the manifold is the Calabi-Yau one.

CONCLUDING REMARKS

In the recent paper [10] our construction of the Wick-type star-product is generalized to the case of the almost Kähler manifold. In the paper the authors also calculate the order zero term of the Fedosov class for the respective star-product and show that it is proportional to the first Chern class of the almost Kähler manifold.

We think that it should be very interesting to generalize the construction presented in the paper [10] to the case of the almost para-Kähler manifold [8] and to calculate the complete Fedosov classes for the both the star-products. We postpone these questions to a future work.

ACKNOWLEDGMENTS

I acknowledge S. L. Lyakhovich and A. A. Sharapov for collaboration in the work [1]. I am also grateful to A. Borowiec for pointing out necessary references, concerning the paracomplex geometry.

REFERENCES

1. Dolgushev, V.A., Lyakhovich, S.L., and Sharapov, A.A., to appear in *Nucl. Phys. B*, hep-th/0101032.
2. Nest, R., and Tsygan, B., *Advances in Math.*, **113**, 151 (1995).
3. Bertelson, M., Cahen, M., and Gutt, S., *Class. Quant. Grav.*, **14**, A93-A107 (1997)
4. Deligne, P., *Selecta Mathematica*, New Series, **1**, 667 (1995).
5. Fedosov, B.V., *J.Diff.Geom.*, **40**, 213 (1994).
6. Xu, P., *Comm. Math. Phys.*, **197**, 167 (1998).
7. Besse, A., *Einstein Manifolds*, Springer-Verlag, Berlin, 1987.
8. Cruceanu, V., Fortuny P., and Gadea, P.M., *Rocky Mt. J. Math.*, **26**, 1, 83 (1996).
9. Sternheimer, D., Proc. of the 1998 Lodz conference "Particles, Fields and Gravitation", math/9809056.
10. Karabegov, A.V., and Schlichenmaier, M., *Almost Kähler deformation quantization*, math.QA /0102169.

The Cartan Covering and Complete Integrability of the KdV-MKdV System

P.H.M. Kersten

University of Twente, Faculty of Mathematical Sciences, P.O.Box 217, 7500 AE Enschede, The Netherlands, email: kersten@math.utwente.nl

Abstract. The coupled KdV–mKdV system arises as the classical part of one of superextensions of the KdV equation. For this system, we prove its complete integrability, i.e., existence of a recursion operator and of infinite series of symmetries.After giving a short introduction into the theory of symmetries,coverings, and the notion of Cartan-covering, the recursion operator will be constructed as a symmetry in the Cartan covering of the KdV–mKdV system.

INTRODUCTION

There are several supersymmetric extensions of the classical Korteweg–de Vries equation (KdV) [5, 9, 10]. One of them is of the form (the so-called $N = 2$, $A = 1$ extension [2])

$$\begin{aligned} u_t &= -u_3 + 6uu_1 - 3\varphi\varphi_2 - 3\psi\psi_2 - 3ww_3 - 3w_1w_2 + 3u_1w^2 + 6uww_1 \\ &\quad + 6\psi\varphi_1 w - 6\varphi\psi_1 w - 6\varphi\psi w_1, \\ \varphi_t &= -\varphi_3 + 3\varphi u_1 + 3\varphi_1 u - 3\psi_2 w - 3\psi_1 w_1 + 3\varphi_1 w^2 + 6\varphi ww_1, \\ \psi_t &= -\psi_3 + 3\psi u_1 + 3\psi_1 u + 3\varphi_2 w + 3\varphi_1 w_1 + 3\psi_1 w^2 + 6\psi ww_1, \\ w_t &= -w_3 + 3w^2 w_1 + 3uw_1 + 3u_1 w, \end{aligned}$$

where u and w are classical (even) independent variables while φ and ψ are odd ones (here and below the numerical subscript at an unknown variable denotes its derivative over x of the corresponding order). Being completely integrable itself, this system gives rise to an interesting system of even equations

$$\begin{aligned} u_t &= -u_3 + 6uu_1 - 3ww_3 - 3w_1w_2 + 3u_1w^2 + 6uww_1, \\ w_t &= -w_3 + 3w^2 w_1 + 3uw_1 + 3u_1 w, \end{aligned} \qquad (1)$$

which can be considered as a sort of coupling between the KdV (with respect to u) and the modified KdV (with respect to w) equations. In fact, setting $w = 0$, we obtain

$$u_t = -u_3 + 6uu_1,$$

while for $u = 0$ we have

$$w_t = -w_3 + 3w^2 w_1.$$

CP589, *Fundamental Interaction Theories*, edited by J. Lukierski and J. Rembieliński

The above indicates why we call (1) the KdV–mKdV system.
In what follows, we prove complete integrability, cf. [1], of system (1) by establishing existence of infinite series of symmetries and/or conservation laws. Towards this end we construct a recursion operator using the techniques of deformation theory introduced in [4] and extensively described and exemplified in [5].
In practical situations the construction of a deformation of the equation structure boils down to the construction of symmetries in an augmented setting of the equation or system at hand.
In Section 1 of this lecture we shall set out the nonlocal setting for differential equations and describe the notion of nonlocal symmetries in this setting.
Section 2 deals with a particular type of nonlocality, the Cartan covering of an equation. Section 3 combines the two previous types of coverings, and it is this covering where the recursion operator for symmetries is obtained as a symmetry in this covering.
In these first three sections the classical KdV equations acts as the main example. Finally in Section 4 we shall present symmetries, conservation laws, nonlocalities and the recursion operator for symmetries for the coupled KdV-mKdV system (1).

NONLOCAL SETTING FOR DIFFERENTIAL EQUATIONS

As standard example , to illustrate the notions, we take KdV-equation

$$u_t = uu_x + u_{xxx}\,. \tag{2}$$

We consider $Y \subset J^\infty(x,t;u)$ the infinite prolongation of (2), c.f.[7, 8], where coordinates in the infinite jet bundle $J^\infty(x,t;u)$ are given by $(x,t,u,u_x,u_t,\cdots)$ and Y is formally described as the submanifold of $J^\infty(x,t;u)$ defined by

$$\begin{aligned} u_t &= uu_x + u_{xxx}\,, \\ u_{xt} &= uu_{xx} + u_x^2 + u_{xxxx}\,, \\ &\vdots \end{aligned} \tag{3}$$

As internal coordinates in Y one chooses $(x,t,u,u_x,u_{xx},\cdots)$ while $u_t, u_{xt},\cdots$ are obtained from (3).

The Cartan distribution on Y is given by the total partial derivative vector fields

$$\begin{aligned} \widetilde{D}_x &= \partial_x + \sum_{n\geq 0} u_{n+1}\partial_{u_n}\,, \\ \widetilde{D}_t &= \partial_t + \sum_{n\geq 0} u_{nt}\partial_{u_n}\,, \end{aligned} \tag{4}$$

where $u_1 = u_x, u_2 = u_{xx},\cdots \quad ; u_{1t} = u_{xt}; u_{2t} = u_{xxt}\cdots$.

Classically the notion of a *generalized* or *higher symmetry* of a differential equation $F = 0$ is defined as a vertical vector field V

$$V = \partial_f = f\partial_u + \widetilde{D}_x(f)\partial_{u_1} + \widetilde{D}_x^2(f)\partial_{u_2} + \ldots, \tag{5}$$

where $f \in C^\infty(Y)$ such that,

$$\ell_F(f) = 0, \tag{6}$$

where in (6) ℓ_F is the *universal linearisation operator* [11, 7] which reads in the case of KdV-equation (3)

$$\widetilde{D}_t(f) - \widetilde{D}_x(f) - u_1 \cdot f - (\widetilde{D}_x)^3(f) = 0. \tag{7}$$

Let now $W \subset \mathbb{R}^m$ with coordinates $(w_1, \cdots w_m)$.
The Cartan distribution on $Y \otimes W$ is given by

$$\begin{aligned} \overline{D}_x &= \widetilde{D}_x + \sum_{j=1}^{m} X^j \frac{\partial}{\partial w_j}, \\ \overline{D}_t &= \widetilde{D}_t + \sum_{j=1}^{m} T^j \partial_{w_j}, \end{aligned} \tag{8}$$

where $X^j, T^j \in C^\infty(Y \otimes W)$ such that

$$[\overline{D}_x, \overline{D}_t] = 0, \tag{9}$$

which yields the socalled *covering condition*

$$D_x(T) - D_t(X) + [X,T] = 0,$$

whereas in (9) [*,*] is the Lie bracket for vector fields $X = \sum_{j=1}^m X^j \partial_{w_j}, T = \sum_{j=1}^m T^j \partial_{w_j}$ defined on W.

A *nonlocal symmetry* is a vertical vector field on $Y \otimes W$ i.e. of the form (5), which satisfies $(f \in C^\infty(Y \otimes W))$

$$\bar{\ell}_F(f) = 0, \tag{10}$$

which for KdV results in

$$\overline{D}_t(f) - u\overline{D}_x(f) - u_1 f - (\overline{D}_x)^3(f) - 0. \tag{11}$$

Formally this is just what is called the *shadow* of the symmetry , i.e., not bothering about the $\partial_{wj}, j = 1 \ldots m$ components.
The construction of the associated $\partial_{wj}, j = 1 \ldots m$ components is called the *reconstruction problem* [6] For reasons of simplicity, we omit this reconstruction problem, i.e. reconstructing the vector field from its shadow.

The classical Lenard recursion operator $\mathcal{R}$ for KdV equation,

$$\mathcal{R} = D_x^2 + \frac{2}{3}u + \frac{1}{3}u_1 D_x^{-1}, \tag{12}$$

which is just such, that

$$\begin{aligned} f_0 &= u_1, \\ \mathcal{R}f_0 = f_1 &= uu_1 + u_3, \\ \mathcal{R}f_1 = f_2 &= u_5 + \frac{5}{3}u_3 u + \frac{10}{3}u_2 u_1 + \frac{5}{6}u_1 u^2, \end{aligned} \tag{13}$$

i.e., creating the (x,t)-independent hierarchy of higher symmetries, has an action on vertical symmetry $\vartheta_{\overline{f}_{-1}}$ (Gallilei-boost)

$$\begin{aligned}\overline{f}_{-1} &= (1+tu_1)/3\,,\\ \mathcal{R}\overline{f}_{-1} &= \overline{f}_0 = 2u + xu_1 + 3t(u_3+uu_1)\,,\\ \overline{f}_1 = \mathcal{R}\overline{f}_0 &= 3t(f_2) + x(f_0) + 4u_2 + \frac{4}{3}u^2 + \frac{1}{3}u_1 D_x^{-1}(u)\,.\end{aligned} \tag{14}$$

If we introduce the variable $p(=w_1)$ through

$$\begin{aligned}p_x &= u\,,\\ p_t &= u_2 + \frac{1}{2}u^2\,,\\ \text{i.e.}\quad D_t(u) &= D_x(u_2 + \frac{1}{2}u^2)\,,\end{aligned} \tag{15}$$

then $\vartheta_{\overline{f}_1}$ is the shadow of a nonlocal symmetry in the one-dimensional covering of KdV-equation by

$$p = w_1\,,\quad X_1 = u\,,\quad T_1 = u_2 + \frac{1}{2}u^2\,.$$

So, by its action the Lenard recursion operator creates nonlocal symmetries in a natural way.
More applications of nonlocal symmetries can be found in e.g.[5].

A SPECIAL TYPE OF COVERING: THE CARTAN-COVERING

We discuss a special type of the nonlocal setting indicated in the previous section, the socalled Cartan-covering. As mentioned before we shall illustrate this by the KdV-equation.
Let $Y \subset J^\infty(x,t;u)$ be the infinite prolongation of KdV-equation (3). Contact one forms on $TJ^\infty(x,t;u)$ are given by

$$\begin{aligned}\alpha_0 &= du - u_1 dx - u_t dt\,,\\ \alpha_1 &= du_1 - u_2 dx - u_{1t} dt\,,\\ \alpha_2 &= du_2 - u_3 dx - u_{2t} dt\,.\end{aligned} \tag{16}$$

From the total partial derivative operators of the previous section we have

$$\widetilde{D}_x(\alpha_1) = \alpha_1\,,\quad \widetilde{D}_x(\alpha_1) = \alpha_2, \ldots$$

$$\begin{aligned}\widetilde{D}_t(\alpha_0) &= \alpha_0 u_x + \alpha_1 u + \alpha_3 = \alpha_t\,,\\ \widetilde{D}_t(\alpha_i) &= (\widetilde{D}_x)^i(\alpha_t)\,.\end{aligned} \tag{17}$$

We now define the Cartan-covering of Y by $Y \otimes \mathbb{R}^\infty$, where local coordinates are given $(x,t,u,u_1,\ldots,\alpha_0,\alpha_1,\ldots)$ by

$$D_x^C = \widetilde{D}_x + \sum_i (\alpha_{i+1}) \frac{\partial}{\partial \alpha_i},$$
$$D_t^C = \widetilde{D}_t + \sum_i (\widetilde{D}_x)^i \alpha_t \frac{\partial}{\partial \alpha_i}. \tag{18}$$

It is a straightforward check, and obvious that

$$[D_x^C, D_t^C] = 0, \tag{19}$$

i.e. they form a Cartan distribution on $Y \otimes \mathbb{R}^\infty$.

Note 1:

Since at first α_i $(i = 0, \ldots)$ are contact forms, they constitute a Grassmann algebra (graded commutative algebra) $\Lambda(\alpha)$, where

$$\alpha_i \wedge \alpha_j = -\alpha_j \wedge \alpha_i, \tag{20}$$

i.e.,

$$xy = (-1)^{|x||y|} yx,$$

where x, y are contact $(*)$-forms of degree $|x|$ and $|y|$ respectively. So in effect we are dealing with a *graded* covering.

Note 2:

Once we have introduced the Cartan-covering by (18) we can forget about the specifics of α_i $(i = 0, \ldots)$ and just treat them as (odd) ordinary variables, associated with their differentiation rules

One can discuss nonlocal symmetries in this type of covering just as in the previous section, the only difference being:

$$f \in V^\infty(Y) \otimes \Lambda(\alpha).$$

In the next section we shall combine constructions of the previous section and this section, in order to construct the recursion operator for symmetries.

THE RECURSION OPERATOR AS SYMMETRY IN THE CARTAN-COVERING

We shall discuss the recursion operator for symmetries of KdV-equation as a geometrical object, i.e., a symmetry in the Cartan-covering.

Our starting point is the four dimensional covering of the KdV-equation in $Y \otimes \mathbb{R}^4$ where

$$\begin{aligned}
\overline{D}_x &= D_x + u\partial_{w_1} + \frac{1}{2}u^2\partial_{w_2} + (u^3 - 3u_1^2)\partial_{w_3} + w_1\partial_{w_4}, \\
\overline{D}_t &= D_t + (\frac{1}{2}u^2 + u_2)\partial_{w_1} + (\frac{1}{3}u^3 - \frac{1}{2}u_1^2 + uu_2)\partial_{w_2} \\
&\quad + (\frac{3}{4}u^4 - 6u_1u_3 + 3u^2u_2 - 6uu_1^2 + 3u_2^2)\partial_{w_3} + (u_1 + w_2)\partial_{w_4},
\end{aligned} \tag{21}$$

$\overline{D}_x, \overline{D}_t$ satisfy the covering condition (9), and note that due to the fact that the coefficients of ∂_{w_i} $(i = 1,2,3)$ in (21) are independent of w_j $(j = 1,2,3)$. These coefficients constitute conservation laws for the KdV-equation. We have the following "formal" variables.

$$\begin{aligned} w_1 &= \int u dx, \\ w_2 &= \int \frac{1}{2} u^2 dx, \\ w_3 &= \int (u^3 - 3u_1^2) dx, \\ w_4 &= \int w_1 dx, \end{aligned} \tag{22}$$

where in (22) w_4 is of a higher nonlocality.

We now build the Cartan-covering of the previous section on the covering given by (21) by introduction of the contact forms $\alpha_0, \alpha_1, \alpha_2, \ldots$ (16) and

$$\begin{aligned} \alpha_{-1} &= dw_1 - udx - (\frac{1}{2}u^2 + u_2)dt, \\ \alpha_{-2} &= dw_2 - \frac{1}{2}u^2 dx - (\frac{1}{3}u^3 - \frac{1}{2}u_1^2 + uu_2)dt, \end{aligned} \tag{23}$$

and similarly for α_{-3}, α_{-4}. It is straightforward to prove the following relations

$$\begin{aligned} \overline{D}_x(\alpha_{-1}) &= \alpha_0\,, \quad \overline{D}_t(\alpha_{-1}) = u\alpha_0 + \alpha_2\,, \\ \overline{D}_x(\alpha_{-2}) &= u\alpha_0\,, \quad \overline{D}_t(\alpha_{-2}) = u^2\alpha_0 - u_1\alpha_1 + u\alpha_2 + u_2\alpha_0\,, \\ \overline{D}_x(\alpha_{-3}) &= 3u^2\alpha_0 - 6u_1\alpha_1\,, \quad \ldots. \end{aligned} \tag{24}$$

We are now constructing symmetries in this Cartan-covering of KdV-equation which are linear w.r.t. α_i $(i = -4, \ldots, 0, 1, \ldots)$.

The symmetry condition for $f \in C^\infty(Y \otimes \mathbb{R}^4) \otimes \Lambda^1(\alpha)$ is just given by (7)

$$\ell_F^C(f) = 0, \tag{25}$$

which for the KdV equation results in

$$\overline{D}_t^C(f) - u\overline{D}_x^C(f) - u_x f - (\overline{D}_x^C)^3 f = 0.$$

As solutions of these equations we obtained

$$\begin{aligned} f^0 &= \alpha_0\,, \\ f^1 &= (\frac{2}{3}u)\alpha_0 + \alpha_2 + (\frac{1}{3}u_1)\alpha_{-1}\,, \\ f^2 &= (\frac{4}{9}u^2 + \frac{4}{3}u_2)\alpha_0 + (2u_1)\alpha_1 + (\frac{4}{3}u)\alpha_2 + \alpha_4 \\ &\quad + \frac{1}{3}(uu_1 + u_3)\alpha_{-1} + \frac{1}{9}(u_1)\alpha_{-2}\,. \end{aligned} \tag{26}$$

As we mentioned above we are working in effect with form-valued vector fields $\vartheta_{f^0}, \vartheta_{f^1} \vartheta_{f^2}$. For these objects one can define Frölicher-Nijenhuis and (by contraction) Richardson-Nijenhuis brackets [4],[5] Without going into details, for which the reader is referred to [4], we can construct the contraction of a (generalized) symmetry and a form valued symmetry p.e.

$$R = (\frac{2}{3}u\alpha_0 + \alpha_2 + \frac{1}{3}u_1\alpha_{-1})\frac{\partial}{\partial u} + \dots. \tag{27}$$

The contraction being defined by

$$(V_1 \lrcorner R) = (V \lrcorner R_u)\partial_u + \overline{D}_x^C(V \lrcorner R_u)\partial_{u_1} + \dots. \tag{28}$$

Start now with

$$V_1 = u_1\frac{\partial}{\partial u} + u_2\frac{\partial}{\partial u_1} + \dots, \tag{29}$$

whose prolongation in the setting $Y \otimes \mathbb{R}^4$ is

$$\overline{V}_1 = u_1\frac{\partial}{\partial u} + u_2\frac{\partial}{\partial u_1} + \dots + u\frac{\partial}{\partial w_1} + \frac{1}{2}u^2\frac{\partial}{\partial w_2} + (u^3 - 3u_1^2)\frac{\partial}{\partial w_3} + w_1\frac{\partial}{\partial w_4}, \tag{30}$$

then

$$\begin{aligned}(\overline{V} \lrcorner R) &= [(\frac{2}{3}u)u_1 + 1 \cdot u_3 + \frac{1}{3}u_1 \cdot u]\frac{\partial}{\partial u} + \dots \\ &= (u_3 + uu_1)\frac{\partial}{\partial u} + \dots = V_3\end{aligned} \tag{31}$$

and similarly

$$(\overline{V}_3 \lrcorner R) = (u_5 + \frac{5}{3}u_3u + \frac{10}{3}u_2u_1 + \frac{5}{6}u^2u_1)\frac{\partial}{\partial u} + \dots = V_5. \tag{32}$$

The result given above means that the well known Lenard recursion operator for symmetries of KdV-equation is represented as a *symmetry* , ϑ_{f_1} , in the Cartan-covering of this equation and in effect is a geometrical object.

THE COUPLED KDV-MKDV SYSTEM

In this section we shall discuss the complete integrability of the KdV–mKdV system $\mathcal{E}$, given in (1), i.e.,

$$\begin{aligned} u_t &= -u_3 + 6uu_1 - 3ww_3 - 3w_1w_2 + 3u_1w^2 + 6uww_1, \\ w_t &= -w_3 + 3w^2w_1 + 3uw_1 + 3u_1w. \end{aligned} \tag{33}$$

In order to demonstrate the complete integrability of this system, we shall construct the recursion operator for symmetries of this coupled system, leading to infinite hierarchies of symmetries and, most probably, of conservation laws.
Due to the very special form of the final results, it seems that integrability of this system, which looks at first glance quite ordinary, has not been discussed before. In order to discuss complete integrability, we shall start to discuss conservation laws in Subsection *Conservation Laws and Nonlocal Variables* leading to the necessary nonlocal variables. In Subsection *Local and Nonlocal Symmetries* we shall discuss local and nonlocal symmetries of the system, while in Subsection *Recursion Operator* we construct the recursion operator or deformation [4], by the construction of a symmetry in the Cartan covering of the equation (33).

Conservation Laws and Nonlocal Variables

Here we shall construct conservation laws for (33) in order to arrive at an abelian covering of the coupled KdV–mKdV system as was shown KdV equation (2).
So we construct $X = X(x,t,u,\ldots,w\ldots), T = T(x,t,u,\ldots,w\ldots)$ such that

$$D_x(T) = D_t(X)\,, \tag{34}$$

where D_x, D_t are defined as the total partial derivative operators on the infinite jetbundle associated to the equation (33)and in a similar way we construct nonlocal conservation laws by the requirement

$$\bar{D}_x(\bar{T}) = \bar{D}_t(\bar{X})\,, \tag{35}$$

where $\bar{D}_*$ is defined as the prolongation of D_* towards the covering of the equation by nonlocal variables arising from local conservation laws; moreover $\bar{X}$, $\bar{T}$ are dependent on local variables x, t, $u,\ldots,w,\ldots$ as well as the already determined nonlocal variables, denoted here by p_* or $p_{*,*}$, which are associated to the conservation laws (X,T) by the formal definition

$$D_x(p_*) = (p_*)_x = X\,,$$
$$D_t(p_*) = (p_*)_t = T\,.$$

Proceeding in this way, we obtained the following set of nonlocal variables

$$p_{0,1},\ p_{0,2},\ p_1,\ p_{1,1},\ p_{1,2},\ p_{2,1},\ p_3,\ p_{3,1},\ p_{3,2},\ p_{4,1},\ p_5\,, \tag{36}$$

where their defining equations are given by

$$(p_1)_x = u\,,$$
$$(p_1)_t = 3u^2 + 3uw^2 - u_2 - 3ww_2\,,$$
$$(p_{0,1})_x = w\,,$$
$$(p_{0,1})_t = 3uw + w^3 - w_2\,,$$

$$
\begin{aligned}
(p_{0,2})_x &= p_1\,,\\
(p_{0,2})_t &= -6p_3 - u_1\,,\\
(p_{1,1})_x &= \cos(2p_{0,1})p_1 w + \sin(2p_{0,1})w^2\,,\\
(p_{1,1})_t &= \cos(2p_{0,1})(3p_1 uw + p_1 w^3 - p_1 w_2 + uw_1 - u_1 w - w^2 w_1)\\
&\quad + \sin(2p_{0,1})(4uw^2 + w^4 - 2ww_2 + w_1^2)\,,\\
(p_{1,2})_x &= \cos(2p_{0,1})w^2 - \sin(2p_{0,1})p_1 w\,,\\
(p_{1,2})_t &= \cos(2p_{0,1})(4uw^2 + w^4 - 2ww_2 + w_1^2)\\
&\quad + \sin(2p_{0,1})(-3p_1 uw - p_1 w^3 + p_1 w_2 - uw_1 + u_1 w + w^2 w_1)\,,\\
(p_{2,1})_x &= (4\cos(2p_{0,1})p_{1,1}w^2 - 4\sin(2p_{0,1})p_1 p_{1,1} w + w(p_1^2 - 2u + w^2))/2\,,\\
(p_{2,1})_t &= (4\cos(2p_{0,1})p_{1,1}(4uw^2 + w^4 - 2ww_2 + w_1^2)\\
&\quad + 4\sin(2p_{0,1})p_{1,1}(-3p_1 uw - p_1 w^3 + p_1 w_2 - uw_1 + u_1 w + w^2 w_1)\\
&\quad + 3p_1^2 uw + p_1^2 w^3 - p_1^2 w_2 + 2p_1 uw_1 - 2p_1 u_1 w - 2p_1 w^2 w_1 - 8u^2 w\\
&\quad - uw^3 + 2uw_2 - 2u_1 w_1 + 2u_2 w + w^5 + 3w^2 w_2)/2\,,\\
(p_3)_x &= (-u^2 - uw^2 + ww_2)/2\,,\\
(p_3)_t &= (-4u^3 - 9u^2 w^2 + 2uu_2 - 3uw^4 + 11uww_2 - uw_1^2 - u_1^2 + u_1 ww_1\\
&\quad + 4u_2 w^2 + 6w^3 w_2 + 3w^2 w_1^2 - ww_4 + w_1 w_3 - w_2^2)/2\,,\\
(p_{3,1})_x &= (\cos(2p_{0,1})w(p_1^3 - 6p_1 u + 39p_1 w^2 - 24p_{1,1}p_{1,2}w + 12p_3 + 6u_1)\\
&\quad + 2\sin(2p_{0,1})w(12p_1 p_{1,1} p_{1,2} + 18p_1 w_1 + 2w^3 + 3w_2)\\
&\quad + 6p_{1,2}w(-p_1^2 + 2u - w^2))/12\,,\\
(p_{3,2})_x &= (2\cos(2p_{0,1})w(12p_1 p_{1,1} p_{1,2} - 18p_1 w_1 - 2w^3 - 3w_2)\\
&\quad + \sin(2p_{0,1})w(p_1^3 - 6p_1 u + 39p_1 w^2 + 24p_{1,1}p_{1,2}w + 12p_3 + 6u_1)\\
&\quad + 6p_{1,1}w(-p_1^2 + 2u - w^2))/12\,,\\
(p_{4,1})_x &= (8\cos(2p_{0,1})w(p_1^3 p_{1,2} + 12p_1 p_{1,1}^2 p_{1,2} - 6p_1 p_{1,2} u + 3p_1 p_{1,2} w^2\\
&\quad - 12p_{1,1} p_{1,2}^2 w + 18p_{1,1} uw - 4p_{1,1} w^3 - 6p_{1,1} w_2 + 12p_{1,2} p_3 + 6p_{1,2} u_1)\\
&\quad + 8\sin(2p_{0,1})w(p_1^3 p_{1,1} + 12p_1 p_{1,1} p_{1,2}^2 - 6p_1 p_{1,1} u + 3p_1 p_{1,1} w^2\\
&\quad + 12p_{1,1}^2 p_{1,2} w + 12p_{1,1} p_3 + 6p_{1,1} u_1 - 18p_{1,2} uw + 4p_{1,2} w^3 + 6p_{1,2} w_2)\\
&\quad + w(-p_1^4 - 24p_1^2 p_{1,1}^2 - 24p_1^2 p_{1,2}^2 + 12p_1^2 u - 6p_1^2 w^2 - 48p_1 p_3\\
&\quad - 24p_1 u_1 + 48p_{1,1}^2 u - 24p_{1,1}^2 w^2 + 48p_{1,2}^2 u - 24p_{1,2}^2 w^2\\
&\quad - 60u^2 + 44uw^2 + 24u_2 - 13w^4 + 6ww_2))/48\,,\\
(p_5)_x &= (12u^3 + 24u^2 w^2 - 6uu_2 + 6uw^4 - 30uww_2 - 3u_2 w^2 - 8w^3 w_2 + 6ww_4)/6\,.
\end{aligned}
$$

In the previous equations, we skipped explicit formulas for $(p_{3,1})_t$, $(p_{3,2})_t$, $(p_{4,1})_t$, and $(p_5)_t$, because they are too massive, though quite important for the setting to be well defined and in order to avoid ambiguities. The reader is referred to [3] for these

explicit formulas.
It is quite a striking result that functions $\cos(2p_{0,1})$, $\sin(2p_{0,1})$ appear in the presentation of the conservation laws and their associated nonlocal variables.
We should note that p_1, $p_{0,1}$, p_3, p_5 arise from **local conservation laws** and we shall call p_1, $p_{0,1}$, p_3, p_5 *nonlocalities of first order*.
In a similar way we see that $p_{0,2}$, $p_{1,1}$, $p_{1,2}$ arise from **nonlocal conservation laws,** where their x- and t-derivatives are dependent on the first order nonlocalities. For this reason $p_{0,2}$, $p_{1,1}$, $p_{1,2}$ are called *nonlocalities of second order*. Proceeding in this way $p_{2,1}$, $p_{3,1}$, $p_{3,2}$, $p_{4,1}$ constitute *nonlocalities of third order*.

Local and Nonlocal Symmetries

In this section we shall present results for the construction of local and nonlocal symmetries of system (33). In order to construct these symmetries, we consider the system of partial differential equations obtained by the infinite prolongation of (33) together with the covering by the nonlocal variables

$$p_{0,1},\ p_{0,2},\ p_1,\ p_{1,1},\ p_{1,2},\ p_{2,1},\ p_3,\ p_{3,1},\ p_{3,2},\ p_{4,1},\ p_5 .$$

So, in the augmented setting governed by (33), their total derivatives and the equations given in Subsection *The Coupled KdV-mKdV System* we construct symmetries $Y = (Y^u, Y^w)$ which have to satisfy the symmetry condition

$$\bar{\ell}_{\mathcal{E}} Y = 0 .$$

From this condition we obtained the following symmetries

$$Y_{0,1},\ Y_{1,1},\ Y_{1,2},\ Y_{1,3},\ Y_{2,1},\ Y_{3,1},\ Y_{3,2},\ Y_{3,3},$$

where generating functions $Y^u_{*,*}$, $Y^w_{*,*}$ are given as

$$\begin{aligned}
Y^u_{0,1} &= 3t(6uu_1 + 6uww_1 + 3u_1w^2 - u_3 - 3ww_3 - 3w_1w_2) + xu_1 + 2u,\\
Y^w_{0,1} &= 3t(3uw_1 + 3u_1w + 3w^2w_1 - w_3) + xw_1 + w,\\
Y^u_{1,1} &= u_1,\\
Y^w_{1,1} &= w_1,\\
Y^u_{1,2} &= \cos(2p_{0,1})(2uw - w_2) + \sin(2p_{0,1})(u_1 + 2ww_1),\\
Y^w_{1,2} &= -\cos(2p_{0,1})u - \sin(2p_{0,1})w_1,\\
Y^u_{1,3} &= \cos(2p_{0,1})(u_1 + 2ww_1) + \sin(2p_{0,1})(-2uw + w_2),\\
Y^w_{1,3} &= -\cos(2p_{0,1})w_1 + \sin(2p_{0,1})u,\\
Y^u_{2,1} &= (2\cos(2p_{0,1})(p_{1,1}u_1 + 2p_{1,1}ww_1 - 2p_{1,2}uw + p_{1,2}w_2)\\
&\quad + 2\sin(2p_{0,1})(-2p_{1,1}uw + p_{1,1}w_2 - p_{1,2}u_1 - 2p_{1,2}ww_1)\\
&\quad + 2p_1uw - p_1w_2 + 2uw_1 + 3u_1w + 2w^2w_1 - w_3)/2,
\end{aligned}$$

$$
\begin{aligned}
Y^w_{2,1} &= (2\cos(2p_{0,1})(-p_{1,1}w_1 + p_{1,2}u) + 2\sin(2p_{0,1})(p_{1,1}u + p_{1,2}w_1) \\
&\quad - p_1 u + u_1 + ww_1)/2, \\
Y^u_{3,1} &= (6uu_1 + 6uww_1 + 3u_1w^2 - u_3 - 3ww_3 - 3w_1w_2)/3, \\
Y^w_{3,1} &= (3uw_1 + 3u_1w + 3w^2w_1 - w_3)/3, \\
Y^u_{3,2} &= (\cos(2p_{0,1})(-2p_1^2uw + p_1^2w_2 - 4p_1uw_1 - 6p_1u_1w - 4p_1w^2w_1 + 2p_1w_3 \\
&\quad + 8p_{1,1}p_{1,2}u_1 + 16p_{1,1}p_{1,2}ww_1 - 8p_{1,2}^2uw + 4p_{1,2}^2w_2 - 4p_{2,1}u_1 - 8p_{2,1}ww_1 \\
&\quad + 10u^2w + 6uw^3 - 8uw_2 - 14u_1w_1 - 8u_2w - 11w^2w_2 - 14ww_1^2 + 2w_4) \\
&\quad + 2\sin(2p_{0,1})(-8p_{1,1}p_{1,2}uw + 4p_{1,1}p_{1,2}w_2 - 2p_{1,2}^2u_1 - 4p_{1,2}^2ww_1 + 4p_{2,1}uw \\
&\quad - 2p_{2,1}w_2 + 6uu_1 + 10uww_1 + 3u_1w^2 - u_3 + 2w^3w_1 - 3ww_3 - 5w_1w_2) \\
&\quad + 4p_{1,2}(2p_1uw - p_1w_2 + 2uw_1 + 3u_1w + 2w^2w_1 - w_3))/8, \\
Y^w_{3,2} &= (\cos(2p_{0,1})(p_1^2u - 2p_1u_1 - 2p_1ww_1 - 8p_{1,1}p_{1,2}w_1 + 4p_{1,2}^2u + 4p_{2,1}w_1 \\
&\quad - 4u^2 - 3uw^2 + 2u_2 + 4ww_2 + 2w_1^2) \\
&\quad + 2\sin(2p_{0,1})(4p_{1,1}p_{1,2}u + 2p_{1,2}^2w_1 - 2p_{2,1}u - 3uw_1 - 3u_1w - 3w^2w_1 + w_3) \\
&\quad + 4p_{1,2}(-p_1u + u_1 + ww_1))/8, \\
Y^u_{3,3} &= (2\cos(2p_{0,1})(2p_{1,1}^2u_1 + 4p_{1,1}^2ww_1 - 4p_{2,1}uw + 2p_{2,1}w_2 - 6uu_1 \\
&\quad - 10uww_1 - 3u_1w^2 + u_3 - 2w^3w_1 + 3ww_3 + 5w_1w_2) \\
&\quad + \sin(2p_{0,1})(-2p_1^2uw + p_1^2w_2 - 4p_1uw_1 - 6p_1u_1w - 4p_1w^2w_1 + 2p_1w_3 \\
&\quad - 8p_{1,1}^2uw + 4p_{1,1}^2w_2 - 4p_{2,1}u_1 - 8p_{2,1}ww_1 + 10u^2w + 6uw^3 \\
&\quad - 8uw_2 - 14u_1w_1 - 8u_2w - 11w^2w_2 - 14ww_1^2 + 2w_4) \\
&\quad + 4p_{1,1}(2p_1uw - p_1w_2 + 2uw_1 + 3u_1w + 2w^2w_1 - w_3))/8, \\
Y^w_{3,3} &= (2\cos(2p_{0,1})(-2p_{1,1}^2w_1 + 2p_{2,1}u + 3uw_1 + 3u_1w + 3w^2w_1 - w_3) \\
&\quad + \sin(2p_{0,1})(p_1^2u - 2p_1u_1 - 2p_1ww_1 + 4p_{1,1}^2u + 4p_{2,1}w_1 - 4u^2 \\
&\quad - 3uw^2 + 2u_2 + 4ww_2 + 2w_1^2) \\
&\quad + 4p_{1,1}(-p_1u + u_1 + ww_1))/8.
\end{aligned}
$$

Recursion Operator

Here we present the recursion operator $\mathcal{R}$ for symmetries for this case obtained as a higher symmetry in the Cartan covering of system of equations (1) augmented by equations governing the nonlocal variables (36).

As demonstrated there, this symmetry is a form-valued vector field (or a vectorfield-valued one-form) and has to satisfy

$$
\bar{\ell}^{\mathcal{C}}_{\mathcal{E}} \mathcal{R} = 0. \tag{37}
$$

In order to arrive at a nontrivial result as was explained for classical KdV equation (3),(26), we have to introduce nonlocal variables

$$p_{0,1},\ p_{0,2},\ p_1,\ p_{1,1},\ p_{1,2},\ p_{2,1},\ p_3,\ p_{3,1},\ p_{3,2},\ p_{4,1},\ p_5$$

and their associated Cartan contact forms

$$\omega_{p_{0,1}},\ \omega_{p_{0,2}},\ \omega_{p_1},\ \omega_{p_{1,1}},\ \omega_{p_{1,2}},\ \omega_{p_{2,1}},\ \omega_{p_3},\ \omega_{p_{3,1}},\ \omega_{p_{3,2}},\ \omega_{p_{4,1}},\ \omega_{p_5}\,.$$

The final result, which is dependent on the nonlocal Cartan forms

$$\omega_{p_{0,1}},\ \omega_{p_1},\ \omega_{p_{1,1}},\ \omega_{p_{1,2}},$$

is given by

$$\mathcal{R} = R^u\frac{\partial}{\partial u} + R^w\frac{\partial}{\partial w} + \ldots, \tag{38}$$

where the components R^u, R^w are given by

$$\begin{aligned}
R_u &= \omega_{u_2}(-1) + \omega_u(4u+w^2) + \omega_{w_2}(-2w) + \omega_{w_1}(-w_1) + \omega_w(3uw - 2w_2)\\
&\quad + \omega_{p_{1,2}}(-\cos(2p_{0,1})(u_1 + 2ww_1) + \sin(2p_{0,1})(2uw - w_2))\\
&\quad + \omega_{p_{1,1}}(\cos(2p_{0,1})(-2uw + w_2) - \sin(2p_{0,1})(u_1 + 2ww_1))\\
&\quad + \omega_{p_1}(2u_1 + ww_1) + \omega_{p_{0,1}}(2p_1uw - p_1w_2 + 2uw_1 + 3u_1w + 2w^2w_1 - w_3)\,,\\
R_w &= \omega_{w_2}(-1) + \omega_w(2u+w^2) + \omega_u(2w)\\
&\quad + \omega_{p_{1,2}}(\cos(2p_{0,1})w_1 - \sin(2p_{0,1})u)\\
&\quad + \omega_{p_{1,1}}(\cos(2p_{0,1})u + \sin(2p_{0,1})w_1)\\
&\quad + \omega_{p_1}(w_1) + \omega_{p_{0,1}}(-p_1u + u_1 + ww_1)\,.
\end{aligned} \tag{39}$$

We shall now present this result in a more conventional form which appeals to expressions using operators of the form D_x and D_x^{-1}. In order to do this, we first split (39) into the so-called local part and nonlocal parts, consisting of terms associated to ω_{u_2}, ω_u, ω_{w_2}, ω_{w_1}, ω_w and those associated to $\omega_{p_{1,2}}$, $\omega_{p_{1,1}}$, ω_{p_1}, $\omega_{p_{0,1}}$ respectively. The first part will account for D_x presentation, while the second one accounts for the D_x^{-1} part.

Due to the action of contraction $\partial_\varphi \lrcorner \mathcal{R}$, the local part is given by the following matrix operator:

$$\begin{bmatrix} -D_x^2 + 4u + w^2 & -2wD_x^2 - w_1D_x + 3uw - 2w_2 \\ 2w & -D_x^2 + 2u + w^2 \end{bmatrix}.$$

The nonlocal part will be split into parts associated to ω_{p_1}, $\omega_{p_{0,1}}$ and $\omega_{p_{1,2}}$, $\omega_{p_{1,1}}$, respectively. The first one is given as

$$\begin{bmatrix} (2u_1 + ww_1)D_x^{-1} & (2p_1uw - p_1w_2 + 2uw_1 + 3u_1w + 2w^2w_1 - w_3)D_x^{-1} \\ w_1D_x^{-1} & (-p_1u + u_1 + ww_1)D_x^{-1} \end{bmatrix}.$$

To deal with the last part, let us introduce the notation:

$$\begin{aligned}
A_1 &= \cos(2p_{0,1})(-2uw+w_2)-\sin(2p_{0,1})(u_1+2ww_1)\,,\\
A_2 &= \cos(2p_{0,1})u+\sin(2p_{0,1})w_1\,,\\
B_1 &= -\cos(2p_{0,1})(u_1+2ww_1)+\sin(2p_{0,1})(2uw-w_2)\,,\\
B_2 &= \cos(2p_{0,1})w_1-\sin(2p_{0,1})u\,,
\end{aligned}$$

being the coefficients at $\omega_{p_{1,1}}$ and $\omega_{p_{1,2}}$ in (39).

According to the presentations of $(p_{1,1})_x$ and $(p_{1,2})_x$, i.e.,

$$\begin{aligned}
(p_{1,1})_x &= \cos(2p_{0,1})p_1w+\sin(2p_{0,1})w^2\,,\\
(p_{1,2})_x &= \cos(2p_{0,1})w^2-\sin(2p_{0,1})p_1w\,,
\end{aligned}$$

we introduce their partial derivatives with respect to $p_{0,1}$, p_1, and w as

$$\begin{aligned}
\alpha_1 &= -2p_1w\sin(2p_{0,1})+2w^2\cos(2p_{0,1})\,,\\
\alpha_2 &= w\cos(2p_{0,1})\,,\\
\alpha_3 &= p_1\cos(2p_{0,1})+2w\sin(2p_{0,1})\,,\\
\beta_1 &= -2w^2\sin(2p_{0,1})-2p_1w\cos(2p_{0,1})\,,\\
\beta_2 &= -w\sin(2p_{0,1})\,,\\
\beta_3 &= 2w\cos(2p_{0,1})-p_1\sin(2p_{0,1})\,.
\end{aligned}$$

From this we arrive in a straightforward way at the last nonlocal part of the recursion operator, i.e.,

$$\begin{bmatrix} A_1D_x^{-1}\alpha_2D_x^{-1} & A_1D_x^{-1}(\alpha_1D_x^{-1}+\alpha_3)\\ A_2D_x^{-1}\alpha_2D_x^{-1} & A_2D_x^{-1}(\alpha_1D_x^{-1}+\alpha_3)\end{bmatrix}+\begin{bmatrix} B_1D_x^{-1}\beta_2D_x^{-1} & B_1D_x^{-1}(\beta_1D_x^{-1}+\beta_3)\\ B_2D_x^{-1}\beta_2D_x^{-1} & B_2D_x^{-1}(\beta_1D_x^{-1}+\beta_3)\end{bmatrix}.$$

So, in the final form we obtain the recursion operator as

$$\begin{aligned}
\mathcal{R} = &\begin{bmatrix} -D_x^2+4u+w^2 & -2wD_x^2-w_1D_x+3uw-2w_2\\ 2w & -D_x^2+2u+w^2\end{bmatrix}\\
&+\begin{bmatrix} (2u_1+ww_1)D_x^{-1} & (2p_1uw-p_1w_2+2uw_1+3u_1w+2w^2w_1-w_3)D_x^{-1}\\ w_1D_x^{-1} & (-p_1u+u_1+ww_1)D_x^{-1}\end{bmatrix}\\
&+\begin{bmatrix} A_1D_x^{-1}\alpha_2D_x^{-1} & A_1D_x^{-1}(\alpha_1D_x^{-1}+\alpha_3)\\ A_2D_x^{-1}\alpha_2D_x^{-1} & A_2D_x^{-1}(\alpha_1D_x^{-1}+\alpha_3)\end{bmatrix}\\
&+\begin{bmatrix} B_1D_x^{-1}\beta_2D_x^{-1} & B_1D_x^{-1}(\beta_1D_x^{-1}+\beta_3)\\ B_2D_x^{-1}\beta_2D_x^{-1} & B_2D_x^{-1}(\beta_1D_x^{-1}+\beta_3)\end{bmatrix}.
\end{aligned}$$

CONCLUSION AND OUTLOOK

We gave an outline of the theory of symmetries of differential equations, leading to the construction of recursion operators for symmetries of such equations. The extension

of this theory to the nonlocal setting of differential equations is essential for getting nontrivial results. The theory has been applied to the construction of the recursion operator for symmetries for a coupled KdV–mKdV system, leading to a highly nonlocal result for this system. Moreover the appearance of nonpolynomial nonlocal terms in all results, e.g., conservation laws, symmetries and recursion operator is striking and reveals some unknown and intriguing underlying structure of the equations.
Work on the construction of Baëcklund transformations for this system is in progress. The construction of all nontrivial couplings of KdV- and MKdV-equation based on symmetry preservation is a challenge.

REFERENCES

1. Dodd, R. K., Eilbeck, J. C., Gibbons, J. D., and Morris, H. C., *Solitons and Nonlinear Wave Equations*. Academic Press, 1982.
2. Kersten, P. H. M., Supersymmetries and recursion operators for $N = 2$ supersymmetric KdV-equation, *RIMS Kokyuroku*, **1150**, Kyoto Uniniv., Kyoto, Japan, 2000, pp. 153–161.
3. Kersten, P. H. M., and Krasil′shchik, I. S., *Complete integrability of the soupled KdV–mKdV system*, to appear in Advanced Studies in Pure Mathematics, Mathematical Society of Japan, 2001. `arXiv:nlin.SI/0010041`(25-10-2000).
4. Krasil′shchik, I. S., *Differential Geom. Appl.*, **2**, 307–350 (1992).
5. Krasil'shchik, I. S., and Kersten, P. H. M. *Symmetries and Recursion Operators for Classical and Supersymmetric Differential Equations*, Kluwer Acad. Publ., Dordrecht/Boston/London, 2000.
6. Khor′kova, N.G., *Mat. Zametki*, **44** (1), 134–144, 157 (1988), translation in *Math. Notes*, **44**, (1-2), 562–568 (1989).
7. Krasil′shchik, I. S., Lychagin, V. V., and Vinogradov, A. M., *Geometry of Jet Spaces and Nonlinear Partial Differential Equations*, Gordon and Breach, New York, 1986.
8. Krasil′shchik, I. S., and Vinogradov, A. M., *Acta Appl. Math.*, **15**, 161–209 (1989).
9. Krivonos, S., and Sorin, A., *Phys. Lett.* **A251**, 109 (1999).
10. Mathieu, P., Open problems for the super KdV equation. *AARMS-CRM Workshop on Baecklund and Darboux transformations. The Geometry of Soliton Theory.* June 4–9, 1999. Halifax, Nova Scotia.
11. Vinogradov, A. M., *Acta Appl. Math.* **3**, 21–78 (1984).

Division Algebras and Extended SuperKdVs

F. Toppan

CBPF, Rua Dr. Xavier Sigaud 150, cep 22290-180 Rio de Janeiro (RJ), Brazil,
email: toppan@cbpf.br

Abstract. The division algebras **R**, **C**, **H**, **O** are used to construct and analyze the $N = 1,2,4,8$ supersymmetric extensions of the KdV hamiltonian equation. In particular a global $N = 8$ super-KdV system is introduced and shown to admit a Poisson bracket structure given by the "Non-Associative $N = 8$ Superconformal Algebra".

INTRODUCTION

In the last several years integrable hierarchies of non-linear differential equations in $1+1$ dimensions have been intensely explored, mainly in connection with the discretization of the two-dimensional gravity (see [1]).

Supersymmetric extensions of such equations have also been largely investigated [2]-[7] using a variety of different methods. Unlike the bosonic theory, many questions have not yet been answered in the supersymmetric case. In this talk I report a recent result, obtained in collaboration with H.L. Carrion and M. Rojas [8], concerning the formulation of the N-extended supersymmetric versions of the bosonic integrable equations (for $N = 1,2,4,8$) in terms of the division algebras **R**, **C**, **H**, **O** respectively.

To be precise we focused our investigation on the supersymmetric extensions of KdV. At first the standard results of Mathieu [3] concerning $N = 2$ KdV are reviewed in the language of division algebras, while a full analysis of the global $N = 4$ extensions of KdV is performed. The Delduc and Ivanov [9] result is recovered as a special case. Later it is proven that a unique $N = 8$ hamiltonian extension of KdV can be found. It admits as a Poisson brackets structure the so-called "Non-Associative $N = 8$ Superconformal Algebra" introduced for the first time in [10]. It is quite a special superconformal algebra, since it does not satisfy the Jacobi property (this is why it was named "non-associative"). In order to have an $N = 8$ KdV, the "non-associativity" of the underlining superconformal algebra of Poisson brackets is mandatory. This is due to the fact that no central extension, which is responsible for the inhomogeneous character of KdV, is allowed by jacobian superconformal algebras.

It is worth noticing that in the following the problem of the integrability of the hamiltonian super-KdVs system is not addressed, only the issue of their global supersymmetric invariances is of concern here.

CP589, *Fundamental Interaction Theories*, edited by J. Lukierski and J. Rembieliński

DIVISION ALGEBRAS AND EXTENDED SUPERCONFORMAL ALGEBRAS

The N-extended superconformal algebras, for $N = 1,2,4,8$, can be recovered from division algebras. Let us here present the largest of such conformal algebras, the "$N = 8$ Non-associative SCA" of reference [10]. The $N = 1,2,4$ SCA's are recovered as subalgebras. The "$N = 8$ Non-associative SCA" can be defined via octonionic structure constants. A generic octonion x is expressed as $x = x_a\tau_a$ (throughout the text the convention over repeated indices is understood), where x_a are real numbers while τ_a denote the basic octonions, with $a = 0,1,2,...,7$.

$\tau_0 \equiv \mathbf{1}$ is the identity, while τ_α, for $\alpha = 1,2,...,7$, denote the imaginary octonions. In the following a Greek index is employed for imaginary octonions, a Latin index for the whole set of octonions (identity included).

The octonionic multiplication can be introduced through

$$\tau_\alpha \cdot \tau_\beta = -\delta_{\alpha\beta}\tau_0 + C_{\alpha\beta\gamma}\tau_\gamma, \tag{1}$$

with $C_{\alpha\beta\gamma}$ a set of totally antisymmetric structure constants which, without loss of generality, can be taken to be

$$C_{123} = C_{147} = C_{165} = C_{246} = C_{257} = C_{354} = C_{367} = 1. \tag{2}$$

and vanishing otherwise.

When α, β, γ are restricted to, let's say, the values $1,2,3$ we recover the quaternionic subalgebra, which is associative. The $N = 8$ extension of the Virasoro algebra is constructed in terms of the above structure constants. Besides the spin-2 Virasoro field, it contains eight fermionic spin-$\frac{3}{2}$ fields Q, Q_α and 7 spin-1 bosonic currents J_α. It is explicitly given by the following Poisson brackets

$$\begin{aligned}
\{T(x),T(y)\} &= -\frac{1}{2}\partial_y{}^3\delta(x-y) + 2T(y)\partial_y\delta(x-y) + T'(y)\delta(x-y),\\
\{T(x),Q(y)\} &= \frac{3}{2}Q(y)\partial_y\delta(x-y) + Q'(y)\delta(x-y) + (X1),\\
\{T(x),Q_\alpha(y)\} &= \frac{3}{2}Q_\alpha(y)\partial_y\delta(x-y) + Q_\alpha{}'(y)\delta(x-y),\\
\{T(x),J_\alpha(y)\} &= J_\alpha(y)\partial_y\delta(x-y) + J_\alpha{}'(y)\delta(x-y),\\
\{Q(x),Q(y)\} &= -\frac{1}{2}\partial_y{}^2\delta(x-y) + +\frac{1}{2}T(y)\delta(x-y),\\
\{Q(x),Q_\alpha(y)\} &= -J_\alpha(y)\partial_y\delta(x-y) - \frac{1}{2}J_\alpha{}'(y)\delta(x-y),\\
\{Q(x),J_\alpha(y)\} &= -\frac{1}{2}Q_\alpha(y)\delta(x-y),\\
\{Q_\alpha(x),Q_\beta(y)\} &= -\frac{1}{2}\delta_{\alpha\beta}\partial_y{}^2\delta(x-y) + C_{\alpha\beta\gamma}J_\gamma(y)\partial_y\delta(x-y)\\
&\quad + \frac{1}{2}(\delta_{\alpha\beta}T(y) + C_{\alpha\beta\gamma}J_\gamma{}'(y))\delta(x-y),
\end{aligned}$$

$$\{Q_\alpha(x), J_\beta(y)\} = \frac{1}{2}(\delta_{\alpha\beta}Q(y) - C_{\alpha\beta\gamma}Q_\gamma(y))\delta(x-y),$$
$$\{J_\alpha(x), J_\beta(y)\} = \frac{1}{2}\delta_{\alpha\beta}\partial_y\delta(x-y) - C_{\alpha\beta\gamma}J_\gamma(y)\delta(x-y). \quad (3)$$

This superconformal algebra can be recovered via Sugawara construction of the affine octonionic algebra (see [11] for details). The failure in closing the Jacobi identity is in consequence of the non-associativity of the multiplication between octonions.

EXTENDED SUPERKDVS

A natural question to be raised is whether the (3) SCA (and its superconformal subalgebras) can be regarded as a Poisson brackets structure for a supersymmetric hamiltonian extension of KdV. This amounts to determine the most general globally supersymmetric hamiltonian of a given dimension 4 (i.e. the second hamiltonian). In the case of $N = 2$, this result is known since the works of Mathieu [3]. In [8] the extension to the $N = 4$ case has been completely worked out. The moduli space of inequivalent $N = 4$ KdV equations has been classified. The special integrable point of Delduc-Ivanov [9] has been recovered in this framework. The complete solution to the $N = 8$ case was given as well.

To get these results an extensive use of computer algebra (with Mathematica and the Thielemans' package for classical OPEs computations) was required. The final results are presented here.

The most general $N = 4$-supersymmetric hamiltonian for the $N = 4$ KdV depends on 5 parameters (plus an overall normalization constant). However, if the hamiltonian is further assumed to be invariant under the involutions of the $N = 4$ minimal SCA, three of the parameters have to be set equal to 0. I recall here that the involutions of the $N = 4$ minimal SCA are induced by the involutions of the quaternionic algebra. Three such involutions exist (any two of them can be assumed as generators), the α-th one (for $\alpha = 1,2,3$) is given by leaving τ_α (together with the identity) invariant and flipping the sign of the two remaining τ's.

What is left is the most general hamiltonian, invariant under $N = 4$ and the involutions of the algebra. It is given by

$$\begin{aligned} H_2 = \; & T^2 + Q'_a Q_a - J''_\alpha J_\alpha + x_\alpha T {J_\alpha}^2 \\ & + 2x_\alpha Q_0 Q_\alpha J_\alpha - C_{\alpha\beta\gamma} x_\alpha J_\alpha Q_\beta Q_\gamma \\ & + 2x_1 J_1 J'_2 J_3 - 2x_2 J'_1 J_2 J_3 \,. \end{aligned} \quad (4)$$

here $a = 0,1,2,3$ ($Q_0 \equiv Q$) and $\alpha, \beta, \gamma \equiv 1,2,3$.

The $N = 4$ global supersymmetry requires the three parameters x_a to satisfy the condition

$$x_1 + x_2 + x_3 = 0, \quad (5)$$

so that only two of them are independent. Since any two of them, at will, can be plotted in a real $x - y$ plane, it can be proven that the fundamental domain of the moduli space of inequivalent $N=4$ KdV equations can be chosen to be the region of the plane comprised between the real axis $y=0$ and the $y=x$ line (boundaries included). There are five other regions of the plane (all such regions are related by an S_3-group transformation) which could be equally well chosen as fundamental domain.

In the region of our choice the $y=x$ line corresponds to an extra global $U(1)$-invariance, while the origin, for $x_1=x_2=x_3=0$, is the most symmetric point (it corresponds to a global $SU(2)$ invariance associated to the generators $\int dx J_\alpha(x)$).

The involutions associated to each given imaginary quaternion allows to consistently reduce the $N=4$ KdV equation to an $N=2$ KdV, by setting simultaneously equal to 0 all the fields associated with the τ's which flip the sign, e.g. the fields $J_2=J_3=Q_2=Q_3=0$ for the first involution (and similarly for the other couples of values $1,3$ and $1,2$). After such a reduction we recover the $N=2$ KdV equation depending on the free parameter x_1 (or, respectively, x_2 and x_3).

The integrability is known for $N=2$ KdV to be ensured for three specific values $a=-2,1,4$, discovered by Mathieu [3], of the free parameter a. We are therefore in the position to determine for which points of the fundamental domain the $N=4$ KdV is mapped, after any reduction, to one of the three Mathieu's integrable $N=2$ KdVs. It turns out that in the fundamental domain only two such points exist. Both of them lie on the $y=x$ line. One of them produces, after inequivalent $N=2$ reductions, the $a=-2$ and the $a=4$ $N=2$ KdV hierarchies. The second point, which produces the $a=1$ and the $a=-2$ $N=2$ KdV equations, however, does not admit at the next order an $N=4$ hamiltonian which is in involution w.r.t. H_2. This has been explicitly proven in [8]. The treatment of [8] is more complete than the one in [9] since it is based on an exhaustive component-fields analysis, rather than on an extended superfield formalism.

The most general equations of motion of the $N=4$ KdV are directly obtained from the hamiltonian (4) together with the (3) Poisson brackets.

THE $N=8$ SUPERKDV

A similar analysis can be extended to the $N=8$ case based on the full $N=8$ non-associative SCA. At first the most general hamiltonian with the right dimension has been written down. Later, some constraints on it have been imposed. The first set of constraints requires the invariance under all the 7 involutions of the algebra. In the case of octonions the total number of involutions is 7 (with 3 generators) each one being associated to one of the seven combinations appearing in (2). In the case, e.g., of the 123 combination the corresponding τ_α's are left invariant, while the remaining four τ's, living in the complement, have the sign flipped.

The second set of constraints requires the invariance under the whole set of $N=8$ global supersymmetries. Under this condition there exists only one hamiltonian, up to the normalization factor, which is $N=8$ invariant. It does not contain any free parameter and is quadratic in the fields. It is explicitly given by

$$H_2 \quad = \quad T^2+Q'_aQ_a-J''_\alpha J_\alpha \tag{6}$$

(here $a=0,1,...,7$ and $\alpha=1,2,...,7$).

The hamiltonian corresponds to the origin of coordinates (confront the previous case) which is also, just like the $N=4$ case, the point of maximal symmetry. This means that the hamiltonian is invariant under the whole set of seven global charges $\int dxJ_\alpha(x)$, obtained by integrating the currents J_α's.

Despite its apparent simplicity, it gives an $N=8$ extension of KdV which does not reduce (for any $N=2$ reduction) to the three Mathieu's values for integrability. Nevertheless it is a highly non-trivial fact that an $N=8$ extension of the KdV equation indeed exists and that it is unique. Explicitly, the associated equations of motion are given by

$$\begin{aligned}
\dot{T} &= T''' + 12T'T + 6Q_a''Q_a - 4J_\alpha''J_\alpha\,, \\
\dot{Q} &= Q''' + 6T'Q + 6TQ' + 4Q_\alpha''J_\alpha - 2Q_\alpha J_\alpha''\,, \\
\dot{Q}_\alpha &= Q_\alpha''' + 2QJ_\alpha'' + 6TQ_\alpha' + 6T'Q_\alpha - 2Q'J_\alpha' - 4Q''J_\alpha \\
&\quad + C_{\alpha\beta\gamma}(Q_\beta J_\gamma'' - Q_\beta' J_\gamma' - 2Q_\beta''J_\gamma)\,, \\
\dot{J}_\alpha &= J_\alpha''' + 4T'J_\alpha + 4TJ_\alpha' + 2QQ_\alpha' + C_{\alpha\beta\gamma}(2J_\alpha J_\beta J_\gamma'' + Q_\beta Q_\gamma')\,.
\end{aligned} \tag{7}$$

Besides the $N=8$ KdV, the fields entering (3) admit globally $N=4$ supersymmetric invariant hamiltonians, which depend on free parameters. Two such classes of hamiltonians are individuated. The first class consists of the hamiltonians invariant under supersymmetries related with the quaternionic subalgebra which, without loss of generality, can be assumed to be given by $\int dxQ_i(x)$, for $i=0,1,2,3$. The second class of invariances is associated to the $N=4$ supersymmetries associated to the remaining generators, i.e. those living in the complement (in the following, without loss of generality, these supersymmetries are labeled by $i=1,2,4,5$). For completeness we report here the results. The first class of $N=4$-invariant hamiltonians is given by

$$\begin{aligned}
H_2 &= T^2 + Q'Q + Q_p'Q_p + xQ_r'Q_r - J_p''J_p - xJ_r''J_r + 2x_pQQ_pJ_p + x_pTJ_pJ_p \\
&\quad - 2x_3Q_1Q_2J_3 + 2x_2Q_1Q_3J_2 - 2x_1Q_2Q_3J_1 + 2x_1J_1J_2'J_3 - 2x_2J_1'J_2J_3\,.
\end{aligned} \tag{8}$$

with x_1, x_2, x free parameters, while $x_3 = -x_1 - x_2$.

In the above expression $p=1,2,3$ and $r=4,5,6,7$.

The second class of $N=4$ invariant hamiltonians is explicitly given by

$$\begin{aligned}
H_2 &= T^2 + x(Q'Q + Q_3'Q_3 + Q_6'Q_6 + Q_7'Q_7) + Q_1'Q_1 + Q_2'Q_2 + Q_4'Q_4 + Q_5'Q_5 \\
&\quad - x(J_1''J_1 + J_2''J_2 + J_4''J_4 + J_5''J_5) - J_3''J_3 - J_6''J_6 - J_7''J_7 \\
&\quad + y(TJ_7J_7 - TJ_3J_3) + 2yJ_3J_6'J_7 + 2y(Q_1Q_2J_3 - Q_1Q_4J_7 - Q_2Q_5J_7 - Q_4Q_5J_3)
\end{aligned} \tag{9}$$

where in this case two free parameters, x and y, appear.

CONCLUSIONS

In this paper I have presented some new results concerning an explicit connection of division algebras and the extended supersymmetrizations of the KdV equation.

In particular it has been proven, following [8], the existence of a unique $N = 8$ KdV equation of hamiltonian type based on the $N = 8$ non-associative SCA as a generalized classical Poisson brackets structure.

Division algebras are a natural ingredient when dealing with extended supersymmetries. It is therefore likely that supersymmetric extensions of other classes of bosonic integrable equations could be studied with the tools furnished by division algebras. Besides KdV, the next simplest equations to be investigated are the mKdV and the NLS. In view of the results of [11], where the $N = 8$ Non-associative SCA is recovered from a singular limit of a Sugawara construction based on the superaffine octonionic algebra of superMalcev type, it is almost for granted that such extensions indeed exist.

Whether such constructions could be applied to other classes of integrable equations is still an open problem. In any case division algebras look as a promising and elegant tool to unveil some of the mysterious features still surrounding the supersymmetrization of the bosonic hierarchies.

ACKNOWLEDGMENTS

I am grateful to the organizers of the Karpacz Winter School for the invitation. It is a pleasure for me to thank my collaborators H.L. Carrion and M. Rojas.

REFERENCES

1. Di Francesco, P., Ginsparg P., and Zinn-Justin, J., *Phys. Rep.*, **254**, 1 (1995).
2. Manin, Y.I., and Radul, A.O., *Comm. Math. Phys.* **98**, 65 (1985).
3. Mathieu, P., *Phys. Lett.*, **B 203**, 65 (1988); Laberge, C.A., and Mathieu, P., *Phys. Lett.* **B 215**, 718 (1988); Labelle, P., and Mathieu, P., *J. Math. Phys.* **89**, 923 (1991).
4. Inami, T., and Kanno, H., *Comm. Math. Phys.*, **136**, 519 (1991); *Int. J. Mod. Phys.* **A 7**, Suppl. 1A, 419 (1992).
5. Popowicz, Z., *Phys. Lett.*, **A 194**, 375 (1994); *J. Phys.* **A 29**, 1281 (1996); *ibid.* **39**, 7935 (1997); *Phys. Lett.* **B 459**, 150 (1999).
6. Brunelli, J.C., and Das, A., *J. Math. Phys.* **36**, 268 (1995).
7. Toppan, F., *Int. J. Mod. Phys.* **A 10**, 895 (1995).
8. Carrion, H.L., Rojas, M., and Toppan, F., Division Algebras and the N=1,2,4,8 Extensions of KdV, Preprint CBPF-NF-012/01, 2001.
9. Delduc, F., and Ivanov, E., Phys. Lett. **B 309**, 312 (1993); Delduc, F., Ivanov, E., and Krivonos, S., *J. Math. Phys.* **37**, 1356 (1996).
10. Englert, F., Sevrin, A., Troost, W., van Proeyen, A., and Spindel, P., *J. Math. Phys.* **29**, 281 (1988).
11. Carrion, H.L., Rojas, M., and Toppan, F., An N=8 Superaffine Malcev Algebra and Its N=8 Sugawara, Preprint CBPF-NF-011/01, 2001.

Participants of XXXVII Winter School in Karpacz

Amelino-Camelia, Giovanni (Roma)
Giovanni.Amelino-Camelia@cern.ch

Arodź, Henryk (Kraków)
ufarodz@thrisc.if.uj.edu.pl

Arzano, Michele (Roma)
Michele.arzano@roma1.infn.it

Azcarraga, JoseA.de (Valencia)
azcarrag@lie1.ific.uv.es

Bagnoud, Maxime (Lousanne)
Maxime.Bagnoud@unine.ch

Bandos, Igor (Valencia)
bandos@ific.uv.es

Bars, Itzak (Los Angeles)
bars@physics1.usc.edu

Bazunowa, Nagegda (Tartu)
nadegda@math.ut.ee

Berghshoeff, Eric (Groningen)
E.A.Bergshoeff@phys.ru

Błaut, Arkadiusz (Wrocław)
blaut@ift.uni.wroc.pl

Bonora, Loriano (Trieste)
bonora@he.sissa.it

Borowiec, Andrzej (Wrocław)
borow@ift.uni.wroc.pl

Broda, Bogusław (Łódź)
bbroda@krysia.uni.lodz.pl

Burdik, Cestmir (Praha)
burdik@siduri.fjfi.cvut.cz

Carlevaro, Luca (Lousanne)
Luca.Carlevaro@unine.ch

Cederwall, Martin (Goeteborg)
martin.cederwall@fy.chalmers.se

Chu,Chong-Sun (Durham)
Chong-sun.chu@iph.unine.ch

Cisło, Jerzy (Wrocław)
cislo@ift.uni.wroc.pl

Coquereaux, Robert (Marseille)
coque@cpt.univ-mrs.fr

Czerhoniak, Piotr (Zielona Góra)
P.Czerhoniak@proton.if.wsp.zgora.pl

Das, Ashok (Rochester)
das@pas.rochester.edu

Dolgushev, Vasily (Tomsk)
vald@phys.tsu.ru

Doplicher, Sergio (Roma)
doplicher@mat.uniroma1.it

Dragovic, Branco (Belgrad)
dragovic@phy.bg.ac.yu

Falkowski, Adam (Kraków)
Adam.Falkowski@fuw.edu.pl

Forys, Marek (Łódź)
williams@krysia.uni.lodz.pl

Frydryszak, Andrzej (Wrocław)
amfry@ift.uni.wroc.pl

Gibbon, Garry W. (Cambridge)
g.w.gibbons@damtp.cam.ac.uk

Graczyk, Krzysztof (Wrocław)
kgracz@ift.uni.wroc.pl

Grosse, Harald (Vienna)
grosse@ap.univie.ac.at

Haba, Zbigniew (Wrocław)
zhab@ift.uni.wroc.pl

Hegerfeldt, Gerhard (Goettingen)
hegerf@Theorie.Physik.UNI-Goettingen.DE

Isaev, Alexei (Dubna)
isaevap@thsun1.jinr.ru

Ivanov,Evgenij (Dubna)
eivanov@thsun1.jinr.dubna.su

Jancewicz, Bernard (Wrocław)
bjan@ift.uni.wroc.pl

Jaramillo, Jose Luis (Granada)
jarama@iaa.es

Jurco, Branislav (Muenchen)
jurco@theorie.physik.uni-muenchen.de

Kerner, Richard (Paris)
rk@ccr.jussieu.fr

Kersten, Paul M. (Enschede)
kersten@math.utwente.nlKersten

Klimek, Małgorzata (Częstochowa)
klimek@matinf.pcz.czest.pl

Knyazev, Michael (Minsk)
knyazev@iaph.bas-net.by

Kowalski-Glikman, Jerzy (Wrocław)
jurekk@ift.uni.wroc.pl

Krivonos, Sergei (Dubna)
krivonos@thsun1.jinr.ru

Lalak, Zygmunt (Warszawa)
Zygmunt.Lalak@fuw.edu.pl

Lukierski, Jerzy (Wrocław)
lukier@ift.uni.wroc.pl

Madore, John (Paris)
madore@theorie.physik.uni-muenchen.de

Marcinek, Władysław (Wrocław)
wmar@ift.uni.wroc.pl

Mozrzymas, Marek (Wrocław)
marmoz@ift.uni.wroc.pl

Nowak, Dobromiła (Wrocław)
dobno@ift.uni.wroc.pl

Nowicki, Anatol (Zielona Góra)
anowicki@wsp.zgora.pl

Pawełczyk, Jacek (Warszawa)
Jacek.pawelczyk@net.fuw.edu.pl

Pilch, Krzysztof (Los Angeles)
pilch@physics1.usc.edu

Popowicz, Ziemowit (Wrocław)
ziemek@ift.uni.wroc.pl

Rembieliński, Jakub (Łódź)
jaremb@krysia.uni.lodz.pl

Ruehl, Werner (Kaiserslautern)
ruehl@physik.uni-kl.de

Serov, Valery, S. (Moscow)
serov@cs.msu.su

Sheikh-Jabbari, Mohammad (Trieste)
jabbari@ictp.trieste.it

Skulimowski, Marek (Łódź)
mskulim@krysia.uni.lodz.pl

Smolin, Lee (London)
l.smolin@ic.ac.uk

Sobczyk, Jan (Wrocław)
jsobczyk@ift.uni.wroc.pl

Sorokin, Dmitri (Padova)
dmitri.sorokin@pd.infn.it

Steinacker, Harold (Paris)
Harold.Steinacker@th.u-psud.fr

Stelle, Kellog (London)
k.stelle@ic.ac.uk

Suszek, Rafał (Warszawa)
suszek@fuw.edu.pl

Ślusarczyk, Maciej (Kraków)
mslus@thrisc.if.uj.edu.pl

Tolstoy, Valerij N. (Moscow)
tolstoy@nucl-th.npi.msu.su

Toppan, Francesco (Rio de Janeiro)
toppan@cbpf.br

Uvarov, Dmytro (Charkov)
uvarov@kipt.kharkov.ua

Van Proyen, Antoine (Leuven)
Antoine.VanProeyen@fys.kuleuven.ac.be

Woronowicz, Leszek (Warszawa)
slworono@fuw.edu.pl

Wereszczyński, Andrzej (Kraków)
wereszcz@thrisc.if.uj.edu.pl

Zakrzewski, Wojtek, J. (Durham)
W.J.Zakrzewski@durham.ac.uk

Zoupanos, George (Athens)
George.Zoupanos@cern.ch

AUTHOR INDEX